Roman Sikorski, *Advanced calculus. Functions of several variables*

ERRATA

Page line	For	Read
62^2	$\mathfrak{A}\cdot\mathfrak{A}^*$	$\mathfrak{A}\bullet\mathfrak{A}^*$
83^3	*isskew*	*is skew*
147_7	$\varphi''_{\mathfrak{a},\mathfrak{b}}(0,0)$ that $\varphi''_{\mathfrak{b},\mathfrak{a}}(0,0)$	$\varphi''_{\mathfrak{a},\mathfrak{b}}(0,0) = \varphi''_{\mathfrak{b},\mathfrak{a}}(0,0)$
156_1	$\psi'^{(q)}_{\mathfrak{a}}(q)$	$\psi'_{\mathfrak{a}}(q)$
414_{10}	Stokes,	Stokes'

INSTYTUT MATEMATYCZNY POLSKIEJ AKADEMII NAUK

MONOGRAFIE MATEMATYCZNE

TOM 51

PWN — POLISH SCIENTIFIC PUBLISHERS

ROMAN SIKORSKI

ADVANCED CALCULUS
FUNCTIONS OF SEVERAL VARIABLES

WARSZAWA 1969

PAŃSTWOWE WYDAWNICTWO NAUKOWE
(PWN — POLISH SCIENTIFIC PUBLISHERS)

WARSZAWA (Poland), ul. Miodowa 10

PRINTED IN POLAND

DRUKARNIA UNIWERSYTETU JAGIELLOŃSKIEGO W KRAKOWIE

CONTENTS

CHAPTER IV

DIFFERENTIATION OF REAL FUNCTIONS

CHAPTER V

DIFFERENTIABLE MAPPINGS

CHAPTER VI

MEASURE THEORY

CHAPTER VII

GENERAL THEORY OF THE INTEGRAL

CHAPTER VIII

THE INTEGRAL IN EUCLIDEAN SPACES

CHAPTER IX

INTEGRALS ON HYPERSURFACES AND POLYHEDRA

CHAPTER X

SUPPLEMENTARY INFORMATION

PREFACE TO THE ENGLISH EDITION

This book is a slight modification of the lectures I gave at Warsaw University in 1963/64 in place of the traditional course in the calculus of functions of several variables. More precisely, only Chapters IV-IX which are the main part of the book, were the subject of my lectures. Chapters I-III, containing a necessary background from set theory, linear algebra and topology, are added for completeness, to avoid the necessity of referring to other books. Chapter X contains information, without proofs, about the further development or generalization of some material discussed in Chapters IV-IX. The book contains more theorems than my lectures which contained more non-precise intuitive explanations than the book.

The lectures were designed for students of mathematics in the second school year. In the first school year they studied elements of higher algebra, analytic geometry, set theory and the calculus of functions of one real variable. It is a tradition at Warsaw University that analytic geometry is presented in general k-dimensional Euclidean spaces. My students knew such fundamental notions as vector, hyperplane, etc. I could also assume that they knew fundamental notions from set theory, such as set-theoretical operations on sets, countable sets, etc. My students also had a course in the theory of metric spaces, which was concurrent with my course. They learned there all the fondamental topological theorems and notions I needed. The correlation between the course on metric spaces and my course was satisfactory. Consequently I was able to start my course with the theory of differentiation of functions defined on subsets of multi-dimensional Euclidean spaces without any introduction from set theory, topology and geometry. Unfortunately, the program of the course of algebra did not include all the notions from linear algebra that I needed. Therefore I had to explain certain parts of the material from Chapter II when I had to make use of it.

At this time a reform of the study of mathematics at Polish universities was under way. One objective of the reform was to avoid the

Riemann integral in the university course in calculus. In the first school year the integral should be defined, for continuous functions only, as the increase of a primitive function. In the second school year the Lebesgue integral in k-dimensional Euclidean spaces should be presented.

Since my lectures on the integral had to be "k-dimensional", I also presented the theory of differentiation in a k-dimensional form. Using suitable notation, such lectures are not more difficult than a traditional course of differentiation of functions in the plane or in the three-dimensional space. After the "k-dimensional" course of differentiation and integration it was necessary to present the theory of surface integrals in a "multi-dimensional" way in order to preserve the general character of my course.

My tendency was to use modern mathematical notation like $f\colon X \to Y$, $f \circ g$ for the composition of functions, etc. and not to use unnecessary symbols. For instance, I wrote $\int\limits_A f$ instead of $\int\limits_A f(x)\,dx$, etc. Consequently, a real function defined on a subset of the k-dimensional Euclidean space was denoted by f only, not by $f(x_1, \ldots, x_k)$. Since I eliminated the variables in symbols of functions, I also had to eliminate them in symbols for partial derivatives. Therefore I introduced for partial derivatives a new notation taken from differential geometry and used also, in a similar form, in some modern books on calculus. However the jargon of the traditional notation is so common that I could not completely ignore it. Therefore the traditional notation is quoted in this book but only in the auxiliary text (small print) and, sometimes, in exercises. The notation and the language of linear algebra is frequently used. One of the reasons is to give many formulas in the multi-dimensional case the same form as in the one-dimensional case (see, e.g., the Taylor formula, the formula for the derivative of the composition of mappings, the formula for the integration by substitution, etc.).

The archaic term "implicit functions" is never used in this book although the theory of systems of function equations is developed.

The main difficulty in preparing my lectures was to choose one of the various possible ways of presenting particular parts of the course.

For instance, one of the problems was how to introduce the notion of the derivative of a function of several variables. It was not my aim to present the theory of differentiation under weak hypotheses. The whole theory of differentiation is developed under the hypothesis that all functions under consideration are at least of class C_1. Under this hypothesis the notion of the total differential does not play an essential part. On the other hand, it seemed to be convenient to introduce a very general notion of derivative.

Another problem was the definition of the k-dimensional measure on k-dimensional hypersurfaces in l-dimensional Euclidean spaces. The definition might be based on the notion of the k-dimensional Hausdorff measure, but then it would have been necessary to prove the well-known integral formula for the area of a surface. To omit the rather difficult proof, the integral formula was assumed as the basis of the definition.

Some notions introduced in Chapter IX are basic notions of algebraic topology. However they are introduced in a different way from the usual one. The notion of a multi-dimensional interval rather than the notion of simplex was assumed as the primitive notion because of a simplification in the proof of Stokes' theorem.

The definitions of an oriented cell and of a chain are not traditional; they are formulated in a way more suitable from the point of view of the theory of integration on polyhedra.

The definition of a hypersurface does not contain the condition of connectedness because this hypothesis is not necessary. Moreover, without this hypothesis, the theorem that a system of equations defines a hypersurface has a simpler formulation.

A tendency to abstraction or generality was not the aim of my course. The subject of my course was only the differential and integral calculus of functions defined on subsets of Euclidean spaces. Therefore the theory of differentiation was not presented in a general form as, for instance, in the well-known book of Dieudonné [1].

In this book the k-dimensional Euclidean space is defined as the Cartesian product of k copies of the set of all real numbers. The product structure in the Euclidean space determines a "natural" coordinate system to which all notions are referred. Consequently, linear and multilinear mappings are in a fixed natural one-to-one correspondence with matrices. Both notions are used by me as equivalent.

For instance, beside the notion of the directional derivative of order m (the mth differential) of a function, which is an m-linear mapping, I introduce and investigate the notion of the mth derivative of the function, which is defined as the matrix of all mth partial derivates. Sometimes I use only the matrix language, for instance in the case of polyvectors.

The Polish version of the book was submitted to the printer in February 1965. That year the book [3] by W. M. Fleming and the book [4] by C. Goffmann appeared, which are similar modern approaches to the classical calculus of functions of several variables.

The English edition of the book is a translation of the second revised Polish edition, which was submitted to the printer in 1967.

I would like to express my gratitude to A. J. Lohwater, who has been kind enough to read the text and to offer valuable suggestions for improvements. I also wish to thank B. P. Schweitzer and W. Waliszewski for a revision of the manuscript.

A few exercises in the book were communicated to me by A. Birkholc, K. Krzyżewski and Z. Sawoń.

Roman Sikorski

Los Angeles–Cleveland–Warsaw
1966–1967

INFORMATION FOR THE READER

To understand this book it is enough to know the fundamental notions from the calculus of functions of one real variable, and the basic properties of determinants from algebra.

A knowledge of analytic geometry is also assumed. All necessary notions from analytic geometry in k-dimensional Euclidean spaces are recollected in Chapter II, §§ 1-2. The student who is not familiar with the notion of multi-dimensional Euclidean spaces can always assume that the dimension of the space under consideration is 1, 2 or 3. However, he can learn the general case from Chapter II, §§ 1-2.

Some authors identify vectors with points in Euclidean spaces. The identification is not assumed in this book. However, it is admissible.

All set-theoretical, algebraical and topological notions needed to understand the main part of the book (Chapters IV-IX) are given in Chapters I-III. The reader familiar with the most fundamental notions from set theory and topology can omit Chapters I and III. The reader familiar with linear algebra should look through Chapter II because of the terminology and notation.

Many examples illustrating the main topics are given in exercises which are added at the end of almost every section. There are only a few exercises involving calculation of multiple integrals and finding extremal points of functions because traditional calculus texts contain enough problems of this type.

The reader familiar with a traditional course in calculus, who wishes to read only selected portions of the book, should use the List of Symbols and the Subject Index at the end of the book; this is necessary because the notation used here differs essentially from the traditional. Sometimes the terminology is also different. For instance, instead of "gradient of a real function" the term "derivative of a function" is used. The term "k-vector" stands for what is usually called "simple k-vector". The term "skew derivative" is used instead of "exterior derivative", etc.

The symbol 0 has in this book many different meanings (the number zero, the zero vector or the zero matrix, the empty set, etc.) but this does not lead to misunderstandings since it is always clear in what sense the symbol 0 is used.

Theorems from the same chapter are quoted by giving their numbers only; otherwise the capital Roman number of the chapter is added. Formulas from the same section are quoted by giving their numbers, those from other section in the same chapter — by giving their numbers and the number of the section, and for formulas from other chapters also the number of the chapter is added. The same holds for exercises. For instance, III 4.1 denotes the theorem 4.1 in Chapter III; III § 1 (10) denotes the formula (10) in Chapter III, section 1; and V § 2, 3 denotes the exercise 3 in Chapter V, section 2, etc.

CHAPTER I

FUNDAMENTAL SET-THEORETICAL NOTIONS

§ 1. Sets. The notion of *set* is the fundamental notion of modern mathematics. The nature of sets which are the subjects of mathematical investigation is diverse. For instance, we examine sets of real numbers (like the set of all rational numbers, the set of all integers, intervals, etc.) sets of points in Euclidean spaces (spheres, balls, multidimensional intervals, etc.), as well as sets whose elements are sets (e.g. the set of all spheres, the set of all triangles in the plane), etc. The part of mathematics that examines arbitrary sets, without looking at the nature of their elements, is called *set theory*. In this chapter we shall introduce a few fundamental notions of set theory that are necessary to understand the other chapters of this book.

We write

$$a \in A$$

if a is an element of a set A (i.e., if a belongs to A), and

$$a \notin A$$

if a is not an element of A (i.e., if a does not belong to the set A). Elements of sets are often called *points*. Thus the phrases "element of a set A" and "point of a set A" have the same meaning.

We say that a set A is a *subset* (or: a *part*) of a set B if every element a of A belongs to B. Then we write

$$A \subset B \quad \text{or} \quad B \supset A .$$

If $A \subset B$ but A is not identical with B, we say that A is a *proper subset* of B.

The relation $\subset$ is called *inclusion*. It has the following properties:

(1) $$A \subset A ;$$

(2) $$\text{if } A \subset B \text{ and } B \subset C, \text{ then } A \subset C ;$$

(3) $$\text{if } A \subset B \text{ and } B \subset A, \text{ then } A = B .$$

Of course, $A = B$ means that the sets A and B are identical, i.e., that they have the same elements.

Property (2), called the *transitivity* of the inclusion, is very often used in calculation with sets. Property (3) is often used in proofs that certain sets are identical. It states that in order to prove that $A = B$ it suffices to show that every element in A belongs to B and, conversely, every element in B is in A.

The set which does not contain any element is said to be *empty* (or *void*) and is denoted by 0.

It is necessary to introduce the notion of the empty set since we often define the set of elements satisfying a condition without a verification that there exists an element satisfying this condition. For instance, in the theory of polynomials we introduce the notion of the set of all real roots of an algebraic equation, in spite of the fact that we do not know, in general, whether or not such roots exist.

It follows from the definition that the empty set is a subset of every set A:

(4) $$0 \subset A\,.$$

A set may contain only one element (this can happen, for instance, for the set of all real roots of an algebraic equation).

The finite set composed of elements $a_1, \ldots, a_n$ only, will be denoted by $\{a_1, \ldots, a_n\}$. In particular, $\{a\}$ stands for the one-element set whose only element is a.

Sometimes we shall examine sets A whose elements are sets. In that case, A will often be called a *class of sets*.

Suppose a set A_n is assigned to every positive integer n. We say then that we have defined a *sequence of sets*. The sequence is said to be

$$\begin{aligned} &\textit{increasing} \quad \text{if} \quad A_n \subset A_{n+1} \quad \text{for} \quad n = 1, 2, \ldots, \\ &\textit{decreasing} \quad \text{if} \quad A_{n+1} \subset A_n \quad \text{for} \quad n = 1, 2, \ldots \end{aligned}$$

Let $\alpha(x)$ be a condition imposed on elements of a set X. In other words, $\alpha(x)$ is an expression which becames a (true or false) sentence if the name of a concrete element in X is substituted for x. The set of all elements x in X which satisfy the condition $\alpha(x)$ will be denoted by the symbol

$$\{x \in X\colon\ \alpha(x)\}$$

or by the shorter symbol

$$\{x\colon\ \alpha(x)\}$$

if it is evident which set X is under consideration.

For instance, $\{x\colon x^2 = 9\}$ is the set composed of the numbers 3 and -3, and $\{x\colon x^2 < 2x+8\}$ is the open interval $-2 < x < 4$, i.e.

$$\{x\colon\ x^2 < 2x+8\} = \{x\colon\ -2 < x < 4\}\,.$$

If $\alpha(x)$ and $\beta(x)$ are conditions imposed on elements x in X, we shall write for brevity

$$\alpha(x) \& \beta(x) \quad \text{instead of} \quad \alpha(x) \textit{ and } \beta(x),$$

$$\alpha(x) \Rightarrow \beta(x) \quad \text{instead of} \quad \textit{if } \alpha(x), \textit{ then } \beta(x).$$

We shall also write

$$\bigvee_x \alpha(x) \quad \text{instead of} \quad \textit{there exists an } x \textit{ such that } \alpha(x),$$

$$\bigwedge_x \alpha(x) \quad \text{instead of} \quad \textit{for every } x, \ \alpha(x).$$

The symbols $\bigvee$ and $\bigwedge$ are called *quantifiers*. Sometimes we write under the quantifiers certain additional conditions which the element x has to satisfy. For instance, the expression

$$\bigwedge_{x>0} \bigvee_{y<0} (x^2 = y^2)$$

should be read: for every positive number x there exists a negative number y such that $x^2 = y^2$.

EXERCISE

1. Prove that

$$\{x: \bigvee_y \bigwedge_z \alpha(x, y, z)\} \subset \{x: \bigwedge_z \bigvee_y \alpha(x, y, z)\}.$$

§ 2. Set-theoretical operations. By the *union* $A \cup B$ of sets A, B we mean the set of all elements x which belong to at least one of the sets A, B. By definition,

$$A \cup B = \{x: x \in A \text{ or } x \in B\}.$$

By the *intersection* (or: *common part*) $A \cap B$ of sets A, B we mean the set of all elements x which belong to each of the sets A, B. By definition,

$$A \cap B = \{x: (x \in A) \& (x \in B)\}.$$

By the *difference* $A - B$ of sets A, B we mean the set of all elements x that belong to A but do not belong to B. By definition,

$$A - B = \{x: (x \in A) \& (x \notin B)\} = \{x \in A: x \notin B\}.$$

If $A \cap B = 0$, i.e., if the sets A, B have no common elements, we say that the sets A and B are *disjoint*.

2.1. *For any sets A, B, C, X,*

(1) $$A \cup B = B \cup A, \quad A \cap B = B \cap A;$$

(2) $$(A \cup B) \cup C = A \cup (B \cup C), \quad (A \cap B) \cap C = A \cap (B \cap C);$$

(3) $$0 \cup A = A\,, \quad 0 \cap A = 0\,;$$

(4) $$A \cup A = A\,, \quad A \cap A = A\,;$$

(5) $$A \cap (B \cup C) = (A \cap B) \cup (A \cap C)\,, \quad A \cup (B \cap C) = (A \cup B) \cap (A \cup C)\,;$$

(6) $$(A - B) \cap B = 0\,, \quad A = (A \cap B) \cup (A - B)\,, \quad A - B = A - (A \cap B)\,;$$

(7) *if* $A \subset X$, *then* $(X - A) \cup A = X$ *and* $X - (X - A) = A$;

(8) *if* $A \subset X$, *then* $A - B = A \cap (X - B)$;

(9) $$X - (A \cup B) = (X - A) \cap (X - B)\,, \quad X - (A \cap B) = (X - A) \cup (X - B)\,;$$

(10) *if* $A \subset X$ *and* $B \subset X$, *then*

$$A \cup B = X - \big((X - A) \cap (X - B)\big) \text{ and } A \cap B = X - \big((X - A) \cup (X - B)\big).$$

Identities (1) and (2) are *commutative* and *associative* laws. Identities (5) are *distributive* laws. Identities (9) and (10) are called the *de Morgan formulas*.

The proofs of all the above equations are based on § 1 (3). One proves that if a point is in the set on the left-hand side of an equation, then it is in the set on the right-hand side, and conversely. Hence it follows, by § 1 (3), that the sets are equal.

As an example we prove the first of the identities (5).

If $x \in A \cap (B \cup C)$, then $x \in A$ and $x \in B \cup C$, i.e. $x \in B$ or $x \in C$. In the first case, $x \in A$ and $x \in B$, i.e. $x \in A \cap B$. In the second case, $x \in A$ and $x \in C$, i.e. $x \in A \cap C$. In each of the two cases, $x \in (A \cap B) \cup (A \cap C)$. This proves that

$$A \cap (B \cup C) \subset (A \cap B) \cap (A \cap C)\,.$$

Conversely, if $x \in (A \cap B) \cup (A \cap C)$, then $x \in A \cap B$ or $x \in A \cap C$. In the first case, $x \in A$ and $x \in B$, and consequently $x \in A$ and $x \in B \cup C$, i.e. $x \in A \cap (B \cup C)$. In the second case, $x \in A$ and $x \in C$, and consequently $x \in A$ and $x \in B \cup C$, i.e. $x \in A \cap (B \cup C)$. This proves that

$$(A \cap B) \cup (A \cap C) \subset A \cap (B \cup C)\,.$$

The first of the identities (5) follows directly from the two inclusions just proved, by § 1 (3).

2.2. *For any sets* A, B, C, D,

(11) $A \subset B$ *if and only if* $A \cup B = B$;

(12) $A \subset B$ *if and only if* $A \cap B = A$;

(13) $A \subset B$ *if and only if* $A - B = 0$;

(14) $$A \subset A \cup B\,, \quad A \cap B \subset A\,,$$

(15) *if* $A \subset C$ *and* $B \subset D$, *then* $A \cup B \subset C \cup D$ *and* $A \cap B \subset C \cap D$;

(16) *if* $A \subset C$ *and* $B \subset C$, *then* $A \cup B \subset C$;

(17) *if* $C \subset A$ *and* $C \subset B$, *then* $C \subset A \cap B$;

(18) *if* $A \subset B$, *then* $C - B \subset C - A$.

The notions of union and intersection of sets can be generalized to the case of an arbitrary, finite or infinite, number of sets. Suppose that A_t is a set for every t in a non-empty set T. By the *union* $\bigcup_{t \in T} A_t$ of all the sets A_t ($t \in T$) we mean the set of all points x which belong to at least one of the sets A_t. By the *intersection* (or: *common part*) $\bigcap_{t \in T} A_t$ of all the sets A_t ($t \in T$) we mean the set of all elements x that belong to each of the sets A_t. In symbols,

$$\bigcup_{t \in T} A_t = \{x \colon \bigvee_{t \in T} x \in A_t\} , \qquad \bigcap_{t \in T} A_t = \{x \colon \bigwedge_{t \in T} x \in A_t\} .$$

Sometimes the shall omit "$t \in T$" under the signs $\bigcup$ and $\bigcap$ if it is clear what values for t are taken under consideration.

In the case where T is a subset of the set of all integers, we shall also use the following notation. The symbols

$$\bigcup_{j=m}^{n} A_j \quad \text{and} \quad A_m \cup \ldots \cup A_n \quad (m \leqslant n)$$

will denote the union of the sets $A_m, A_{m+1}, \ldots, A_n$, and the symbols

$$\bigcap_{j=m}^{n} A_j \quad \text{and} \quad A_m \cap \ldots \cap A_n \quad (m \leqslant n)$$

will denote the intersection of these sets. Similarly the symbols

$$\bigcup_{j=m}^{\infty} A_j \quad \text{or} \quad A_m \cup A_{m+1} \cup \ldots$$

will denote the union of the sets $A_m, A_{m+1}, \ldots$, and the symbols

$$\bigcap_{j=m}^{\infty} A_j \quad \text{and} \quad A_m \cap A_{m+1} \cap \ldots$$

will denote the intersection of these sets.

The union $\bigcup_{t \in T} A_t$ and the intersection $\bigcap_{t \in T} A_t$ are also commutative and associative, i.e., roughly speaking, they do not depend on the arrangement of the sets A_t. The exact formulation of the general commutative and associative laws is given in exercise 1 and in § 3, exercise 1.

2.3. *For any sets,*

(19) *if* $A_t \subset B_t$ *for every* $t \in T$, *then* $\bigcup_{t \in T} A_t \subset \bigcup_{t \in T} B_t$ *and* $\bigcap_{t \in T} A_t \subset \bigcap_{t \in T} B_t$;

(20) $$\text{if } A_t \subset C \text{ for every } t \in T, \text{ then } \bigcup_{t \in T} A_t \subset C;$$

(21) $$\text{if } C \subset A_t \text{ for every } t \in T, \text{ then } C \subset \bigcap_{t \in T} A_t;$$

(22) $$\bigcap_{t \in T} A_t \subset A_{t_0} \subset \bigcup_{t \in T} A_t \text{ for every } t_0 \in T;$$

(23) $$A \cap \bigcup_{t \in T} A_t = \bigcup_{t \in T} (A \cap A_t), \qquad A \cup \bigcap_{t \in T} A_t = \bigcap_{t \in T} (A \cup A_t);$$

(24) $$\bigcup_{t \in T} A_t \cup \bigcup_{t \in T} B_t = \bigcup_{t \in T} (A_t \cup B_t), \qquad \bigcap_{t \in T} A_t \cap \bigcap_{t \in T} B_t = \bigcap_{t \in T} (A_t \cap B_t);$$

(25) $$(\bigcup_{t \in T} A_t) \cap (\bigcup_{s \in S} B_s) = \bigcup_{t \in T} \bigcup_{s \in S} (A_t \cap B_s),$$
$$(\bigcap_{t \in T} A_t) \cup (\bigcap_{s \in S} B_s) = \bigcap_{t \in T} \bigcap_{s \in S} (A_t \cup B_s);$$

(26) $$X - \bigcup_{t \in T} A_t = \bigcap_{t \in T} (X - A_t), \qquad X - \bigcap_{t \in T} A_t = \bigcup_{t \in T} (X - A_t);$$

(27) *if* $A_t \subset X$ *for every* $t \in T$, *then*

$$\bigcup_{t \in T} A_t = X - \bigcap_{t \in T} (X - A_t) \quad \text{and} \quad \bigcap_{t \in T} A_t = X - \bigcup_{t \in T} (X - A_t).$$

Identities (23) and (25) are *distributive laws*. Identities (26) and (27) are called the *de Morgan formulas*.

We often examine only subsets of a fixed set X. Then the set X is called a *space*. For every set $A \subset X$, the set $X - A$ is then called the *complement* of A. The de Morgan formula can then be expressed as follows: the complement of the union is equal to the intersection of the complements; the complement of the intersection is equal to the union of the complements.

It follows from (27) that the union (intersection) of subsets of a space can be defined by means of the intersection (the union) and the complement only.

The identities in 2.3 can be proved, as in the case of 2.1, by a simple verification that if a point belongs to the set on the left-hand side of an equation, then it belongs to the set on the right-hand side, and conversely.

The following properties of union and intersection will be often used in this book (especially in Chapter VI).

2.4. *If* $B_n = A_1 \cup \ldots \cup A_n$ *for* $n = 1, 2, \ldots,$ *then* $B_n \subset B_{n+1}$ *for* $n = 1, 2, \ldots,$ *and* $\bigcup_{n=1}^{\infty} A_n = \bigcup_{n=1}^{\infty} B_n$.

If $B_n = A_1 \cap \ldots \cap A_n$ *for* $n = 1, 2, \ldots$, *then* $B_{n+1} \subset B_n$ *for* $n = 1, 2, \ldots$, *and* $\bigcap_{n=1}^{\infty} A_n = \bigcap_{n=1}^{\infty} B_n$.

2.5. *If* $B_1 = A_1$ *and* $B_n = A_n - (A_1 \cup \ldots \cup A_{n-1})$ *for* $n = 2, 3, \ldots$ *then the sets* B_n *are disjoint and*

$$A_1 \cup \ldots \cup A_n = B_1 \cup \ldots \cup B_n \quad \textit{for} \quad n = 1, 2, \ldots, \tag{28}$$

$$\bigcup_{n=1}^{\infty} A_n = \bigcup_{n=1}^{\infty} B_n. \tag{29}$$

2.6. *If* $A_n \subset A_{n+1}$ *for* $n = 1, 2, \ldots$ *and* $B_1 = A_1$, $B_n = A_n - A_{n-1}$ *for* $n = 2, 3, \ldots$, *then the sets* B_n *are disjoint and*

$$A_n = B_1 \cup \ldots \cup B_n \quad \textit{for} \quad n = 1, 2, \ldots, \tag{30}$$

$$\bigcup_{n=1}^{\infty} A_n = \bigcup_{n=1}^{\infty} B_n. \tag{31}$$

To prove the first part of 2.4, let us observe that the inclusion $B_n \subset B_{n+1}$ follows directly from the definition of the sets B_n. Since $A_n \subset B_n$ for $n = 1, 2, \ldots$, we have $\bigcup_{n=1}^{\infty} A_n \subset \bigcup_{n=1}^{\infty} B_n$ by (19). Since $B_m \subset \bigcup_{n=1}^{\infty} A_n$ for $m = 1, 2, \ldots$, we have $\bigcup_{n=1}^{\infty} B_n \subset \bigcup_{n=1}^{\infty} A_n$ by (20). The two inclusions just proved imply the equality of the two infinite unions.

The proof of the second part of 2.4 is analogous.

To prove 2.5 let us observe that if $m < n$, then $B_m \cap B_n \subset A_m \cap (A_n - A_m) = 0$, which proves that the sets B_n are disjoint. Let $C_n = A_1 \cup \ldots \cup A_n$. Since $B_j \subset A_j$, we have $B_1 \cup \ldots \cup B_n \subset C_n$. Conversely, if $x \in C_n$, then $x \in A_j$ for some $j \leqslant n$. Let j_0 be the smallest integer such that $x \in A_{j_0}$. If $j_0 = 1$, then $x \in B_1 = A_1$. If $j_0 > 1$, then $x \in A_{j_0} - (A_1 \cup \ldots \cup A_{j_0 - 1}) = B_{j_0}$. Thus $C_n \subset B_1 \cup \ldots \cup B_n$ which completes the proof of (28). Applying twice the first part of 2.4, we infer that

$$\bigcup_{n=1}^{\infty} A_n = \bigcup_{n=1}^{\infty} C_n = \bigcup_{n=1}^{\infty} B_n,$$

since $C_n = A_1 \cup \ldots \cup A_n = B_1 \cup \ldots \cup B_n$.

Theorem 2.6 follows directly from 2.5 since the hypothesis that the sequence (A_n) is increasing implies that

$$A_n = A_1 \cup \ldots \cup A_n \quad \text{and} \quad A_n - A_{n-1} = A_n - (A_1 \cup \ldots \cup A_{n-1}).$$

EXERCISES

1. If $T = \bigcup_{u \in U} T_u$, then

$$\bigcup_{t \in T} A_t = \bigcup_{u \in U} \bigcup_{t \in T_u} A_t \quad \text{and} \quad \bigcap_{t \in T} A_t = \bigcap_{u \in U} \bigcap_{t \in T_u} A_t$$

(the *associative laws* for union and intersection).

2. For any sequence of sets (A_n) the set

$$\limsup_{n\to\infty} A_n = \bigcap_{n=1}^{\infty} \bigcup_{m=0}^{\infty} A_{n+m}$$

is called the *limit superior* of (A_n), and the set

$$\liminf_{n\to\infty} A_n = \bigcup_{n=1}^{\infty} \bigcap_{m=0}^{\infty} A_{n+m}$$

is called the *limit inferior* of (A_n).

Prove that

$$\limsup_{n\to\infty} A_n = \{x\colon \bigwedge_{n} \bigvee_{m>n} (x \in A_m)\}\ ;$$

$$\liminf_{n\to\infty} A_n = \{x\colon \bigvee_{n} \bigwedge_{m>n} (x \in A_m)\}\ ;$$

$$\limsup_{n\to\infty} (A_n \cup B_n) = \limsup_{n\to\infty} A_n \cup \limsup_{n\to\infty} B_n\ ;$$

$$\limsup_{n\to\infty} (A_n \cap B_n) \subset \limsup_{n\to\infty} A_n \cap \limsup_{n\to\infty} B_n\ ;$$

$$\liminf_{n\to\infty} (A_n \cup B_n) \supset \liminf_{n\to\infty} A_n \cup \liminf_{n\to\infty} B_n\ ;$$

$$\liminf_{n\to\infty} (A_n \cap B_n) = \liminf_{n\to\infty} A_n \cap \liminf_{n\to\infty} B_n\ ;$$

$$\liminf_{n\to\infty} A_n \subset \limsup_{n\to\infty} A_n\ .$$

3. If $\limsup_{n\to\infty} A_n = \liminf_{n\to\infty} A_n = A$ we say that the sequence (A_n) *converges* to A (or: A is the *limit* of (A_n)). We write then

$$A = \lim_{n\to\infty} A_n\ .$$

Prove that if $A = \lim_{n\to\infty} A_n$ and $B = \lim_{n\to\infty} B_n$, then

$$A \cup B = \lim_{n\to\infty} (A_n \cup B_n)\ , \qquad A \cap B = \lim_{n\to\infty} (A_n \cap B_n)\ , \qquad A - B = \lim_{n\to\infty} (A_n - B_n)\ .$$

4. If $A_n \subset A_{n+1}$ and $B_n \subset B_{n+1}$ for $n = 1, 2, \ldots$, then

$$\Big(\bigcup_{n=1}^{\infty} A_n\Big) \cap \Big(\bigcup_{n=1}^{\infty} B_n\Big) = \bigcup_{n=1}^{\infty} (A_n \cap B_n)\ .$$

If $A_{n+1} \subset A_n$ and $B_{n+1} \subset B_n$ for $n = 1, 2, \ldots$, then

$$\Big(\bigcap_{n=1}^{\infty} A_n\Big) \cup \Big(\bigcap_{n=1}^{\infty} B_n\Big) = \bigcap_{n=1}^{\infty} (A_n \cup B_n)\ .$$

§ 3. Functions. The second fundamental mathematical notion, besides the notion of set, is that of function.

Let X and Y be sets. By a *function* on (or: from) X into Y we mean a rule which assigns an element in Y to each element x in X. The words *mapping* and *transformation* are often used as synonyms for "function". Functions will be denoted in this book by letters f, g, φ, ψ, Φ, Ψ, etc. If f is a function on X into Y, and $x \in X$, then $f(x)$ denotes the element in Y which is assigned to the element x. The element $f(x)$ is called the *value* of the function f at x (or: the *image* of x under f). The symbol $f(x)$ is the standard notation for the value of a function f at a point x. However, sometimes we shall also use another notation; this will be always explicitly stated.

According to the above convention, the symbols f and $f(x)$ have different meanings. The first symbol denotes the function, i.e., the rule that assigns elements in Y to elements in X. The second symbol denotes the element in Y which was assigned to a given element x in X. However, the above convention is not rigorously applied in practice. Often the symbol $f(x)$, where x denotes an arbitrary element in X (we say then that x is a *variable* running through X), is also used to denote the function f. This notation does not lead, in general, to any misunderstanding and it is convenient in many cases. Tradition is also one of the reasons for using it. For instance, the function which assigns the number $\cos x$ to a real number x is traditionally denoted by $\cos x$ instead of cos. Similarly, we talk about the function $x/(1+x^2)$ (which assigns the number $x/(1+x^2)$ to any real x), especially if we do not want to introduce any symbol to denote this function. In this book, except in the excercises, functions will be denoted by symbols without variables.

The notation $f(x)$ instead of f for functions is applied in older or more elementary mathematical books. In such books another notation is also used: sometimes certain letters are reserved to denote values of a function at arbitrary points. If y is used to denote the values of a function f at any point x, one often says that $y = f(x)$ is a function of x. As a consequence of this non-precise notation we mention here the traditional notation dy/dx for the derivative of a function f, instead of the correct notation f'. The traditional notation $\partial y/\partial x$ for partial derivatives (see p. 134) is another consequence. This non-precise notation for derivatives will not be used in this book in the main text. Since it is generally used in almost all books and since it is sometimes convenient in calculation, it will be used in some of the exercises and in the supplementary text in fine print.

Let f be a mapping from X into Y. The set X on which f is defined is called the *domain* of f. The set of all elements $f(x)$ is called the *range* of f. The range of f is a subset of Y. If Y is the range of f, i.e. if for every $y \in Y$ there is an $x \in X$ such that $y = f(x)$, we say that f is a mapping *onto* Y. For instance, the set R of all real numbers is the domain of the function cos, and the interval $I = \{y: -1 \leqslant y \leqslant 1\}$ is its range; cos is a mapping from R into R but a mapping from R onto I. The domain of the derivative f' of a function f defined on an interval of real numbers is the set of all numbers x at which the function f is differentiable.

We write

$$f\colon X \to Y \tag{1}$$

if we want to indicate that f is a mapping from X into Y.

Let $f\colon X \to Y$. Sometimes we have to examine the behaviour of the function f only on a subset $A \subset X$. More precisely, we examine then a new function $g\colon A \to Y$ defined as follows:

$$g(x) = f(x) \quad \text{for} \quad x \in A .$$

The new function g will be denoted by $f|A$ and will be called the *restriction* of f to A. The domain of $f|A$ is A.

If f_1, f_2: $X \to Y$ and $A \subset X$, then $f_1|A = f_2|A$ if and only if $f_1(x) = f_2(x)$ for every $x \in A$.

Let $A \subset X$ and g: $A \to Y$. Any function f: $X \to Y$ such that $f|A = g$ is said to be an *extension* of g to X.

Suppose f is a function from X onto Y. By hypothesis, for every $y \in Y$ there is at least one $x \in X$ such that $y = f(x)$. If, for every $y \in Y$, there exists exactly one point $x \in X$ such that $y = f(x)$, we say that the function f is *one-to-one*. If f is one-to-one, the function that assigns to $y \in Y$ the unique $x \in X$ such that $y = f(x)$ is denoted by f^{-1} and is called the *inverse* of f. By definition, for any elements $x \in X$ and $y \in Y$

$$y = f(x) \text{ if and only if } x = f^{-1}(y)\,,$$

$$f^{-1}(f(x)) = x\,, \quad f(f^{-1}(y)) = y\,.$$

The inverse f^{-1} of a one-to-one function f is also one-to-one and

$$(f^{-1})^{-1} = f\,.$$

Let f: $X \to Y$ and let $\alpha(x)$ be a condition defined on X. Generalizing the notation introduced in § 1 we shall denote by the symbol

$$\{f(x)\colon\ \alpha(x)\}$$

the set of all elements $f(x)$ for which $\alpha(x)$ holds. For instance, for any set $A \subset X$,

$$\{f(x)\colon\ x \in A\}$$

is the set of all values of the function f on the set A (i.e. the range of the function $f|A$). This set is said to be the *image* of the set A under f and is denoted by $f(A)$. By definition,

$$f(A) = \{f(x)\colon\ x \in A\}\,.$$

More generally, for any condition $\alpha(x)$,

$$\{f(x)\colon\ \alpha(x)\} = f(\{x\colon\ \alpha(x)\})\,.$$

For any function f: $X \to Y$ and for any set B the symbol $f^{-1}(B)$ will denote the set of all $x \in X$ such that $f(x) \in B$. The set $f^{-1}(B)$ is called the *inverse image* of B. By definition

$$f^{-1}(B) = \{x\colon\ f(x) \in B\}\,.$$

3.1. *For any function* f: $X \to Y$,

(2) $$\textit{if } A_1 \subset A_2, \textit{ then } f(A_1) \subset f(A_2)\,;$$

(3) $$\textit{if } B_1 \subset B_2, \textit{ then } f^{-1}(B_1) \subset f^{-1}(B_2)\,;$$

(4) $f(A_1 \cup A_2) = f(A_1) \cup f(A_2)$; *more generally* $f(\bigcup_{t\in T} A_t) = \bigcup_{t\in T} f(A_t)$;

(5) $f^{-1}(B_1 \cup B_2) = f^{-1}(B_1) \cup f^{-1}(B_2)$; *more generally*

$$f^{-1}(\bigcup_{t\in T} A_t) = \bigcup_{t\in T} f^{-1}(A_t);$$

6) $f^{-1}(B_1 \cap B_2) = f^{-1}(B_1) \cap f^{-1}(B_2)$; *more generally*

$$f^{-1}(\bigcap_{t\in T} A_t) = \bigcap_{t\in T} f^{-1}(A_t);$$

(7) $$f^{-1}(B_1 - B_2) = f^{-1}(B_1) - f^{-1}(B_2);$$

in particular

(8) $$f^{-1}(Y - B) = X - f^{-1}(B).$$

If f is one-to-one, then

(9) $f(A_1 \cap A_2) = f(A_1) \cap f(A_2)$; *more generally* $f(\bigcap_{t\in T} A_t) = \bigcap_{t\in T} f(A_t)$;

(10 $$f(A_1 - A_2) = f(A_1) - f(A_2).$$

The proof of each of these identities is by verification that a point belonging to the set on one side of the identity also belongs to the set on the other side.

If $f: X \to Y$ and $g: Y \to Z$, then the function $h: X \to Z$ defined by

$$h(x) = g(f(x))$$

is called the *composition* of g and f and is denoted by the symbol $g \circ f$. By definition

(11) $$g \circ f(x) = g(f(x)) \quad \text{for every } x \in X.$$

Observe that

$$g \circ f(A) = g(f(A)) \quad \text{and} \quad (g \circ f)^{-1}(B) = f^{-1}(g^{-1}(B)).$$

If f and g are one-to-one, so is $g \circ f$ and

(12) $$(g \circ f)^{-1} = f^{-1} \circ g^{-1}.$$

In the traditional notation the formula for the derivative of the composition $z = g(f(x))$ of the functions $z = g(y)$ and $y = f(x)$ is written as follows

$$\frac{dz}{dx} = \frac{dz}{dy} \cdot \frac{dy}{dx}.$$

Using the notation $g \circ f$ for the composition of functions, this formula should be written as follows

13) $$(g \circ f)' = (g' \circ f) \cdot f'.$$

This identity at any point x can be written in the form

(13') $$(g \circ f)'(x) = (g' \circ f)(x) \cdot f'(x)$$

or

(13'') $$(g \circ f)'(x) = g'(f(x)) \cdot f'(x) .$$

Sometimes the following generalization of the definition of the composition of functions is convenient. If $f: X_1 \to Y_1$ and $g: X_2 \to Y_2$, then the composition $g \circ f$ is the function defined on the set $A = f^{-1}(X_2)$ by the formula

$$g \circ f(x) = g(f(x)) \quad \text{for} \quad x \in A .$$

Since it was not supposed that the range of f is a subset of the domain of g, the domain A of the composition $g \circ f$ is a subset of the domain X_1 of f; in general, A is not identical with X_1. Observe that

$$g \circ f = g \circ (f|A) ;$$

the composition on the right-hand side satisfies the hypotheses of definition (11) since $f|A: A \to X_2$ and $g: X_2 \to Y_2$.

We do not exclude the case where the domain of a function is the empty set; then the function is said to be *empty*. There is only one empty function (i.e. all empty functions are equal). We are not interested in examination of the empty function and normally we suppose that domains of functions under consideration are not empty. However, the empty function appears often in mathematical investigation. For instance, if in the generalized definition of the composition $g \circ f$ the sets Y_1 and X_2 are disjoint, then the domain A of $g \circ f$ is empty, i.e. $g \circ f$ is the empty function. Another example will be taken from algebra. Consider a fixed real polynomial. Let X be the set of all real roots of the polynomial and let f be a function on X that assigns to every root $x \in X$, its multiplicity. The function f is well defined in spite of the fact that we do not know if X is empty or not. If X is empty, then f is the empty function.

Let $f: X \to Y$.

If Y is the set of all real numbers, then f is said to be a *real function*. If Y is the set of all real numbers including $\pm\infty$, then f is said to be an *extended real function*. Sometimes we say *a finite real function* instead of "a real function" to emphasize that $\pm\infty$ are not values of f.

If X is a set a real numbers, then f is said to be a *function of a real variable* (or *of one real variable*).

If X is the set of all positive integers or of all non-negative integers (Y being arbitrary), then f is said to be an *infinite sequence*, or simply a *sequence*. The element $f(n)$ is said to be the nth *term* of the sequence. In general we use for sequences a notation different from the standard one. The nth term of a sequence is usually denoted by a symbol of the form a_n etc., and the sequence (i.e. function) is denoted by (a_n) or $(a_1, a_2, \ldots)$. If (a_n) is a sequence of elements, and (m_n) is an increasing sequence of positive integers, then the sequence (a_{m_n}), i.e. the sequence $(a_{m_1}, a_{m_2}, \dots)$, is said to be a *subsequence* of (a_n).

If X is the set of positive integers $1, 2, \dots, k$, then any function on X is called a *finite sequence*, more precisely: *k-term sequence*. We use for finite sequences a notation similar to that for infinite sequences. A *k-term* sequence, the nth term of which is a_n $(n = 1, 2, \dots, k)$, is denoted by $(a_1, \dots, a_k)$. Two-term sequences are called *ordered pairs*, or simply *pairs*.

EXERCISES

1. If f is a one-to-one mapping of a set T onto itself, then

$$\bigcup_{t\in T} A_t = \bigcup_{t\in T} A_{f(t)} \quad \text{and} \quad \bigcap_{t\in T} A_t = \bigcap_{t\in T} A_{f(t)}$$

(the *commutative laws* for union and intersection).

2. If $f\colon X \to Y$ and $A, A_1, A_2 \subset X$, $B \subset Y$, then

$$f(A_1 \cap A_2) \subset f(A_1) \cap f(A_2)\,, \quad f(A_1) - f(A_2) \subset f(A_1 - A_2)\,,$$

$$A \subset f^{-1}(f(A))\,, \quad f(f^{-1}(B)) \subset B\,.$$

Give examples showing that $\subset$ cannot be replaced by $=$. The equation $f(f^{-1}(B)) = B$ holds if and only if $B \subset f(X)$.

Prove that

$$f(A) \cap B = f(A \cap f^{-1}(B))\,;$$

$$\text{if } f^{-1}(f(A_1)) = A_1, \text{ then } f(A_1 \cap A_2) = f(A_1) \cap f(A_2)\,;$$

$$\text{if } g = f|A, \text{ then } g^{-1}(B) = A \cap f^{-1}(B)\,;$$

$$\text{if } X = \bigcup_{t\in T} X_t \text{ and } f_t = f|X_t, \text{ then } f^{-1}(B) = \bigcup_{t\in T} f_t^{-1}(B)\,.$$

3. Suppose the sets X_t are disjoint and $X = \bigcup_{t\in T} X_t$. A mapping $f\colon X \to Y$ is one-to-one if and only if all the mappings $f|X_t$ are one-to-one and the sets $f(X_t)$ are disjoint.

§ 4. Real numbers. Sequences and series of real numbers. By a *real number* we shall always mean a finite number. The set of all real numbers will always be denoted by R. However, the infinite real numbers ∞ and $-\infty$ will also appear in our investigations. Finite and infinite real numbers will be called *extended real numbers*.

Of course we assume that

$$-\infty < a < \infty \quad \text{for every } a \in R\,.$$

We admit the following natural operations on the infinite real numbers (where $a \in R$)

$$\infty + a = a + \infty = a - (-\infty) = \infty + \infty = \infty - (-\infty) = \infty\,,$$

$$-\infty+a = a+(-\infty) = a-\infty = (-\infty)-\infty = (-\infty)+(-\infty) = -\infty\,,$$

$$\infty\cdot a = a\cdot\infty = \infty\cdot\infty = (-\infty)(-\infty) = (-a)(-\infty) = (-\infty)(-a) = \infty \quad \text{for} \quad a>0\,,$$

$$(-\infty)\cdot a = a\cdot(-\infty) = \infty\cdot(-\infty) = (-\infty)\cdot\infty = (-a)\cdot\infty = \infty\cdot(-a) = -\infty \quad \text{for} \quad a>0\,,$$

$$-(-\infty) = \infty\,,$$

$$\frac{a}{\infty} = \frac{a}{-\infty} = 0\,,$$

$$\frac{a}{0} = \infty \quad \text{and} \quad \frac{-a}{0} = -\infty \quad \text{for} \quad a>0\,.$$

We also admit the following convention

$$0\cdot(\pm\infty) = 0\,,$$

which will be used in Chapter VII. Other operations are assumed to be non-feasible. In particular, the operations $\infty-\infty$ and $\infty+(-\infty)$ are not feasible.

If Z is a non-empty set of extended real numbers, then

$$\sup Z \quad \text{and} \quad \inf Z$$

will denote respectively the least upper bound and the greatest lower bound of Z, respectively. By definition, $\sup Z$ is the smallest extended real number a such that $a \geqslant x$ for every $x \in Z$, and $\inf Z$ is the greatest extended real number a such that $a \leqslant x$ for every $x \in Z$.

If $(a_1, a_2, \ldots)$ is any sequence of extended real numbers, the following shorter notation will be used

$$\sup a_n = \sup_n a_n = \sup\{a_n\colon\ n = 1, 2, \ldots\}\,,$$

$$\sup_{n\geqslant m} a_n = \sup\{a_n\colon\ n = m, m+1, \ldots\}\,,$$

$$\inf a_n = \inf_n a_n = \inf\{a_n\colon\ n = 1, 2, \ldots\}\,,$$

$$\inf_{n\geqslant m} a_n = \inf\{a_n\colon\ n = m, m+1, \ldots\}\,.$$

Let

(1) $$u_m = \sup_{n\geqslant m} a_n\,, \qquad v_m = \inf_{n\geqslant m} a_n \qquad (m = 1, 2, \ldots)\,.$$

The numbers

(2) $$\inf u_n\,, \qquad \sup v_n$$

are called the *limit superior* and the *limit inferior* of (a_n) and denoted by

$$\limsup_{n\to\infty} a_n\,, \qquad \liminf_{n\to\infty} a_n$$

respectively. By definition,

$$-\infty \leqslant \liminf_{n\to\infty} a_n \leqslant \limsup_{n\to\infty} a_n \leqslant \infty\,.$$

If the limit superior and the limit inferior are equal, their common values a is called the *limit* of (a_n) and denoted by

$$\lim_{n\to\infty} a_n\,.$$

We also write

$$a_n \to a\,.$$

If the limit $a = \lim\limits_{n\to\infty} a_n$ exists and is finite, we say that the sequence (a_n) *converges* to a. The sequence (a_n) converges to a if and only if for every $\varepsilon > 0$ there exists an integer n_0 such that

$$|a_n - a| < \varepsilon \qquad \text{for every } n > n_0\,.$$

Every monotonic sequence (a_n) has a limit. We write

$$a_n \nearrow a$$

if a is the limit of a non-decreasing sequence (a_n); then $a = \sup a_n$. We write

$$a_n \searrow a$$

if a is the limit of a non-increasing sequence (a_n); then $a = \inf a_n$.

For every sequence (a_n), the sequences (u_n) and (v_n) defined by (1) are respectively non-increasing and non-decreasing. Thus

$$\limsup_{n\to\infty} a_n = \lim_{m\to\infty} \sup_{n\geqslant m} a_n\,, \qquad \liminf_{n\to\infty} a_n = \lim_{m\to\infty} \inf_{n\geqslant m} a_n\,.$$

If an infinite series $\sum\limits_{n=1}^{\infty} a_n$ is examined, we shall always assume that either all the terms a_n are non-negative extended real numbers, or all the terms a_n are finite. In any case the sequence

$$s_n = a_1 + \ldots + a_n \qquad (n = 1, 2, \ldots)$$

of partial sums is defined. The limit $s = \lim\limits_{n\to\infty} s_n$, if exists, is said to be the *sum* of the series; in symbols

$$s = \sum_{n=1}^{\infty} a_n\,.$$

If s is finite, the series is said to *converge*.

If $a_n \geqslant 0$ for $1, 2, \ldots$, the sum s of the series always exists $(0 \leqslant s \leqslant \infty)$ and it does not depend on the arrangement of the terms, i.e., the addition of an infinite sequence of non-negative numbers is commutative. It is also associative, i.e., the result is the same, however the terms are grouped in finite or infinite collections.

Sometimes we shall deal with double series $\sum_{m,n=1}^{\infty} a_{m,n}$ but only in the case where $a_{m,n} \geqslant 0$ for $m, n = 1, 2, \ldots$ The sum s of this series is defined to be the limit of the sequence

$$s_n = \sum_{i,j=1}^{n} a_{i,j} \qquad (n = 1, 2, \ldots)$$

of partial sums. The sum s always exists $(0 \leqslant s \leqslant \infty)$. Also in this case the addition is associative and commutative. In particular,

$$\sum_{m,n=1}^{\infty} a_{m,n} = \sum_{m=1}^{\infty} \sum_{n=1}^{\infty} a_{m,n} = \sum_{n=1}^{\infty} \sum_{m=1}^{\infty} a_{m,n} .$$

For any extended real numbers a, b, let

$$\max(a, b) = \begin{cases} a & \text{if} \quad a \geqslant b , \\ b & \text{if} \quad b \geqslant a , \end{cases}$$

$$\min(a, b) = \begin{cases} a & \text{if} \quad a \leqslant b , \\ b & \text{if} \quad b \leqslant a . \end{cases}$$

We define analogously $\max(a_1, \ldots, a_n)$ and $\min(a_1, \ldots, a_n)$ for any finite sequence of extended real numbers.

EXERCISES

1. Verify that

$$\min(a, b) = \tfrac{1}{2}(a + b - |b - a|) , \qquad \max(a, b) = \tfrac{1}{2}(a + b + |b - a|) , \qquad |a| = \max(a, -a) ,$$

$$\min(a, \max(b, c)) = \max(\min(a, b), \min(a, c)) ,$$

$$\max(a, \min(b, c)) = \min(\max(a, b), \max(a, c)) .$$

2. Prove that

$$\min(\sup A, \sup B) = \sup\{\min(a, b) \colon (a \in A) \& (b \in B)\} ;$$

$$\max(\inf A, \inf B) = \inf\{\max(a, b) \colon (a \in A) \& (b \in B)\} .$$

§ 5. Operations on functions. Let f and g be extended real functions. By the *sum*, *difference*, *product* and *quotient* of f and g we shall mean the functions, denoted by symbols $f+g$, $f-g$, fg (or $f\cdot g$), $\frac{f}{g}$ (or f/g) respectively, defined by the equations

$$(f+g)(x) = f(x)+g(x)\,,$$

$$(f-g)(x) = f(x)-g(x)\,,$$

$$fg(x) = f(x)g(x)\,,$$

$$\frac{f}{g}(x) = \frac{f(x)}{g(x)}\,.$$

The domain of $f+g$ is the set of the points x at which $f(x)$ and $g(x)$ are defined and the addition $f(x)+g(x)$ is feasible. Thus if f and g are finite, the domain of $f+g$ is the intersection of the domains of f and g, but if f and g assume infinite values, the domain of $f+g$ can be a proper subset of the common part of the domains of f and g.

The same remarks hold for the domains of the difference, product and quotient of f and g.

Of course, the only interesting case is when the domain of $f+g$ (or $f-g$, fg, f/g) is not empty. Most often we shall assume that the domains of f and g are the same set which is also the domain of the result of the operation on f and g in question.

By the *absolute value* of an extended function f we mean the function denoted by $|f|$ and defined by

$$|f|(x) = |f(x)|\,.$$

The domain of $|f|$ coincides with the domain of f.

If f, g are extended real functions, then $\max(f, g)$ is the function whose value at a point x is equal to $\max\big(f(x), g(x)\big)$. Similarly we define the functions $\min(f, g)$, $\max(f_1, \dots, f_n)$, $\min(f_1, \dots, f_n)$, $\sup f_n$, $\inf g_n$, $\limsup\limits_{n\to\infty} f_n$, $\liminf\limits_{n\to\infty} f_n$, $\lim\limits_{n\to\infty} f_n$. It is obvious what the domains of these functions are. As a rule, we shall suppose in all the above cases that all the functions f, g or $f_1, f_2, \dots$ have a common domain X. Then the set X is the domain of the results of the all the above operations, except perhaps for $\lim\limits_{n\to\infty} f_n$, whose domain is the set of all the points $x \in X$ such that $\lim\limits_{n\to\infty} f_n(x)$ (the value of the function $\lim\limits_{n\to\infty} f_n$ at the point x) exists. If all the functions f, f_n have the same domain and $f = \lim\limits_{n\to\infty} f_n$, we write

(1) $$f_n \to f\,.$$

We write

$$f_n \nearrow f \quad \text{or} \quad f_n \searrow f$$

respectively, if (1) holds and the sequence (f_n) is non-decreasing or non-increasing, i.e., if

$$f_n \leqslant f_{n+1} \ (n = 1, 2, \ldots) \quad \text{or} \quad f_n \geqslant f_{n+1} \ (n = 1, 2, \ldots)$$

respectively.

Of course, if f and g are extended real functions defined on the same set X, we write $f < g$ or $f \leqslant g$ provided that

$$f(x) < g(x) \text{ for every } x \in X, \quad \text{or} \quad f(x) \leqslant g(x) \text{ for every } x \in X,$$

respectively.

In order to simplify the notation, a constant extended real function, i.e. a function defined on a set X, assuming only one value c, will be denoted by the same letter c. In any particular case the domain of the constant function c is usually apparent.

Interpreting an extended real number c as a function defined on the domain of an extended real function f, it is easy to see the meaning of the symbols

$$f+c, \quad f-c, \quad cf, \quad f/c.$$

Also it is obvious what we mean by the inequalities

$$f < c, \quad f > c, \quad f \leqslant c, \quad f \geqslant c.$$

The first inequality holds if $f(x) < c$ for every x in the domain of f, and similarly for the remaining ones.

If $c_1 \leqslant f \leqslant c_2$ for some finite real numbers c_1, c_2, the function f is said to be *bounded*.

Besides real functions, in this book we shall also examine functions whose values are vectors or matrices (for details, see II § 7) on which certain operations like addition, subtraction, multiplication, etc., can be performed. The reader will guess without any difficulty what we shall mean by the sum (difference, product, etc.) of such functions: this is a function whose value at a given point is equal to the sum (difference, product, etc.) of the values of the summands, etc.

§ 6. Cartesian products. By the *Cartesian product* (or simply *product*) $X_1 \times \ldots \times X_k$ of sets $X_1, \ldots, X_k$ we mean the set of all finite sequences

$$(x_1, \ldots, x_k)$$

where $x_i \in X_i$ for $i = 1, \ldots, k$. In particular, the product $X \times Y$ of two sets X, Y is the set of all pairs (x, y) where $x \in X$ and $y \in Y$. In symbols

$$X \times Y = \{(x, y): (x \in X) \& (y \in Y)\}.$$

Note that $(x, y) \neq (y, x)$ if $x \neq y$.

6.1. *For any sets,*

(1) $$A \times B = 0 \text{ if and only if } A = 0 \text{ or } B = 0\,;$$

(2) $$\text{if } A \subset C \text{ and } B \subset D, \text{ then } A \times B \subset C \times D\,;$$

(3) $$\begin{aligned} A \times (B \cup C) &= A \times B \cup A \times C\,, \\ A \times (B \cap C) &= A \times B \cap A \times C\,, \\ A \times (B - C) &= A \times B - A \times C\,; \end{aligned}$$

(3′) $$\begin{aligned} (A \cup B) \times C &= A \times C \cup B \times C\,, \\ (A \cap B) \times C &= A \times C \cap B \times C\,, \\ (A - B) \times C &= A \times C - B \times C\,; \end{aligned}$$

(4) $$(A \times B) \cap (C \times D) = (A \cap C) \times (B \cap D)\,;$$

(5) $$A \times B - C \times D = \big(A \times (B - D)\big) \cup \big((A - C) \times B\big)\,;$$

(6) $$A \times \bigcup_{t \in T} B_t = \bigcup_{t \in T} (A \times B_t)\,, \qquad A \times \bigcap_{t \in T} B_t = \bigcap_{t \in T} (A \times B_t)\,;$$

(6′) $$\Big(\bigcup_{t \in T} A_t\Big) \times B = \bigcup_{t \in T} (A_t \times B)\,, \qquad \Big(\bigcap_{t \in T} A_t\Big) \times B = \bigcap_{t \in T} (A_t \times B)\,.$$

One proves these identities by showing that a point (x, y) belongs to the set on the left-hand side of the equation if and only if it belongs to the set on the right-hand side.

The Cartesian product

$$X \times X \times \ldots \times X \qquad (k \text{ times})$$

of k copies of a set X will be denoted by X^k. By definition, X^k is the set of all finite sequences

$$(x_1, \ldots, x_k) \quad \text{where} \quad x_i \in X \quad \text{for } i = 1, \ldots, k\,.$$

If $k = m + n$, where m and n are positive numbers, the Cartesian product X^k can be conceived of as the Cartesian product of X^m and X^n. This will be done in this book in the following way. Fix two sequences

$$\mathbf{i} = (i_1, \ldots, i_m)\,, \qquad \mathbf{j} = (j_1, \ldots, j_n)$$

of positive integers such that $i_1, \ldots, i_m, j_1, \ldots, j_n$ is a permutation of the integers $1, \ldots, k$. By definition, every positive integer $\leqslant k$ appears exactly once in one of the sequences $\mathbf{i}, \mathbf{j}$. Any point

(7) $$p = (x_1, \ldots, x_k) \in X^k$$

will be identified with the point

(8) $$(p_1, p_2) \in X^m \times X^n\,,$$

where

$$(9) \qquad p_1 = (x_{i_1}, \ldots, x_{i_m}) \in X^m ,$$

$$(10) \qquad p_2 = (x_{j_1}, \ldots, x_{j_n}) \in X^n .$$

We say then that X^k is interpreted as the $(\mathbf{i}, \mathbf{j})$-*product* of X^m and X^n. We then write

$$(11) \qquad p = (p_1, p_2) .$$

The same identification can be made in the case of products $X_1 \times X_2 \times \ldots \times X_k$ of different sets $X_1, \ldots, X_k$; it is not described here because it will not be applied in this book.

In particular, the product X^{k+1} will be often conceived of as the product $X^k \times X$ by identification of any point $(x_1, \ldots, x_k, x) \in X^{k+1}$ with the point $(p, x) \in X^k \times X$ where $p = (x_1, \ldots, x_k)$. This is the case where $\mathbf{i} = (1, \ldots, k)$ and $\mathbf{j} = (k+1)$.

If f is a function defined on a subset A of a product $X_1 \times \ldots \times X_k$, then the value $f(p)$ of f at a point $p = (x_1, \ldots, x_k)$ will also be denoted by the symbol

$$(12) \qquad f(x_1, \ldots, x_k) .$$

In this case the function f is often said to be a *function of k variables* $x_1, \ldots, x_k$. This terminology is far from precise; its purpose is to emphasize that f is defined on a subset of the product of k sets.

If we fix the values of the $k-1$ variables $x_1, \ldots, x_{i-1}, x_{i+1}, \ldots, x_k$, then the expression (12) can be conceived of as a function of one variable x_i defined on the set

$$A_i = \{x_i \colon (x_1, \ldots, x_{i-1}, x_i, x_{i+1}, \ldots, x_k) \in A\} \subset X_i .$$

Of course, the only interesting case is where the set A_i is not empty.

Let f be a function defined on a set $A \subset X^k$. If X^k is interpreted as an $(\mathbf{i}, \mathbf{j})$-product of X^m and X^n, then the value of f at the point (11) will also be denoted by

$$(13) \qquad f(p_1, p_2) .$$

If a point $p_2 \in X^n$ is fixed, then (13) can be considered as a function of p_1, defined on the set

$$A_1 = \{p_1 \in X^m \colon (p_1, p_2) \in A\} .$$

This function is denoted by the symbol

$$f(\cdot, p_2) .$$

Similarly, if we fix a point $p_1 \in X^m$, then (13) can be considered as a function of p_2 on the set

$$A_2 = \{p_2 \in X^n \colon (p_1, p_2) \in A\} .$$

This function is denoted by

$$f(p_1, \cdot)\,.$$

By definition,

$$f(\cdot, p_2)(p_1) = f(p_1, p_2)\,, \qquad f(p_1, \cdot)(p_2) = f(p_1, p_2)\,.$$

Now let f be a mapping of a set $A \subset X$ into Y. The set

$$\{(x, f(x)):\ x \in A\} \subset X \times Y$$

is called the *graph* of f.

EXERCISE

1. Prove that

$$\limsup_{n\to\infty} (A_n \times B_n) \subset \limsup_{n\to\infty} A_n \times \limsup_{n\to\infty} B_n\,;$$

$$\liminf_{n\to\infty} (A_n \times B_n) = \liminf_{n\to\infty} A_n \times \liminf_{n\to\infty} B_n\,.$$

If (A_n) and (B_n) are convergent, then so is $(A_n \times B_n)$ and

$$\lim_{n\to\infty} (A_n \times B_n) = \lim_{n\to\infty} A_n \times \lim_{n\to\infty} B_n\,.$$

§ 7. Countable sets. A set X is said to be *countable* if there is a sequence (x_n) containing all the elements of X (and conceivably some other elements), i.e. such that for every $x \in X$ there is an n such that $x = x_n$. A set X is *uncountable* if it is not countable.

By definition, every finite set (in particular, the empty set) is countable. For if $x_1, \ldots, x_m$ ($m \geqslant 0$) are all the elements in X, then defining x_n for $n > m$ in an arbitrary way we get a sequence containing all elements in X.

The set of all positive integers is an example of an infinite countable set (define $x_n = n$ for $n = 1, 2, \ldots$). The set of all non-negative integers is also countable and infinite (define $x_n = n-1$ for $n = 1, 2, \ldots$). The set of all integers is countable too, since the sequence

$$(0, 1, -1, 2, -2, 3, -3, \ldots)$$

contains all the integers.

It is less obvious that

7.1. *The set of all rational numbers is countable.*

Indeed, the sequence

$$\left(\frac{0}{1}, -\frac{1}{1}, \frac{0}{2}, \frac{1}{1}, -\frac{2}{1}, -\frac{1}{2}, \frac{0}{3}, \frac{1}{2}, \frac{2}{1}, \ldots\right)$$

obtained by writing all the fractions (with positive denominators) for which the sum of the absolute value of the numerator and of the denominator is equal to one, then those for which the sum is equal to 2, 3, etc., contains all rational numbers.

However the interval $I = \{x\colon -1 < x < 1\}$ is uncountable. For suppose that I is countable, i.e. there is a sequence (x_n) containing all real numbers in I. We may suppose that all the terms x_n are in I. Let

$$0, a_{n,1}\, a_{n,2} \ldots$$

be the decimal expansion of the number x_n. The number x whose decimal expansion is

$$0, a_1 a_2 \ldots$$

where

$$a_n = \begin{cases} 4 & \text{if} \quad a_{n,n} \geqslant 5\,, \\ 5 & \text{if} \quad a_{n,n} < 5\,, \end{cases}$$

is in I but $x \neq x_n$ for $n = 1, 2, \ldots$ This contradicts the hypothesis that (x_n) contains all numbers from I.

It follows directly from the definition that

7.2. *Any subset of a countable set is countable.*

Thus if $A \subset B$ and A is uncountable, so is B. Hence it follows that the set R of all real numbers is uncountable.

7.3. *The union of a finite or infinite sequence of countable sets is countable.*

It is enough to prove the theorem for an infinite sequence of sets (if the sequence is finite, repeat one term infinitely many times).

Suppose the sequence $(x_{n,1}, x_{n,2}, \ldots)$ contains all elements of X_n. The sequence

$$(x_{1,1}, x_{2,1}, x_{1,2}, x_{3,1}, x_{2,2}, x_{1,3}, x_{4,1}, \ldots)$$

obtained by writing first all the terms for which the sum of indices is equal to 2, then those for which the sum is equal to 3, 4, etc., contains all elements of $X_1 \cup X_2 \cup \ldots$

7.4. *If A is a countable set, f is a function on A and $B \subset f(A)$, then B is countable.*

If (x_n) contains all elements in A, and (y_n) is a sequence such that $y_n = f(x_n)$ if $x_n \in A$, then (y_n) contains all elements of B.

7.5. *If there is a one-to-one mapping g from B into a countable set A, then B is countable.*

The function $f = g^{-1}$ maps the set $A_0 = g(B)$ onto B. Since $A_0 \subset A$, the set A_0 is countable. By 7.4, the set $B = f(A_0)$ is countable.

7.6. *The Cartesian product of a finite number of countable sets is countable.*

Suppose that $(x_{j,1}, x_{j,2}, .)$ contains all elements of $X_j, j = 1, \ldots, m$. The sequence whose terms are finite sequences

(1) $$(x_{1,n_1}, \ldots, x_{m,n_m})$$

and which is obtained by writing all the terms (1) for which the sum $n_1 + \ldots + n_m$ is equal to m, then those for which the sum is equal to $m+1$, $m+2$, etc., contains all the elements of $X_1 \times \ldots \times X_m$.

Let N be the set of all positive integers. By 7.6 the set N^2 of all ordered pairs of positive integers is countable, i.e. there exists an infinite sequence

(2) $$((m_1, n_1), (m_2, n_2), (m_3, n_3), \ldots)$$

which contains all pairs of positive integers. We may assume that every ordered pair of positive integers appears in (2) exactly once (for if not, we may cancel the pairs which are a repeatition of a pair). Now let $(a_{m,n})$ be a double sequence the terms of which are completely arbitrary. The infinite sequence

(3) $$(a_{m_1,n_1}, a_{m_2,n_2}, a_{m_3,n_3}, \ldots)$$

contains all the terms of the double sequence $(a_{m,n})$. We say that the sequence (3) was obtained by an *arrangement of the double sequence* $(a_{m,n})$ *into a single sequence.*

EXERCISES

1. Prove that the set of all infinite sequences of positive integers and the set of all infinite sequences of numbers 0,1 are uncountable.

Hint. Apply a method similar to that used in the proof that the unit interval is uncountable.

2. The set of all real numbers

(4) $$x = \sum_{n=1}^{\infty} \frac{2a_n}{3^n} \quad \text{where } a_n = 0 \text{ or } 1$$

is called the *Cantor set.* Prove that the mapping f which assigns the number (4) to every infinite sequence (a_n) of numbers 0, 1 is one-to-one and maps the set of all these sequences onto the Cantor set. Thus the Cantor set is uncountable.

CHAPTER II

FUNDAMENTAL GEOMETRIC AND ALGEBRAIC NOTIONS

§ 1. Points and vectors in Euclidean spaces. By the *k-dimensional Euclidean space* ($k = 1, 2, \ldots$) we understand the Cartesian product R^k of k copies of the set R of all real numbers, i.e. the set of all k-term sequences

$$p = (x_1, \ldots, x_k) \tag{1}$$

where $x_1, \ldots, x_k$ are finite real numbers. Sequences (1) are called *points* in R^k. The numbers $x_1, \ldots, x_k$ are called the *coordinates* of the point (1); more precisely, x_i is the *i-th coordinate* of the point p, $i = 1, \ldots, k$. The one-dimensional Euclidean space R^1 is the set R of all real numbers.

Points in R^k will be denoted by the letters p, q (with indices).

One of the most fundamental notions of geometry of Euclidean spaces is that of *vector*. The notion of a vector in R^k can be defined in various ways. It is immaterial what definition is assumed. For instance, a vector $\mathfrak{a}$ in R^k can be defined as a mapping (from R^k onto itself) which assigns to every point (1) the point

$$q = (x_1 + a_1, \ldots, x_k + a_k) \tag{2}$$

where $a_1, \ldots, a_k$ is a given fixed sequence of real numbers. The vector $\mathfrak{a}$ is denoted by

$$[a_1, \ldots, a_k] .$$

The numbers $a_1, \ldots, a_k$ are called the *coordinates* of $\mathfrak{a}$. More precisely, a_i is the *i-th coordinate* of $\mathfrak{a}$.

The point (2) is usually denoted by $p + \mathfrak{a}$ (instead of the standard notation $\mathfrak{a}(p)$ for values of mappings). By definition,

(3) if $p = (x_1, \ldots, x_k)$ and $\mathfrak{a} = [a_1, \ldots, a_k]$, then

$$p + \mathfrak{a} = (x_1 + a_1, \ldots, x_k + a_k) .$$

Graphically the vector $\mathfrak{a}$ is represented by an arrow whose initial point is a point $p \in R^k$ and whose end-point is the point $q = p + \mathfrak{a}$. It is immaterial what point p is chosen to be the initial point. However, sometimes there are reasons for selecting the initial point p in a special way. For instance, if we speak about a vector tangent to a curve C at a point $p \in C$ (see p. 170 for the definition) we usually take the point p as the initial point in the graphical representation of the vector, etc.

The reader who knows another definition of a vector in R^k can use it (then (3) should be treated as a definition of $p+\mathfrak{a}$). The only use we shall make of any definition of vectors is that every vector in R^k is uniquely determined by the sequence of its coordinates, which can be arbitrary finite real numbers.

Instead of "vector in R^k" we shall also say *k-dimensional vector*. The symbol V_k will always stand for the set of all vectors in R^k. According to I § 6, the symbol V_k^m will stand for the Cartesian product $V_k \times \ldots \times V_k$ m-times. Vectors in Euclidean spaces will be denoted by small German letters $\mathfrak{a}, \mathfrak{b}, \mathfrak{c}, \ldots, \mathfrak{w}, \mathfrak{v}, \ldots$

Note that for any given points

(4) $$p = (x_1, \ldots, x_k), \qquad q = (y_1, \ldots, y_k)$$

in R^k there is exactly one vector $\mathfrak{a}$ such that $q = p+\mathfrak{a}$, viz. the vector whose coordinates are the numbers $y_i - x_i$, $i = 1, \ldots, k$. The vector $\mathfrak{a}$ is called the *difference* of the points q and p and is denoted by $q-p$. By definition,

(5) $$q-p = [y_1-x_1, \ldots, y_k-x_k] \quad \text{and} \quad p+(q-p) = q$$

for any points (4).

If all the coordinates of a vector $\mathfrak{a}$ in R^k are equal to zero, then $\mathfrak{a}$ is said to be the *zero vector* in R^k and is denoted by 0. By definition,

$$0 = [0, \ldots, 0].$$

For instance, $p-p = 0$ for every point $p \in R^k$.

We shall perform some algebraical operations on vectors in R^k.

The *product* of a vector

(6) $$\mathfrak{a} = [a_1, \ldots, a_k]$$

and a real number a is the vector

(7) $$a\mathfrak{a} = [aa_1, \ldots, aa_k].$$

We write $-\mathfrak{a}$ instead of $(-1)\mathfrak{a}$, and $\mathfrak{a}/a$ instead of $a^{-1}\mathfrak{a}$ $(a \neq 0)$. Sometimes we write $\mathfrak{a}a$ instead of $a\mathfrak{a}$.

The *sum* of a vector (6) and a vector

(8) $$\mathfrak{b} = [b_1, \ldots, b_k]$$

is the vector

(9) $$\mathfrak{a}+\mathfrak{b} = [a_1+b_1, \ldots, a_k+b_k],$$

and the *difference* of the vectors (6), (8) is the vector

(10) $$\mathfrak{a}-\mathfrak{b} = [a_1-b_1, \ldots, a_k-b_k].$$

Note the following fundamental properties of addition of vectors and multiplication by numbers:

(11) $$\mathfrak{a}+\mathfrak{b} = \mathfrak{b}+\mathfrak{a}\,;$$

(12) $$\mathfrak{a}+(\mathfrak{b}+\mathfrak{c}) = (\mathfrak{a}+\mathfrak{b})+\mathfrak{c}\,;$$

(13) $$\text{if } \mathfrak{a}+\mathfrak{b} = \mathfrak{a}+\mathfrak{c}, \text{ then } \mathfrak{b} = \mathfrak{c}\,;$$

(14) $$a(\mathfrak{a}+\mathfrak{b}) = a\mathfrak{a}+a\mathfrak{b}\,;$$

(15) $$(a+b)\mathfrak{a} = a\mathfrak{a}+b\mathfrak{a}\,;$$

(16) $$a(b\mathfrak{a}) = (ab)\mathfrak{a}\,;$$

(17) $$1\mathfrak{a} = \mathfrak{a}\,.$$

The difference of vectors can be defined by means of addition and multiplication:

$$\mathfrak{a}-\mathfrak{b} = \mathfrak{a}+(-\mathfrak{b})\,.$$

Note that

(18) $$\mathfrak{a}+(\mathfrak{b}-\mathfrak{a}) = \mathfrak{b}\,,$$

(19) $$(a-b)\mathfrak{a} = a\mathfrak{a}-a\mathfrak{b}\,, \qquad a(\mathfrak{a}-\mathfrak{b}) = a\mathfrak{a}-a\mathfrak{b}\,,$$

(20) $$\mathfrak{a}\pm 0 = \mathfrak{a}\,, \qquad a0 = 0\,, \qquad 0\mathfrak{a} = 0\,, \qquad \mathfrak{a}-\mathfrak{a} = 0\,.$$

For any points $p_1, \dots, p_m \in R^k$,

(21) $$(p_1-p_2)+(p_2-p_3)+\dots+(p_{m-1}-p_m) = p_1-p_m\,.$$

A vector $\mathfrak{a}$ is said to be a *linear combination* of vectors $\mathfrak{a}_1, \dots, \mathfrak{a}_m$ if there exist real numbers $a_1, \dots, a_m$ such that

$$\mathfrak{a} = a_1\mathfrak{a}_1+\dots+a_m\mathfrak{a}_m\,.$$

Vectors $\mathfrak{a}_1, \dots, \mathfrak{a}_m$ are said to be *linearly independent* if the equation

(22) $$a_1\mathfrak{a}_1+\dots+a_m\mathfrak{a}_m = 0$$

holds if and only if $a_1 = a_2 = \dots = a_m = 0$. Otherwise they are *linearly dependent*.

If vectors $\mathfrak{a}_1, \dots, \mathfrak{a}_m$ are linearly independent, and $\mathfrak{a}, \mathfrak{a}_1, \dots, \mathfrak{a}_m$ are linearly dependent, then $\mathfrak{a}$ is a linear combination of $\mathfrak{a}_1, \dots, \mathfrak{a}_m$. Indeed, we have $a\mathfrak{a}+a_1\mathfrak{a}_1+\dots+a_m\mathfrak{a}_m = 0$ where not all numbers $a, a_1, \dots, a_m$ are equal to 0. Hence $a \neq 0$ by the linear independence of $\mathfrak{a}_1, \dots, \mathfrak{a}_m$. Consequently

$$\mathfrak{a} = a_1\mathfrak{a}_m+\dots+a_m\mathfrak{a}_m \qquad \text{where} \qquad a_i = -a^{-1}a_i \qquad \text{for} \qquad i = 1, \dots, m\,.$$

Suppose that

$$(23)\qquad \begin{aligned} \mathfrak{a}_1 &= [a_{1,1}, \ldots, a_{1,k}]\,, \\ &\ldots\ldots\ldots \\ \mathfrak{a}_m &= [a_{m,1}, \ldots, a_{m,k}]\,. \end{aligned}$$

Then (22) holds if and only if

$$(24)\qquad \begin{aligned} a_1 a_{1,1} + \ldots + a_m a_{m,1} &= 0\,, \\ \ldots\ldots\ldots\ldots \\ a_1 a_{1,k} + \ldots + a_m a_{m,k} &= 0\,. \end{aligned}$$

It follows from the theory of systems of linear equations that in order that

$$a_1 = a_2 = \ldots = a_m = 0$$

be the only solution of the system (24) it is necessary and sufficient that $m \leqslant k$ and at least one of the determinants

$$(25)\qquad \begin{vmatrix} a_{1,i_1} & \cdots & a_{m,i_1} \\ \cdots & \cdots & \cdots \\ a_{1,i_m} & \cdots & a_{m,i_m} \end{vmatrix} \qquad (i_1 < i_2 < \ldots < i_m)$$

do not vanish. This proves that

1.1. *The vectors* (22) *are linearly independent if and only if* $m \leqslant k$ *and at least one of the determinants* (25) *does not vanish.*

By the *scalar product* of the vectors (6) and (8) we mean the number

$$(26)\qquad \mathfrak{a}\mathfrak{b} = a_1 b_1 + \ldots + a_k b_k$$

denoted sometimes by $\mathfrak{a}\cdot\mathfrak{b}$. In particular, for any vector (6)

$$(27)\qquad \mathfrak{a}\mathfrak{a} = a_1^2 + \ldots + a_k^2\,.$$

In many books the scalar product $\mathfrak{a}\mathfrak{a}$ of a vector $\mathfrak{a}$ by itself is denoted by $\mathfrak{a}^2$. We shall never use the notation $\mathfrak{a}^2$ for $\mathfrak{a}\mathfrak{a}$ in this book because the symbol $\mathfrak{a}^2$ will be reserved to denote another operation on a vector $\mathfrak{a}$ — see § 4, p. 55.

It is easy to verify that the scalar product has the following properties:

$$(28)\qquad \mathfrak{a}\mathfrak{b} = \mathfrak{b}\mathfrak{a}\,;$$

$$(29)\qquad \mathfrak{a}\cdot(\mathfrak{b}+\mathfrak{c}) = \mathfrak{a}\mathfrak{b}+\mathfrak{a}\mathfrak{c}\,; \qquad (\mathfrak{a}+\mathfrak{b})\cdot\mathfrak{c} = \mathfrak{a}\mathfrak{c}+\mathfrak{b}\mathfrak{c}\,;$$

$$(30)\qquad a\mathfrak{a}\cdot\mathfrak{b} = a\cdot\mathfrak{a}\mathfrak{b} = \mathfrak{a}\cdot a\mathfrak{b}\,;$$

$$(31)\qquad 0\cdot\mathfrak{a} = \mathfrak{a}\cdot 0 = 0\,;$$

$$(32)\qquad \mathfrak{a}\mathfrak{a} > 0 \text{ for } \mathfrak{a} \neq 0\,; \qquad \mathfrak{a}\mathfrak{a} = 0 \text{ for } \mathfrak{a} = 0\,.$$

It follows from the above properties that for any vectors $\mathfrak{a}$, $\mathfrak{b}$ in R^k and for every real number x

$$\mathfrak{a}\mathfrak{a}\cdot x^2+2\mathfrak{a}\mathfrak{b}\cdot x+\mathfrak{b}\mathfrak{b}=(x\mathfrak{a}+\mathfrak{b})\cdot(x\mathfrak{a}+\mathfrak{b})\geqslant 0,$$

which proves, by known properties of polynomials of degree 2, that

$$(\mathfrak{a}\mathfrak{b})^2\leqslant(\mathfrak{a}\mathfrak{a})\cdot(\mathfrak{b}\mathfrak{b}). \tag{33}$$

By the *length* (or the *norm*) of a vector (6) we understand the number denoted by $|\mathfrak{a}|$ or by $\|\mathfrak{a}\|$, defined by

$$|\mathfrak{a}|=\|\mathfrak{a}\|=\sqrt{\mathfrak{a}\mathfrak{a}}=\sqrt{a_1^2+\ldots+a_k^2}. \tag{34}$$

The length has the following properties:

$$|\mathfrak{a}|\geqslant 0;\qquad |\mathfrak{a}|=0 \text{ if and only if } \mathfrak{a}=0; \tag{35}$$

$$|a\mathfrak{a}|=|a|\cdot|\mathfrak{a}|\qquad(\text{in particular } |-\mathfrak{a}|=|\mathfrak{a}|); \tag{36}$$

$$|\mathfrak{a}+\mathfrak{b}|\leqslant|\mathfrak{a}|+|\mathfrak{b}|. \tag{37}$$

Only (37) requires a proof. It follows from (34) and (33) that

$$|\mathfrak{a}+\mathfrak{b}|^2=(\mathfrak{a}+\mathfrak{b})(\mathfrak{a}+\mathfrak{b})=\mathfrak{a}\mathfrak{a}+2\mathfrak{a}\mathfrak{b}+\mathfrak{b}\mathfrak{b}\leqslant|\mathfrak{a}|^2+2|\mathfrak{a}|\cdot|\mathfrak{b}|+|\mathfrak{b}|^2=(|\mathfrak{a}|+|\mathfrak{b}|)^2$$

which yields (37).

The following generalization of (37) can be proved by induction on m:

$$|\mathfrak{a}_1+\ldots+\mathfrak{a}_m|\leqslant|\mathfrak{a}_1|+\ldots+|\mathfrak{a}_m| \tag{37'}$$

for any vectors $\mathfrak{a}_1, \ldots, \mathfrak{a}_m$.

A vector $\mathfrak{a}$ is said to be a *unit vector* if $|\mathfrak{a}|=1$. For instance, all the vectors

$$\begin{aligned}\mathfrak{e}_1&=[1,0,0,\ldots,0],\\ \mathfrak{e}_2&=[0,1,0,\ldots,0],\\ &\ldots\ldots\ldots\ldots\\ \mathfrak{e}_k&=[0,0,\ldots,0,1]\end{aligned} \tag{38}$$

are unit vectors. They are called the *unit vectors of the coordinate axes* in R^k. More precisely, the vector $\mathfrak{e}_i$ is called the *unit vector of the i-th coordinate axis*.

For any vector $\mathfrak{a}=[a_1,\ldots,a_k]$,

$$\mathfrak{a}=a_1\mathfrak{e}_1+\ldots+a_k\mathfrak{e}_k. \tag{39}$$

This implies, together with (36) and (37′), that

$$|\mathfrak{a}|\leqslant|a_1|+\ldots+|a_k| \tag{40}$$

for every vector (6). Note that by (34)

$$|a_i|\leqslant|\mathfrak{a}|\quad\text{for}\quad i=1,\ldots,k. \tag{41}$$

We say that a *vector* $\mathfrak{a} \in V_k$ *is parallel to a vector* $\mathfrak{b} \in V_k$ (or that the *vectors* $\mathfrak{a}, \mathfrak{b}$ *are parallel*) if $\mathfrak{a} = c\mathfrak{b}$ or $\mathfrak{b} = c\mathfrak{a}$ for a number c. According to this definition, the zero vector in R^k is parallel to every vector in R^k. For every vector $\mathfrak{a} \neq 0$ in R^k there exist exactly two unit vectors parallel to $\mathfrak{a}$, viz. the vectors

$$\frac{\mathfrak{a}}{|\mathfrak{a}|} \quad \text{and} \quad -\frac{\mathfrak{a}}{|\mathfrak{a}|}.$$

By the *distance* between points (4) we mean the length of the vector $q-p$, i.e. the number

$$(42) \qquad |q-p| = \sqrt{(y_1-x_1)^2+\ldots+(y_k-x_k)^2}\,.$$

It follows from (35), (36) and (37) that

$$(43) \qquad |q-p| \geqslant 0; \qquad |q-p| = 0 \text{ if and only if } q = p\,;$$

$$(44) \qquad |q-p| = |p-q|\,;$$

$$(45) \qquad |p_1-p_3| \leqslant |p_1-p_2|+|p_2-p_3|\,.$$

More generally, by (37′),

$$(45') \qquad |p_1-p_m| \leqslant |p_1-p_2|+|p_2-p_3|+\ldots+|p_{m-1}-p_m|$$

for any points $p_1, \ldots, p_m \in R^k$.

We say that vectors $\mathfrak{a}, \mathfrak{b}$ in R^k are *orthogonal* (or that $\mathfrak{a}$ *is orthogonal to* $\mathfrak{b}$) if $\mathfrak{a}\mathfrak{b} = 0$ (or equivalently $\mathfrak{b}\mathfrak{a} = 0$), i.e. if their scalar product vanishes. For instance, the zero vector is orthogonal to every vector, by (31). It follows from (29) and (30) that if a vector $\mathfrak{a}$ is orthogonal to each of the vectors $\mathfrak{a}_1, \ldots, \mathfrak{a}_m$, then it is orthogonal to every linear combination $a_1\mathfrak{a}_1+\ldots+a_m\mathfrak{a}_m$.

1.2. *Let* $\mathfrak{a}_1, \ldots, \mathfrak{a}_m$ *be linearly independent vectors in* R^k, $1 \leqslant m < k$. *There exist linearly independent vectors* $\mathfrak{b}_1, \ldots, \mathfrak{b}_{k-m}$ *orthogonal to all the vectors* $\mathfrak{a}_1, \ldots, \mathfrak{a}_m$, *that is,*

$$(46) \qquad \mathfrak{a}_i\mathfrak{b}_j = 0 \quad \textit{for} \quad i = 1, \ldots, m \textit{ and } j = 1, \ldots, k-m\,.$$

1.3. *Suppose that* $\mathfrak{a}_1, \ldots, \mathfrak{a}_m$ *are linearly independent vectors in* R^k, $\mathfrak{b}_1, \ldots, \mathfrak{b}_{k-m}$ *are linearly independent vectors in* R^k *and* (46) *holds. Then a vector* $\mathfrak{b}$ *is orthogonal to all the vectors* $\mathfrak{a}_1, \ldots, \mathfrak{a}_m$ *if and only if it is a linear combination of* $\mathfrak{b}_1, \ldots, \mathfrak{b}_{k-m}$, *that is,*

$$(47) \qquad \mathfrak{b} = \beta_1\mathfrak{b}_1+\ldots+\beta_{k-m}\mathfrak{b}_{k-m}\,.$$

A vector $\mathfrak{b} = [b_1, \ldots, b_k]$ is orthogonal to all the vectors (23) if and only if its coordinates satisfy the following system of homogeneous linear equations:

$$\begin{aligned} a_{1,1}b_1 + \ldots + a_{1,k}b_k &= 0 , \\ \cdots\cdots\cdots & \\ a_{m,1}b_1 + \ldots + a_{m,k}b_k &= 0 . \end{aligned} \tag{48}$$

Since the vectors (23) are linearly independent, at least one of the determinants (25) is not equal to zero.

It is known from algebra that then there exist $k-m$ solutions

$$\begin{aligned} & b_{1,1}, \ldots, b_{1,k}, \\ & \cdots\cdots\cdots \\ & b_{k-m,1}, \ldots, b_{k-m,k} \end{aligned} \tag{49}$$

of the system (48) of linear equations, such that at least one of the determinants

$$\begin{vmatrix} b_{1,j_1} & \ldots & b_{1,j_{k-m}} \\ \cdots & \cdots & \cdots \\ b_{k-m,j_1} & \ldots & b_{k-m,j_m} \end{vmatrix} \qquad (j_1 < j_2 < \ldots < j_{k-m}) \tag{50}$$

is not equal to zero. This means that the vectors

$$\begin{aligned} \mathfrak{b}_1 &= [b_{1,1}, \ldots, b_{1,k}] , \\ \cdots\cdots & \cdots\cdots\cdots \\ \mathfrak{b}_{k-m} &= [b_{k-m,1}, \ldots, b_{k-m,k}] \end{aligned} \tag{51}$$

satisfy (46) and are linearly independent. This proves 1.2.

Suppose now that vectors (23) are linearly independent, vectors (51) are linearly independent and (46) holds. In other words, we have $k-m$ solutions (49) of the system (48) of linear equations, such that at least one of the determinants (50) does not vanish. Moreover at least one of the determinants (25) does not vanish. It is known from algebra that, under these hypotheses, $b_1, \ldots, b_k$ form a solution of (48) if and only if there exist numbers $\beta_1, \ldots, \beta_{k-m}$ such that

$$\begin{aligned} b_1 &= \beta_1 b_{1,1} + \ldots + \beta_{k-m} b_{k-m,1} , \\ \cdots & \cdots\cdots\cdots \\ b_k &= \beta_1 b_{k,1} + \ldots + \beta_{k-m} b_{k-m,k} . \end{aligned}$$

This means that a vector $\mathfrak{b} = [b_1, \ldots, b_k]$ is orthogonal to all the vectors (23) if and only if it is of the form (47), i.e. 1.3 is true.

Every set of the form

$$S = \{p_0 + t\mathfrak{a}\colon\ a \leqslant t \leqslant b\} \subset R^k ,$$

where $p_0 \in R^k$, $0 \neq \mathfrak{a} \in V_k$ and $-\infty < a < b < \infty$, is said to be the *segment* joining the points $p_0 + a\mathfrak{a}$ and $p_0 + b\mathfrak{a}$. In particular, the set

$$S = \{p + t(q-p) \colon 0 \leqslant t \leqslant 1\}$$

is the segment joining the points p and q.

A set $C \subset R^k$ is said to be *convex* if, for any $p_1, p_2 \in C$, the segment joining p_1 and p_2 is a subset of C.

By an *open ball* (or simply, *ball*) of radius $\eta > 0$ and center $p_0 \in R^k$ we mean the set

$$\mathrm{B}(p_0, \eta) = \{p \colon |p - p_0| < \eta\}\,.$$

The ball $\mathrm{B}(p_0, \eta)$ is convex. Indeed if $p_1, p_2 \in \mathrm{B}(p_0, \eta)$ and $p = p_1 + t(p_2 - p_1)$ (where $0 \leqslant t \leqslant 1$) is any point of the segment joining p_1 and p_2, then

$$p - q = (1-t)(p_1 - p_0) + t(p_2 - p_0)$$

and consequently, by (36) and (37),

$$|p - p_0| \leqslant (1-t)|p_1 - p_0| + t|p_2 - p_0| < (1-t)\eta + t\eta = \eta$$

which proves that $p \in \mathrm{B}(p_0, \eta)$.

According to a general convention from I § 6 (p. 34), every function defined on a set $A \subset R^k$ will be called a *function of k real variables*.

EXERCISES

1. Are the vectors

$$\mathfrak{a}_1 = [1, 2, 3, 4, 5]\,, \qquad \mathfrak{a}_2 = [5, 4, 3, 2, 1]\,, \qquad \mathfrak{a}_3 = [6, 6, 6, 6, 0]$$

linearly independent?

2. Find two linearly independent vectors $\mathfrak{b}_1, \mathfrak{b}_2$ which are orthogonal to each of the vectors $\mathfrak{a}_1, \mathfrak{a}_2, \mathfrak{a}_3$ in exercise 1.

§ 2. Hyperplanes in Euclidean spaces. By an *m-dimensional hyperplane* in R^k we mean every set $H \subset R^k$ of the form

$$(1) \qquad H = \{p_0 + t_1\mathfrak{a}_1 + \ldots + t_m\mathfrak{a}_m \colon -\infty < t_i < \infty \text{ for } i = 1, \ldots, m\}\,,$$

where p_0 is a fixed point in R^k and $\mathfrak{a}_1, \ldots, \mathfrak{a}_m$ are linearly independent vectors in R^k, $1 \leqslant m \leqslant k$. More precisely, we say that (1) is the *m-dimensional hyperplane through p_0 parallel to the vectors* $\mathfrak{a}_1, \ldots, \mathfrak{a}_m$. By definition,

$$p \in H \text{ if and only if } p = p_0 + t_1\mathfrak{a}_1 + \ldots + t_m\mathfrak{a}_m\,,$$

i.e. if and only if $p - p_0 = t_1\mathfrak{a}_1 + \ldots + t_m\mathfrak{a}_m$ for certain real numbers $t_1, \ldots, t_m$. Thus the m-dimensional hyperplane (1) through p_0 parallel to the linearly independent vectors $\mathfrak{a}_1, \ldots, \mathfrak{a}_m$ is the set of all points p such that the vector $p - p_0$ is a linear combination of the vectors $\mathfrak{a}_1, \ldots, \mathfrak{a}_m$. It is easy to verify that every hyperplane is a convex set.

Two-dimensional hyperplanes are called *planes*.

One-dimensional hyperplanes are called *straight lines*. In other words, a straight line is a set of the form

$$(2) \qquad \{p_0 + t\mathfrak{a}:\ -\infty < t < \infty\} \subset R^k$$

where $p_0 \in R^k$ and $\mathfrak{a} \neq 0$; we say then that (2) is the *line through* p_0 *parallel to* $\mathfrak{a}$.

The general notion of hyperplane is a multi-dimensional analogue of the notions of line and plane.

Suppose H is the m-dimensional hyperplane (1) and $q_0 \in H$. Then also

$$H = \{q_0 + t_1\mathfrak{a}_1 + \ldots + t_m\mathfrak{a}_m:\ -\infty < t_i < \infty \text{ for } i = 1, \ldots, m\},$$

i.e. H is the m-dimensional hyperplane through q_0 parallel to $\mathfrak{a}_1, \ldots, \mathfrak{a}_m$. In fact, we have

$$q_0 = p_0 + \tau_1\mathfrak{a}_1 + \ldots + \tau_m\mathfrak{a}_m$$

for certain real numbers $\tau_1, \ldots, \tau_m$. If $p = p_0 + t_1\mathfrak{a}_1 + \ldots + t_m\mathfrak{a}_m$, then

$$p = q_0 + (t_1 - \tau_1)\mathfrak{a}_1 + \ldots + (t_m - \tau_m)\mathfrak{a}_m .$$

Conversely, if $p = q_0 + t_1\mathfrak{a}_1 + \ldots + t_m\mathfrak{a}_m$, then

$$p = p_0 + (t_1 + \tau_1)\mathfrak{a}_1 + \ldots + (t_m + \tau_m)\mathfrak{a}_m .$$

A non-zero vector $\mathfrak{a}$ is said to be *parallel* to an m-dimensional hyperplane $H \subset R^k$ provided that for every point $p_0 \in H$ the line through p_0 parallel to $\mathfrak{a}$ is a subset of H. We assume, moreover, that the zero vector in R^k is parallel to every hyperplane in R^k. It follows from the above definition that a non-zero vector $\mathfrak{a}$ is parallel to the hyperplane (1) if and only if

$$p_0 + t\mathfrak{a} = p_0 + t_1\mathfrak{a}_1 + \ldots + t_m\mathfrak{a}_m ,$$

i.e. if for every t there exist numbers $t_1, \ldots, t_m$ such that

$$t\mathfrak{a} = t_1\mathfrak{a}_1 + \ldots + t_m\mathfrak{a}_m .$$

This is possible if and only if the vector $\mathfrak{a}$ is a linear combination of the vectors $\mathfrak{a}_1, \ldots, \mathfrak{a}_m$, i.e. if

$$\mathfrak{a} = a_1\mathfrak{a}_1 + \ldots + a_m\mathfrak{a}_m$$

for certain real numbers $a_1, \ldots, a_m$. Including the case of $\mathfrak{a} = 0$ we get the following statement.

2.1. *A vector* $\mathfrak{a}$ *is parallel to the hyperplane* (1) *if and only if it is a linear combination of the vectors* $\mathfrak{a}_1, \ldots, \mathfrak{a}_m$.

In particular, the vectors $\mathfrak{a}_1, \ldots, \mathfrak{a}_m$ are parallel to the hyperplane (1).

Two m-dimensional hyperplanes H_1 and H_2 in R^k are said to be *parallel* if every vector parallel to H_1 is parallel to H_2, and conversely.

We say that a vector $\mathfrak{b}$ is *orthogonal* or *normal* to a hyperplane H (or that H is *orthogonal* to $\mathfrak{b}$) if $\mathfrak{b}$ is orthogonal to every vector $\mathfrak{a}$ which is parallel to H.

Let H be an m-dimensional hyperplane in R^k, and let H' be an m'-dimensional hyperplane in R^k, $m+m' \leqslant k$. We say that the hyperplanes H and H' are *orthogonal* (or that H is *orthogonal* to H') if and only if every vector parallel to H is orthogonal to every vector parallel to H'.

2.2. *A vector* $\mathfrak{b}$ *is orthogonal to the* m*-dimensional hyperplane through* p_0 *parallel to linearly independent vectors* $\mathfrak{a}_1, \ldots, \mathfrak{a}_m$ *if and only if*

$$\mathfrak{a}_i \mathfrak{b} = 0 \quad \textit{for} \quad i = 1, \ldots, m .$$

By 2.1, a vector $\mathfrak{b}$ is orthogonal to H if and only if

$$(a_1 \mathfrak{a}_1 + \ldots + a_m \mathfrak{a}_m)\mathfrak{b} = 0 ,$$

i.e. if and only if

$$a_1 \mathfrak{a}_1 \mathfrak{b} + \ldots + a_m \mathfrak{a}_m \mathfrak{b} = 0$$

for all real numbers $a_1, \ldots, a_m$. This is possible if and only if $\mathfrak{a}_i \mathfrak{b} = 0$ for $i = 1, \ldots, m$.

2.3. *A hyperplane* H' *is orthogonal to the* m*-dimensional hyperplane* H *through* p_0 *parallel to linearly independent vectors* $\mathfrak{a}_1, \ldots, \mathfrak{a}_m$ *if and only if all the vectors* $\mathfrak{a}_1, \ldots, \mathfrak{a}_m$ *are orthogonal to* H'.

This follows directly from the definition and 2.2.

2.4. *Let* $p_0 \in R^k$, $q_0 \in R^k$, *let* $\mathfrak{a}_1, \ldots, \mathfrak{a}_m$ *be linearly independent vectors in* R^k, *let* $\mathfrak{b}_1, \ldots, \mathfrak{b}_{k-m}$ *be linearly vectors in* R^k *and let*

$$\mathfrak{a}_i \mathfrak{b}_j = 0 \quad \textit{for} \quad i = 1, \ldots, m \textit{ and } j = 1, \ldots, k-m$$

$(1 \leqslant m < k)$. *The* m*-dimensional hyperplane*

$$H = \{p_0 + t_1 \mathfrak{a}_1 + \ldots + t_m \mathfrak{a}_m \colon -\infty < t_i < \infty \textit{ for } i = 1, \ldots, m\}$$

through p_0 *parallel to* $\mathfrak{a}_1, \ldots, \mathfrak{a}_m$ *is orthogonal to the* $(k-m)$*-dimensional hyperplane*

$$H' = \{q_0 + t_1 \mathfrak{b}_1 + \ldots + t_{k-m} \mathfrak{b}_{k-m} \colon -\infty < t_j < \infty \textit{ for } j = 1, \ldots, k-m\}$$

through q_0 *parallel to* $\mathfrak{b}_1, \ldots, \mathfrak{b}_{k-m}$. *Moreover,*

$$H = \{p \in R^k \colon \mathfrak{b}_j(p - p_0) = 0 \textit{ for } j = 1, \ldots, k-m\} \tag{3}$$

and

$$H' = \{q \in R^k \colon \mathfrak{a}_i(q - q_0) = 0 \textit{ for } i = 1, \ldots, m\} . \tag{4}$$

The hyperplane H' *is the only* $(k-m)$*-dimensional hyperplane through* q_0 *orthogonal to* H. *The hyperplane* H *is the only* k*-dimensional hyperplane through* p_0 *orthogonal to* H'.

H' is the set of all points q such that the vector $q-q_0$ is a linear combination of $\mathfrak{b}_1, \ldots, \mathfrak{b}_{k-m}$, i.e. $q-q_0$ is orthogonal to all the vectors $\mathfrak{a}_1, \ldots, \mathfrak{a}_m$ (see 1.3). This proves (4). The proof of (3) is analogous.

If vectors $\mathfrak{a}$, $\mathfrak{b}$ are parallel to H and H' respectively, then

$$\mathfrak{a} = a_1\mathfrak{a}_1 + \ldots + a_m\mathfrak{a}_m, \qquad \mathfrak{b} = b_1\mathfrak{b}_1 + \ldots + b_{k-m}\mathfrak{b}_{k-m}$$

by 2.1. Hence

$$\mathfrak{a}\mathfrak{b} = \sum_{i=1}^{m}\sum_{j=1}^{k-m} a_i b_j \mathfrak{a}_i \mathfrak{b}_j = 0 .$$

This proves that H and H' are orthogonal.

Suppose that

$$H'' = \{q_0 + t_1\mathfrak{c}_1 + \ldots + t_{k-m}\mathfrak{c}_{k-m}\colon -\infty < t_j < \infty \text{ for } j = 1, \ldots, k-m\},$$

where $\mathfrak{c}_1, \ldots, \mathfrak{c}_{k-m}$ are linearly independent vectors, is a $(k-m)$-dimensional hyperplane through q_0 orthogonal to H. Thus

$$\mathfrak{a}_i\mathfrak{c}_j = 0 \quad \text{for} \quad i = 1, \ldots, m \text{ and } j = 1, \ldots, k-m .$$

It follows from the part of 2.4 just proved (where $\mathfrak{b}_1, \ldots, \mathfrak{b}_{k-m}$ should now be replaced by $\mathfrak{c}_1, \ldots, \mathfrak{c}_{k-m}$) that

$$H'' = \{q \in R^k\colon \mathfrak{a}_i(q-q_0) = 0 \text{ for } i = 1, \ldots, k-m\} .$$

Thus $H'' = H'$. Analogously we prove that every m-dimensional hyperplane through p_0 orthogonal to H' coincides with H.

EXERCISES

1. Let $p_0 \in R^k$ and let $\mathfrak{a}_1, \ldots, \mathfrak{a}_m$ be any vectors in R^k. Prove that the set H of all points

(5) $$p = p_0 + t_1\mathfrak{a}_1 + \ldots + t_m\mathfrak{a}_m$$

where $t_1, \ldots, t_m$ are any real numbers is an s-dimensional hyperplane, s being the maximal number of linearly independent vectors in the set $\{\mathfrak{a}_1, \ldots, \mathfrak{a}_m\}$. All the vectors $\mathfrak{a}_1, \ldots, \mathfrak{a}_m$ are parallel to H.

In the case where all the vectors $\mathfrak{a}_1, \ldots, \mathfrak{a}_m$ are linearly independent we call (5) a *parametric equation of the hyperplane* H.

2. Write a parametric equation of the plane through $(-2, 0, 2, 0, -2)$ orthogonal to the vectors $\mathfrak{a}_1, \mathfrak{a}_2, \mathfrak{a}_3$ from § 1, exercise 1.

3. The set H of all points $p = (x_1, \ldots, x_k) \in R^k$ whose coordinates satisfy all the linear equations

(6) $$\begin{aligned} b_{1,1}x_1 + \ldots + b_{1,k}x_k + b_1 &= 0 , \\ \ldots\ldots\ldots\ldots\ldots& \\ b_{s,1}x_1 + \ldots + b_{s,k}x_k + b_k &= 0 \end{aligned}$$

is a $(k-r)$-dimensional hyperplane, r being the maximal number of linearly independent vectors in the set of vectors

$$\mathfrak{b}_1 = [b_{1,1}, \dots, b_{1,k}], \quad \dots\dots\dots\dots, \quad \mathfrak{b}_s = [b_{s,1}, \dots, b_{s,k}]. \tag{7}$$

All the vectors (7) are orthogonal to H.

Thus, if the vectors (7) are linearly independent, then H is a $(k-s)$-dimensional hyperplane. In this case we say that (6) is a *system of linear equations for the hyperplane H*.

4. Write a system of linear equations for the 3-dimensional hyperplane H through $(-2, 0, 2, 0, -2)$ parallel to the vectors $\mathfrak{a}_1, \mathfrak{a}_2, \mathfrak{a}_3$ from § 1, exercise 1.

§ 3. Intervals. By an *open interval* we mean every subset of R^k of the form

$$\{(x_1, \dots, x_k): a_i < x_i < b_i \text{ for } i = 1, \dots, k\} \tag{1}$$

where $a_1, \dots, a_k, b_1, \dots, b_k$ are finite numbers such that

$$a_i < b_i \quad \text{for} \quad i = 1, \dots, k. \tag{2}$$

By a *closed interval* we mean every set of the form

$$\{(x_1, \dots, x_k): a_i \leqslant x_i \leqslant b_i \text{ for } i = 1, \dots, k\} \tag{3}$$

where $a_1, \dots, a_k, b_1, \dots, b_k$ are finite numbers such that (2) holds. The point

$$\left(\frac{a_1+b_1}{2}, \dots, \frac{a_k+b_k}{2}\right) \tag{4}$$

is said to be the *center* of the interval (1) or (3). The positive numbers

$$b_1-a_1, \dots, b_k-a_k \tag{5}$$

are the *lengths of the edges* of the interval (1) or (3). If all the numbers (5) are equal, the closed interval (3) is said to be a k-dimensional *cube*.

In the case where $k = 1$, the above definition of an interval is the usual one. In the case where $k = 2$, open and closed intervals are rectangles without or with the boundary, respectively, such that their edges are parallel to the coordinate axes. In the case where $k = 3$, open and closed intervals are rectangular parallelepipeds without or with the boundary, respectively, such that their edges are parallel to the coordinate axes.

Under the hypothesis (2), every set of the form

$$[(x_1, \dots, x_k): a_i \leqslant x_i < b_i \text{ for } i = 1, \dots, k\}$$

is said to be an interval *closed from the left* or *open from the right*, and every set of the form

$$\{(x_1, \dots, x_k): a_i < x_i \leqslant b_i \text{ for } i = 1, \dots, k\}$$

is said to be an interval *closed from the right* or *open from the left.* Each of these intervals is said to be *half-open* or *half-closed.*

In the case $k = 1$, the following notation for intervals will be used:

$$(a; b) = \{x \in R\colon a < x < b\}, \quad \langle a; b\rangle = \{x \in R\colon a \leqslant x \leqslant b\},$$

$$\langle a; b) = \{x \in R\colon a \leqslant x < b\}, \quad (a; b\rangle = \{x \in R\colon a < x \leqslant b\}.$$

The same terminology and notation is applied if one of the numbers a, b in these formulas, or if some of the numbers $a_1, \ldots, a_k, b_1, \ldots, b_k$ in (1) or (3) are infinite; the intervals in question are then said to be *unbounded.* If all the numbers a, b or $a_1, \ldots, a_k, b_1, \ldots, b_k$ are finite, then the intervals are said to be *bounded.*

Any set of the form (3), where $a_1, \ldots, a_k, b_1, \ldots, b_k$ are finite numbers,

$$a_i \leqslant b_i \quad \text{for} \quad i = 1, \ldots, k \tag{6}$$

and $a_j = b_j$ for at least one integer j, is called a *degenerate interval.*

In the case where $k = 1$, degenerate intervals are one-point sets. In the case where $k = 2$, degenerate intervals are one-point sets or segments parallel to the coordinate axes. In the case where $k = 3$, degenerate intervals are one-point sets, or segments parallel to the coordinate axes, or rectangles whose edges are parallel to the coordinate axes.

If I is a non-degenerate interval (1) or (3), then the sets

$$\begin{aligned} I_{j,1} &= \{(x_1, \ldots, x_k)\colon a_i \leqslant x_i \leqslant b_i \text{ for } i \neq j \text{ and } x_j = b_j\}, \\ I_{j,-1} &= \{(x_1, \ldots, x_k)\colon a_i \leqslant x_i \leqslant b_i \text{ for } i \neq j \text{ and } x_j = a_j\} \end{aligned} \tag{7}$$

are degenerate intervals. They are called *faces* of I, more precisely $(k-1)$-*dimensional closed faces* of I.

Observe that the closed interval (3) is the union of the open interval (1) and all the faces (7).

It is easy to check that open, closed, half-open and degenerate intervals are convex.

In what follows the word *interval* without any adjectives will always mean a bounded non-degenerate interval.

Sometimes it is convenient to consider the empty set as an open, closed or degenerate interval.

We say that an interval (1) or (3) is *rational* if all the numbers $a_1, \ldots, a_k, b_1, \ldots, b_k$ are rational.

3.1. *The set of all rational open* (*or closed*) *intervals is countable.*

The function which assigns to every rational interval (1) the finite sequence $(a_1, \ldots, a_k, b_1, \ldots, b_k)$ is a one-to-one mapping of the set of all open rational intervals into the Cartesian product Q^{2k}, where Q is the set of all rational numbers. Since the set Q^{2k} is countable (see I 7.1 and

I 7.6), the set of all open rational intervals is countable by I 7.5. The proof for closed rational intervals is the same.

§ 4. Matrices. Let $k_1, \ldots, k_m$ be fixed positive integers. Suppose that we assign a real number $a_{i_1, \ldots, i_m}$ to every finite sequence $i_1, \ldots, i_m$ of positive integers such that

$$(1) \qquad i_1 \leqslant k_1, \ \ldots, \ i_m \leqslant k_m .$$

We say then that we have defined an *m-indexed matrix*

$$(2) \qquad [a_{i_1, \ldots, i_m}]_{i_1 \leqslant k_1, \ldots, i_m \leqslant k_m}$$

or, more precisely, that we have defined a *matrix of type* $k_1, \ldots, k_m$. From the point of view of the general definition in I § 3, a *matrix* is a real function defined on the set of all finite sequences of positive integers $i_1, \ldots, i_m$ satisfying (1).

As a rule, matrices will be denoted by capital German letters $\mathfrak{A}, \mathfrak{B}, \ldots$ If

$$(3) \qquad \mathfrak{A} = [a_{i_1, \ldots, i_m}]_{i_1 \leqslant k_1, \ldots, i_m \leqslant k_m} ,$$

then the numbers $a_{i_1, \ldots, i_m}$ are said to be the *coordinates* (or *terms*) of the matrix $\mathfrak{A}$; more precisely, the term $a_{i_1, \ldots, i_m}$ is said to be the $(i_1, \ldots, i_m)$*-th coordinate* of the matrix $\mathfrak{A}$.

Sometimes we shall use a shorter notation instead of (2). For instance if it is evident that (1) holds, we shall write

$$[a_{i_1, \ldots, i_m}]$$

instead of (2). If $k_1 = k_2 = \ldots = k_m = k$, then we shall write

$$[a_{i_1, \ldots, i_m}]_{i_1, \ldots, i_m \leqslant k}$$

instead of (2). In that case we shall say that the matrix is *k-dimensional.* The symbol

$$[a_{i, i_1, \ldots, i_m}]_{i \leqslant l, i_1, \ldots, i_m \leqslant k}$$

will denote the matrix, where the first index runs over the integers $1, \ldots, l$, and the remaining indices over the integers $1, \ldots, k$.

The integer m is the *number of indices* of the matrix (2). The integer $k_1 \cdot \ldots \cdot k_m$ is the *number of terms* of the matrix (2).

Any k-dimensional one-indexed matrix $[a_i]_{i \leqslant k}$ will be identified with the k-dimensional vector $[a_1, \ldots, a_k]$.

We shall now define some algebraic operations on matrices.

The *product* $a\mathfrak{A}$ of a matrix (3) by a real number a is the matrix

$$a\mathfrak{A} = [aa_{i_1, \ldots, i_m}]_{i_1 \leqslant k_1, \ldots, i_m \leqslant k_m}$$

of the same type. Instead of $(-1)\mathfrak{A}$ we shall write $-\mathfrak{A}$. Instead of $a^{-1}\mathfrak{A}$ we shall also write $\mathfrak{A}/a$. Sometimes we write $\mathfrak{A}a$ or $a \cdot \mathfrak{A}$ instead of $a\mathfrak{A}$.

By the *sum* $\mathfrak{A}+\mathfrak{B}$ of two matrices of the same type

(4) $$\mathfrak{A} = [a_{i_1,\dots,i_m}]_{i_1\leqslant k_1,\dots,i_m\leqslant k_m}, \qquad \mathfrak{B} = [b_{i_1,\dots,i_m}]_{i_1\leqslant k_1,\dots,i_m\leqslant k_m}$$

we shall mean the matrix

$$\mathfrak{A}+\mathfrak{B} = [a_{i_1,\dots,i_m}+b_{i_1,\dots,i_m}]_{i_1\leqslant k_1,\dots,i_m\leqslant k_m}$$

which is also of the same type. Similarly we define the *difference* $\mathfrak{A}-\mathfrak{B}$ of the matrices (4):

$$\mathfrak{A}-\mathfrak{B} = [a_{i_1,\dots,i_m}-b_{i_1,\dots,i_m}]_{i_1\leqslant k_1,\dots,i_m\leqslant k_m},$$

which is also of the same type.

The *scalar product* of matrices (4) of the same type is the number

$$\mathfrak{A}\mathfrak{B} = \sum_{i_1=1}^{k_1}\dots\sum_{i_m=1}^{k_m} a_{i_1,\dots,i_m} b_{i_1,\dots,i_m}.$$

The operations on matrices which we have just defined have the same properties as the analogous operations on vectors. In particular,

(5) $$\mathfrak{A}+\mathfrak{B} = \mathfrak{B}+\mathfrak{A}, \qquad (\mathfrak{A}+\mathfrak{B})+\mathfrak{C} = \mathfrak{A}+(\mathfrak{B}+\mathfrak{C}),$$

(6) the difference $\mathfrak{C} = \mathfrak{A}-\mathfrak{B}$ is the only solution of the equation

$$\mathfrak{B}+\mathfrak{C} = \mathfrak{A},$$

(7) $$a(\mathfrak{A}+\mathfrak{B}) = a\mathfrak{A}+a\mathfrak{B}, \qquad (a+b)\mathfrak{A} = a\mathfrak{A}+b\mathfrak{A},$$

(8) $$a(b\mathfrak{A}) = (ab)\mathfrak{A}, \qquad 1\mathfrak{A} = \mathfrak{A},$$

(9) $$\mathfrak{A}\mathfrak{B} = \mathfrak{B}\mathfrak{A}, \qquad (a\mathfrak{A})\mathfrak{B} = a(\mathfrak{A}\mathfrak{B}) = \mathfrak{A}(a\mathfrak{B}),$$

(10) $$\mathfrak{A}(\mathfrak{B}+\mathfrak{C}) = \mathfrak{A}\mathfrak{B}+\mathfrak{A}\mathfrak{C}, \qquad (\mathfrak{A}+\mathfrak{B})\mathfrak{C} = \mathfrak{A}\mathfrak{C}+\mathfrak{B}\mathfrak{C}.$$

The matrix of type $k_1, \dots, k_m$ whose terms are all equal to 0 is called the *zero matrix* and is denoted by 0. The following identities hold:

$$\mathfrak{A}+0 = \mathfrak{A}, \qquad 0\mathfrak{A} = 0.$$

The last identity is true in the case where the symbol 0 on the left-hand side is the number zero, as well as in the case where it is the zero matrix of the same type as $\mathfrak{A}$.

The *norm* $\|\mathfrak{A}\|$ of a matrix (3) is defined by

(11) $$\|\mathfrak{A}\| = \sqrt{\sum_{i_1=1}^{k_1}\dots\sum_{i_m=1}^{k_m} a^2_{i_1,\dots,i_m}}.$$

In other symbols,

(11′) $$\|\mathfrak{A}\| = \sqrt{\mathfrak{A}\mathfrak{A}}.$$

The norm of a matrix has the same properties as the norm of a vector (see § 1 (35)-(37) and § 1 (33)):

(12) $$\|\mathfrak{A}\| \geqslant 0\,; \qquad \|\mathfrak{A}\| = 0 \text{ if and only if } \mathfrak{A} = 0\,;$$

(13) $$\|a\mathfrak{A}\| = |a|\cdot\|\mathfrak{A}\|; \qquad \|\mathfrak{A}+\mathfrak{B}\| \leqslant \|\mathfrak{A}\|+\|\mathfrak{B}\|\,;$$

(14) $$|\mathfrak{A}\mathfrak{B}| \leqslant \|\mathfrak{A}\|\cdot\|\mathfrak{B}\|\,.$$

The proof is the same as for vectors.

Observe that in the case of one-indexed matrices, i.e. of vectors, all the above definitions coincide with the definitions assumed in § 1 for vectors.

On the other hand, every matrix (3) of type $k_1, \ldots, k_m$ can be interpreted as a vector in the $k_1\cdot\ldots\cdot k_m$-dimensional Euclidean space whose coordinates are numbered by sequences $i_1, \ldots, i_m$ of integers instead of by the integers $1, \ldots, k_1\cdot\ldots\cdot k_m$. If we so interpret matrices as vectors, the above definitions for matrices coincide with those for vectors in § 1.

The definition of the product $\mathfrak{A}\mathfrak{B}$ can be generalized to the case where the matrices $\mathfrak{A}$, $\mathfrak{B}$ have different numbers of indices. The product $\mathfrak{A}\mathfrak{B}$ is then a matrix, not a number. Namely, if

$$\mathfrak{A} = [a_{i_1,\ldots,i_m,j_1,\ldots,j_n}]_{i_1\leqslant k_1,\ldots,i_m\leqslant k_m,j_1\leqslant l_1,\ldots,j_n\leqslant l_n}$$

and

$$\mathfrak{B} = [b_{j_1,\ldots,j_n}]_{j_1\leqslant l_1,\ldots,j_n\leqslant l_n}\,,$$

then

(15) $$\mathfrak{A}\mathfrak{B} = [c_{i_1,\ldots,i_m}]_{i_1\leqslant k_1,\ldots,i_m\leqslant k_m}$$

where

$$c_{i_1,\ldots,i_m} = \sum_{j_1=1}^{l_1}\cdots\sum_{j_n=1}^{l_n} a_{i_1,\ldots,i_m,j_1,\ldots,j_n} b_{j_1,\ldots,j_n}\,.$$

Analogously, if

$$\mathfrak{A} = [a_{i_1,\ldots,i_m}]_{i_1\leqslant k_1,\ldots,i_m\leqslant k_m}$$

and

$$\mathfrak{B} = [b_{i_1,\ldots,i_m,j_1,\ldots,j_n}]_{i_1\leqslant k_1,\ldots,i_m\leqslant k_m,j_1\leqslant l_1,\ldots,j_n\leqslant l_n}\,,$$

then

(15′) $$\mathfrak{A}\mathfrak{B} = [c_{j_1,\ldots,j_n}]_{j_1\leqslant l_1,\ldots,j_n\leqslant l_n}$$

where

$$c_{j_1,\ldots,j_n} = \sum_{i_1=1}^{k_1}\cdots\sum_{i_m=1}^{k_m} a_{i_1,\ldots,i_m} b_{i_1,\ldots,i_m,j_1,\ldots,j_n}\,.$$

The first of the formulas (9) is not true, in general, for the generalized product. However, the second formula of (9) and the formulas (10) remains true.

Except for the cases discussed above, the product $\mathfrak{A}\mathfrak{B}$ is not defined. Sometimes we write $\mathfrak{A}\cdot\mathfrak{B}$ instead of $\mathfrak{A}\mathfrak{B}$.

By the *tensor product* $\mathfrak{A}\times\mathfrak{B}$ of an m-indexed matrix

$$\mathfrak{A} = [a_{i_1,\dots,i_m}]_{i_1\leqslant k_1,\dots,i_m\leqslant k_m}$$

and an n-indexed matrix

$$\mathfrak{B} = [b_{j_1,\dots,j_n}]_{j_1\leqslant l_1,\dots,j_n\leqslant l_n}$$

we mean the $(m+n)$-indexed matrix

$$\mathfrak{A}\times\mathfrak{B} = [c_{i_1,\dots,i_m,j_1,\dots,j_n}]_{i_1\leqslant k_1,\dots,i_m\leqslant k_m,j_1\leqslant l_1,\dots,j_n\leqslant l_n} \tag{16}$$

where

$$c_{i_1,\dots i_m,j_1,\dots,j_n} = a_{i_1,\dots,i_m} b_{j_1,\dots,j_n}\,.$$

By definition,

$$(\mathfrak{A}+\mathfrak{A}')\times\mathfrak{B} = \mathfrak{A}\times\mathfrak{B}+\mathfrak{A}'\times\mathfrak{B}\,,\qquad \mathfrak{A}\times(\mathfrak{B}+\mathfrak{B}') = \mathfrak{A}\times\mathfrak{B}+\mathfrak{A}\times\mathfrak{B}', \tag{16'}$$

$$(a\mathfrak{A})\times\mathfrak{B} = a(\mathfrak{A}\times\mathfrak{B}) = \mathfrak{A}\times(a\mathfrak{B})\,,\qquad (\mathfrak{A}\times\mathfrak{B})\times\mathfrak{C} = \mathfrak{A}\times(\mathfrak{B}\times\mathfrak{C})\,. \tag{16''}$$

Of course, $\mathfrak{A}'$ is here a matrix of the same type as $\mathfrak{A}$, and $\mathfrak{B}'$ is of the same type as $\mathfrak{B}$. It follows from the last associative law that the tensor product of several matrices $\mathfrak{A}_1,\dots,\mathfrak{A}_m$ can be denoted by the symbol $\mathfrak{A}_1\times\dots\times\mathfrak{A}_m$ without any parentheses.

The tensor multiplication is not commutative, i.e., the equation $\mathfrak{A}\times\mathfrak{B} = \mathfrak{B}\times\mathfrak{A}$ does not hold in general.

In this book we shall examine mainly tensor products of one-indexed matrices, i.e. of vectors. By definition, if

$$\mathfrak{a} = [a_1,\dots,a_k]\,,\qquad \mathfrak{b} = [b_1,\dots,b_k]\,,\qquad \mathfrak{c} = [c_1,\dots,c_k]\,,\qquad \dots$$

then

$$\mathfrak{a}\times\mathfrak{b} = [a_i b_j]_{i,j\leqslant k}\,,\qquad \mathfrak{a}\times\mathfrak{b}\times\mathfrak{c} = [a_i b_j c_s]_{i,j,s\leqslant k}\,,\qquad \dots$$

More generally, if

$$\begin{aligned} \mathfrak{a}_1 &= [a_{1,1},\dots,a_{1,k}]\,,\\ &\dots\dots\dots\dots\dots\\ \mathfrak{a}_m &= [a_{m,1},\dots,a_{m,k}]\,, \end{aligned} \tag{17}$$

then

$$\mathfrak{a}_1\times\dots\times\mathfrak{a}_m = [a_{1,i_1}a_{2,i_2}\cdots a_{m,i_m}]_{i_1,i_2,\dots,i_m\leqslant k}\,. \tag{18}$$

The tensor product $\mathfrak{a}\times\mathfrak{a}\times\dots\times\mathfrak{a}$ m-times will be denoted by the symbol $\mathfrak{a}^m$. By $\mathfrak{a}^1$ we shall mean the vector $\mathfrak{a}$. By definition, if

$$\mathfrak{a} = [a_1,\dots,a_k]\,,$$

then

(18′) $$\mathfrak{a}^m = [a_{i_1} a_{i_2} \dots a_{i_m}]_{i_1, i_2, \dots, i_m \leqslant k} .$$

In particular,

$$\mathfrak{a}^2 = [a_i a_j]_{i,j \leqslant k} , \quad \mathfrak{a}^3 = [a_i a_j a_s]_{i,j,s \leqslant k} , \quad \dots$$

The following identities are true

(19) $$a \cdot (\mathfrak{a}_1 \times \dots \times \mathfrak{a}_m) = \mathfrak{a}_1 \times \dots \times \mathfrak{a}_{r-1} \times a\mathfrak{a}_r \times \mathfrak{a}_{r+1} \times \dots \times \mathfrak{a}_m ,$$

(19′) $$\mathfrak{a}_1 \times \dots \times \mathfrak{a}_{r-1} \times (\mathfrak{a}_r + \mathfrak{a}_r') \times \mathfrak{a}_{r+1} \times \dots \times \mathfrak{a}_m = \mathfrak{a}_1 \times \dots \times \mathfrak{a}_{r-1} \times \mathfrak{a}_r \times \mathfrak{a}_{r+1} \times \dots \times \mathfrak{a}_m + \mathfrak{a}_1 \times \dots \times \mathfrak{a}_{r-1} \times \mathfrak{a}_r' \times \mathfrak{a}_{r+1} \times \dots \times \mathfrak{a}_m ,$$

(19″) $$\|\mathfrak{a}_1 \times \dots \times \mathfrak{a}_m\| = |\mathfrak{a}_1| \cdot \dots \cdot |\mathfrak{a}_m| , \quad \|\mathfrak{a}^m\| = |\mathfrak{a}|^m .$$

Observe the following formula for the scalar product of a matrix

(20) $$\mathfrak{A} = [a_{i_1, \dots, i_m}]_{i_1, \dots, i_m \leqslant k}$$

and the tensor product (18):

(21) $$\mathfrak{A} \cdot (\mathfrak{a}_1 \times \dots \times \mathfrak{a}_m) = \sum_{i_1, \dots, i_m = 1}^{k} a_{i_1, \dots, i_m} a_{1,i_1} \dots a_{m,i_m} .$$

In particular, we have the following formula for the scalar product of (20) and the tensor power (18):

(22) $$\mathfrak{A}\mathfrak{a}^m = \sum_{i_1, \dots, i_m = 1}^{k} a_{i_1, \dots, i_m} a_{i_1} \dots a_{i_m} .$$

Observe also that

(23) $$\mathfrak{A} \cdot (\mathfrak{a}_1 \times \dots \times \mathfrak{a}_m) = (\mathfrak{A} \cdot \mathfrak{a}_m) \cdot (\mathfrak{a}_1 \times \dots \times \mathfrak{a}_{m-1}) .$$

If

$$\mathfrak{B} = [b_{i,i_1, \dots, i_m}]_{i \leqslant l, i_1, \dots, i_m \leqslant k} ,$$

then for any vectors (17)

(24) $$\mathfrak{B} \cdot (\mathfrak{a}_1 \times \dots \times \mathfrak{a}_m) = \Big[\sum_{i_1, \dots, i_m = 1}^{k} b_{i,i_1, \dots, i_m} a_{1,i_1} \dots a_{m,i_m} \Big]_{i \leqslant l} .$$

In particular for any vector $\mathfrak{a} = [a_1, \dots, a_k]$

(25) $$\mathfrak{B}\mathfrak{a}^m = \Big[\sum_{i_1, \dots, i_m = 1}^{k} b_{i,i_1, \dots, i_m} a_{i_1} \dots a_{i_m} \Big]_{i \leqslant l} .$$

The products (24) and (25) are l-dimensional vectors.

We say that a matrix (20) is *symmetric* if

$$a_{i_{r_1}, \dots, i_{r_m}} = a_{i_1, \dots, i_m}$$

for any permutation $r_1, \ldots, r_m$ of the integers $1, \ldots, m$. Since every permutation is the result of a finite number of transpositions, in order that a matrix (20) be symmetric it is necessary and sufficient that

$$a_{i_1, \ldots, i_{r-1}, i_{r+1}, i_r, i_{r+2}, \ldots, i_m} = a_{i_1, \ldots, i_{r-1}, i_r, i_{r+1} i_{r+2}, \ldots, i_m} .$$

More generally, we say that a matrix

$$\mathfrak{A} = [a_{j_1, \ldots, j_n, i_1, \ldots, i_m}]_{j_1 \leqslant l_1, \ldots, j_n \leqslant l_n, i_1, \ldots, i_m \leqslant k}$$

is *symmetric in the last m indices* if

$$a_{j_1, \ldots, j_n, i_{r_1}, \ldots, i_{r_m}} = a_{j_1, \ldots, j_n, i_1, \ldots, i_m}$$

for any permutation $r_1, \ldots, r_m$ of the integers $1, \ldots, m$, and for any indices $j_1, \ldots, j_n, i_1, \ldots, i_m$.

The following simple remark will be useful in what follows. Let

$$\mathfrak{A} = [a_{i_1, \ldots, i_m}]_{i_1 \leqslant k_1, \ldots, i_m \leqslant k_m}$$

be any matrix. Suppose that one of the numbers $k_1, \ldots, k_m$ is equal to one, say $k_r = 1$. Then the index i_r can assume only one value, viz. the number 1, i.e. the terms of the matrix $\mathfrak{A}$ does not depend on the rth index. It is natural, and sometimes very convenient, to cancel the rth index i_r and to consider $\mathfrak{A}$ to be the $(m-1)$-indexed matrix

$$\mathfrak{A} = [a_{i_1, \ldots, i_{r-1}, 1, i_{r+1}, \ldots, i_m}]_{i_1 \leqslant k_1, \ldots, i_{r-1} \leqslant k_{r-1}, i_{r+1} \leqslant k_{r+1}, \ldots, i_m \leqslant k_m},$$

i.e. the matrix

$$\mathfrak{A} = [b_{j_1, \ldots, j_{m-1}}]_{j_1 \leqslant l_1, \ldots, j_{m-1} \leqslant l_{m-1}}$$

where

$$b_{j_1, \ldots, j_{m-1}} = a_{j_1, \ldots, j_{r-1}, 1, j_r, \ldots, j_{m-1}}$$

and

$$l_1 = k_1, \ \ldots, \ l_{r-1} = k_{r-1}, \ l_r = k_{r+1}, \ \ldots, \ l_{m-1} = k_m .$$

EXERCISES

1. Calculate the norm of the matrix $[i+j+r]_{i,j,r \leqslant k}$ (i.e. of the matrix $[a_{i,j,r}]_{i,j,r \leqslant k}$, where $a_{i,j,r} = i+j+r$).

2. Calculate the scalar product of the matrices

$$[i+j+r]_{i,j,r \leqslant k}, \quad [(-1)^{i+j+r}]_{i,j,r \leqslant k}.$$

3. Calculate the product of the matrices

$$[i+j+r]_{i,j,r \leqslant k}, \quad [(-1)^{j+r}]_{j,r \leqslant k} .$$

4. For any matrix (20),

$$\mathfrak{U} = \sum_{i_1, \ldots, i_m = 1}^{k} a_{i_1, \ldots, i_m} \mathfrak{e}_{i_1} \times \ldots \times \mathfrak{e}_{i_m},$$

where $\mathfrak{e}_1, \ldots, \mathfrak{e}_k$ are unit vectors of coordinate axes in R^k.

5. For any vectors $\mathfrak{a}_1, \ldots, \mathfrak{a}_m$ and $0 < r < m$,

$$(\mathfrak{a}_1 \times \ldots \times \mathfrak{a}_m)(\mathfrak{a}_{r+1} \times \ldots \times \mathfrak{a}_m) = |\mathfrak{a}_{r+1}|^2 \ldots |\mathfrak{a}_m|^2 \cdot (\mathfrak{a}_1 \times \ldots \times \mathfrak{a}_r) .$$

6. The equation $\mathfrak{a}_1 \times \ldots \times \mathfrak{a}_m = 0$ holds if and only if one of the vectors $\mathfrak{a}_1, \ldots, \mathfrak{a}_m$ is the zero vector.

7. By the *sum* $\sum_{n=1}^{\infty} \mathfrak{U}_n$ of a series of matrices of the same type

$$\mathfrak{U}_n = [a_{n,i_1, \ldots, i_m}]_{i_1 \leqslant k_1, \ldots, i_m \leqslant k_m}$$

we mean the matrix $\mathfrak{U} = [a_{i_1, \ldots, i_m}]_{i_1 \leqslant k_1, \ldots, i_m \leqslant k_m}$ where $a_{i_1, \ldots, i_m} = \sum_{n=1}^{\infty} a_{n,i_1, \ldots, i_m}$, provided all the series converge; then the series $\sum_{n=1}^{\infty} \mathfrak{U}_n$ is said to *converge*.

A series $\sum_{n=1}^{\infty} \mathfrak{A}_n$ of matrices of the same type is said to *converge absolutely* if $\sum_{n=1}^{\infty} \|\mathfrak{A}_n\| < \infty$. Prove that every absolutely convergent series converges.

§ 5. Rectangular matrices. Any two-indexed matrix

$$\mathfrak{A} = [a_{i,j}]_{i \leqslant l, j \leqslant k} \tag{1}$$

is called a *rectangular matrix*. If $k = l$, it is called a *square matrix*. Instead of the notation (1) we also write

$$\mathfrak{A} = \begin{bmatrix} a_{1,1} & \ldots & a_{1,k} \\ \cdot & \cdot\,\cdot\,\cdot & \cdot \\ a_{l,1} & \ldots & a_{l,k} \end{bmatrix} . \tag{1'}$$

The vectors

$$\begin{aligned} \mathfrak{w}_1 &= [a_{1,1}, \ldots, a_{1,k}] , \\ & \ldots\ldots\ldots \\ \mathfrak{w}_l &= [a_{l,1}, \ldots, a_{l,k}] \end{aligned} \tag{2}$$

are said to be the *rows* of the matrix $\mathfrak{A}$, and the vectors

$$\begin{aligned} \mathfrak{v}_1 &= [a_{1,1}, \ldots, a_{l,1}] , \\ & \ldots\ldots\ldots \\ \mathfrak{v}_k &= [a_{1,k}, \ldots, a_{l,k}] \end{aligned} \tag{3}$$

are said to be the *columns* of $\mathfrak{A}$.

By a general definition from p. 53, if $\mathfrak{A}$ is the matrix (1) of type l, k and if

$$\mathfrak{a} = [a_1, \ldots, a_k]$$

is a k-dimensional vector, then the product $\mathfrak{A}\mathfrak{a}$ is the l-dimensional vector

$$\mathfrak{A}\mathfrak{a} = a_1 \mathfrak{v}_1 + \ldots + a_k \mathfrak{v}_k = \Big[\sum_{j=1}^{\infty} a_{i,j} a_j\Big]_{i \leqslant l} . \tag{4}$$

Similarly, if $\mathfrak{A}$ is a matrix (1) of type l, k and

$$\mathfrak{a} = [a_1, \ldots, a_l]$$

is an l-dimensional vector, then the product $\mathfrak{a}\mathfrak{A}$ is the k-dimensional vector

$$(5) \qquad \mathfrak{a}\mathfrak{A} = a_1\mathfrak{w}_1 + \ldots + a_l\mathfrak{w}_l = \Big[\sum_{i=1}^{l} a_i a_{i,j}\Big]_{j \leqslant k}.$$

Instead of $\mathfrak{A}\mathfrak{a}$ and $\mathfrak{a}\mathfrak{A}$ we write sometimes $\mathfrak{A}\cdot\mathfrak{a}$ and $\mathfrak{a}\cdot\mathfrak{A}$ respectively.

If $\mathfrak{A}$ is a square matrix, the symbol $/\mathfrak{A}/$ will denote its determinant. In the case of rectangular matrices the number $|\mathfrak{A}|$ defined below and called the *modulus* of $\mathfrak{A}$ is an incomplete substitute for the notion of determinant of a square matrix.

If $\mathfrak{A}$ is the matrix (1) and $l \geqslant k$, then

$$(6) \qquad |\mathfrak{A}| = \sqrt{\sum_{i_1 < i_2 < \ldots < i_k} \begin{vmatrix} a_{i_1,1} & \ldots & a_{i_1,k} \\ \cdot & \cdot & \cdot \\ a_{i_k,1} & \ldots & a_{i_k,k} \end{vmatrix}^2}.$$

If $\mathfrak{A}$ is the matrix (1) and $l \leqslant k$, then

$$(6') \qquad |\mathfrak{A}| = \sqrt{\sum_{j_1 < j_2 < \ldots < j_l} \begin{vmatrix} a_{1,j_1} & \ldots & a_{1,j_l} \\ \cdot & \cdot & \cdot \\ a_{l,j_1} & \ldots & a_{l,j_l} \end{vmatrix}^2}.$$

Hence it follows that $|\mathfrak{A}| \neq 0$ if and only if at least one of the determinants of the maximal order which we can form from $\mathfrak{A}$ by cancelling certain rows or columns is not equal to 0. Consequently (see 1.1), in the case where $l \geqslant k$,

(7) $|\mathfrak{A}| \neq 0$ if and only if the columns of $\mathfrak{A}$ are linearly independent vectors.

In other words (see (4)), in the case where $l \geqslant k$,

(8) $|\mathfrak{A}| \neq 0$ if and only if $\mathfrak{A}\mathfrak{a} \neq 0$ for every k-dimensional vector $\mathfrak{a} \neq 0$.

Similarly, in the case where $l \leqslant k$,

(7′) $|\mathfrak{A}| \neq 0$ if and only if the rows of $\mathfrak{A}$ are linearly independent vectors.

In other words (see (5)), in the case where $l \leqslant k$,

(8′) $|\mathfrak{A}| \neq 0$ if and only if $\mathfrak{a}\mathfrak{A} \neq 0$ for every l-dimensional vector $\mathfrak{a} \neq 0$.

If $\mathfrak{A}$ is a square matrix, the modulus $|\mathfrak{A}|$ of $\mathfrak{A}$ is the absolute value of the determinant of $\mathfrak{A}$:

$$|\mathfrak{A}| = |/\mathfrak{A}/|.$$

If one of the numbers k or l is equal to 1, then by the remark at the end of § 4, the matrix (1) can be thought of as a one-indexed matrix, i.e. as a vector. Then the modulus $|\mathfrak{A}|$ of the rectangular matrix $\mathfrak{A}$ is equal to the length $|\mathfrak{A}|$ of the vector $\mathfrak{A}$ (see § 1 (34)).

By the *matrix product* of two rectangular matrices

$$\mathfrak{A} = [a_{i,j}]_{i\leqslant l, j\leqslant k}, \qquad \mathfrak{B} = [b_{j,r}]_{j\leqslant k, r\leqslant \varkappa} \tag{9}$$

we understand the rectangular matrix

$$\mathfrak{A}\cdot\mathfrak{B} = [c_{i,r}]_{i\leqslant l, r\leqslant \varkappa}, \tag{10}$$

where

$$c_{i,r} = \sum_{j=1}^{k} a_{i,j}\, b_{j,r}\,. \tag{10'}$$

It follows easily from this definition that

$$(a\mathfrak{A})\cdot\mathfrak{B} = a(\mathfrak{A}\cdot\mathfrak{B}) = \mathfrak{A}\cdot(a\mathfrak{B}),$$

$$(\mathfrak{A}+\mathfrak{A}')\cdot\mathfrak{B} = \mathfrak{A}\cdot\mathfrak{B}+\mathfrak{A}'\cdot\mathfrak{B}, \qquad \mathfrak{A}\cdot(\mathfrak{B}+\mathfrak{B}') = \mathfrak{A}\cdot\mathfrak{B}+\mathfrak{A}\cdot\mathfrak{B}',$$

where $\mathfrak{A}'$ is a matrix of the same type as $\mathfrak{A}$, and $\mathfrak{B}'$ is a matrix of the same type as $\mathfrak{B}$.

If $\mathfrak{A}$ and $\mathfrak{B}$ are the matrices (9) and

$$\mathfrak{C} = [c_{r,s}]_{r\leqslant\varkappa, s\leqslant\lambda},$$

then

$$\mathfrak{A}\cdot(\mathfrak{B}\cdot\mathfrak{C}) = (\mathfrak{A}\cdot\mathfrak{B})\cdot\mathfrak{C}$$

since both sides of the equation are the rectangular matrix

$$[d_{i,s}]_{i\leqslant l, s\leqslant\lambda}, \qquad \text{where} \qquad d_{i,s} = \sum_{j=1}^{k}\sum_{r=1}^{\varkappa} a_{i,j}\, b_{j,r}\, c_{r,s}\,.$$

It follows from the associative law that we can omit the parentheses in the expressions on both sides of the equation and we can write simply $\mathfrak{A}\cdot\mathfrak{B}\cdot\mathfrak{C}$.

If $\mathfrak{A}$ and $\mathfrak{B}$ are matrices (9) and $\mathfrak{a}$ is a $\varkappa$-dimensional vector, then

$$(\mathfrak{A}\cdot\mathfrak{B})\mathfrak{a} = \mathfrak{A}(\mathfrak{B}\mathfrak{a})\,. \tag{11}$$

Similarly, if $\mathfrak{A}$ and $\mathfrak{B}$ are matrices (9) and $\mathfrak{a}$ is an l-dimensional vector, then

$$\mathfrak{a}(\mathfrak{A}\cdot\mathfrak{B}) = (\mathfrak{a}\mathfrak{A})\mathfrak{B}\,. \tag{11'}$$

The easy verification of (11) and (11′) is left to the reader. The vectors (11) and (11′) will be denoted simply by

$$\mathfrak{A}\bullet\mathfrak{B}\cdot\mathfrak{a} \quad \text{and} \quad \mathfrak{a}\cdot\mathfrak{A}\bullet\mathfrak{B}$$

respectively.

In the case where the matrices (9) are square matrices (i.e. $l = k = \varkappa$), one of the fundamental theorems on determinants says that

$$(12) \qquad /\mathfrak{A}\bullet\mathfrak{B}/ = /\mathfrak{A}/\cdot/\mathfrak{B}/ .$$

Consequently if the determinants of $\mathfrak{A}$ and $\mathfrak{B}$ are not equal to zero, then the determinant of $\mathfrak{A}\bullet\mathfrak{B}$ is not equal to zero. Conversely, if the determinant of $\mathfrak{A}\bullet\mathfrak{B}$ is different from zero, so are the determinants of $\mathfrak{A}$ and $\mathfrak{B}$. The next theorem contains a generalization of these remarks for the case of rectangular matrices of the form (9).

5.1. *If* $l\geqslant k\geqslant\varkappa$ *and* $|\mathfrak{A}|\neq 0\neq|\mathfrak{B}|$, *then* $|\mathfrak{A}\bullet\mathfrak{B}|\neq 0$. *Conversely if* $l\geqslant k\geqslant\varkappa$ *and* $|\mathfrak{A}\bullet\mathfrak{B}|\neq 0$, *then* $|\mathfrak{B}|\neq 0$. *If* $l\geqslant k=\varkappa$, *then*

$$(13) \qquad |\mathfrak{A}\bullet\mathfrak{B}| = |\mathfrak{A}|\cdot|\mathfrak{B}| .$$

Similarly, if $l\leqslant k\leqslant\varkappa$ *and* $|\mathfrak{A}|\neq 0\neq|\mathfrak{B}|$, *then* $|\mathfrak{A}\bullet\mathfrak{B}|\neq 0$. *Conversely, if* $l\leqslant k\leqslant\varkappa$ *and* $|\mathfrak{A}\bullet\mathfrak{B}|\neq 0$, *then* $|\mathfrak{A}|\neq 0$. *If* $l=k\leqslant\varkappa$, *then* (13) *holds.*

Suppose $l\geqslant k\geqslant\varkappa$. Let $\mathfrak{a}$ be any $\varkappa$-dimensional non-zero vector. If $|\mathfrak{B}|\neq 0$, then $\mathfrak{B}\mathfrak{a}\neq 0$ by (8). If also $|\mathfrak{A}|\neq 0$, then $(\mathfrak{A}\bullet\mathfrak{B})\mathfrak{a} = \mathfrak{A}(\mathfrak{B}\mathfrak{a})\neq 0$ by the same argument (see also (11)). This proves that $|\mathfrak{A}\bullet\mathfrak{B}|\neq 0$.

Conversely, if $|\mathfrak{A}\bullet\mathfrak{B}|\neq 0$, then $\mathfrak{A}(\mathfrak{B}\mathfrak{a}) = (\mathfrak{A}\bullet\mathfrak{B})\mathfrak{a}\neq 0$. Hence $\mathfrak{B}\mathfrak{a}\neq 0$. This proves that $|\mathfrak{B}|\neq 0$.

Suppose now that $l\geqslant k=\varkappa$. By definition, $\mathfrak{A}\bullet\mathfrak{B} = [c_{i,r}]$ where the numbers $c_{i,r}$ are defined by (10′). Since

$$\begin{vmatrix} c_{i_1,1} & \dots & c_{i_1,k} \\ \dots & \dots & \dots \\ c_{i_k,1} & \dots & c_{i_k,k} \end{vmatrix} = \begin{vmatrix} a_{i_1,1} & \dots & a_{i_1,k} \\ \dots & \dots & \dots \\ a_{i_k,1} & \dots & a_{i_k,k} \end{vmatrix} \cdot \begin{vmatrix} b_{1,1} & \dots & b_{1,k} \\ \dots & \dots & \dots \\ b_{k,1} & \dots & b_{k,k} \end{vmatrix}$$

by the theorem on multiplication of determinants (see (12)), applying the definition of $|\mathfrak{A}\bullet\mathfrak{B}|$ we easily get (13).

The proof of the second part of 5.1 is analogous.

Let $\mathfrak{A} = [a_{i,j}]_{i,j\leqslant k}$ be a square matrix. A square matrix $\mathfrak{B} = [b_{i,j}]_{i,j\leqslant k}$ is said to be the *inverse* of $\mathfrak{A}$ if

$$(14) \qquad \mathfrak{A}\bullet\mathfrak{B} = \mathfrak{I} = \mathfrak{B}\bullet\mathfrak{A}$$

where $\mathfrak{I}$ is the *unit matrix*, i.e.

$$\mathfrak{I} = [\delta_{i,j}]_{i,j\leqslant k}, \qquad \delta_{i,j} = \begin{cases} 1 & \text{if} \quad i=j, \\ 0 & \text{if} \quad i\neq j. \end{cases}$$

It follows from the theory of determinants that the inverse matrix $\mathfrak{B}$ exists if and only if $/\mathfrak{A}/ \neq 0$. Then

$$b_{i,j} = \frac{A_{j,i}}{/\mathfrak{A}/}$$

where $A_{i,j}$ is the minor corresponding to the term $a_{i,j}$, i.e. $A_{i,j}$ is the product of $(-1)^{i+j}$ and the determinant of the matrix obtained from the matrix $\mathfrak{A}$ by canceling the ith row and the jth column.

The inverse of $\mathfrak{A}$ will be denoted, if exists, by $\mathfrak{A}^{-1}$. By definition

(14′) $$\mathfrak{A} \cdot \mathfrak{A}^{-1} = \mathfrak{J} = \mathfrak{A}^{-1} \cdot \mathfrak{A}.$$

$\mathfrak{A}$ is said to be *invertible* if $\mathfrak{A}^{-1}$ exists.

If $\mathfrak{A}$ and $\mathfrak{B}$ are invertible matrices of the same dimension, then $\mathfrak{A} \cdot \mathfrak{B}$ is also invertible and

(15) $$(\mathfrak{A} \cdot \mathfrak{B})^{-1} = \mathfrak{B}^{-1} \cdot \mathfrak{A}^{-1}.$$

Note the following characteristic property of the unit matrix $\mathfrak{J}$:

(16) $$\mathfrak{A} \cdot \mathfrak{J} = \mathfrak{J} \cdot \mathfrak{A} = \mathfrak{A}$$

for every square matrix $\mathfrak{A}$ of the same dimension as $\mathfrak{J}$.

If $\mathfrak{A}$ is a rectangular matrix (1), then $\mathfrak{A}^*$ denotes the *transpose* of $\mathfrak{A}$, i.e. the matrix obtained from $\mathfrak{A}$ by replacing columns by rows and conversely. In other words,

(17) $$\mathfrak{A}^* = [a^*_{i,j}]_{i \leqslant k, j \leqslant l} \quad \text{where} \quad a^*_{i,j} = a_{j,i}.$$

Observe that if $\mathfrak{v}_j$ denotes the jth column of the matrix $\mathfrak{A}$ (see (1) and (3)), then

(18) $$\mathfrak{A}^* \cdot \mathfrak{A} = \begin{bmatrix} \mathfrak{v}_1 \mathfrak{v}_1 & \dots & \mathfrak{v}_1 \mathfrak{v}_k \\ \cdot & \cdots & \cdot \\ \mathfrak{v}_k \mathfrak{v}_1 & \dots & \mathfrak{v}_k \mathfrak{v}_k \end{bmatrix} = [\mathfrak{v}_r \mathfrak{v}_s]_{r,s \leqslant k},$$

i.e. the terms of the square matrix $\mathfrak{A}^* \cdot \mathfrak{A}$ are the scalar products of columns of $\mathfrak{A}$

(19) $$\mathfrak{v}_r \mathfrak{v}_s = \sum_{i=1}^{l} a_{i,r} a_{i,s}.$$

Consequently,

(20) $$/\mathfrak{A}^* \cdot \mathfrak{A}/ = \begin{vmatrix} \mathfrak{v}_1 \mathfrak{v}_1 & \dots & \mathfrak{v}_1 \mathfrak{v}_k \\ \cdot & \cdots & \cdot \\ \mathfrak{v}_k \mathfrak{v}_1 & \dots & \mathfrak{v}_k \mathfrak{v}_k \end{vmatrix}.$$

The matrix (18) and the determinant (20) are said to be the *Gram matrix* and the *Gram determinant* of vectors $\mathfrak{v}_1, \dots, \mathfrak{v}_k$.

If $\mathfrak{w}_i$ denotes the ith row of $\mathfrak{A}$ (see (1) and (2)), then similarly

$$(18')\qquad \mathfrak{A}\cdot\mathfrak{A}^* = \begin{bmatrix} \mathfrak{w}_1\mathfrak{w}_1 & \dots & \mathfrak{w}_1\mathfrak{w}_l \\ \dots & \dots & \dots \\ \mathfrak{w}_l\mathfrak{w}_1 & \dots & \mathfrak{w}_l\mathfrak{w}_l \end{bmatrix} = [\mathfrak{w}_r\mathfrak{w}_s]_{r,s\leqslant l},$$

i.e. the terms of $\mathfrak{A}\cdot\mathfrak{A}^*$ are the scalar products of the rows of $\mathfrak{A}$:

$$(19')\qquad \mathfrak{w}_r\mathfrak{w}_s = \sum_{j=1}^{k} a_{r,j}a_{s,j}.$$

Consequently,

$$(20')\qquad /\mathfrak{A}\cdot\mathfrak{A}^*/ = \begin{vmatrix} \mathfrak{w}_1\mathfrak{w}_1 & \dots & \mathfrak{w}_1\mathfrak{w}_l \\ \dots & \dots & \dots \\ \mathfrak{w}_l\mathfrak{w}_1 & \dots & \mathfrak{w}_l\mathfrak{w}_l \end{vmatrix}.$$

Thus $\mathfrak{A}\cdot\mathfrak{A}^*$ and $/\mathfrak{A}\cdot\mathfrak{A}^*/$ are respectively the Gram matrix and the Gram determinant of the vectors $\mathfrak{w}_1, \dots, \mathfrak{w}_l$.

5.2. *For any rectangular matrix $\mathfrak{A}$ of type l, k $(l \geqslant k)$*

$$(21)\qquad |\mathfrak{A}|^2 = /\mathfrak{A}^*\cdot\mathfrak{A}/,$$

that is,

$$(21')\qquad |\mathfrak{A}| = \sqrt{/\mathfrak{A}^*\cdot\mathfrak{A}/}.$$

Consequently, if $|\mathfrak{A}| \neq 0$, then the inverse matrix $(\mathfrak{A}^*\cdot\mathfrak{A})^{-1}$ exists.

If $|\mathfrak{A}| \neq 0$, $\mathfrak{B}$ is a rectangular matrix of type $k, \varkappa$ and $\mathfrak{C} = \mathfrak{A}\cdot\mathfrak{B}$, then

$$(22)\qquad \mathfrak{B} = (\mathfrak{A}^*\cdot\mathfrak{A})^{-1}\cdot\mathfrak{A}^*\cdot\mathfrak{C}.$$

The formulation of the analogous theorem for $k \geqslant l$ is left to the reader (it will not be used in the sequel).

To prove (21), let us substitute into (20) the values of the scalar products from (19). Applying the known properties of determinants (the additivity and homogeneity with respect to columns, etc.), we observe that (21) holds (see the definition (6) of $|\mathfrak{A}|$). A more detailed proof can be found in every textbook in higher algebra.

If $|\mathfrak{A}| \neq 0$ and $\mathfrak{C} = \mathfrak{A}\cdot\mathfrak{B}$, then

$$\mathfrak{A}^*\cdot\mathfrak{C} = (\mathfrak{A}^*\cdot\mathfrak{A})\cdot\mathfrak{B}.$$

Multiplying the last equation by $(\mathfrak{A}^*\cdot\mathfrak{A})^{-1}$ from the left-hand side, we get (22).

If $i_1, \dots, i_l$ are positive integers $\leqslant l$, then the symbol

$$(23)\qquad \operatorname{sgn}(i_1, \dots, i_l)$$

denotes the number 1 if $i_1, \dots, i_l$ is an even permutation of the numbers $1, \dots, l$, it denotes the number -1 if $i_1, \dots, i_l$ is an odd permutation of the numbers $1, \ldots, l$, and it denotes the number 0 if it is not a permutation, i.e. if $i_r = i_s$ for some $r, s, r \neq s$.

The next theorem from algebra is not very well known, therefore a complete proof is given.

5.3. *Let* $k < l$,

$$\mathfrak{A} = [a_{i,j}]_{i\leqslant k, j\leqslant l}, \qquad \mathfrak{B} = [b_{i,j}]_{i\leqslant l-k, j\leqslant l}.$$

If each of the rows

$$\begin{aligned} \mathfrak{a}_1 &= [a_{1,1}, \ldots, a_{1,l}], \\ &\ldots\ldots\ldots \\ \mathfrak{a}_k &= [a_{k,1}, \ldots, a_{k,l}] \end{aligned}$$

of $\mathfrak{A}$ *is orthogonal to each of the rows*

$$\begin{aligned} \mathfrak{b}_1 &= [b_{1,1}, \ldots, b_{1,l}], \\ &\ldots\ldots\ldots \\ \mathfrak{b}_{l-k} &= [b_{l-k,1}, \ldots, b_{l-k,1}] \end{aligned}$$

of $\mathfrak{B}$, *then for every permutation* $i_1, \ldots, i_k, j_1, \ldots, j_{l-k}$ *of the integers* $1, \ldots, l$

$$|\mathfrak{A}| \cdot \begin{vmatrix} b_{1,j_1} & \ldots & b_{1,j_{l-k}} \\ \ldots & \ldots & \ldots \\ b_{l-k,j_1} & \ldots & b_{l-k,j_{l-k}} \end{vmatrix} = \eta \cdot \operatorname{sgn}(i_1, \ldots, i_k, j_1, \ldots, j_{l-k}) \cdot |\mathfrak{B}| \cdot \begin{vmatrix} a_{1,i_1} & \ldots & a_{1,i_k} \\ \ldots & \ldots & \ldots \\ a_{k,i_1} & \ldots & a_{k,i_k} \end{vmatrix} \tag{24}$$

where $\eta = 1$ *if the determinant of the square matrix*

$$\mathfrak{C} = \begin{bmatrix} a_{1,1} & \ldots & a_{1,l} \\ \ldots & \ldots & \ldots \\ a_{k,1} & \ldots & a_{k,l} \\ b_{1,1} & \ldots & b_{1,l} \\ \ldots & \ldots & \ldots \\ b_{l-k,1} & \ldots & b_{l-k,l} \end{bmatrix}$$

is non-negative, and $\eta = -1$ *in the opposite case.*

It suffices to prove theorem 5.3 in the case where $i_1 = 1, \ldots, i_k = k$, $j_1 = k+1, \ldots, j_{l-k} = l$. The remaining cases can be reduced to this particular case by a common permutation of the columns of $\mathfrak{A}$ and $\mathfrak{B}$.

Let

$$\alpha = \begin{vmatrix} a_{1,1} & \ldots & a_{1,k} \\ \ldots & \ldots & \ldots \\ a_{k,1} & \ldots & a_{k,k} \end{vmatrix}, \qquad \beta = \begin{vmatrix} b_{1,k+1} & \ldots & b_{1,l} \\ \ldots & \ldots & \ldots \\ b_{l-k,k+1} & \ldots & b_{l-k,l} \end{vmatrix}$$

and

$$\mathfrak{D} = \begin{bmatrix} a_{1,1} & \dots & a_{k,1} & 0 & 0 & \dots & 0 \\ \cdot & \cdot & \cdot & \cdot & \cdot & \cdot & \cdot \\ a_{1,k} & \dots & a_{k,k} & 0 & 0 & \dots & 0 \\ a_{1,k+1} & \dots & a_{k,k+1} & 1 & 0 & \dots & 0 \\ \cdot & \cdot & \cdot & \cdot & \cdot & \cdot & \cdot \\ a_{1,l} & \dots & a_{k,l} & 0 & 0 & \dots & 1 \end{bmatrix}.$$

In other words, the vectors

$$\mathfrak{a}_1, \dots, \mathfrak{a}_k, \mathfrak{e}_{k+1}, \dots, \mathfrak{e}_l$$

(see § 1 (38)) are the consecutive columns of $\mathfrak{D}$.

Since $\mathfrak{a}_i \mathfrak{b}_j = 0$ for $i = 1, \dots, k$ and $j = 1, \dots, l-k$, we easily verify that

$$\mathfrak{C} \cdot \mathfrak{D} = \begin{bmatrix} \mathfrak{a}_1\mathfrak{a}_1 & \dots & \mathfrak{a}_1\mathfrak{a}_k & a_{1,k+1} & \dots & a_{1,l} \\ \cdot & \cdot & \cdot & \cdot & \cdot & \cdot \\ \mathfrak{a}_k\mathfrak{a}_1 & \dots & \mathfrak{a}_k\mathfrak{a}_k & a_{k,k+1} & \dots & a_{k,l} \\ 0 & \dots & 0 & b_{1,k+1} & \dots & b_{1,l} \\ \cdot & \cdot & \cdot & \cdot & \cdot & \cdot \\ 0 & \dots & 0 & b_{l-k,k+1} & \dots & b_{l-k,l} \end{bmatrix}.$$

Thus the determinant of the matrix $\mathfrak{C} \cdot \mathfrak{D}$ is equal to the product of the number β and the Gram determinant of the vectors $\mathfrak{a}_1, \dots, \mathfrak{a}_k$, i.e. the number $|\mathfrak{A}|^2$ (see (21)). The determinant of $\mathfrak{D}$ is equal to a. Hence

$$(25) \qquad |\mathfrak{A}|^2 \cdot \beta = /\mathfrak{C}/ \cdot a .$$

On the other hand,

$$\mathfrak{C} \cdot \mathfrak{C}^* = \begin{bmatrix} \mathfrak{a}_1\mathfrak{a}_1 & \dots & \mathfrak{a}_1\mathfrak{a}_k & 0 & \dots & 0 \\ \cdot & \cdot & \cdot & \cdot & \cdot & \cdot \\ \mathfrak{a}_k\mathfrak{a}_1 & \dots & \mathfrak{a}_k\mathfrak{a}_k & 0 & \dots & 0 \\ 0 & \dots & 0 & \mathfrak{b}_1\mathfrak{b}_1 & \dots & \mathfrak{b}_1\mathfrak{b}_{l-k} \\ \cdot & \cdot & \cdot & \cdot & \cdot & \cdot \\ 0 & \dots & 0 & \mathfrak{b}_{l-k}\mathfrak{b}_1 & \dots & \mathfrak{b}_{l-k}\mathfrak{b}_{l-k} \end{bmatrix}$$

and consequently $/\mathfrak{C}/^2 = /\mathfrak{C} \cdot \mathfrak{C}^*/ = |\mathfrak{A}|^2|\mathfrak{B}|^2$ (see (21)). Hence it follows, by (25), that

$$(26) \qquad |\mathfrak{A}| \cdot \beta = \eta \cdot |\mathfrak{B}| \cdot a ,$$

provided $|\mathfrak{A}| \neq 0$.

If $|\mathfrak{A}| = 0$, then (26) also holds since then $a = 0$. The identity (26) is the particular case of (24) we wished to prove.

EXERCISES

1. Prove that

$$\|\mathfrak{A} \bullet \mathfrak{B}\| \leqslant \|\mathfrak{A}\| \cdot \|\mathfrak{B}\|$$

for any rectangular matrices $\mathfrak{A}$, $\mathfrak{B}$ for which the matrix product is defined.

For any square matrix $\mathfrak{A}$,

$$\|\mathfrak{A}^n\| \leqslant \|\mathfrak{A}\|^n \quad (n = 0, 1, 2, \ldots),$$

where

(27) $$\mathfrak{A}^0 = \mathfrak{I}, \quad \mathfrak{A}^n = \mathfrak{A} \bullet \mathfrak{A} \bullet \ldots \bullet \mathfrak{A} \quad n\text{-times},$$

$\mathfrak{I}$ is the unit matrix of the same type as $\mathfrak{A}$.

2. Calculate the matrix product

$$\begin{bmatrix} 1 & 7 & -1 & 0 \\ 2 & -1 & 3 & -5 \end{bmatrix} \bullet \begin{bmatrix} 1 & 0 & -2 \\ 2 & -1 & 3 \\ 3 & 1 & 0 \\ 4 & -1 & 0 \end{bmatrix}.$$

3. Let

$$\mathfrak{A} = \begin{bmatrix} -6 & 3 & 2 \\ 0 & 5 & 4 \\ 1 & 1 & 1 \end{bmatrix}.$$

Calculate $\mathfrak{A}^{-1}$.

4. Calculate the Gram matrix of the vectors

$$\mathfrak{v}_1 = [2, 1, -3, 6, 5],$$
$$\mathfrak{v}_2 = [-1, 1, 1, -1, 0],$$
$$\mathfrak{v}_3 = [0, 1, 2, 0, -1].$$

5. For any rectangular matrix $\mathfrak{A}$ of type l, k $(l \geqslant k)$, if $|\mathfrak{A}| \neq 0$ and $\mathfrak{b} = \mathfrak{A}\mathfrak{a}$ ($\mathfrak{a}, \mathfrak{b}$ — vectors), then

$$\mathfrak{a} = (\mathfrak{A}^* \bullet \mathfrak{A})^{-1} \bullet \mathfrak{A} \cdot \mathfrak{b}.$$

6. The *trace* $\operatorname{tr}\mathfrak{A}$ of a square matrix $\mathfrak{A} = [a_{i,j}]_{i,j \leqslant k}$ is the number

$$\operatorname{tr}\mathfrak{A} = a_{1,1} + a_{2,2} + \ldots + a_{k,k}.$$

Prove that

$$\operatorname{tr} a\mathfrak{A} = a \operatorname{tr}\mathfrak{A}, \quad \operatorname{tr}(\mathfrak{A} + \mathfrak{B}) = \operatorname{tr}\mathfrak{A} + \operatorname{tr}\mathfrak{B},$$

$$\operatorname{tr}\mathfrak{A} \bullet \mathfrak{B} = \mathfrak{A}^*\mathfrak{B} = \operatorname{tr}\mathfrak{B} \bullet \mathfrak{A} = \mathfrak{B}^*\mathfrak{A}, \quad |\operatorname{tr}\mathfrak{A} \bullet \mathfrak{B}| \leqslant \|\mathfrak{A}\| \cdot \|\mathfrak{B}\|,$$

$\mathfrak{B}$ being any matrix of the same type. If a series $\sum_{n=1}^{\infty} \mathfrak{A}_n$ of square matrices converges, then

$$\operatorname{tr} \sum_{n=1}^{\infty} \mathfrak{A}_n = \sum_{n=1}^{\infty} \operatorname{tr}\mathfrak{A}_n.$$

7. Let $\mathfrak{A}$ and $\mathfrak{B}$ be square matrices of the same type. Prove the following formula for the derivative of the function

$$f(t) = |\mathfrak{A} + t\mathfrak{B}|$$

at the point 0:

$$f'(0) = |\mathfrak{A}| \cdot (\mathfrak{B}^* \mathfrak{A}^{-1}) = |\mathfrak{A}| \operatorname{tr}(\mathfrak{B} \bullet \mathfrak{A}^{-1})$$

provided $\mathfrak{A}^{-1}$ exists.

8. Let $\mathfrak{A}$ be a square matrix, let $\mathfrak{I}$ be the unit square matrix of the same type and let

$$f(t) = |\mathfrak{I} + t\mathfrak{A}| .$$

Prove the following formula for the derivative of f:

$$f'(t) = |\mathfrak{I} + t\mathfrak{A}| \cdot \operatorname{tr}(\mathfrak{A} \bullet (\mathfrak{I} + t\mathfrak{A})^{-1})$$

for every t such that $|\mathfrak{I} + t\mathfrak{A}| \neq 0$.

9. If $\mathfrak{A}$ is a square matrix and $\|\mathfrak{A}\| < 1$, then the series

$$\sum_{n=0}^{\infty} (-1)^n \mathfrak{A}^n \quad \text{and} \quad \sum_{n=1}^{\infty} (-1)^{n-1} \frac{\mathfrak{A}^n}{n}$$

(see (27)) converge absolutely, and therefore converge. The sum of the first series is the matrix $(\mathfrak{I} + \mathfrak{A})^{-1}$, $\mathfrak{I}$ being the unit matrix of the same type as $\mathfrak{A}$. The sum of the second series is denoted by $\log(\mathfrak{I} + \mathfrak{A})$ by an analogy with the development of $\log(1+x)$ into a power series.

Prove that the real function

$$f(t) = \operatorname{tr} \log(\mathfrak{I} + t\mathfrak{A}) \quad \text{for} \quad |t| < \|\mathfrak{A}\|^{-1}$$

is differentiable and

$$f'(t) = \operatorname{tr}(\mathfrak{A} \bullet (\mathfrak{I} + t\mathfrak{A})^{-1}) .$$

10. Prove that

$$|\mathfrak{I} + t\mathfrak{A}| = e^{\operatorname{tr} \log(\mathfrak{I} + t\mathfrak{A})} \quad \text{for} \quad |t| < \|\mathfrak{A}\|^{-1} .$$

11. For any square matrix $\mathfrak{A}$, the series

$$e^{\mathfrak{A}} = \sum_{n=0}^{\infty} \frac{\mathfrak{A}^n}{n!}$$

(see (27)) converges absolutely. Prove that for every matrix $\mathfrak{B}$ of the same type,

$$\text{if } \mathfrak{A} \bullet \mathfrak{B} = \mathfrak{B} \bullet \mathfrak{A}, \text{ then } e^{\mathfrak{A}+\mathfrak{B}} = e^{\mathfrak{A}} \bullet e^{\mathfrak{B}} .$$

Prove that

$$e^{\log(\mathfrak{I} + \mathfrak{A})} = \mathfrak{I} + \mathfrak{A} .$$

§ 6. Linear and multilinear forms. Forms of degree m. Let F be a real function defined on the set V_k of all vectors in R^k. F is said to be *homogeneous of degree m* provided

(1) $$F(a\mathfrak{a}) = a^m F(\mathfrak{a})$$

for every vector $\mathfrak{a} \in V_k$ and every real number a. F is said to be *homogeneous* if it is homogeneous of degree 1, i.e. if

(2) $$F(a\mathfrak{a}) = aF(\mathfrak{a}) .$$

F is said to be *additive* provided

(3) $$F(\mathfrak{a}+\mathfrak{b}) = F(\mathfrak{a})+F(\mathfrak{b})$$

for all vectors $\mathfrak{a}, \mathfrak{b} \in V_k$.

F is said to be a *linear form* (or a *linear function*) if F is homogeneous and additive. If F is a linear form, then besides (2) and (3) the following identities hold:

(4) $$F(-\mathfrak{a}) = -F(\mathfrak{a}) ,$$

(5) $$F(\mathfrak{a}-\mathfrak{b}) = F(\mathfrak{a})-F(\mathfrak{b}) ,$$

(6) $$F(0) = 0 ,$$

(7) $$F(a_1\mathfrak{a}_1+\ldots+a_m\mathfrak{a}_m) = a_1F(\mathfrak{a}_1)+\ldots+a_mF(\mathfrak{a}_m) .$$

Suppose F is a linear form on V_k. For any vector $\mathfrak{a} = [a_1, \ldots, a_k]$ we have

$$\mathfrak{a} = a_1\mathfrak{e}_1+\ldots+a_k\mathfrak{e}_k ,$$

where $\mathfrak{e}_1, \ldots, \mathfrak{e}_k$ are the unit vectors of the coordinate axes (see § 1 (38)). Hence by (7)

$$F(\mathfrak{a}) = a_1F(\mathfrak{e}_1)+\ldots+a_kF(\mathfrak{e}_k) ,$$

that is,

(8) $$F(\mathfrak{a}) = \mathfrak{w}\mathfrak{a} ,$$

where

(9) $$\mathfrak{w} = [F(\mathfrak{e}_1), \ldots, F(\mathfrak{e}_k)] .$$

Conversely, if $\mathfrak{w}$ is any k-dimensional vector, then (8) defines a linear form on V_k and (9) holds. Consequently,

6.1. *A function F: $V_k \to R$ is a linear form if and only if it is of the form* (8) *where $\mathfrak{w}$ is a k-dimensional vector. Then* (9) *holds.*

Now let F be a real function defined on the Cartesian product $V_k \times V_k$. By definition, for any vectors $\mathfrak{a}, \mathfrak{b} \in V_k$, $F(\mathfrak{a}, \mathfrak{b})$ is a finite real number. F is said to be a *bilinear form on $V_k \times V_k$* provided that

1° for every fixed $\mathfrak{b} \in V_k$, $F(\cdot, \mathfrak{b})$ is a linear form;

2° for every fixed $\mathfrak{a} \in V_k$, $F(\mathfrak{a}, \cdot)$ is a linear form.

In other words, the expression $F(\mathfrak{a}, \mathfrak{b})$ considered as a function of one of the variables $\mathfrak{a}, \mathfrak{b}$ only (the remaining being fixed) is a linear form. Thus F is bilinear if and only if

(10) $$F(a\mathfrak{a}, \mathfrak{b}) = aF(\mathfrak{a}, \mathfrak{b}) = F(\mathfrak{a}, a\mathfrak{b}) ,$$

(11) $$F(\mathfrak{a}+\mathfrak{a}_1, \mathfrak{b}) = F(\mathfrak{a}, \mathfrak{b})+F(\mathfrak{a}_1, \mathfrak{b}) ,$$

(11′) $$F(\mathfrak{a}, \mathfrak{b}+\mathfrak{b}_1) = F(\mathfrak{a}, \mathfrak{b})+F(\mathfrak{a}, \mathfrak{b}_1) ,$$

for any real number a and any vectors $\mathfrak{a}, \mathfrak{a}_1, \mathfrak{b}, \mathfrak{b}_1 \in V_k$. The following identities are immediate consequences of (10), (11), (11'):

(12) $$F(-\mathfrak{a}, \mathfrak{b}) = -F(\mathfrak{a}, \mathfrak{b}) = F(\mathfrak{a}, -\mathfrak{b}),$$

(13) $$F(\mathfrak{a}-\mathfrak{a}_1, \mathfrak{b}) = F(\mathfrak{a}, \mathfrak{b}) - F(\mathfrak{a}_1, \mathfrak{b}),$$

(14) $$F(\mathfrak{a}, \mathfrak{b}-\mathfrak{b}_1) = F(\mathfrak{a}, \mathfrak{b}) - F(\mathfrak{a}, \mathfrak{b}_1),$$

(15) $$F(0, \mathfrak{b}) = 0 = F(\mathfrak{a}, 0),$$

(16) $$F\Big(\sum_{i=1}^{m} a_i \mathfrak{a}_i, \sum_{j=1}^{n} b_j \mathfrak{b}_j\Big) = \sum_{i=1}^{m} \sum_{j=1}^{n} a_i b_j F(\mathfrak{a}_i, \mathfrak{b}_j).$$

For any vectors

$$\mathfrak{a} = [a_1, \ldots, a_k], \qquad \mathfrak{b} = [b_1, \ldots, b_k]$$

we have

$$\mathfrak{a} = a_1 \mathfrak{e}_1 + \ldots + a_k \mathfrak{e}_k, \qquad \mathfrak{b} = b_1 \mathfrak{e}_1 + \ldots + b_k \mathfrak{e}_k,$$

where $\mathfrak{e}_1, \ldots, \mathfrak{e}_k$ are the unit vectors of the coordinate axes. Hence by (16)

(17) $$F(\mathfrak{a}, \mathfrak{b}) = \sum_{i=1}^{k} \sum_{j=1}^{k} F(\mathfrak{e}_i, \mathfrak{e}_j) a_i b_j,$$

that is,

(17') $$F(\mathfrak{a}, \mathfrak{b}) = \mathfrak{A} \cdot (\mathfrak{a} \times \mathfrak{b}),$$

where

(18) $$\mathfrak{A} = [F(\mathfrak{e}_i, \mathfrak{e}_j)]_{i,j \leqslant k}.$$

Conversely, if $\mathfrak{A}$ is any two-indexed k-dimensional matrix, then (17') is a bilinear form on $V_k \times V_k$ (see § 4 (9), (10), (19), (19')), and (18) holds.

Consequently,

6.2. *A function F: $V_k \times V_k \to R$ is a bilinear form if and only if it is of the form* (17') *where $\mathfrak{A}$ is a two-indexed k-dimensional matrix. Then* (18) *holds.*

The definition of linear and bilinear forms can be generalized as follows. A real function F on the Cartesian product V_k^m of m replicas of V_k ($m = 1, 2, \ldots$) is said to be an *m-linear form* (or less precisely, a *multilinear form*) provided that for any fixed vectors $\mathfrak{a}_1, \ldots, \mathfrak{a}_{i-1}, \mathfrak{a}_{i+1}, \ldots, \mathfrak{a}_m \in V_k$, the expression

$$F(\mathfrak{a}_1, \ldots, \mathfrak{a}_{i-1}, \mathfrak{a}_i, \mathfrak{a}_{i+1}, \ldots, \mathfrak{a}_m)$$

interpreted as a function of $\mathfrak{a}_i$ only, is linear, $i = 1, \ldots, m$. In other words, F is an m-linear form on V_k^m provided that

$$F(\mathfrak{a}_1, \ldots, \mathfrak{a}_{i-1}, a\mathfrak{a}_i, \mathfrak{a}_{i+1}, \ldots, \mathfrak{a}_m) = aF(\mathfrak{a}_1, \ldots, \mathfrak{a}_m), \tag{19}$$

$$\begin{aligned} &F(\mathfrak{a}_1, \ldots, \mathfrak{a}_{i-1}, \mathfrak{a}_i + \mathfrak{a}_i', \mathfrak{a}_{i+1}, \ldots, \mathfrak{a}_m) \\ &\quad = F(\mathfrak{a}_1, \ldots, \mathfrak{a}_{i-1}, \mathfrak{a}_i, \mathfrak{a}_{i+1}, \ldots, \mathfrak{a}_m) + F(\mathfrak{a}_1, \ldots, \mathfrak{a}_{i-1}, \mathfrak{a}_i', \mathfrak{a}_{i+1}, \ldots, \mathfrak{a}_m) \end{aligned} \tag{20}$$

for any real number a and any vectors $\mathfrak{a}_1, \ldots, \mathfrak{a}_m, \mathfrak{a}_i' \in V_k$, $i = 1, \ldots, m$. It follows immediately from (19) and (20) that

$$F(\mathfrak{a}_1, \ldots, \mathfrak{a}_{i-1}, -\mathfrak{a}_i, \mathfrak{a}_{i+1}, \ldots, \mathfrak{a}_m) = -F(\mathfrak{a}_1, \ldots, \mathfrak{a}_{i-1}, \mathfrak{a}_i, \mathfrak{a}_{i+1}, \ldots, \mathfrak{a}_m), \tag{21}$$

$$\begin{aligned} &F(\mathfrak{a}_1, \ldots, \mathfrak{a}_{i-1}, \mathfrak{a}_i - \mathfrak{a}_i', \mathfrak{a}_{i+1}, \ldots, \mathfrak{a}_m) \\ &\quad = F(\mathfrak{a}_1, \ldots, \mathfrak{a}_{i-1}, \mathfrak{a}_i, \mathfrak{a}_{i+1}, \ldots, \mathfrak{a}_m) - F(\mathfrak{a}_1, \ldots, \mathfrak{a}_{i-1}, \mathfrak{a}_i', \mathfrak{a}_{i+1}, \ldots, \mathfrak{a}_m), \end{aligned} \tag{22}$$

$$F(\mathfrak{a}_1, \ldots, \mathfrak{a}_{i-1}, 0, \mathfrak{a}_{i+1}, \ldots, \mathfrak{a}_m) = 0, \tag{23}$$

$$\begin{aligned} &F\Big(\sum_{i_1=1}^{l_1} a_{1,i_1}\mathfrak{a}_{1,i_1}, \ldots, \sum_{i_m=1}^{l_m} a_{m,i_m}\mathfrak{a}_{m,i_m}\Big) \\ &\qquad\qquad = \sum_{i_1=1}^{l_1} \ldots \sum_{i_m=1}^{l_m} F(\mathfrak{a}_{1,i_1}, \ldots, \mathfrak{a}_{m,i_m}) \cdot a_{1,i_1} \ldots a_{m,i_m}. \end{aligned} \tag{24}$$

Let $\mathfrak{e}_1, \ldots, \mathfrak{e}_k$ be the unit vectors of the coordinate axes. If F is an m-linear form on V_k^m, then for any vectors

$$\mathfrak{a}_1 = [a_{1,1}, \ldots, a_{1,k}] = \sum_{i_1=1}^{k} a_{1,i_1}\mathfrak{e}_{i_1} \in V_k,$$

$$\ldots\ldots\ldots\ldots\ldots\ldots\ldots\ldots\ldots$$

$$\mathfrak{a}_m = [a_{m,1}, \ldots, a_{m,k}] = \sum_{i_m=1}^{k} a_{m,i_m}\mathfrak{e}_{i_m} \in V_k$$

we have, by (24),

$$F(\mathfrak{a}_1, \ldots, \mathfrak{a}_m) = \sum_{i_1=1}^{k} \ldots \sum_{i_m=1}^{k} F(\mathfrak{e}_{i_1}, \ldots, \mathfrak{e}_{i_m}) \cdot a_{1,i_1} \ldots a_{m,i_m}, \tag{25}$$

that is,

$$F(\mathfrak{a}_1, \ldots, \mathfrak{a}_m) = \mathfrak{A} \cdot (\mathfrak{a}_1 \times \ldots \times \mathfrak{a}_m), \tag{25'}$$

where

$$\mathfrak{A} = [F(\mathfrak{e}_{i_1}, \ldots, \mathfrak{e}_{i_m})]_{i_1, \ldots, i_m \leqslant k}. \tag{26}$$

Conversely, if $\mathfrak{A}$ is any m-indexed k-dimensional matrix, then (25′) defines an m-linear form on V_k^m and (26) holds. Consequently,

6.3. *A function* $F\colon V_k^m \to R$ *is an* m*-linear form if and only if it is of the form* (25′), *where* $\mathfrak{A}$ *is an* m*-indexed* k*-dimensional matrix. Then* (26) *holds.*

It follows from 6.3 that there is a natural one-to-one correspondence between m-linear forms on V_k^m and k-dimensional m-indexed matrices. Roughly speaking, both notions express the same mathematical content. The terms of the matrix $\mathfrak{A}$ in (25′) uniquely determine the form F. Therefore they will be called the *coordinates* of F; more precisely, the $(i_1, \ldots, i_m)$-th coordinate of $\mathfrak{A}$ will also be called the $(i_1, \ldots, i_m)$*-th coordinate* of F. The matrix $\mathfrak{A}$ will be called the *matrix of the coordinates* of the multi-linear form F.

A generalization of Theorem 6.3 to the case of arbitrary matrices is given in exercise 3.

An m-linear form F on V_k^m is said to be *symmetric* if

$$F(\mathfrak{a}_{i_1}, \ldots, \mathfrak{a}_{i_m}) = F(\mathfrak{a}_1, \ldots, \mathfrak{a}_m) \tag{27}$$

for any permutation $i_1, \ldots, i_m$ of the integers $1, \ldots, m$ and for any vectors $\mathfrak{a}_1, \ldots, \mathfrak{a}_m \in V_k$. It is easy to verify that an m-linear form (25′) is symmetric if and only if the matrix (26) is symmetric.

A function $G\colon V_k \to R$ is said to be a *form of degree* m if there exists a symmetric form F on the Cartesian product V_k^m such that

$$G(\mathfrak{a}) = F(\mathfrak{a}, \mathfrak{a}, \ldots, \mathfrak{a}) \quad \text{for} \quad \mathfrak{a} \in V_k . \tag{28}$$

In other words, G is a form of degree m if

$$G(\mathfrak{a}) = \mathfrak{A}\cdot\mathfrak{a}^m \quad \text{for} \quad \mathfrak{a} \in V_k , \tag{29}$$

where $\mathfrak{A}$ is a symmetric m-indexed k-dimensional matrix ($\mathfrak{a}^m$ denotes here the tensor product $\mathfrak{a} \times \mathfrak{a} \times \ldots \times \mathfrak{a}$ m-times; see § 4 (18′), p. 54). It follows from the definition that the form G of degree m is homogeneous of degree m, i.e.,

$$G(a\mathfrak{a}) = a^m G(\mathfrak{a}) . \tag{30}$$

Observe that

$$G(0) = 0 .$$

Forms of degree 1 coincide with linear forms.

If $\mathfrak{A}$ is the zero matrix, then the form G given by (29) is identically equal to zero. The converse theorem is also true:

6.4. *If the form* (29) *of degree* m *is identically equal to zero, then the symmetric matrix* $\mathfrak{A}$ *is the zero matrix*:

$$\mathfrak{A} = 0 .$$

A proof of 6.4 will be given in IV § 5, p. 155-156.

It follows from 6.4 that, for every form G of degree m, there exists exactly one symmetric matrix $\mathfrak{A}$ such that (29) holds. For if $\mathfrak{A}'$ is another symmetric matrix such that $G(\mathfrak{a}) = \mathfrak{A}'\cdot\mathfrak{a}^m$ for all $a \in V_k$, then $(\mathfrak{A}-\mathfrak{A}')\cdot\mathfrak{a}^m = 0$ for all $\mathfrak{a} \in V_k$ and, consequently, $\mathfrak{A}-\mathfrak{A}' = 0$ by 6.4, i.e. $\mathfrak{A} = \mathfrak{A}'$.

Let $G\colon V_k \to R$ be a form of degree m. We say that the form G

is *positive* if $G(\mathfrak{a}) > 0$ for every vector $\mathfrak{a} \neq 0$;

is *negative* if $G(\mathfrak{a}) < 0$ for every vector $\mathfrak{a} \neq 0$;

is *non-negative* if $G(\mathfrak{a}) \geqslant 0$ for every vector $\mathfrak{a}$;

is *non-positive* if $G(\mathfrak{a}) \leqslant 0$ for every vector $\mathfrak{a}$;

has no sign if there exist vectors $\mathfrak{a}_1$ and $\mathfrak{a}_2$ such that $G(\mathfrak{a}_1) > 0$ and $G(\mathfrak{a}_2) < 0$.

It follows from (30) (where $a = -1$) that if m is odd and $G \neq 0$, then G has no sign.

If m is an even positive integer, then an m-indexed symmetric matrix $\mathfrak{A}$ is said to be *positive*, *negative*, *non-negative* or *non-positive* whenever the corresponding form (29) of degree m is positive, negative, non-negative or non-positive respectively; similarly $\mathfrak{A}$ is said *to have no sign* if (29) has no sign.

If $k = 1$, then $\mathfrak{A}$ has only one coordinate: $A = a_{1,\ldots,1}$. For every vector $\mathfrak{a} = [a]$, $G(\mathfrak{a}) = Aa^m$. If m is even, then $\mathfrak{A}$ (i.e. G) is positive, negative, non-negative, or non-positive if and only if the number A is positive, negative, non-negative or non-positive, respectively.

EXERCISES

1. For any m-indexed symmetric matrix $\mathfrak{A}$,

$$\mathfrak{A}\cdot(\mathfrak{a}+\mathfrak{b})^m = \mathfrak{A}\cdot\sum_{n=1}^{m}\binom{m}{n}\mathfrak{a}^{m-n}\times\mathfrak{b}^n = \sum_{n=0}^{m}\binom{m}{n}\mathfrak{A}\cdot(\mathfrak{a}^{m-n}\times\mathfrak{b}^n).$$

2. For any matrix $\mathfrak{A} = [a_{i_1,\ldots,i_m}]_{i_1,\ldots,i_m \leqslant k}$, the matrix

$$\mathfrak{B} = [b_{i_1,\ldots,i_m}]_{i_1,\ldots,i_m \leqslant k},$$

where

$$b_{i_1,\ldots,i_m} = \frac{1}{m!}\sum a_{i_{r_1},\ldots,i_{r_m}}$$

and the summation is extended over all permutations $r_1, \ldots, r_m$ of the integers $1, \ldots, m$, is symmetric and

$$\mathfrak{A}\mathfrak{a}^m = \mathfrak{B}\mathfrak{a}^m$$

for every vector $\mathfrak{a} \in V_k$.

3. Generalizing the definition on p. 68 we say that a function

$$F\colon V_{k_1}\times\ldots\times V_{k_m} \to R \tag{31}$$

is an *m-linear form* (or, less precisely, a *multi-linear form*) if (19) and (20) hold. Verify that conditions (21)-(24) are also satisfied.

Let $\mathfrak{e}_{i,j}$ be the unit vector of the jth coordinate axis in R^{k_i}. If (31) is an m-linear form, then

(32) $$F(\mathfrak{a}_1, \ldots, \mathfrak{a}_m) = \mathfrak{U}\cdot(\mathfrak{a}_1 \times \ldots \times \mathfrak{a}_m) \qquad (\mathfrak{a}_i \in V_{k_i},\ i = 1, \ldots, m)$$

where

(33) $$\mathfrak{U} = [F(\mathfrak{e}_{1,j_1}, \ldots, \mathfrak{e}_{m,j_m})]_{j_1 \leqslant k_1, \ldots, j_m \leqslant k_m} .$$

Conversely, if $\mathfrak{U}$ is matrix of type $k_1, \ldots, k_m$, then formula (32) defines an m-linear form (31), and (33) holds.

Thus, in order that a function (31) be an m-linear form it is necessary and sufficient that F be given by (32). Then $\mathfrak{U}$ satisfies (33).

The matrix $\mathfrak{U}$ is called the *matrix of the coordinates* of F.

If $\mathfrak{U}$ is the matrix of coordinates of F, then $a\mathfrak{U}$ is the matrix of coordinates of the multi-linear form aF ($a \in R$). If $\mathfrak{U}$ and $\mathfrak{B}$ are the matrices of coordinates of multi-linear forms F and G defined on the product $V_{k_1} \times \ldots \times V_{k_m}$, then $\mathfrak{U} + \mathfrak{B}$ (then $\mathfrak{U} - \mathfrak{B}$) is the matrix of coordinates of $F+G$ (of $F-G$).

4. The *tensor product* of multilinear forms

$$F\colon V_{k_1} \times \ldots \times V_{k_m} \to R, \qquad G\colon V_{l_1} \times \ldots \times V_{l_n} \to R$$

is the multilinear form $F \times G$ on the product $V_{k_1} \times \ldots \times V_{k_m} \times V_{l_1} \times \ldots \times V_{l_n}$ given by

$$F \times G(\mathfrak{a}_1, \ldots, \mathfrak{a}_m, \mathfrak{b}_1, \ldots, \mathfrak{b}_n) = F(\mathfrak{a}_1, \ldots, \mathfrak{a}_m)\cdot G(\mathfrak{b}_1, \ldots, \mathfrak{b}_n)$$

($\mathfrak{a}_i \in V_{k_i}$ for $i = 1, \ldots, m$ and $\mathfrak{b}_j \in V_{l_j}$ for $j = 1, \ldots, n$).

Verify that if $\mathfrak{U}$ and $\mathfrak{B}$ are matrices of the coordinates of the forms F and G, then the tensor product $\mathfrak{U} \times \mathfrak{B}$ is the matrix of the coordinates of the tensor product $F \times G$.

5. Let ψ be a real function defined on the set V of all unit vectors in R^k, such that

$$\psi(-\mathfrak{a}) = -\psi(\mathfrak{a}) \quad \text{for} \quad \mathfrak{a} \in V .$$

The formula

$$F(\mathfrak{a}) = \begin{cases} 0 & \text{if} \quad \mathfrak{a} = 0 \in V_k , \\ |\mathfrak{a}|\psi\left(\dfrac{\mathfrak{a}}{|\mathfrak{a}|}\right) & \text{if} \quad 0 \neq \mathfrak{a} \in V_k \end{cases}$$

defines a homogeneous function on V_k. In general, the function F is not additive.

§ 7. The main types of mappings. We can now describe the main types of functions which we shall examine in this book.

First of all we shall examine real functions defined on certain sets. As a rule, the domains of these functions will be subsets of Euclidean spaces.

We shall also examine mappings $\Phi\colon X \to R^l$ of a set X into the l-dimensional Euclidean space. As a rule, X will be a subset of a Euclidean space. This is not essential for the following remark on these mappings.

Let $\Phi\colon X \to R^l$ and let φ_i be the real function which assigns to every $p \in X$ the ith coordinate of the point $\Phi(p)$, $i = 1, \ldots, l$. By definition,

(1) $$\Phi(p) = (\varphi_1(p), \ldots, \varphi_l(p)) \quad \text{for every } p \in X .$$

The function φ_i is called the *i-th coordinate* of the mapping Φ. Instead of (1) we write

$$(1') \qquad \Phi = (\varphi_1, \ldots, \varphi_l)$$

for short. Conversely, if $\varphi_1, \ldots, \varphi_l$ are real functions on X, then the formula (1), or the equivalent formula (1′), defines a mapping Φ: $X \to R^l$. We see that an examination of mappings Φ: $X \to R^l$ reduces to an examination of a finite sequence of real functions, the coordinates of Φ, and conversely.

We shall also examine mappings Φ: $X \to V_l$ of a set X into the set V_l of all l-dimensional vectors. If Φ: $X \to V_l$ is a mapping of this kind, let φ_i be the real function which assigns to every $p \in X$ the ith coordinate of the vector $\Phi(p)$, $i = 1, \ldots, l$.

By definition,

$$(2) \qquad \Phi(p) = [\varphi_1(p), \ldots, \varphi_l(p)] \quad \text{for every } p \in X .$$

The function φ_i is called the *i-th coordinate* of Φ. Instead of (2) we shall write

$$(2') \qquad \Phi = [\varphi_1, \ldots, \varphi_l]$$

for short. Conversely, if $\varphi_1, \ldots, \varphi_l$ are real functions on X, then the formula (2), or the equivalent formula (2′), defines a mapping Φ: $X \to V_l$. We see that an examination of mappings Φ: $X \to V_l$ reduces to an examination of a finite sequence of real functions, and conversely.

Thus, from an analytic point of view, the study of mappings Φ: $X \to R^l$ coincides with the study of mappings Φ: $X \to V_l$ since each case reduces to the study of the sequence of their coordinates $\varphi_1, \ldots, \varphi_l$. However the geometrical sense of the mappings Φ: $X \to R^l$ and Φ: $X \to V_l$ is different because of the diversity of the notions of point and of vector.

Mappings Φ: $X \to V_l$ are often called *vector fields* (more precisely, *l-dimensional vector fields*) on X. This name will be used especially in the case where X is a subset of a Euclidean space.

The second case considered in this book is that where $X = V_k$ or $X = V_k^m$; in this case we shall examine only linear and multilinear mappings.

We say that a mapping Φ: $V_k \to V_l$ is *linear* if it is *homogeneous*

$$(3) \qquad \Phi(a\mathfrak{a}) = a\Phi(\mathfrak{a})$$

and *additive*

$$(4) \qquad \Phi(\mathfrak{a}+\mathfrak{b}) = \Phi(\mathfrak{a})+\Phi(\mathfrak{b}) .$$

In order that a mapping $\Phi = [\varphi_1, \ldots, \varphi_l]$: $V_k \to V_l$ be homogeneous (or additive) it is necessary and sufficient that all its coordinates $\varphi_1, \ldots, \varphi_l$ be homogeneous (or additive) in the sense defined in § 6. Thus, in order that a mapping $\Phi = [\varphi_1, ..., \varphi_l]$: $V_k \to V_l$ be linear it is necessary and sufficient that all its coordinates $\varphi_1, ..., \varphi_l$ be linear forms.

Suppose that $\Phi = [\varphi_1, ..., \varphi_l]$: $V_k \to V_l$ is linear. Let $\mathfrak{e}_1, ..., \mathfrak{e}_k$ be the unit vectors of the coordinate axes in R^k. For any vector

$$\mathfrak{a} = [a_1, ..., a_k] = a_1 \mathfrak{e}_1 + ... + a_k \mathfrak{e}_k$$

we have

$$\varphi_i(\mathfrak{a}) = \sum_{j=1}^{k} \varphi_i(\mathfrak{e}_j) a_j \quad \text{for} \quad i = 1, ..., l,$$

that is,

$$\Phi(\mathfrak{a}) = \Big[\sum_{j=1}^{k} \varphi_1(\mathfrak{e}_j) a_j, ..., \sum_{j=1}^{k} \varphi_l(\mathfrak{e}_j) a_j\Big].$$

Denoting

(5) $$\mathfrak{A} = [\varphi_i(\mathfrak{e}_j)]_{i \leqslant l,\, j \leqslant k},$$

we obtain the formula

(6) $$\Phi(\mathfrak{a}) = \mathfrak{A}\mathfrak{a} \quad \text{for} \quad \mathfrak{a} \in V_k.$$

Conversely, if $\mathfrak{A}$ is any rectangular matrix of type l, k, then formula (6) defines a linear mapping Φ: $V_k \to V_l$ and (5) holds. We see that the following analogue of Theorem 6.1 is true.

7.1. *A mapping Φ: $V_k \to V_l$ is linear if and only if it is of the form* (6), *where $\mathfrak{A}$ is a rectangular matrix of type l, k. Then* (5) *holds.*

Identities (3) and (4) are analogous to § 6 (2) and (3) respectively. It is easy to see that the identities analogous to § 6 (4)-(7) also hold for any linear mapping Φ: $V_k \to V_l$.

A mapping Φ: $V_k^m \to V_l$ is said to be *m-linear* if for any fixed vectors $\mathfrak{a}_1, ..., \mathfrak{a}_{i-1}, \mathfrak{a}_{i+1}, ..., \mathfrak{a}_m \in V_k$, the expression

$$\Phi(\mathfrak{a}_1, ..., \mathfrak{a}_{i-1}, \mathfrak{a}_i, \mathfrak{a}_{i+1}, ..., \mathfrak{a}_m),$$

interpreted as a function of $\mathfrak{a}_i \in V_k$, is linear ($i = 1, ..., m$). In other words, Φ is m-linear if the identities analogous to § 6 (19)-(20) hold. Then identities analogous to § 6 (21)-(24) are also true. In order that a mapping $\Phi = [\varphi_1, ..., \varphi_l]$: $V_k^m \to V_l$ be m-linear it is necessary and sufficient that all its coordinates $\varphi_1, ..., \varphi_l$ be m-linear forms in the sense defined in § 6.

Suppose that $\Phi = [\varphi_1, \dots, \varphi_l]: V_k^m \to V_l$ is m-linear. Let $\mathfrak{e}_1, \dots, \mathfrak{e}_k$ be unit vectors of the coordinate axes in R^k. For any vectors

$$\mathfrak{a}_1 = [a_{1,1}, \dots, a_{1,k}] = \sum_{j_1=1}^{k} a_{1,j_1} \mathfrak{e}_{j_1} \in V_k ,$$

$$\dots\dots\dots\dots\dots\dots\dots\dots$$

$$\mathfrak{a}_m = [a_{m,1}, \dots, a_{m,k}] = \sum_{j_m=1}^{k} a_{m,j_m} \mathfrak{e}_{j_m} \in V_k$$

we have

$$\varphi_i(\mathfrak{a}_1, \dots, \mathfrak{a}_m) = \sum_{j_1=1}^{k} \dots \sum_{j_m=1}^{k} \varphi_i(\mathfrak{e}_{j_1}, \dots, \mathfrak{e}_{j_m}) \cdot a_{1,j_1} \dots a_{m,j_m} .$$

Denoting

(7) $$\mathfrak{A} = [\varphi_i(\mathfrak{e}_{j_1}, \dots, \mathfrak{e}_{j_m})]_{i \leqslant l, j_1, \dots, j_m \leqslant k}$$

we infer that

(8) $$\Phi(\mathfrak{a}_1, \dots, \mathfrak{a}_m) = \mathfrak{A} \cdot (\mathfrak{a}_1 \times \dots \times \mathfrak{a}_m) \quad \text{for} \quad \mathfrak{a}_1, \dots, \mathfrak{a}_m \in V_k .$$

Conversely, if $\mathfrak{A}$ is any $(m+1)$-indexed matrix of type $l, k, \dots, k$, then formula (8) defines an m-linear mapping $\Phi: V_k^m \to V_l$ and (7) holds. Thus the following analogue of Theorem 6.3 is true.

7.2. *A mapping* $\Phi: V_k^m \to V_l$ *is* m-*linear if and only if it is of the form* (8), *where* $\mathfrak{A}$ *is an* $(m+1)$-*indexed matrix of type* $l, k, \dots, k$. *Then* (7) *holds.*

In this book we shall also examine mappings whose values are multi-indexed matrices of a fixed type $l_1, \dots, l_n$, i.e. mappings Φ which assign to every point p from a set X a matrix

(9) $$\Phi(p) = [\varphi_{i_1, \dots, i_n}(p)]_{i_1 \leqslant l_1, \dots, i_n \leqslant l_n} .$$

The real function $\varphi_{i_1, \dots, i_n}$ which assigns to every point $x \in X$ the $(i_1, \dots, i_n)$-th coordinate of the matrix $\Phi(p)$ is said to be the $(i_1, \dots, i_n)$-*th coordinate* of Φ. Instead of (9) we also write

(9′) $$\Phi = [\varphi_{i_1, \dots, i_n}]_{i_1 \leqslant l_1, \dots, i_n \leqslant l_n}$$

for short. Conversely, if $\varphi_{i_1, \dots, i_n}$ is any given system of real functions on X ($i_1 \leqslant l_1, \dots, i_n \leqslant l_n$), then the formula (9), or the equivalent formula (9′), defines a mapping on X, whose values are matrices of type $l_1, \dots, l_n$. We see that the study of mappings Φ of this type reduces to the study of a system $\varphi_{i_1, \dots, i_n}$ of real functions, the coordinates of Φ.

Mappings (9) are often called *matrix fields* (more precisely, *matrix fields of type* $l_1, \dots, l_n$). This name will be used especially in the case where X is a subset of a Euclidean space.

A mapping (9′) defined on V_k is said to be *linear* provided it is *homogeneous* and *additive*, i.e. it satisfies conditions (3) and (4). A necessary and sufficient condition for (9′) to be linear is that each of its coordinates be linear.

Denoting

$$\mathfrak{A} = [\varphi_{i_1,\ldots,i_n}(\mathfrak{e}_j)]_{i_1\leqslant l_1,\ldots,i_n\leqslant l_n,j\leqslant k}\,, \tag{10}$$

where as always $\mathfrak{e}_1, \ldots, \mathfrak{e}_k$ are unit vectors of the coordinate axes in R^k, we infer that

$$\Phi(\mathfrak{a}) = \mathfrak{A}\mathfrak{a} \quad \text{for} \quad \mathfrak{a} \in V_k\,. \tag{11}$$

Conversely, if $\mathfrak{A}$ is a matrix of type $l_1, \ldots, l_n, k$, then (11) defines a linear mapping of the form (9′) defined on V_k and (10) holds. Thus we get the following generalization of Theorem 7.1.

7.3. *A mapping Φ of V_k into the set of matrices of type $l_1, \ldots, l_n$ is linear if and only if it is of the form* (11), *where $\mathfrak{A}$ is a matrix of type $l_1, \ldots, l_n, k$. Then* (10) *holds.*

We say that a mapping (9′) from the Cartesian product V_k^m into the set of all matrices of type $l_1, \ldots, l_n$ is *m-linear* if for any fixed vectors $\mathfrak{a}_1, \ldots, \mathfrak{a}_{i-1}, \mathfrak{a}_{i+1}, \ldots, \mathfrak{a}_m \in V_k$ the expression

$$\Phi(\mathfrak{a}_1, \ldots, \mathfrak{a}_{i-1}, \mathfrak{a}_i, \mathfrak{a}_{i+1}, \ldots, \mathfrak{a}_m) \tag{12}$$

interpreted as a function of a vector $\mathfrak{a}_i \in V_k$ is linear ($i = 1, \ldots, m$). In other words, Φ is m-linear if the identities analogous to § 6 (19)-(20) hold. Observe that identities analogous to § 6 (21)-(24) also hold. It is easy to verify that a mapping (9′) is m-linear if and only if all its coordinates $\varphi_{i_1,\ldots,i_n}$ are m-linear forms in the sense defined in § 6. The proof of the following theorem is analogous to the proofs of 7.1-7.3.

7.4. *A mapping* (9′) *from V_k^m into the set of all matrices of type $l_1, \ldots, l_n$ is m-linear if and only if it is of the form*

$$\Phi(\mathfrak{a}_1, \ldots, \mathfrak{a}_m) = \mathfrak{A}\cdot(\mathfrak{a}_1 \times \ldots \times \mathfrak{a}_m)\,,$$

where $\mathfrak{A}$ is a matrix of type $l_1, \ldots, l_n, \underbrace{k, \ldots, k}_{m\text{-times}}$. *Then*

$$\mathfrak{A} = [\varphi_{i_1,\ldots,i_l}(\mathfrak{e}_{j_1}, \ldots, \mathfrak{e}_{j_m})]_{i_1\leqslant l_1,\ldots,i_n\leqslant l_n,j_1,\ldots,j_m\leqslant k}\,.$$

We recall the convention introduced at the end of I § 5. If

$$\Phi = (\varphi_1, \ldots, \varphi_k)\colon X \to R^k \quad \text{and} \quad \Psi = [\psi_1, \ldots, \psi_k]\colon X \to V_k\,,$$

then $\Phi + \Psi\colon X \to R^k$ is defined by

$$(\Phi+\Psi)(p) = \Phi(p) + \Psi(p) \quad \text{for} \quad p \in X\,,$$

i.e.

$$\Phi+\Psi = (\varphi_1+\psi_1, \ldots, \varphi_k+\psi_k).$$

If

$$\Phi = [\varphi_1, \ldots, \varphi_k]: X \to V_k \quad \text{and} \quad \Psi = [\psi_1, \ldots, \psi_k]: X \to V_k,$$

then $\Phi+\Psi: X \to V_k$ is defined by

$$(\Phi+\Psi)(p) = \Phi(p)+\Psi(p) \quad \text{for} \quad p \in X,$$

i.e.

$$\Phi+\Psi = [\varphi_1+\psi_1, \ldots, \varphi_k+\psi_k].$$

With the same notation, $\Phi\Psi: X \to R$ is defined by

$$(\Phi\Psi)(p) = \Phi(p)\Psi(p) \quad \text{for} \quad p \in X,$$

i.e.

$$\Phi\Psi = \varphi_1\psi_1+\ldots+\varphi_k\psi_k.$$

A similar notation is applied for other mappings into a Euclidean space or for vector fields and matrix fields. For instance, if

$$\Phi = [\varphi_{i,j}]_{i\leqslant k, j\leqslant l} \quad \text{and} \quad \Psi = [\psi_{j,r}]_{j\leqslant l, r\leqslant s}$$

are matrix fields on X, then $\Phi\cdot\Psi$ is the matrix field on X, defined by

$$(\Phi\cdot\Psi)(p) = \Phi(p)\cdot\Psi(p) \quad \text{for} \quad p \in X,$$

i.e.

$$\Phi\cdot\Psi = \Big[\sum_{j=1}^{l} \varphi_{i,j}\psi_{j,r}\Big]_{i\leqslant k,\, r\leqslant s}.$$

If

$$\Phi = [\varphi_{i,j}]_{i\leqslant l, j\leqslant k} \quad \text{and} \quad \Psi = [\psi_1, \ldots, \psi_k]$$

are a matrix field and a vector field on X respectively, then the vector field $\Phi\Psi: X \to V_l$ is defined by

$$(\Phi\Psi)(p) = \Phi(p)\Psi(p) \quad \text{for} \quad p \in X,$$

i.e.

$$\Phi\Psi = \Big[\sum_{j=1}^{k} \varphi_{1,j}\psi_j, \ldots, \sum_{j=1}^{k} \varphi_{l,j}\psi_j\Big].$$

If $\Phi = [\varphi_{i_1 \ldots, i_m}]$ is a matrix field on X and φ is a real function on X, then $\varphi\Phi$ is the matrix field on X defined by

$$(\varphi\Phi)(p) = \varphi(p)\Phi(p) \quad \text{for} \quad p \in X,$$

i.e.

$$\varphi\Phi = [\varphi\varphi_{i_1, \dots, i_m}] .$$

The reader will guess, without any difficulty, the meaning of other operations on mappings and vector or matrix fields which will appear in this book.

EXERCISES

1. Generalizing the definitions from this section, we say that a mapping Φ of the product $V_{k_1} \times \dots \times V_{k_m}$ into the set of matrices of the type $l_1, \dots, l_n$ is *m-linear* if the expression (12) interpreted as the function of $\mathfrak{a}_i \in V_{k_i}$ is a linear mapping, $i = 1, \dots, m$. Formulate and prove the analogue of Theorem 7.4 for m-linear mappings just defined.

2. The composition $\Psi \circ \Phi$: $V_k \to V_r$ of linear mappings Φ: $V_k \to V_l$ and Ψ: $V_l \to V_r$ is linear. If

$$\Phi(\mathfrak{a}) = \mathfrak{A}\mathfrak{a} \quad \text{for} \quad \mathfrak{a} \in V_k \quad \text{and} \quad \Psi(\mathfrak{b}) = \mathfrak{B}\mathfrak{b} \quad \text{for} \quad \mathfrak{b} \in V_l ,$$

then

$$(\Psi \circ \Phi)(\mathfrak{a}) = (\mathfrak{B} \bullet \mathfrak{A})\mathfrak{a} \quad \text{for} \quad \mathfrak{a} \in V_k .$$

§ 8. Skew-symmetric forms. We recall that

$$\operatorname{sgn}(r_1, \dots, r_k)$$

is equal to 1 if $r_1, \dots, r_k$ is an even permutation of $1, \dots, k$ and is equal to -1 if it is an odd permutation of those integers. Moreover, this expression is equal to 0 if the sequence $r_1, \dots, r_k$ of positive integers $\leqslant k$ is not any permutation, i.e. if $r_i = r_j$ for some $i \neq j$.

A k-linear form

$$F\colon V_l^k \to R \tag{1}$$

is said to be *skew-symmetric* provided that

$$F(\mathfrak{a}_{r_1}, \dots, \mathfrak{a}_{r_k}) = \operatorname{sgn}(r_1, \dots, r_k) F(\mathfrak{a}_1, \dots, \mathfrak{a}_k) \tag{2}$$

for every permutation $r_1, \dots, r_k$ of the integers $1, \dots, k$ and for arbitrary vectors $\mathfrak{a}_1, \dots, \mathfrak{a}_k \in V_l$. Since every permutation is the result of a finite number of transpositions of consecutive terms, a k-linear form (1) is skew-symmetric if and only if

$$\begin{aligned} F(\mathfrak{a}_1, \dots, \mathfrak{a}_{r-1}, \mathfrak{a}_{r+1}, \mathfrak{a}_r, \mathfrak{a}_{r+2}, \dots, \mathfrak{a}_k) \\ = -F(\mathfrak{a}_1, \dots, \mathfrak{a}_{r-1}, \mathfrak{a}_r, \mathfrak{a}_{r+1}, \mathfrak{a}_{r+2}, \dots, \mathfrak{a}_k) \end{aligned} \tag{3}$$

for any vectors $\mathfrak{a}_1, \dots, \mathfrak{a}_k \in V_l$ and $r = 1, \dots, k-1$. Thus a k-linear form (1) is skew-symmetric if and only if

$$\begin{aligned} F(\mathfrak{a}_1, \dots, \mathfrak{a}_{r-1}, \mathfrak{a}_s, \mathfrak{a}_{r+1}, \dots, \mathfrak{a}_{s-1}, \mathfrak{a}_r, \mathfrak{a}_{s+1}, \dots, \mathfrak{a}_k) \\ = -F(\mathfrak{a}_1, \dots, \mathfrak{a}_{r-1}, \mathfrak{a}_r, \mathfrak{a}_{r+1}, \dots, \mathfrak{a}_{s-1}, \mathfrak{a}_s, \mathfrak{a}_{s+1}, \dots, \mathfrak{a}_k) \end{aligned} \tag{3'}$$

for any vectors $\mathfrak{a}_1, \dots, \mathfrak{a}_k \in V_l$ and for any integers r, s $(1 \leqslant r < s \leqslant k)$.

It follows from (3′) that for every skew-symmetric form (1)

(4) if $\mathfrak{a}_r = \mathfrak{a}_s$ for $r \neq s$, then $F(\mathfrak{a}_1, \ldots, \mathfrak{a}_k) = 0$.

This proves that (2) holds also in the case where the sequence $r_1, \ldots, r_k$ of positive integers $\leqslant k$ is not any permutation, i.e. if $r_i = r_j$ for some $i \neq j$.

It follows easily from (4) and § 6 (19)-(20) that for any skew-symmetric form F,

(4′) if a vector $\mathfrak{a}_i$ is a linear combination of some of the vectors $\mathfrak{a}_1, \ldots, \mathfrak{a}_{i-1}, \mathfrak{a}_{i+1}, \ldots, \mathfrak{a}_k$, then $F(\mathfrak{a}_1, \ldots, \mathfrak{a}_k) = 0$.

If $\mathfrak{a}_1, \ldots, \mathfrak{a}_k$ are vectors in R^l and $k > l$, then at least one of these vectors is a linear combination of the remaining ones. Thus if $k > l$ and (1) is skew-symmetric, then (1) is identically equal to 0.

It follows directly from the definition that if F is skew-symmetric and a is a number, then aF is skew-symmetric. If F and F_1 are skew-symmetric forms on V_l^k, then $F+F_1$ and $F-F_1$ are skew-symmetric.

In the case of $k = 1$, every linear (i.e. 1-linear) form $F\colon V_l \to R$ is skew-symmetric. Sometimes it is convenient to consider every fixed real number as a skew-symmetric form (the case of $k = 0$).

The following lemma will be useful in IX § 11.

8.1. *If F is a $(k-1)$-linear skew-symmetric form on V_l^{k-1}, and $\mathfrak{a}, \mathfrak{a}_1, \ldots, \mathfrak{a}_k \in V_l$, then*

$$(5)\quad \sum_{j=1}^{k}(-1)^{j-1}\Big(\sum_{r=1}^{j-1} F(\mathfrak{a}_1, \ldots, \mathfrak{a}_{r-1}, \mathfrak{a}, \mathfrak{a}_{r+1}, \ldots, \mathfrak{a}_{j-1}, \mathfrak{a}_{j+1}, \ldots, \mathfrak{a}_k) + \sum_{r=j+1}^{k} F(\mathfrak{a}_1, \ldots, \mathfrak{a}_{j-1}, \mathfrak{a}_{j+1}, \ldots, \mathfrak{a}_{r-1}, \mathfrak{a}, \mathfrak{a}_{r+1}, \ldots, \mathfrak{a}_k)\Big) = 0 .$$

Denote by L the left-hand side of (5). By definition,

$$L = \sum_{r<j} (-1)^{j-1} F(\mathfrak{a}_1, \ldots, \mathfrak{a}_{r-1}, \mathfrak{a}, \mathfrak{a}_{r+1}, \ldots, \mathfrak{a}_{j-1}, \mathfrak{a}_{j+1}, \ldots, \mathfrak{a}_k) + \sum_{j<r} (-1)^{j-1} F(\mathfrak{a}_1, \ldots, \mathfrak{a}_{j-1}, \mathfrak{a}_{j+1}, \ldots, \mathfrak{a}_{r-1}, \mathfrak{a}, \mathfrak{a}_{r+1}, \ldots, \mathfrak{a}_k) .$$

Replacing in the second sum the letters j and r by r and j respectively, we get

$$L = \sum_{r<j} \Big((-1)^{j-1} F(\mathfrak{a}_1, \ldots, \mathfrak{a}_{r-1}, \mathfrak{a}, \mathfrak{a}_{r+1}, \ldots, \mathfrak{a}_{j-1}, \mathfrak{a}_{j+1}, \ldots, \mathfrak{a}_k) + (-1)^{r-1} F(\mathfrak{a}_1, \ldots, \mathfrak{a}_{r-1}, \mathfrak{a}_{r+1}, \ldots, \mathfrak{a}_{j-1}, \mathfrak{a}, \mathfrak{a}_{j+1}, \ldots, \mathfrak{a}_k)\Big) .$$

It follows from the skew symmetry of F that the two terms under the sign $\sum_{r<j}$ differ in sign only, i.e. their sum is equal to 0. This proves that $L = 0$.

For every k-linear form (1) the symbol $\mathrm{sk}F$ will stand for the k-linear form

(6) $$\mathrm{sk}F\colon V_l^k \to R$$

defined by

(6′) $$\mathrm{sk}F(\mathfrak{a}_1, \ldots, \mathfrak{a}_k) = \sum \mathrm{sgn}(r_1, \ldots, r_k) F(\mathfrak{a}_{r_1}, \ldots, \mathfrak{a}_{r_k}) \quad \text{for any } \mathfrak{a}_1, \ldots, \mathfrak{a}_k \in V_l,$$

the summation being extended over all permutations $r_1, \ldots, r_k$ of the positive integers $\leqslant k$. It follows directly from the definition that the k-linear form $\mathrm{sk}F$ is skew-symmetric. Indeed, if we perform any permutation of vectors $\mathfrak{a}_1, \ldots, \mathfrak{a}_k$ in (6′), the right-hand side of (6′) will be multiplied by the sign of the permutation.

Suppose that $k = s_1 + s_2$, s_1 and s_2 being positive integers. A k-linear form (1) is said to be *skew-symmetric in the first s_1 variables* if for any fixed vectors $\mathfrak{a}_{s_1+1}, \ldots, \mathfrak{a}_k \in V_l$ the expression

(7) $$F(\mathfrak{a}_1, \ldots, \mathfrak{a}_{s_1}, \mathfrak{a}_{s_1+1}, \ldots, \mathfrak{a}_k)$$

considered as a function of the variables $\mathfrak{a}_1, \ldots, \mathfrak{a}_{s_1}$ only is skew-symmetric (clearly it is s_1-linear). Similarly we say that a k-linear form (1) is *skew-symmetric in the last s_2 variables* if for any fixed vectors $\mathfrak{a}_1, \ldots, \mathfrak{a}_{s_1}$ the expression (7) considered as a function of variables $\mathfrak{a}_{s_1+1}, \ldots, \mathfrak{a}_k$ only is skew-symmetric (clearly it is s_2-linear). If $s_1 = 1$ (if $s_2 = 1$) every form (1) is skew-symmetric in the first s_1 variables (in the last s_2 variables).

If a k-linear form (1) is skew-symmetric in the first s_1 variables and in the last s_2 variables, $s_1 + s_2 = k$, then the symbol $\mathrm{sk}_{s_1,s_2}F$ will stand for the k-linear form

(8) $$\mathrm{sk}_{s_1,s_2}F\colon V_l^k \to R$$

defined by

(8′) $$\mathrm{sk}_{s_1,s_2}F(\mathfrak{a}_1, \ldots, \mathfrak{a}_k) = \sum \mathrm{sgn}(r_1, \ldots, r_k) F(\mathfrak{a}_{r_1}, \ldots, \mathfrak{a}_{r_k}) \quad \text{for } \mathfrak{a}_1, \ldots, \mathfrak{a}_k \in V_l,$$

the summation being extended over all permutations $r_1, \ldots, r_k$ of the positive integers $1, \ldots, k$, such that

(9) $$r_1 < r_2 < \ldots < r_{s_1} \quad \text{and} \quad r_{s_1+1} < r_{s_1+2} < \ldots < r_k .$$

It follows from the skew symmetry in the first s_1 variables and in the last s_2 variables that

(10) $$\mathrm{sk}F = s_1!\, s_2!\, \mathrm{sk}_{s_1,s_2}F ,$$

that is,

$$(10') \qquad \mathrm{sk}_{s_1,s_2}F = \frac{1}{s_1!\,s_2!}\,\mathrm{sk}F\,.$$

It follows from (10′) that $\mathrm{sk}_{s_1,s_2}F$ is a skew-symmetric k-linear form on V_l^k.

To prove (10), divide all permutations of the integers $1, \ldots, k$ into disjoint classes such that two permutations

$$r_1, \ldots, r_k \quad \text{and} \quad \bar{r}_1, \ldots, \bar{r}_k$$

are in the same class if and only if

$$\bar{r}_1, \ldots, \bar{r}_{s_1} \text{ is a permutation of } r_1, \ldots, r_{s_1}$$

and

$$\bar{r}_{s_1+1}, \ldots, \bar{r}_k \text{ is a permutation of } r_{s_1+1}, \ldots, r_k\,.$$

Every class contains $s_1!\,s_2!$ permutations. It follows from the skew-symmetry in the first s_1 variables and in the last s_2 variables that the number

$$\mathrm{sgn}(r_1, \ldots, r_k)F(\mathfrak{a}_{r_1}, \ldots, \mathfrak{a}_{r_k})$$

is the same for all permutations in the same class. Every class contains exactly one permutation satisfying condition (9). Grouping all the summands in (6′) which correspond to permutations in the same class, we see that the right-hand side of (6′) is equal to the right-hand side of (8′) multiplied by $s_1!\,s_2!$.

The following particular case of $\mathrm{sk}_{s_1,s_2}F$ will be useful in § 9 and in IX § 10.

If a $(k+1)$-linear form

$$(11) \qquad F\colon V_l^{k+1} \to R$$

is skew-symmetric in the last k variables, then the $(k+1)$-linear form

$$(12) \qquad \begin{aligned} &\mathrm{sk}_{1,k}F(\mathfrak{a}_1, \ldots, \mathfrak{a}_{k+1}) \\ &= F(\mathfrak{a}_1, \ldots, \mathfrak{a}_{k+1}) - F(\mathfrak{a}_2, \mathfrak{a}_1, \mathfrak{a}_3, \ldots, \mathfrak{a}_{k+1}) + \\ &\quad + F(\mathfrak{a}_3, \mathfrak{a}_1, \mathfrak{a}_2, \mathfrak{a}_4, \ldots, \mathfrak{a}_{k+1}) - \ldots + (-1)^k F(\mathfrak{a}_{k+1}, \mathfrak{a}_1, \ldots, \mathfrak{a}_k) \\ &= \sum_{j=1}^{k+1} (-1)^{j-1} F(\mathfrak{a}_j, \mathfrak{a}_1, \ldots, \mathfrak{a}_{j-1}, \mathfrak{a}_{j+1}, \ldots, \mathfrak{a}_{k+1}) \end{aligned}$$

is skew-symmetric.

EXERCISES

1. The tensor product $F_1 \times F_2\colon V_l^k \to R$ of a skew-symmetric s_1-linear form $F_1\colon V_l^{s_1} \to R$ and a skew-symmetric s_2-linear form $F_2\colon V_l^{s_2} \to R$, $k = s_1 + s_2$, is skew-symmetric in the first s_1 variables and is skew-symmetric in the last s_2 variables. The skew-symmetric k-linear form

$$F_1 \wedge F_2 = \mathrm{sk}_{s_1,s_2} F_1 \times F_2$$

is called the *skew-product* of F_1 and F_2. Verify that

$$F_1 \wedge F_2(\mathfrak{a}_1, \ldots, \mathfrak{a}_k) = \sum \operatorname{sgn}(r_1, \ldots, r_k) F_1(\mathfrak{a}_{r_1}, \ldots, \mathfrak{a}_{r_{s_1}}) F_2(\mathfrak{a}_{r_{s_1+1}}, \ldots, \mathfrak{a}_{r_k}) ,$$

the summation being extended over all permutations $r_1, \ldots, r_k$ of the integers $1, \ldots, k$ such that inequalities (9) holds. Prove that

$$F_2 \wedge F_1 = (-1)^{s_1 s_2} F_1 \wedge F_2 , \quad (F_1 \wedge F_2) \wedge F_3 = F_1 \wedge (F_2 \wedge F_3) ,$$
$$a(F_1 \wedge F_2) = (aF_1) \wedge F_2 = F_1 \wedge (aF_2) , \quad (F_1 + \overline{F}_1) \wedge F_2 = F_1 \wedge F_2 + \overline{F}_1 \wedge F_2 .$$

2. If $F: V_l^{k+1} \to R$ is $(k+1)$-linear and skew-symmetric in the first k variables, then

$$\mathrm{sk}_{k,1} F(\mathfrak{a}_1, \ldots, \mathfrak{a}_{k+1}) = \sum_{j=1}^{k+1} (-1)^{k+j-1} F(\mathfrak{a}_1, \ldots, \mathfrak{a}_{j-1}, \mathfrak{a}_{j+1}, \ldots, \mathfrak{a}_{k+1}, \mathfrak{a}_j) ,$$

and consequently,

$$(-1)^k \mathrm{sk}_{k,1} \mathrm{F}(\mathfrak{a}_1, \ldots, \mathfrak{a}_{k+1}) = \sum_{j=1}^{k+1} (-1)^{j-1} F(\mathfrak{a}_1, \ldots, \mathfrak{a}_{j-1}, \mathfrak{a}_{j+1}, \ldots, \mathfrak{a}_{k+1}, \mathfrak{a}_j)$$
$$= F(\mathfrak{a}_2, \ldots, \mathfrak{a}_{k+1}, \mathfrak{a}_1) - F(\mathfrak{a}_1, \mathfrak{a}_3, \ldots, \mathfrak{a}_{k+1}, \mathfrak{a}_2) + F(\mathfrak{a}_1, \mathfrak{a}_2, \mathfrak{a}_4, \ldots, \mathfrak{a}_{k+1}, \mathfrak{a}_3) -$$
$$- \ldots + (-1)^k F(\mathfrak{a}_1, \ldots, \mathfrak{a}_k, \mathfrak{a}_{k+1}) .$$

§ 9. Skew-symmetric matrices. A k-indexed l-dimensional matrix

(1) $$\mathfrak{A} = [a_{i_1, \ldots, i_k}]_{i_1, \ldots, i_k \leqslant l}$$

is said to be *skew-symmetric* provided that

(2) $$a_{i_{r_1}, \ldots, i_{r_k}} = \operatorname{sgn}(r_1, \ldots, r_k) a_{i_1, \ldots, i_k}$$

for all positive integers $i_1, \ldots, i_k \leqslant l$ and for every permutation $r_1, \ldots, r_k$ of the integers $1, \ldots, k$. Since every permutation is the result of a finite number of transpositions, the matrix (1) is skew-symmetric if and only if

(3) $$a_{i_1, \ldots, i_{r-1}, i_{r+1}, i_r, i_{r+2}, \ldots, i_k} = -a_{i_1, \ldots, i_{r-1}, i_r, i_{r+1}, i_{r+2}, \ldots, i_k} .$$

Thus (1) is skew-symmetric if and only if for $r < s$

(3′) $$a_{i_1, \ldots, i_{r-1}, i_s, i_{r+1}, \ldots, i_{s-1}, i_r, i_{s+1}, \ldots, i_k} = -a_{i_1, \ldots, i_k} .$$

Hence it follows that if (1) is skew-symmetric and

(4) if $i_r = i_s$ for $r \neq s$, then $a_{i_1, \ldots, i_k} = 0$.

This proves that (2) holds also in the case where the sequence $r_1, \ldots, r_k$ of positive integers $\leqslant k$ is not a permutation, i.e. if $r_i = r_j$ for some $i \neq j$.

If follows from (4) that if the matrix (1) is skew-symmetric and $k > l$, then all terms of $\mathfrak{A}$ are equal to 0, i.e. $\mathfrak{A} = 0$. Therefore it suffices to examine only the case where $k \leqslant l$.

Every one-dimensional matrix (i.e. every vector) is skew-symmetric.

The following theorem explains the connexion between skew-symmetric matrices and skew-symmetric forms.

9.1. *A matrix* (1) *is skew-symmetric if and only if the k-linear form*

(5) $$F(\mathfrak{a}_1, \ldots, \mathfrak{a}_k) = \mathfrak{A}\cdot(\mathfrak{a}_1 \times \ldots \times \mathfrak{a}_k) \qquad (\mathfrak{a}_1, \ldots, \mathfrak{a}_k \in V_l)$$

isskew-symmetric.

In other words,

A k-linear form F is skew-symmetric if and only if the matrix $\mathfrak{A}$ *of the coordinates of F is skew-symmetric.*

Thus Theorem 9.1 states that in the natural one-to-one correspondence between multi-linear forms and matrices, the skew-symmetric forms correspond to the skew-symmetric matrices, i.e. both notion express the same mathematical content.

To prove 9.1 let us recall that (5) means that

$$a_{i_1, \ldots, i_k} = F(\mathfrak{e}_{i_1}, \ldots, \mathfrak{e}_{i_k})$$

(see 6.3) where $\mathfrak{e}_1, \ldots, \mathfrak{e}_l$ are the unit vectors of the coordinate axes in R^l.

If F is skew-symmetric, then by § 8 (2)

$$F(\mathfrak{e}_{i_{r_1}}, \ldots, \mathfrak{e}_{i_{r_k}}) = \operatorname{sgn}(r_1, \ldots, r_k) F(\mathfrak{e}_{i_1}, \ldots, \mathfrak{e}_{i_k})$$

ior every permutation $r_1, \ldots, r_k$ of the integers $1, \ldots, k$. Thus (2) holds, f.e. $\mathfrak{A}$ is skew-symmetric.

Suppose now that $\mathfrak{A}$ is skew-symmetric. If

$$\mathfrak{a}_1 = [a_{1,1}, \ldots, a_{1,l}],$$
$$\ldots\ldots\ldots\ldots$$
$$\mathfrak{a}_k = [a_{k,1}, \ldots, a_{k,l}]$$

are any vectors in R^l and $r_1, \ldots, r_k$ is a permutation of the integers $1, \ldots, k$, then by (5)

$$F(\mathfrak{a}_{r_1}, \ldots, \mathfrak{a}_{r_k}) = \sum_{i_1, \ldots, i_k = 1}^{l} a_{i_1, \ldots, i_k} a_{r_1, i_1} \ldots a_{r_k, i_k}.$$

Replacing indices $i_1, \ldots, i_k$ by $j_{r_1}, \ldots, j_{r_k}$ respectively, we obtain

$$\begin{aligned} F(\mathfrak{a}_{r_1}, \ldots, \mathfrak{a}_{r_k}) &= \sum_{j_1, \ldots, j_k = 1}^{l} a_{j_{r_1}, \ldots, j_{r_k}} a_{1, j_1} \ldots a_{k, j_k} \\ &= \operatorname{sgn}(r_1, \ldots, r_k) \sum_{j_1, \ldots, j_k = 1}^{l} a_{j_1, \ldots, j_k} a_{1, j_1} \ldots a_{k, j_k} \\ &= \operatorname{sgn}(r_1, \ldots, r_k) F(\mathfrak{a}_1, \ldots, \mathfrak{a}_k) \end{aligned}$$

by (2) and (5). This proves that F is skew-symmetric.

For any k-indexed l-dimensional matrix $\mathfrak{B} = [b_{i_1, \ldots, i_k}]_{i_1, \ldots, i_k \leqslant l}$ let

(6) $$\operatorname{sk} \mathfrak{B} = [c_{i_1, \ldots, i_k}]_{i_1, \ldots, i_k \leqslant l}$$

where

(6′) $$c_{i_1,\dots,i_k} = \sum \operatorname{sgn}(r_1,\dots,r_k)\, b_{i_{r_1},\dots,i_{r_k}},$$

the summation being extended over all permutations $r_1, \dots, r_k$ of the integers $1, \dots, k$.

9.2. *If* $\mathfrak{B}$ *is the matrix of coordinates of a* k*-linear form* $F\colon V_l^k \to R$, *then* $\operatorname{sk}\mathfrak{B}$ *is the matrix of coordinates of the skew-symmetric* k*-linear form* $\operatorname{sk}F\colon V_l^k \to R$. *Thus* $\operatorname{sk}\mathfrak{B}$ *is skew-symmetric.*

By hypothesis, $b_{i_1,\dots,i_k} = F(\mathfrak{e}_{i_1},\dots,\mathfrak{e}_{i_k})$. The $(i_1,\dots,i_k)$-coordinate of $\operatorname{sk}F$ is equal to

$$\begin{aligned}\operatorname{sk}F(\mathfrak{e}_{i_1},\dots,\mathfrak{e}_{i_k}) &= \sum \operatorname{sgn}(r_1,\dots,r_k) F(\mathfrak{e}_{i_{r_1}},\dots,\mathfrak{e}_{i_{r_k}})\\ &= \sum \operatorname{sgn}(r_1,\dots,r_k)\, b_{i_{r_1},\dots,i_{r_k}} = c_{i_1,\dots,i_k}.\end{aligned}$$

This proves the first part of 9.2. The second part follows from the first and Theorem 9.1 (where $\mathfrak{A}$ should be replaced by $\operatorname{sk}\mathfrak{B}$).

By the *skew product* of an l-dimensional vector

$$\mathfrak{a} = [a_1, \dots, a_l]$$

and an l-dimensional k-indexed skew-symmetric matrix (1) we mean the following l-dimensional $(k+1)$-indexed matrix

(7) $$\mathfrak{a} \wedge \mathfrak{A} = [c_{i_1,\dots,i_{k+1}}]_{i_1,\dots,i_{k+1} \leq l},$$

where

(7′) $$\begin{aligned}c_{i_1,\dots,i_{k+1}} &= \sum_{j=1}^{k+1} (-1)^{j-1} a_{i_j} a_{i_1,\dots,i_{j-1},i_{j+1},\dots,i_{k+1}}\\ &= a_{i_1} a_{i_2,\dots,i_{k+1}} - a_{i_2} a_{i_1,i_3,\dots,i_{k+1}} + a_{i_3} a_{i_1,i_2,i_4,\dots,i_{k+1}} - \dots +\\ &\quad + (-1)^k a_{i_{k+1}} a_{i_1,\dots,i_k}.\end{aligned}$$

9.3. *The skew product* $\mathfrak{a} \wedge \mathfrak{A}$ *is the matrix of the coordinates of the skew-symmetric* $(k+1)$*-linear form* $\operatorname{sk}_{1,k}F$ *where*

(8) $$F(\mathfrak{a}_1,\dots,\mathfrak{a}_{k+1}) = (\mathfrak{a}\mathfrak{a}_1)\cdot\big(\mathfrak{A}\cdot(\mathfrak{a}_2 \times \dots \times \mathfrak{a}_{k+1})\big) \quad \textit{for} \quad \mathfrak{a}_1,\dots,\mathfrak{a}_{k+1} \in V_l.$$

Thus the matrix $\mathfrak{a} \wedge \mathfrak{A}$ *is skew-symmetric.*

It follows directly from the definition of F and the skew symmetry of $\mathfrak{A}$ that the $(k+1)$-linear form $F\colon V_l^{k+1} \to R$ is skew-symmetric in the last k variables. The $(i_1,\dots,i_{k+1})$-coordinate of $\operatorname{sk}_{1,k}F$ is equal to

$$\begin{aligned}\operatorname{sk}_{1,k}F(\mathfrak{e}_{i_1},\dots,\mathfrak{e}_{i_{k+1}}) &= \sum_{j=1}^{k+1} (-1)^{j-1} F(\mathfrak{e}_{i_j},\mathfrak{e}_{i_1},\dots,\mathfrak{e}_{i_{j-1}},\mathfrak{e}_{i_{j+1}},\dots,\mathfrak{e}_{i_{k+1}})\\ &= \sum_{j=1}^{k+1} (-1)^{j-1} a_{i_j} a_{i_1,\dots,i_{j-1},i_{j+1},\dots,i_{k+1}} = c_{i_1,\dots,i_{k+1}}\end{aligned}$$

by § 8 (12). This proves the first part of the theorem. The second part follows from the first and 9.1.

Suppose (1) is skew-symmetric. It follows from (2) and (4) that, practically, it suffices to consider only those terms $a_{i_1, \dots, i_k}$ of $\mathfrak{A}$ for which

(9) $$i_1 < i_2 < \dots < i_k .$$

In fact, for any sequence $j_1, \dots, j_k$ of integers $\leqslant l$, either $j_r = j_s$ for some $r \neq s$ and then $a_{j_1, \dots, j_k} = 0$, or $j_1, \dots, j_k$ is a permutation of a sequence (9) and then

(10) $$a_{j_1, \dots, j_n} = \pm a_{i_1, \dots, i_k} ,$$

according to whether the permutation is even or odd. Since any sequence (9) has $k!$ permutations, every term of the matrix $\mathfrak{A}$ is repeated $k!$ times, roughly speaking. This implies the following consequences.

If

(11) $$\mathfrak{B} = [b_{i_1, \dots, i_k}]_{i_1, \dots, i_k \leqslant l}$$

is another skew-symmetric matrix of the same type, then

(12) $$\mathfrak{A}\mathfrak{B} = \sum_{j_1, \dots, j_k = 1}^{l} a_{j_1, \dots, j_k} b_{j_1, \dots, j_k} = k! \sum_{i_1 < \dots < i_k} a_{i_1, \dots, i_k} b_{i_1, \dots, i_k} .$$

In particular,

$$\|\mathfrak{A}\| = \sqrt{\sum_{j_1, \dots, j_k = 1}^{l} a^2_{j_1, \dots, j_k}} = \sqrt{k!} \sqrt{\sum_{i_1 < \dots < i_k} a^2_{i_1, \dots, i_k}} .$$

The $k!$-fold repetition of every term in the formulas for $\mathfrak{A}\mathfrak{B}$ and $\|\mathfrak{A}\|$ seems to be superfluous. Therefore, for skew-symmetric matrices (1), we shall use the *reduced norm*

(13) $$\|\mathfrak{A}\|^* = \sqrt{\sum_{i_1 < \dots < i_k} a^2_{i_1, \dots, i_k}}$$

instead of the norm $\|\mathfrak{A}\|$. The reduced norm $\|\mathfrak{A}\|^*$ also has all the properties of the norm $\|\mathfrak{A}\|$ mentioned on p. 53, (12)-(13). By definition

(14) $$\|\mathfrak{A}\|^* = \frac{\|\mathfrak{A}\|}{\sqrt{k!}} .$$

Instead of the scalar product $\mathfrak{A}\mathfrak{B}$ of the skew-symmetric matrices (1), (11) we shall use the *reduced scalar product*

(15) $$\frac{1}{k!}\mathfrak{A}\mathfrak{B} = \sum_{i_1 < \dots < i_k} a_{i_1, \dots, i_k} b_{i_1, \dots, i_k} ,$$

although we do not introduce any new symbol for the reduced scalar product.

Observe that in the case where $k=1$, i.e. when the skew-symmetric matrix under consideration is a vector $\mathfrak{a}$, then

(16) $$\|\mathfrak{a}\|^* = \|\mathfrak{a}\| = |\mathfrak{a}| .$$

Note that if $\mathfrak{A}$ is a k-indexed l-dimensional skew-symmetric matrix and $\mathfrak{B}$ is any k-indexed l-dimensional matrix, then

(17) $$\frac{1}{k!}\mathfrak{A}\cdot \operatorname{sk}\mathfrak{B} = \mathfrak{A}\mathfrak{B} .$$

Let $\mathfrak{A}$ be the skew-symmetric matrix (1) and let $k<l$. We say that a skew-symmetric matrix

(18) $$\mathfrak{B} = [b_{j_1,\ldots,j_{l-k}}]_{j_1,\ldots,j_{l-k}\leqslant l}$$

is *dual to* $\mathfrak{A}$ provided that, for any permutation $j_1,\ldots,j_{l-k},i_1,\ldots,i_k$ of the integers $1,\ldots,l$,

(19) $$a_{i_1,\ldots,i_k} = \operatorname{sgn}(j_1,\ldots,j_{l-k},i_1,\ldots,i_k)\cdot b_{j_1,\ldots,j_{l-k}} ,$$

or equivalently,

(20) $$b_{j_1,\ldots,j_{l-k}} = \operatorname{sgn}(j_1,\ldots,j_{l-k},i_1,\ldots,i_k)\cdot a_{i_1,\ldots,i_k} .$$

Formula (20) uniquely defines the terms of the dual matrix $\mathfrak{B}$. Indeed, if $j_1,\ldots,j_{l-k}$ is a sequence of different positive integers $\leqslant l$, then taking as $i_1,\ldots,i_k$ a sequence formed of all the remaining positive integers $\leqslant l$ we calculate $b_{j_1,\ldots,j_{l-k}}$ from (20) (the sequence $i_1,\ldots,i_k$ is determined uniquely up to a permutation but the right-hand side of (20) is independent of the permutation because of the skew-symmetric properties of $a_{i_1,\ldots,i_k}$ and $\operatorname{sgn}(j_1,\ldots,i_k)$). If some of the integers $j_1,\ldots,j_{l-k}$ are equal, then $b_{j_1,\ldots,j_{l-k}}=0$ by the hypothesis of skew symmetry. It follows from (20) that the matrix $\mathfrak{B}$ just defined is skew-symmetric.

We see that for a given skew-symmetric matrix $\mathfrak{A}$ there exists exactly one dual skew-symmetric matrix $\mathfrak{B}$. We shall denote by $\mathfrak{A}^\vee$ the matrix dual to $\mathfrak{A}$. Roughly speaking, the dual matrix $\mathfrak{A}^\vee$ has the same terms as $\mathfrak{A}$ but indexed in a different way.

Conversely, for any given skew-symmetric matrix $\mathfrak{B}$ of the form (18) there is exactly one skew-symmetric matrix $\mathfrak{A}$ such that $\mathfrak{B}$ is dual to $\mathfrak{A}$; the coordinates of $\mathfrak{A}$ are defined by (19) and the hypothesis of skew symmetry. The unique matrix $\mathfrak{A}$ will be denoted by $\mathfrak{B}^\wedge$.

By definition

(21) $$\mathfrak{A}^{\vee\wedge} = \mathfrak{A} , \quad \mathfrak{B}^{\wedge\vee} = \mathfrak{B} ,$$

(22) $$(c\mathfrak{A})^\vee = c\mathfrak{A}^\vee , \quad (c\mathfrak{B})^\wedge = c\mathfrak{B}^\wedge ,$$

$$(\mathfrak{A}+\mathfrak{A}_1)^{\vee} = \mathfrak{A}^{\vee}+\mathfrak{A}_1^{\vee}, \quad (\mathfrak{B}+\mathfrak{B}_1)^{\wedge} = \mathfrak{B}^{\wedge}+\mathfrak{B}_1^{\wedge}, \tag{23}$$

$$\frac{1}{k!}\mathfrak{A}\mathfrak{A}_1 = \frac{1}{(l-k)!}\mathfrak{A}^{\vee}\mathfrak{A}_1^{\vee}, \quad \frac{1}{(l-k)!}\mathfrak{B}\mathfrak{B}_1 = \frac{1}{k_1}\mathfrak{B}^{\wedge}\mathfrak{B}_1^{\wedge}, \tag{24}$$

$$\|\mathfrak{A}\|^* = \|\mathfrak{A}^{\vee}\|^*, \quad \|\mathfrak{B}\|^* = \|\mathfrak{B}^{\wedge}\|^*; \tag{25}$$

the matrices $\mathfrak{A}_1$ and $\mathfrak{B}_1$ in (23)-(24) must be of the same type as $\mathfrak{A}$ and $\mathfrak{B}$, respectively.

Observe that $\mathfrak{A}^{\vee}$ is a zero matrix if and only if $\mathfrak{A}$ is the zero matrix. Similarly $\mathfrak{B}^{\wedge}$ is a zero matrix if and only if $\mathfrak{B}$ is a zero matrix.

For any given skew-symmetric matrix $\mathfrak{A}$ of the form (1) we can form two matrices: $\mathfrak{A}^{\vee}$ and $\mathfrak{A}^{\wedge}$. They are equal or they differ by sign only. More precisely,

$$\mathfrak{A}^{\vee} = (-1)^{k(l-k)}\mathfrak{A}^{\wedge}. \tag{26}$$

In fact, $\mathfrak{A}^{\vee}$ is the matrix $\mathfrak{B}$ defined by (18) and (20), and $\mathfrak{A}^{\wedge}$ is the skew-symmetric matrix $\mathfrak{C} = [c_{j_1,\dots,j_{l-k}}]_{j_1,\dots,j_{l-k}\leq l}$ such that

$$c_{j_1,\dots,j_{l-k}} = \operatorname{sgn}(i_1,\dots,i_k,j_1,\dots,j_{l-k})\,a_{i_1,\dots,i_k}$$

for any permutation $i_1,\dots,i_k,j_1,\dots,j_{l-k}$ of the integers $1,\dots,l$. Since

$$\operatorname{sgn}(j_1,\dots,j_{l-k},i_1,\dots,i_k) = (-1)^{k(l-k)}\operatorname{sgn}(i_1,\dots,i_k,j_1,\dots,j_{l-k}),$$

we have

$$b_{j_1,\dots,j_{l-k}} = (-1)^{k(l-k)}c_{j_1,\dots,j_{l-k}}$$

which proves (26).

In defining the notion of dual matrix we assumed that $0<k<l$. It is convenient to extend the notion to the case where $k=0$ or $k=l$. For this purpose we have to introduce the notion of a 0-indexed matrix and to explain the structure of l-dimensional l-indexed skew-symmetric matrices.

The matrix

$$\operatorname{sgn} = [\operatorname{sgn}(i_1,\dots,i_l)]_{i_1,\dots,i_l\leq l} \tag{27}$$

is the simplest example of an l-dimensional l-indexed skew-symmetric matrix. Let

$$\mathfrak{A} = [a_{i_1,\dots,i_l}]_{i_1,\dots,i_l\leq l} \tag{28}$$

be any l-dimensional l-indexed skew-symmetric matrix. It follows from (2) that

$$a_{i_1,\dots,i_l} = a_{1,\dots,l}\cdot\operatorname{sgn}(i_1,\dots,i_l),$$

that is,

$$\mathfrak{A} = a_{1,\dots,l}\cdot\operatorname{sgn}.$$

We see that every l-dimensional l-indexed skew-symmetric matrix is of the form

$$b \cdot \mathrm{sgn}$$

where b is a real number.

By a 0-*indexed matrix* $\mathfrak{B}$ we shall mean any real number b; we then write $\mathfrak{B} = [b]$. Each 0-indexed matrix is considered to be skew-symmetric.

Generalizing (19) we shall say that a 0-indexed matrix $\mathfrak{B} = [b]$ is *dual* to a matrix (28) if

$$a_{i_1, \ldots, i_k} = \mathrm{sgn}(i_1, \ldots, i_l) \cdot b \,, \tag{29}$$

i.e. if $\mathfrak{A} = b \cdot \mathrm{sgn}$. The matrix $\mathfrak{B}$ will be denoted by $\mathfrak{A}^\vee$.

According to (20) (where $\mathfrak{A}$ should be replaced by $\mathfrak{B}$ and conversely) we say that a matrix (28) is *dual* to a 0-indexed matrix $\mathfrak{B} = [b]$ if and only if (29) holds, i.e. if $\mathfrak{A} = b \cdot \mathrm{sgn}$. The matrix $\mathfrak{A}$ will be then denoted by $\mathfrak{B}^\vee$.

Similarly we can define $\mathfrak{A}^\wedge$ in the case where $\mathfrak{A}$ is a 0-indexed or l-indexed skew-symmetric matrix. Since then $\mathfrak{A}^\wedge = \mathfrak{A}^\vee$ according to (26), it is not necessary to introduce the notions.

Observe that if

$$\mathfrak{A} = [a_{i_1, \ldots, i_{l-1}}]_{i_1, \ldots, i_{l-1} \leqslant l}$$

is an l-dimensional $(l-1)$-indexed skew-symmetric matrix, then the dual matrix $\mathfrak{A}^\vee$ is an l-dimensional vector, viz.

$$\mathfrak{A}^\vee = [a_1, \ldots, a_l] \,, \tag{30}$$

where

$$a_i = (-1)^{i-1} a_{1, \ldots, i-1, i+1, \ldots, l} \quad \text{for} \quad i = 1, \ldots, l \,. \tag{31}$$

For any permutation $i_1, \ldots, i_l$ of the integers $1, \ldots, l$,

$$a_{i_r} = (-1)^{r-1} a_{i_1, \ldots, i_{r-1}, i_{r+1}, \ldots, i_l} \cdot \mathrm{sgn}(i_1, \ldots, i_l) \,. \tag{32}$$

The following identity holds

$$\mathfrak{A}^\vee \wedge \mathfrak{A} = (\|\mathfrak{A}\|^*)^2 \cdot \mathrm{sgn} \,. \tag{33}$$

In particular,

$$\text{if } \|\mathfrak{A}\|^* = 1, \text{ then } \mathfrak{A}^\vee \wedge \mathfrak{A} = \mathrm{sgn} \,. \tag{34}$$

Indeed, $\mathfrak{A}^\vee \wedge \mathfrak{A} = [b_{i_1, \ldots, i_l}]$, where

$$b_{i_1, \ldots, i_l} = \sum_{r=1}^{l} (-1)^{r-1} a_{i_r} a_{i_1, \ldots, i_{r-1}, i_{r+1}, \ldots, i_l} \,.$$

If $i_1, \ldots, i_l$ is a permutation of the integers $1, \ldots, l$, then

$$b_{i_1, \ldots, i_l} = (\|\mathfrak{A}\|^*)^2 \mathrm{sgn}(i_1, \ldots, i_l)$$

by (32). If $i_r = i_s$ for $r \neq s$, then the equation also holds since $b_{i_1, \dots, i_n}$ and $\operatorname{sgn}(i_1, \dots, i_l)$ are equal to 0.

EXERCISES

1. The product $\mathfrak{A}\mathfrak{B}$ of two skew-symmetric matrices $\mathfrak{A}$, $\mathfrak{B}$ of the same dimension is a skew-symmetric matrix.

2. Prove that, for any skew-symmetric matrix (1),

$$\mathfrak{A}^{\vee} = \frac{1}{k!}\operatorname{sgn}\cdot\mathfrak{A} \quad \text{and} \quad \mathfrak{A}^{\wedge} = \frac{1}{k!}\mathfrak{A}\cdot\operatorname{sgn},$$

i.e. that $\mathfrak{A}^{\vee}$ and $\mathfrak{A}^{\wedge}$ are the reduced products of $\mathfrak{A}$ and sgn, the factors arranged as indicated.

3. Calculate the reduced norm of the skew-symmetric square matrix $\mathfrak{A} = [i-j]_{i,j\leq l}$.

4. Let $s_1 + s_2 = k$. A matrix (1) is said to be *skew-symmetric in the first s_1 indices* if

$$a_{i_{r_1}, \dots, i_{r_{s_1}}, j_1, \dots, j_{s_2}} = \operatorname{sgn}(r_1, \dots, r_{s_1})\, a_{i_1, \dots, i_{s_1}, j_1, \dots, j_{s_2}}$$

for every permutation $r_1, \dots, r_{s_1}$ of the integers $1, \dots, s_1$. Similarly, (1) is said to be *skew-symmetric in the last s_2 indices* if

$$a_{i_1, \dots, i_{s_1}, j_{r_1}, \dots, j_{r_{s_2}}} = \operatorname{sgn}(r_1, \dots, r_{s_2})\, a_{i_1, \dots, i_{s_1}, j_1, \dots, j_{s_2}}$$

for every permutation $r_1, \dots, r_{s_2}$ of the integers $1, \dots, s_2$. Verify that (1) is skew-symmetric in the first s_1 indices (in the last s_2 indices) if and only if the corresponding k-linear form (5) is skew-symmetric in the first s_1 variables (in the last s_2 variables).

If the matrix (1) is skew-symmetric in the first s_1 indices and is skew-symmetric in the last s_2 indices, then $\operatorname{sk}_{s_1,s_2}\mathfrak{A}$ denotes the matrix

$$\operatorname{sk}_{s_1,s_2}\mathfrak{A} = [c_{i_1, \dots, i_k}]_{i_1, \dots, i_k \leq l}$$

where

$$c_{i_1, \dots, i_k} = \sum \operatorname{sgn}(r_1, \dots, r_k)\, a_{i_{r_1}, \dots, i_{r_k}},$$

the summation being extended over all permutations $r_1, \dots, r_k$ of the integers $1, \dots, k$ such that

(35) $$r_1 < r_2 < \dots < r_{s_1} \quad \text{and} \quad r_{s_1+1} < r_{s_1+2} < \dots < r_k.$$

Verify that $\operatorname{sk}_{s_1,s_2}\mathfrak{A}$ is skew-symmetric and

$$\operatorname{sk}_{s_1,s_2}\mathfrak{A} = \frac{1}{s_1!\,s_2!}\operatorname{sk}\mathfrak{A}.$$

Moreover if $\mathfrak{A}$ is the matrix of coordinates of a k-linear form $F\colon V_l^k \to R$, then $\operatorname{sk}_{s_1,s_2}\mathfrak{A}$ is the matrix of the coordinates of the form $\operatorname{sk}_{s_1,s_2} F$.

5. If a $(k+1)$-indexed matrix

(36) $$\mathfrak{A} = [a_{i_1, \dots, i_{k+1}}]_{i_1, \dots, i_{k+1} \leq l}$$

is skew-symmetric in the last k indices, then

$$\operatorname{sk}_{1,k}\mathfrak{A} = [b_{i_1, \dots, i_{k+1}}]_{i_1, \dots, i_{k+1} \leq l}$$

where

$$b_{i_1, \dots, i_{k+1}} = \sum_{j=1}^{k+1} (-1)^{j-1} a_{i_j, i_1, \dots, i_{j-1}, i_{j+1}, \dots, i_{k+1}}.$$

6. If (36) is skew-symmetric in the first k indices, then

$$(-1)^k \mathrm{sk}_{k,1}\mathfrak{A} = [b_{i_1,\dots,i_{k+1}}]_{i_1,\dots,i_{k+1}\leqslant l}$$

where

$$b_{i_1,\dots,i_{k+1}} = \sum_{j=1}^{k+1}(-1)^{j-1}a_{i_1,\dots,i_{j-1},i_{j+1},\dots,i_{k+1},i_j}\,.$$

Verify that

$$(-1)^k \mathbf{sk}_{k,1}\mathfrak{A}\cdot(\mathfrak{a}_1\times\dots\times\mathfrak{a}_{k+1}) = \sum_{j=1}^{k+1}(-1)^{j-1}\mathfrak{A}\cdot(\mathfrak{a}_1\times\dots\times\mathfrak{a}_{j-1}\times\mathfrak{a}_{j+1}\times\dots\times\mathfrak{a}_{k+1}\times\mathfrak{a}_j)\,.$$

7. Let $k = s_1+s_2$. If matrices

$$\mathfrak{A} = [a_{i_1,\dots,i_{s_1}}]_{i_1,\dots,i_{s_1}\leqslant l}\,,\qquad \mathfrak{B} = [b_{i_1,\dots,i_{s_2}}]_{i_1,\dots,i_{s_2}\leqslant l}$$

are skew-symmetric, then their tensor product $\mathfrak{A}\times\mathfrak{B}$ is skew-symmetric in the first s_1 indices and in the last s_2 indices. The skew-symmetric matrix

$$\mathfrak{A}\wedge\mathfrak{B} = \mathrm{sk}_{s_1,s_2}\mathfrak{A}\times\mathfrak{B}$$

is called the *skew product* of $\mathfrak{A}$ and $\mathfrak{B}$. Verify that

$$\mathfrak{A}\wedge\mathfrak{B} = [c_{i_1,\dots,i_k}]_{i_1,\dots,i_k\leqslant l}\,,$$

where

$$c_{i_1,\dots,i_k} = \sum \mathrm{sgn}(r_1,\dots,r_k)\,a_{i_{r_1},\dots,i_{r_{s_1}}}b_{i_{r_{s_1+1}},\dots,i_{r_k}}\,,$$

the summation being extended over all permutations $r_1,\dots,r_k$ of the integers $1,\dots,k$ such that (35) holds. Prove that

$$\mathfrak{B}\wedge\mathfrak{A} = (-1)^{s_1s_2}\mathfrak{A}\wedge\mathfrak{B}\,,\qquad (\mathfrak{A}\wedge\mathfrak{B})\wedge\mathfrak{C} = \mathfrak{A}\wedge(\mathfrak{B}\wedge\mathfrak{C})\,,$$

$$a\cdot(\mathfrak{A}\wedge\mathfrak{B}) = (a\mathfrak{A})\wedge\mathfrak{B} = \mathfrak{A}\wedge(a\mathfrak{B})\,,\qquad (\mathfrak{A}+\mathfrak{A}_1)\wedge\mathfrak{B} = \mathfrak{A}\wedge\mathfrak{B}+\mathfrak{A}_1\wedge\mathfrak{B}\,.$$

If $\mathfrak{A}$ and $\mathfrak{B}$ are the matrices of the coordinates of forms F_1: $V_l^{s_1}\to R$ and F_2: $V_l^{s_2}\to R$ respectively, then $\mathfrak{A}\wedge\mathfrak{B}$ is the matrix of the coordinates of the form $F_1\wedge F_2$: $V_l^k\to R$.

If $\mathfrak{A}$ is a vector, then $\mathfrak{A}\wedge\mathfrak{B}$ coincide with the skew product of a vector and a matrix defined on p. 84.

8. Verify that identities (33), (34) hold for every k-indexed skew-symmetric matrix (1), $1\leqslant k\leqslant l$.

§ 10. Polyvectors.

By the *skew product* of k vectors

$$(1)\qquad \begin{aligned}\mathfrak{a}_1 &= [a_{1,1},\dots,a_{1,l}]\,,\\ &\dots\dots\dots\dots\\ \mathfrak{a}_k &= [a_{k,1},\dots,a_{k,l}]\end{aligned}$$

where $l\geqslant k$ we mean the skew-symmetric matrix

$$(2)\qquad \mathfrak{a}_1\wedge\dots\wedge\mathfrak{a}_k = \mathrm{sk}\,\mathfrak{a}_1\times\dots\times\mathfrak{a}_k\,,$$

i.e. the matrix

(3) $$\mathfrak{a}_1 \wedge \ldots \wedge \mathfrak{a}_k = [a_{i_1,\ldots,i_k}]_{i_1,\ldots,i_k \leqslant l}$$

where

(3′) $$a_{i_1,\ldots,i_k} = \begin{vmatrix} a_{1,i_1} & \ldots & a_{1,i_k} \\ \cdot & \cdot & \cdot \\ a_{k,i_1} & \ldots & a_{k,i_k} \end{vmatrix}.$$

Roughly speaking, the terms of (2) are all determinants of degree k which can be obtained from the rectangular matrix

(4) $$\mathfrak{A} = \begin{bmatrix} a_{1,1} & \ldots & a_{1,l} \\ \cdot & \cdot & \cdot \\ a_{k,1} & \ldots & a_{k,l} \end{bmatrix}$$

by cancelling certain columns.

Matrices of the form (3)-(3′) will be called *polyvectors*, more precisely *l-dimensional k-vectors*.

In the case of $k = 1$, the 1-vector (2) is identical with $\mathfrak{a}_1$. In other words, the notion of 1-vector and the notion of vector coincide.

Developing the determinants (3′) by the first row we get for $k > 1$

(5) $$\mathfrak{a}_1 \wedge \ldots \wedge \mathfrak{a}_k = \mathfrak{a}_1 \wedge (\mathfrak{a}_2 \wedge \ldots \wedge \mathfrak{a}_k),$$

i.e. k-vector $\mathfrak{a}_1 \wedge \ldots \wedge \mathfrak{a}_k$ is the skew product (in the sense of § 9, p. 84) of the vector $\mathfrak{a}_1$ and the $(k-1)$-vector $\mathfrak{a}_2 \wedge \ldots \wedge \mathfrak{a}_k$.

The last remarks constitute an equivalent inductive definition of $\mathfrak{a}_1 \wedge \ldots \wedge \mathfrak{a}_k$.

The following properties of the skew product (2) follow directly from the definition and the known properties of determinants:

(6) $$c(\mathfrak{a}_1 \wedge \ldots \wedge \mathfrak{a}_k) = \mathfrak{a}_1 \wedge \ldots \wedge \mathfrak{a}_{i-1} \wedge c\mathfrak{a}_i \wedge \mathfrak{a}_{i+1} \wedge \ldots \wedge \mathfrak{a}_k,$$

(7) $$\mathfrak{a}_1 \wedge \ldots \wedge \mathfrak{a}_{i-1} \wedge (\mathfrak{a}_i \pm \mathfrak{a}_i') \wedge \mathfrak{a}_{i+1} \wedge \ldots \wedge \mathfrak{a}_k$$
$$= \mathfrak{a}_1 \wedge \ldots \wedge \mathfrak{a}_{i-1} \wedge \mathfrak{a}_i \wedge \mathfrak{a}_{i+1} \wedge \ldots \wedge \mathfrak{a}_k \pm \mathfrak{a}_1 \wedge \ldots \wedge \mathfrak{a}_{i-1} \wedge \mathfrak{a}_i' \wedge \mathfrak{a}_{i+1} \wedge \ldots \wedge \mathfrak{a}_k,$$

(8) $$\mathfrak{a}_1 \wedge \ldots \wedge \mathfrak{a}_{r-1} \wedge \mathfrak{a}_s \wedge \mathfrak{a}_{r+1} \wedge \ldots \wedge \mathfrak{a}_{s-1} \wedge \mathfrak{a}_r \wedge \mathfrak{a}_{s+1} \wedge \ldots \wedge \mathfrak{a}_k$$
$$= -\mathfrak{a}_1 \wedge \ldots \wedge \mathfrak{a}_{r-1} \wedge \mathfrak{a}_r \wedge \mathfrak{a}_{r+1} \wedge \ldots \wedge \mathfrak{a}_{s-1} \wedge \mathfrak{a}_s \wedge \mathfrak{a}_{s+1} \wedge \ldots \wedge \mathfrak{a}_k.$$

The last property expresses the skew symmetry of $\mathfrak{a}_1 \wedge \ldots \wedge \mathfrak{a}_k$. More generally, we have

(9) $$\mathfrak{a}_{j_1} \wedge \ldots \wedge \mathfrak{a}_{j_k} = \operatorname{sgn}(j_1, \ldots, j_k)\, \mathfrak{a}_1 \wedge \ldots \wedge \mathfrak{a}_k$$

for every permutation $j_1, \ldots, j_k$ of the integers $1, \ldots, k$.

It follows from (8) that

(10) $$\mathfrak{a}_1 \wedge \ldots \wedge \mathfrak{a}_k = 0 \text{ if } \mathfrak{a}_r = \mathfrak{a}_s \text{ for } r \neq s\,.$$

This property can also be deduced directly from the definition (3)-(3′).

The next property is a different formulation of the condition 1.1 for the vectors $\mathfrak{a}_1, \ldots, \mathfrak{a}_k$ to be linearly independent:

(11) $\mathfrak{a}_1 \wedge \ldots \wedge \mathfrak{a}_k \neq 0$ if and only if $\mathfrak{a}_1, \ldots, \mathfrak{a}_k$ are linearly independent.

It follows from (2) and § 9 (17) that

(12) $$\frac{1}{k!}\,\mathfrak{C}\cdot(\mathfrak{a}_1 \wedge \ldots \wedge \mathfrak{a}_k) = \mathfrak{C}\cdot(\mathfrak{a}_1 \times \ldots \times \mathfrak{a}_k)$$

for every skew-symmetric k-indexed l-dimensional matrix $\mathfrak{C}$.

Now let

$$\mathfrak{B} = \begin{bmatrix} b_{1,1} & \ldots & b_{1,k} \\ \cdot & \cdot\cdot\cdot & \cdot \\ b_{k,1} & \ldots & b_{k,k} \end{bmatrix}$$

be any square matrix, and let

(13) $$\begin{aligned} \mathfrak{b}_1 &= b_{1,1}\mathfrak{a}_1 + \ldots + b_{1,k}\mathfrak{a}_k\,, \\ &\cdots\cdots\cdots\cdots \\ \mathfrak{b}_k &= b_{k,1}\mathfrak{a}_1 + \ldots + b_{k,k}\mathfrak{a}_k\,. \end{aligned}$$

It follows from the homogeneity (6), distributivity (7) and skew symmetry (9)-(10) that

(14) $$\mathfrak{b}_1 \wedge \ldots \wedge \mathfrak{b}_k = /\mathfrak{B}/\,\mathfrak{a}_1 \wedge \ldots \wedge \mathfrak{a}_k\,.$$

Property (14) can also be deduced from the definition (3)-(3′) on account of the known theorem on multiplication of determinants.

By a general definition § 9 (13)

(15) $$\|\mathfrak{a}_1 \wedge \ldots \wedge \mathfrak{a}_k\|^* = \sqrt{\sum_{i_1<\ldots<i_k} \begin{vmatrix} a_{1,i_1} & \ldots & a_{1,i_k} \\ \cdot & \cdot\cdot\cdot & \cdot \\ a_{k,i_1} & \ldots & a_{k,i_k} \end{vmatrix}^2}\,,$$

i.e.

(16) $$\|\mathfrak{a}_1 \wedge \ldots \wedge \mathfrak{a}_k\|^* = |\mathfrak{A}|$$

where $|\mathfrak{A}|$ is the modulus (see § 5, p. 58) of the matrix (4).

If $\|\mathfrak{a}_1 \wedge \ldots \wedge \mathfrak{a}_k\|^* = 1$, then $\mathfrak{a}_1 \wedge \ldots \wedge \mathfrak{a}_k$ is said to be a *unit k-vector.*

By (6), the product of a k-vector by a number is a k-vector (however, the sum or the difference of two k-vectors is not always a k-vector). We say that two non-zero k-vectors are *parallel* if they differ only by a factor (which is, of course, a number different from 0). Every non-zero

k-vector is parallel to exactly two unit k-vectors which differ only by sign. Namely, the only unit k-vectors parallel to $\mathfrak{a}_1 \wedge \ldots \wedge \mathfrak{a}_k$ are

$$(17) \qquad \frac{\mathfrak{a}_1 \wedge \ldots \wedge \mathfrak{a}_k}{\|\mathfrak{a}_1 \wedge \ldots \wedge \mathfrak{a}_k\|^*}, \qquad -\frac{\mathfrak{a}_1 \wedge \ldots \wedge \mathfrak{a}_k}{\|\mathfrak{a}_1 \wedge \ldots \wedge \mathfrak{a}_k\|^*},$$

that is, the unit k-vectors

$$(18) \qquad \frac{\mathfrak{a}_1 \wedge \ldots \wedge \mathfrak{a}_k}{|\mathfrak{A}|}, \qquad -\frac{\mathfrak{a}_1 \wedge \ldots \wedge \mathfrak{a}_k}{|\mathfrak{A}|},$$

where $|\mathfrak{A}|$ is the modulus of (4).

Let H be a k-dimensional hyperplane in the l-dimensional Euclidean space R^l.

We say that a non-zero k-vector is *parallel* to H if it is of the form $\mathfrak{a}_1 \wedge \ldots \wedge \mathfrak{a}_k$ where $\mathfrak{a}_1, \ldots, \mathfrak{a}_k$ are linearly independent vectors parallel to H. It follows from (6) that a k-vector parallel to a k-vector parallel to H is also parallel to H.

10.1. *Let $\mathfrak{S}$ be a non-zero k-vector parallel to a k-dimensional hyperplane H, and let $\mathfrak{S}_1$ be a non-zero k-vector parallel to a k-dimensional hyperplane H_1. The k-vectors $\mathfrak{S}$ and $\mathfrak{S}_1$ are parallel if and only if the hyperplanes H and H_1 are parallel.*

By hypothesis, $\mathfrak{S} = \mathfrak{a}_1 \wedge \ldots \wedge \mathfrak{a}_k$ where $\mathfrak{a}_1, \ldots, \mathfrak{a}_k$ are linearly independent vectors parallel to H, and similarly $\mathfrak{S}_1 = \mathfrak{b}_1 \wedge \ldots \wedge \mathfrak{b}_k$ where $\mathfrak{b}_1, \ldots, \mathfrak{b}_k$ are linearly independent vectors parallel to H_1.

If H_1 is parallel to H, then the identities (13) hold and, by (14), $\mathfrak{S}_1$ is parallel to $\mathfrak{S}$.

Conversely, if $\mathfrak{S}_1$ is parallel to $\mathfrak{S}$, i.e. $\mathfrak{S}_1 = c\mathfrak{S}$ ($c \neq 0$), then for every vector $\mathfrak{a}$ in R^l

$$\mathfrak{a} \wedge \mathfrak{S} = 0 \text{ if and only if } \mathfrak{a} \wedge \mathfrak{S}_1 = 0,$$

i.e. (see (5))

$$\mathfrak{a} \wedge \mathfrak{a}_1 \wedge \ldots \wedge \mathfrak{a}_k = 0 \text{ if and only if } \mathfrak{a} \wedge \mathfrak{b}_1 \wedge \ldots \wedge \mathfrak{b}_k = 0.$$

By (11) (where k should be replaced by $k+1$), the first equation holds if and only if $\mathfrak{a}$ is a linear combination of $\mathfrak{a}_1, \ldots, \mathfrak{a}_k$, i.e. if and only if $\mathfrak{a}$ is parallel to H. By the same argument, the second equation holds if and only if $\mathfrak{a}$ is parallel to H_1. This proves that H and H_1 are parallel.

In geometry vectors are used to determine directions of straight lines. It follows from 10.1 that, similarly, k-vectors can be used to determine directions of k-dimensional hyperplanes.

For any given k-dimensional hyperplane H in R^l there exist exactly two unit k-vectors parallel to H, differing only in sign. To obtain them, take any linearly independent vectors $\mathfrak{a}_1, \ldots, \mathfrak{a}_k$ parallel to H. Then the formulas (17) (or the equivalent formulas (18)) yield the required unit k-vectors.

By analogy with the case of vectors, the zero k-vector is said to be *parallel* to every k-vector and to every k-dimensional hyperplane H.

10.2. *If* $\mathfrak{S}$ *is a* k*-vector parallel to a* k*-dimensional hyperplane* H *in* R^l $(0<k<l)$, *then the dual matrix* $\mathfrak{S}^\vee$ *and the matrix* $\mathfrak{S}^\wedge$ *are* $(l-k)$*-vectors parallel to every* $(l-k)$*-dimensional hyperplane* H' *which is orthogonal to* H. *Moreover* $\|\mathfrak{S}\|^* = \|\mathfrak{S}^\vee\|^* = \|\mathfrak{S}^\wedge\|^*$. *Thus if* $\mathfrak{S}$ *is a unit* k*-vector parallel to* H, *then* $\mathfrak{S}^\vee$ *and* $\mathfrak{S}^\wedge$ *are unit* $(l-k)$*-vectors parallel to* H'.

The equality of reduced norms follows directly from § 9 (25). Consequently it suffices to prove the first part of 10.2.

If $\mathfrak{S}$ is a zero matrix, then $\mathfrak{S}^\vee$ and $\mathfrak{S}^\wedge$ are zero matrices and the theorem is true.

Suppose that $\mathfrak{S} \neq 0$, i.e. $\mathfrak{S} = \mathfrak{a}_1 \wedge \ldots \wedge \mathfrak{a}_k$ where $\mathfrak{a}_1, \ldots, \mathfrak{a}_k$ are linearly independent vectors (1) parallel to H. Consequently, the coordinates of $\mathfrak{S}$ are given by (3)-(3′). Let

$$\mathfrak{b}_1 = [b_{1,1}, \ldots, b_{1,l}],$$
$$\ldots\ldots\ldots\ldots\ldots$$
$$\mathfrak{b}_{l-k} = [b_{l-k,1}, \ldots, b_{l-k,l}]$$

be linearly independent vectors parallel to H'. We have

$$\mathfrak{b}_1 \wedge \ldots \wedge \mathfrak{b}_{l-k} = [b_{j_1,\ldots,j_{l-k}}]_{j_1,\ldots,j_{l-k} \leqslant l}$$

where

$$b_{j_1,\ldots,j_{l-k}} = \begin{vmatrix} b_{1,j_1} & \ldots & b_{1,j_{l-k}} \\ \ldots & \ldots & \ldots \\ b_{l-k,j_1} & \ldots & b_{l-k,j_{l-k}} \end{vmatrix}.$$

Since each of the vectors $\mathfrak{a}_1, \ldots, \mathfrak{a}_k$ is orthogonal to all the vectors $\mathfrak{b}_1, \ldots, \mathfrak{b}_{l-k}$, it follows from 5.3 that

$$b_{j_1,\ldots,j_{l-k}} = \operatorname{sgn}(i_1, \ldots, i_k, j_1, \ldots, j_{l-k}) \cdot c \cdot a_{i_1,\ldots,i_k}$$

where c is a number different from 0. In other words,

$$\mathfrak{b}_1 \wedge \ldots \wedge \mathfrak{b}_{l-k} = (c\mathfrak{S})^\wedge = c\mathfrak{S}^\wedge.$$

Thus $\mathfrak{S}^\wedge = c^{-1}\mathfrak{b}_1 \wedge \ldots \wedge \mathfrak{b}_{l-k}$ is an $(l-k)$-vector parallel to H'. Since $\mathfrak{S}^\vee = \pm\mathfrak{S}^\wedge$ (see § 9 (26)), $\mathfrak{S}^\vee$ is also an $(l-k)$-vector parallel to H'.

The $(l-k)$-vector $(\mathfrak{a}_1 \wedge \ldots \wedge \mathfrak{a}_k)^\vee$ dual to the k-vector $\mathfrak{a}_1 \wedge \ldots \wedge \mathfrak{a}_k$ will be denoted by the symbol $\mathfrak{a}_1 \vee \ldots \vee \mathfrak{a}_k$. By definition (see § 9 (18)-(20)),

(19) $$\mathfrak{a}_1 \vee \ldots \vee \mathfrak{a}_k = [b_{j_1,\ldots,j_{l-k}}]_{j_1,\ldots,j_{l-k} \leqslant l}$$

where

$$(20) \qquad b_{j_1,\ldots,j_{l-k}} = \operatorname{sgn}(j_1, \ldots, j_{l-k}, i_1, \ldots, i_k) \cdot \begin{vmatrix} a_{1,i_1} & \cdots & a_{1,i_k} \\ \cdots & \cdots & \cdots \\ a_{k,i_1} & \cdots & a_{k,i_k} \end{vmatrix}$$

for any permutation $j_1, \ldots, j_{l-k}, i_1, \ldots, i_k$ of the integers $1, \ldots, l$, and

$$b_{j_1,\ldots,j_{l-k}} = 0 \text{ if } j_r = j_s \text{ for } r \neq s .$$

The $(l-k)$-vector (19) is said to be the *dual skew product* of the vectors $\mathfrak{a}_1, \ldots, \mathfrak{a}_k$.

In the case where $k = 1$, the dual skew product (19) is the $(l-1)$-indexed matrix $\mathfrak{a}_1^{\vee}$ dual to the vector $\mathfrak{a}_1$.

In the case where $k = l-1$, the dual skew product (19) is a 1-vector, i.e. it is a vector. It follows from (19)-(20) that the dual skew product

$$(21) \qquad \mathfrak{a}_1 \vee \ldots \vee \mathfrak{a}_{l-1}$$

of vectors

$$\mathfrak{a}_1 = [a_{1,1}, \ldots, a_{1,l}] ,$$
$$\cdots\cdots\cdots\cdots$$
$$\mathfrak{a}_{l-1} = [a_{l-1,1}, \ldots, a_{l-1,l}]$$

is the vector whose consecutive coordinates are the numbers

$$(21') \qquad \begin{vmatrix} a_{1,2} & \cdots & a_{1,l} \\ \cdots & \cdots & \cdots \\ a_{l-1,2} & \cdots & a_{l-1,l} \end{vmatrix}, \; -\begin{vmatrix} a_{1,1} & a_{1,3} & \cdots & a_{1,l} \\ \cdots & \cdots & \cdots & \cdots \\ a_{l-1,1} & a_{l-1,3} & \cdots & a_{l-1,l} \end{vmatrix},$$

$$\begin{vmatrix} a_{1,1} & a_{1,2} & a_{1,4} & \cdots & a_{1,l} \\ \cdots & \cdots & \cdots & \cdots & \cdots \\ a_{l-1,1} & a_{l-1,2} & a_{l-1,4} & \cdots & a_{l-1,l} \end{vmatrix}, \ldots, (-1)^{l-1} \begin{vmatrix} a_{1,1} & \cdots & a_{1,l-1} \\ \cdots & \cdots & \cdots \\ a_{l-1,1} & \cdots & a_{l-1,l-1} \end{vmatrix}.$$

The vector $\mathfrak{a}_1 \vee \ldots \vee \mathfrak{a}_{l-1}$ is said to be the *vector product* of the vectors $\mathfrak{a}_1, \ldots, \mathfrak{a}_{l-1}$. It follows from the definition that

$$(22) \qquad |\mathfrak{a}_1 \vee \ldots \vee \mathfrak{a}_{l-1}| = \|\mathfrak{a}_1 \wedge \ldots \wedge \mathfrak{a}_{l-1}\|^* .$$

Note the following fundamental property of the vector product: for any vector $\mathfrak{a} = [a_1, \ldots, a_l]$

$$(23) \qquad \mathfrak{a} \cdot (\mathfrak{a}_1 \vee \ldots \vee \mathfrak{a}_{l-1}) = \begin{vmatrix} a_1 & \cdots & a_l \\ a_{1,1} & \cdots & a_{1,l} \\ \cdots & \cdots & \cdots \\ a_{l-1,1} & \cdots & a_{l-1,l} \end{vmatrix},$$

i.e. the scalar product of $\mathfrak{a}$ and the vector product of $\mathfrak{a}_1, \ldots, \mathfrak{a}_{l-1}$ is equal to the determinant formed by all the coordinates of the vectors $\mathfrak{a}, \mathfrak{a}_1, \ldots, \mathfrak{a}_{l-1}$.

It follows from (23) that the vector product $\mathfrak{a}_1\vee\ldots\vee\mathfrak{a}_{l-1}$ is orthogonal to each of the vectors $\mathfrak{a}_1, \ldots, \mathfrak{a}_{l-1}$. In symbols

$$\mathfrak{a}_i\cdot(\mathfrak{a}_1\vee\ldots\vee\mathfrak{a}_{i-1}) = 0 \quad \text{for} \quad i = 1, \ldots, l-1 . \tag{24}$$

It follows directly from the definition that in the case where $l = 3$ the vector product $\mathfrak{a}_1\vee\mathfrak{a}_2$ (of two vectors in R^3) just defined coincides with the vector product defined in text-books in analytic geometry.

It is easy to verify that the dual skew product (19) (in particular the vector product (21)) has properties analogous to (6), (7), (8), (9), (10), (11).

An $(l-k)$-vector is said to be *orthogonal* to a k-dimensional hyperplane H in R^l provided it is parallel to an $(l-k)$-dimensional hyperplane H' orthogonal to H. It follows from 10.2 that a k-vector $\mathfrak{a}_1\wedge\ldots\wedge\mathfrak{a}_k$ is parallel to H if and only if the $(l-k)$-vector $\mathfrak{a}_1\vee\ldots\vee\mathfrak{a}_k$ is orthogonal to H. Then

$$\frac{\mathfrak{a}_1\wedge\ldots\wedge\mathfrak{a}_k}{|\mathfrak{A}|} \text{ is a unit } k\text{-vector parallel to } H , \tag{25}$$

$$\frac{\mathfrak{a}_1\vee\ldots\vee\mathfrak{a}_k}{|\mathfrak{A}|} \text{ is a unit } (l-k)\text{-vector orthogonal to } H , \tag{26}$$

(see (16) and § 9 (25)) provided that the modulus $|\mathfrak{A}|$ of the matrix (4) is not equal to 0.

As has been stated, not every skew-symmetric k-indexed matrix is a k-vector. We have observed that, in the case of $k = 1$, the notions of vector, 1-vector and one-indexed skew-symmetric matrix coincide. This implies that in the case of $k = l-1$ every skew-symmetric $(l-1)$-indexed l-dimensional matrix $\mathfrak{A}$ is an $(l-1)$-vector. Indeed, the dual matrix $\mathfrak{S} = \mathfrak{A}^\vee$ is one-indexed, i.e. it is a vector. By 10.2 the matrix $\mathfrak{A} = \mathfrak{S}^\wedge$ is an $(l-1)$-vector.

Also in the case of $k = l$ every l-indexed l-dimensional skew-symmetric matrix is an l-vector. In fact, we know (see p. 88) that $\mathfrak{A} = b\cdot\mathrm{sgn}$. On the other hand, the matrix sgn is a unit l-vector, viz.

$$\|\mathrm{sgn}\|^* = 1 \quad \text{and} \quad \mathrm{sgn} = \mathfrak{e}_1\wedge\ldots\wedge\mathfrak{e}_l, \tag{27}$$

where $\mathfrak{e}_1, \ldots, \mathfrak{e}_l$ are unit vectors of the coordinate axes in R^l.

According to § 7, any function which, to every point in a set X, assigns an l-dimensional k-vector (or unit k-vector) will be called an *l-dimensional k-vector field* (or *unit k-vector field*) on X.

EXERCISES

1. Calculate the coordinates of the 2-vector $\mathfrak{a}_1\wedge\mathfrak{a}_2$ where

$$\mathfrak{a}_1 = [1, 2, \ldots, l] , \quad \mathfrak{a}_2 = [2, 3, \ldots, l+1] .$$

Calculate the reduced norm of this 2-vector.

2. Calculate the coordinates of the vector product $\mathfrak{a}_1 \vee \mathfrak{a}_2 \vee \mathfrak{a}_3$ of the vectors

$$\mathfrak{a}_1 = [1, 2, 3, 4], \quad \mathfrak{a}_2 = [-1, 3, 5, 1], \quad \mathfrak{a}_3 = [0, 7, 1, -1].$$

3. Let H be the two-dimensional hyperplane in R^4 defined by the equations

$$x_1 + x_2 + x_3 + x_4 = 0, \quad x_1 - x_2 - x_3 + x_4 = 1.$$

Find a unit 2-vector parallel to H and a unit 2-vector orthogonal to H.

4. Verify that the polyvector $\mathfrak{a}_1 \wedge \ldots \wedge \mathfrak{a}_k$ defined on pp. 90-91 is the skew product of the 1-indexed matrices $\mathfrak{a}_1, \ldots, \mathfrak{a}_k$ in the sense defined in § 9, exercise 7.

5. For any skew-symmetric matrix $\mathfrak{U} = [a_{i_1, \ldots, i_k}]_{i_1, \ldots, i_k \leqslant l}$

$$\mathfrak{U} = \sum_{i_1 < i_2 < \ldots < i_k} a_{i_1, \ldots, i_k} \mathfrak{e}_{i_1} \wedge \ldots \wedge \mathfrak{e}_{i_k}$$

where $\mathfrak{e}_1, \ldots, \mathfrak{e}_k$ are the unit vectors of the coordinate axes in R^l.

Thus every k-indexed l-dimensional skew-symmetric matrix is the sum of at most $\binom{l}{k}$ l-dimensional k-vectors.

6. Prove that for every 2-vector $[a_{i,j}]_{i,j \leqslant l}$

$$a_{i,j} a_{m,n} + a_{i,m} a_{n,j} + a_{i,n} a_{j,m} = 0.$$

CHAPTER III

FUNDAMENTAL TOPOLOGICAL NOTIONS

§ 1. Metric spaces. By a *metric space* we mean a set X such that to every pair of points $p, q \in X$ there is assigned a finite real number denoted by the symbol $|q-p|$ so that the following conditions are satisfied:

(1) $$|q-p| \geqslant 0; \quad |q-p| = 0 \text{ if and only if } q = p;$$

(2) $$|q-p| = |p-q|;$$

(3) $$|q-p| \leqslant |q-p_1| + |p_1-p|.$$

The number $|q-p|$ is said to be the *distance* from p to q (or between p and q). The distance $|q-p|$ is a real function defined on $X \times X$ and satisfying (1)-(3); the function is said to be the *metric* of X. Condition (2) expresses the symmetry of the distance. Condition (3) is called the *triangle inequality*.

In this book we shall mainly examine the following examples of metric spaces.

A) The set R of all real numbers with the usual definition of the metric: the distance from p to q is the absolute value of their difference, i.e. $|q-p| = |q-p|$.

B) The k-dimensional Euclidean space R^k with the definition of the metric from II § 1: the distance from p to q is the length of the vector $q-p$, i.e. $|q-p| = |q-p|$.

C) Any subset X of the k-dimensional Euclidean space with the same definition of the distance.

D) The space V_k of all vectors in R^k with the following definition of the metric: the distance from a vector $\mathfrak{a}$ to a vector $\mathfrak{b}$ is the length of the vector $\mathfrak{b}-\mathfrak{a}$, i.e. $|\mathfrak{b}-\mathfrak{a}| = |\mathfrak{b}-\mathfrak{a}|$.

E) The set of all r-indexed matrices of a fixed type $k_1, \ldots, k_r$ with the following definition of the metric: the distance from a matrix $\mathfrak{A}$ to a matrix $\mathfrak{B}$ is the norm of the matrix $\mathfrak{B}-\mathfrak{A}$ (see II § 4 (11)), i.e. $|\mathfrak{B}-\mathfrak{A}| = \|\mathfrak{B}-\mathfrak{A}\|$.

The verification that the example A) is a metric space is very easy. Conditions (1)-(3) follow from the known properties of the absolute value of a real number. In particular, the triangle inequality is another formulation of the inequality $|x+y| \leqslant |x|+|y|$.

The verification of conditions (1)-(3) in the case of B) and C) was done in II §1 (43)-(45).

The verification that the example D) is a metric space is as easy as in the case of A). It is based on the fundamental properties of the length (i.e. the norm) of a vector (see II §1 (35)-(37)) which are a generalization of known properties of the absolute value of a number. In particular, the triangle inequality is another formulation of the property II §1 (37).

The verification of (1)-(3) in the case of E) is the same as in the case of D), on account of the fundamental properties of the norm of a matrix — see II §4 (12)-(13).

Let X be a metric space.

The following inequality is a generalization of the triangle inequality:

$$(3') \qquad |q-p| \leqslant |q-p_1|+|p_1-p_2|+\ldots+|p_{m-1}-p_m|+|p_m-q|$$

for any points $p, p_1, \ldots, p_m, q$ in X. The proof of (3′), which is a generalization of the inequality II §1 (45), is by induction with respect to m. For $m=1$ the inequality (3′) is identical with (3). Suppose it is true for $m-1$, i.e.

$$|q-p| \leqslant |q-p_1|+|p_1-p_2|+\ldots+|p_{m-1}-p| \, .$$

By the triangle inequality, $|p_{m-1}-p| \leqslant |p_{m-1}-p_m|+|p_m-p|$. This implies (3′).

By the *diameter* $\operatorname{dia} A$ of a non-empty subset A of a metric space X we mean the finite, or infinite, positive number

$$(4) \qquad \operatorname{dia} A = \sup\{|q-p| \colon p, q \in A\} \, .$$

For the empty set we assume additionally that

$$\operatorname{dia} 0 = 0 \, .$$

Observe that

$$(5) \qquad \text{if } A \subset B, \text{ then } \operatorname{dia} A \leqslant \operatorname{dia} B \, .$$

We say that a subset A of X is *bounded* if $\operatorname{dia} A < \infty$. Any subset of a bounded set is bounded.

By the *distance* between non-empty subsets A, B of a metric space X we mean the finite number

$$(6) \qquad \operatorname{dist}(B, A) = \inf\{|q-p| \colon p \in A \text{ and } q \in B\} \, .$$

If one of the sets A, B is empty and the other is not empty, we assume additionally that $\operatorname{dist}(B, A) = \infty$. By definition

$$(7) \qquad 0 \leqslant \operatorname{dist}(B, A) \leqslant \infty, \qquad \operatorname{dist}(B, A) = \operatorname{dist}(A, B) \, .$$

If $B = \{q\}$, i.e. if B has only one element q, then instead of $\operatorname{dist}(\{q\}, A)$ we write $\operatorname{dist}(q, A)$. By definition,

(8) $$\operatorname{dist}(q, A) = \inf\{|q-p|\colon p \in A\} < \infty$$

if A is not empty, and

(8′) $$\operatorname{dist}(q, 0) = \infty .$$

The number $\operatorname{dist}(q, A)$ is called the *distance* from q to A. It follows directly from the definition that

(9) $$0 \leqslant \operatorname{dist}(q, A) \leqslant \infty,$$

(10) $$\text{if } A \subset B, \text{ then } \operatorname{dist}(q, B) \leqslant \operatorname{dist}(q, A) ,$$

(11) $$\operatorname{dist}(q, \{p\}) = |q-p| ,$$

(12) $$\operatorname{dist}(q, A \cup B) = \min\bigl(\operatorname{dist}(q, A), \operatorname{dist}(q, B)\bigr) ,$$

(13) $$\operatorname{dist}(q, A) = 0 \quad \text{for every } q \in A .$$

It is a little more difficult to show that

(14) $$|\operatorname{dist}(q_1, A) - \operatorname{dist}(q_2, A)| \leqslant |q_1 - q_2| \quad \text{for} \quad A \neq 0 .$$

Let p be any point in A. Since

$$\operatorname{dist}(q_1, A) \leqslant |q_1 - p| \quad \text{and} \quad |q_1 - p| \leqslant |q_1 - q_2| + |q_2 - p| ,$$

we have

$$\operatorname{dist}(q_1, A) \leqslant |q_1 - q_2| + |q_2 - p| .$$

Since the point $p \in A$ is arbitrary, we infer that

$$\operatorname{dist}(q_1, A) \leqslant |q_1 - q_2| + \operatorname{dist}(q_2, A) .$$

Replacing here q_1 by q_2 and conversely we get

$$\operatorname{dist}(q_2, A) \leqslant |q_1 - q_2| + \operatorname{dist}(q_1, A) .$$

The last two inequalities prove (14).

Generalizing the definition from II § 1, p. 45, we call the set

(15) $$\mathrm{B}(p_0, \eta) = \{p \in X\colon |p - p_0| < \eta\}$$

the *open ball* (or simply *ball*) of radius $\eta > 0$ and center $p_0 \in X$.

If $p_1, p_2 \in \mathrm{B}(p_0, \eta)$, then $|p_1 - p_2| \leqslant |p_1 - p_0| + |p_2 - p_0| < 2\eta$, which proves that

$$\operatorname{dia} \mathrm{B}(p_0, \eta) \leqslant 2\eta .$$

We see that every ball and, consequently, every subset of a ball are bounded sets. Conversely, every bounded set A is a subset of a ball $\mathrm{B}(p_0, \eta)$, the center p_0 being specified in advance. As the radius η it suffices to take any number η greater than $\operatorname{dist}(p_0, A)+\operatorname{dia} A$. In fact, let p and q be any points in A. Since

$$|p-p_0| \leqslant |p-q|+|q-p_0| \leqslant \operatorname{dia} A+|p_0-q|$$

we have

$$|p-p_0| \leqslant \operatorname{dia} A+\operatorname{dist}(p_0, A) < \eta$$

for $p \in A$, i.e. $A \subset \mathrm{B}(p_0, \eta)$.

EXERCISES

1. A function $|q-p|$ is a metric if and only if it satisfies the second part of condition (1) and the condition

$$|q-p| \leqslant |q-p_1|+|p-p_1| .$$

2. Prove that

$$\operatorname{dia}(A \cup B) \leqslant \operatorname{dia} A+\operatorname{dia} B+\operatorname{dist}(A, B) .$$

3. If X and Y are metric spaces, then $X \times Y$ is also a metric space with the metric

$$|(p_1, q_1)-(p_2, q_2)| = \sqrt{|p_1-p_2|^2+|p_2-q_2|^2}$$

for $p_1, p_2 \in X$ and $q_1, q_2 \in Y$.

§ 2. Convergent sequences. A sequence (p_n) of points in a metric space X is said to *converge* to a point $p \in X$ provided $\lim_{n\to\infty} |p_n-p| = 0$, i.e., for every $\varepsilon > 0$ there is an n_0 such that $|p_n-p| < \varepsilon$ for every integer $n > n_0$. We then write

$$p = \lim_{n\to\infty} p_n \quad \text{or} \quad p_n \to p ,$$

and we say that p is the *limit* of (p_n).

A sequence (p_n) of points in X is said to be *convergent* (or *to converge*) if there exists a point $p \in X$ such that $p_n \to p$.

In the case where X is the set R of all real numbers (see § 1 A)) the above definitions of convergence and limit coincide with the ordinary definitions (see I § 4, p. 29). Thus the notion of convergence of sequences of points in metric spaces is a generalization of the same notion for real numbers. It has many properties as in the case of real numbers. For instance,

2.1. *In any metric space X,*

(a) *every sequence has at most one limit;*

(b) *if $p_n = p$ for all integers $n > n_0$, then $p = \lim_{n\to\infty} p_n$;*

(c) *if $p = \lim_{n\to\infty} p_n$ and $m_1 < m_2 < m_3 < \ldots$, then $p = \lim_{n\to\infty} p_{m_n}$, i.e. every subsequence of a sequence convergent to p converges to p;*

(d) *if every subsequence* (p_{m_n}) *of a sequence* (p_n) *contains a subsequence* $(p_{m_{r_n}})$ *convergent to* p, *then* (p_n) *converges to* p.

Properties (b), (c), (d) follow directly from the definition of convergence in X and from the analogous properties of number sequences, applied to $(|p_n-p|)$. To prove (a), suppose that $p_n \to p$ and $p_n \to q$ simultaneously. Thus

$$0 \leqslant |q-p| \leqslant |q-p_n|+|p_n-p| \to 0$$

which proves that $|q-p| = 0$, i.e. $q = p$.

2.2. *If* $q_n \to q$ *and* A *is a non-empty subset of* X, *then*

$$\operatorname{dist}(q_n, A) \to \operatorname{dist}(q, A) .$$

In particular, if $q_n \to q$, *then* $|q_n-p| \to |q-p|$.

The first part of 2.2 follows directly from § 1 (14). The second part follows from the first, where $A = \{p\}$.

A sequence (p_n) of points in a metric space X is said to *satisfy the Cauchy condition* if $\lim\limits_{m,n\to\infty} |p_n-p_m| = 0$, i.e. for every $\varepsilon > 0$ there exists an integer n_0 such that

$$|p_n-p_m| < \varepsilon \quad \text{for all } m, n > n_0 .$$

2.3. *Every convergent sequence satisfies the Cauchy condition.*

Indeed, if $p_n \to p$, then

$$0 \leqslant |p_n-p_m| \leqslant |p_n-p|+|p_m-p|$$

which proves that $|p_n-p_m| \to 0$ for $n, m \to \infty$.

A theorem converse to 2.3 is not true, in general. One can give examples of metric spaces in which there are non-convergent sequences satisfying the Cauchy condition. A metric space X is said to be *complete* provided the theorem converse to 2.3 is true for X, i.e. every sequence (of points in X) satisfying the Cauchy condition converges to a point in X.

All the metric spaces in § 1 A), B), D), E) are complete.

The case of A), i.e. of the space R of all real numbers, is well known: a sequence of real numbers converges if and only if it satisfies the Cauchy condition.

To examine the case of B), i.e. of the k-dimensional Euclidean space R^k, suppose that

$$p_n = (x_{n,1}, \ldots, x_{n,k}) \in R^k, \quad n = 1, 2, \ldots,$$
$$p \;= (x_1, \ldots, x_k) \in R^k.$$

By definition of the distance in R^k,

$$|p_n-p| = \sqrt{\sum_{j=1}^{k} (x_{n,j}-x_j)^2}, \quad |p_n-p_m| = \sqrt{\sum_{j=1}^{k} (x_{n,j}-x_{m,j})^2} .$$

Hence it follows that

$$p_n \to p \text{ if and only if } x_{n,j} \to x_j \text{ for } j = 1, \ldots, k,$$

i.e. if the number sequences

$$(1) \qquad (x_{n,1}), \ldots, (x_{n,k})$$

of the coordinates of the points p_n converge to the corresponding coordinates of the point p. This implies that the sequence (p_n) converges if and only if all the sequences (1) converge. Similarly we infer (see II § 1 (40)-(41)) that the sequence (p_n) satisfies the Cauchy condition if and only if all the sequences (1) satisfy the Cauchy condition, i.e. converge. Thus (p_n) satisfies the Cauchy condition if and only if it converges.

The case D) of the space V_k of all k-dimensional vectors can be examined in the same way. Let

$$\mathfrak{a}_n = [a_{n,1}, \ldots, a_{n,k}], \qquad n = 1, 2, \ldots,$$
$$\mathfrak{a} = [a_1, \ldots, a_k].$$

Then

$$|\mathfrak{a}_n - \mathfrak{a}| = \sqrt{\sum_{j=1}^{k} (a_{n,j} - a_j)^2}, \qquad |\mathfrak{a}_n - \mathfrak{a}_m| = \sqrt{\sum_{j=1}^{k} (a_{n,j} - a_{m,j})^2}.$$

Hence it follows that

$$\mathfrak{a}_n \to \mathfrak{a} \text{ if and only if } a_{n,j} \to a_j \text{ for } j = 1, \ldots, k,$$

i.e. if the number sequences

$$(2) \qquad (a_{n,1}), \ldots, (a_{n,k})$$

of the coordinates of $\mathfrak{a}_n$ converge to the corresponding coordinates of the vector $\mathfrak{a}$. This implies that the sequence $(\mathfrak{a}_n)$ converges if and only if all the sequences (2) converge. Similarly, we infer that $(\mathfrak{a}_n)$ satisfies the Cauchy condition if and only if all the sequences (2) satisfy the Cauchy condition, i.e. they converge. Thus $(\mathfrak{a}_n)$ satisfies the Cauchy condition if and only if it converges.

Similarly in the case E) of the space of all matrices of a fixed type $k_1, \ldots, k_r$, if

$$\mathfrak{A}_n = [a_{n;i_1,\ldots,i_r}]_{i_1 \leqslant k_1, \ldots, i_r \leqslant k_r}, \qquad n = 1, 2, \ldots,$$
$$\mathfrak{A} = [a_{i_1,\ldots,i_r}]_{i_1 \leqslant k_1, \ldots, i_r \leqslant k_r},$$

then

$$|\mathfrak{A}_n - \mathfrak{A}| = \sqrt{\sum_{i_1=1}^{k_1} \cdots \sum_{i_r=1}^{k_r} (a_{n;i_1,\ldots,i_r} - a_{i_1,\ldots,i_r})^2},$$

$$|\mathfrak{A}_n - \mathfrak{A}_m| = \sqrt{\sum_{i_1=1}^{k_1} \cdots \sum_{i_r=1}^{k_r} (a_{n;i_1,\ldots,i_r} - a_{m;i_1,\ldots,i_r})^2}.$$

Hence it follows that $\mathfrak{A}_n \to \mathfrak{A}$ if and only if

$$a_{n;i_1,\ldots,i_r} \to a_{i_1,\ldots,i_r} \quad \text{for} \quad i_1 = 1, \ldots, k_1, \ldots, i_r = 1, \ldots, k_r,$$

i.e. if the number sequences

$$(a_{n;i_1,\ldots,i_r}), \quad i_1 = 1, \ldots, k_1, \ldots, i_r = 1, \ldots, k_r, \tag{3}$$

of the coordinates of $\mathfrak{A}_n$ converge to the corresponding coordinates of $\mathfrak{A}$. This implies that the sequence $(\mathfrak{A}_n)$ converges if and only if all the sequences (3) converge. Similarly, we infer that $(\mathfrak{A}_n)$ satisfies the Cauchy condition if and only if all the sequences (3) satisfy the Cauchy condition, i.e. they converge. Thus $(\mathfrak{A}_n)$ satisfies the Cauchy condition if and only if it converges.

We summarize all these remarks in the form of the following theorem.

2.4. *In the space R^k (in the space V_k, in the space of all matrices of a fixed type) a sequence of points (of vectors, of matrices) converges if and only if it satisfies the Cauchy condition, and if and only if all the sequences of its coordinates converge. Then it converges to the point (vector, matrix) whose coordinates are the limits of the corresponding sequences of coordinates.*

We know that for any sequences (a_n), (b_n) of real numbers

$$\text{if } a_n \to a \text{ and } b_n \to b, \text{ then } a_n \pm b_n \to a \pm b,\ a_n b_n \to ab, \text{ etc.} \tag{4}$$

Analogous statements are true for vectors and matrices and all the operations defined for them. For instance, for any sequences of vectors of the same dimensions,

$$\text{if } a_n \to a \text{ and } \mathfrak{a}_n \to \mathfrak{a}, \text{ then } a_n \mathfrak{a}_n \to a\mathfrak{a}; \tag{5}$$

$$\text{if } \mathfrak{a}_n \to \mathfrak{a} \text{ and } \mathfrak{b}_n \to \mathfrak{b}, \text{ then } \mathfrak{a}_n \pm \mathfrak{b}_n \to \mathfrak{a} \pm \mathfrak{b} \text{ and } \mathfrak{a}_n \mathfrak{b}_n \to \mathfrak{a}\mathfrak{b}. \tag{6}$$

Similarly, for matrices of a fixed type,

$$\text{if } a_n \to a \text{ and } \mathfrak{A}_n \to \mathfrak{A}, \text{ then } a_n \mathfrak{A}_n \to a\mathfrak{A}; \tag{7}$$

$$\text{if } \mathfrak{A}_n \to \mathfrak{A} \text{ and } \mathfrak{B}_n \to \mathfrak{B}, \text{ then } \mathfrak{A}_n \pm \mathfrak{B}_n \to \mathfrak{A} \pm \mathfrak{B} \text{ and } \mathfrak{A}_n\mathfrak{B}_n \to \mathfrak{A}\mathfrak{B}. \tag{8}$$

Analogously, for rectangular matrices,

$$\text{if } \mathfrak{A}_n \to \mathfrak{A} \text{ and } \mathfrak{B}_n \to \mathfrak{B}, \text{ then } \mathfrak{A}_n \cdot \mathfrak{B}_n \to \mathfrak{A} \cdot \mathfrak{B}; \tag{9}$$

$$\text{if } \mathfrak{A}_n \to \mathfrak{A} \text{ and } \mathfrak{a}_n \to \mathfrak{a}, \text{ then } \mathfrak{A}_n \mathfrak{a}_n \to \mathfrak{A}\mathfrak{a} \tag{10}$$

and the same for the left-hand multiplication of rectangular matrices by vectors. Note also that

$$\text{if } \mathfrak{a}_{n,1} \to \mathfrak{a}_1, \ldots, \mathfrak{a}_{n,k} \to \mathfrak{a}_k, \text{ then } \mathfrak{a}_{n,1} \wedge \ldots \wedge \mathfrak{a}_{n,k} \to \mathfrak{a}_1 \wedge \ldots \wedge \mathfrak{a}_k \tag{11}$$

and the same for the dual skew multiplication $\vee$.

All these properties follow from 2.4 and (4) because the operations are defined by means of addition, subtraction and multiplication of coordinates.

EXERCISES

1. Prove that if $p_n \to p$ and $q_n \to q$, then

$$|p_n - q_n| \to |p - q| .$$

2. The set of all rational numbers with the ordinary definition of distance is a non-complete metric space. The same holds for the set of all irrational numbers.

3. Every closed interval in R^k (with the ordinary definition of distance) is a complete metric space but every non-empty open bounded interval is a non-complete metric space.

§ 3. Closed and open sets. Let A be a subset of a metric space X. The set

$$\{q \in X\colon \operatorname{dist}(q, A) = 0\}$$

is called the *closure* of A and denoted by $\operatorname{cl} A$. By definition,

(1) $\quad q \in \operatorname{cl} A$ if and only if $A \cap \mathrm{B}(q, \eta) \neq 0$ for every $\eta > 0$,

or equivalently

2) $\quad q \in \operatorname{cl} A$ if and only if q is the limit of a sequence of points in A .

3.1. *For any subsets* A, B *of* X,

(a) *if* $A \subset B$, *then* $\operatorname{cl} A \subset \operatorname{cl} B$;

(b) $A \subset \operatorname{cl} A$;

(c) $\operatorname{cl}(A \cup B) = \operatorname{cl} A \cup \operatorname{cl} B$;

(d) $\operatorname{cl}\operatorname{cl} A = \operatorname{cl} A$;

(e) *if the set* A *is finite, then* $\operatorname{cl} A = A$.

(a) follows from § 1 (10).

(b) follows from § 1 (13).

It follows from § 1 (12) that $\operatorname{dist}(q, A \cup B) = 0$ if and only if $\operatorname{dist}(q, A) = 0$ or $\operatorname{dist}(q, B) = 0$. In other words, $q \in \operatorname{cl}(A \cup B)$ if and only if $q \in \operatorname{cl} A$ or $q \in \operatorname{cl} B$. This proves (c).

It follows from (b) that $\operatorname{cl} A \subset \operatorname{cl}\operatorname{cl} A$ by (a). Conversely, if $q \in \operatorname{cl}\operatorname{cl} A$, then $q = \lim\limits_{n\to\infty} q_n$ where $q_n \in \operatorname{cl} A$ for $n = 1, 2, \ldots$, i.e. $\operatorname{dist}(q_n, A) = 0$ for $n = 1, 2, \ldots$ By 2.2

$$\operatorname{dist}(q, A) = \lim_{n\to\infty} \operatorname{dist}(q_n, A) = 0 , \quad \text{i.e.} \quad q \in \operatorname{cl} A .$$

This proves (d).

If A is empty, then (e) follows from § 1 (8′). If A is a one-point set, (e) follows from § 1 (1), (11). Hence it follows, by (c), that (e) is true for any finite non-empty set.

We say that a set $A \subset X$ is *closed* if $\mathrm{cl}A = A$. It follows from 3.1 (d) that the closure of any set is a closed set. It follows from 3.1 (e) that every finite set is closed. In particular, the empty set is closed. Since $X \subset \mathrm{cl}X \subset X$ (see 3.1 (b)), we infer that $\mathrm{cl}X = X$, i.e. the whole space X is closed.

It follows from 3.1 (b) and (2) that a set A is closed if and only if it contains the limit of every convergent sequence of points in A.

3.2. *If A and B are closed, then $A \cup B$ is closed. Consequently the union of a finite number of closed sets is closed.*

The intersection of any number of closed sets is closed.

If $\mathrm{cl}A = A$ and $\mathrm{cl}B = B$, then $\mathrm{cl}(A \cup B) = A \cup B$ by 3.1 (c). This proves the first part of 3.2.

Suppose $A = \bigcap_{t \in T} A_t$ and $\mathrm{cl}A_t = A_t$ for every $t \in T$. Since $A \subset A_t$, we have $\mathrm{cl}A \subset \mathrm{cl}A_t = A_t$ (see 3.1 (a)). Hence

$$A \subset \mathrm{cl}A \subset \bigcap_{t \in T} A_t = A ,$$

i.e. $\mathrm{cl}A = A$, which proves the second part of 3.2.

Let $A \subset X$. The set

$$\{q \in X\colon \mathrm{dist}(q, X - A) > 0\}$$

is said to be the *interior* of A and is denoted by $\mathrm{int}A$. Points in $\mathrm{int}A$ are said to be *interior points* of A.

By definition,

(3) $\quad q \in \mathrm{int}A$ if and only if $\mathrm{B}(q, \eta) \subset A$ for a number $\eta > 0$.

It follows directly from the definition of the closure and the interior that

$$X - \mathrm{int}A = \mathrm{cl}(X - A) , \tag{4}$$

$$X - \mathrm{cl}A = \mathrm{int}(X - A) . \tag{4'}$$

In other words,

$$\mathrm{int}A = X - \mathrm{cl}(X - A) , \tag{5}$$

$$\mathrm{cl}A = X - \mathrm{int}(X - A) . \tag{5'}$$

These identities permit one to deduce properties of the interior from properties of the closure, and conversely. For instance, the properties of the interior in the next theorem follow from the properties of the closure, denoted in 3.1 by the same letters (a) ... (d).

3.3. *For any subsets A, B of X,*

(a) *if $A \subset B$, then* $\mathrm{int}A \subset \mathrm{int}B$;

(b) $\operatorname{int} A \subset A$;

(c) $\operatorname{int}(A \cap B) = \operatorname{int} A \cap \operatorname{int} B$;

(d) $\operatorname{int}\operatorname{int} A = \operatorname{int} A$.

Of course Theorem 3.3 can also be proved directly.

A set $A \subset X$ is said to be *open* if $\operatorname{int} A = A$. By (3), a set A is open if and only if every $q \in A$ is an interior point of A or, equivalently, if for every point $q \in A$ there exists a number $\eta > 0$ such that $\mathrm{B}(q, \eta) \subset A$.

It follows from 3.3 (d) that the interior of any set is open.

The following remark is a direct consequence of identities (4), (4′), (5), (5′).

3.4. *A set A is open (closed) if and only if its complement $X - A$ is closed (open).*

Since the empty set and the whole space X are closed, it follows from 3.4 that the whole space X and the empty set are also open.

The following theorem will be deduced from 3.2 and 3.4.

3.5. *If A and B are open, then $A \cap B$ is open. Consequently the intersection of any finite number of open sets is open.*

The union of any number of open sets is open.

If A and B are open, then $X-A$ and $X-B$ are closed. Consequently the set $(X-A) \cup (X-B)$ is closed and the set $A \cap B = X - \big((X-A) \cup (X-B)\big)$ is open.

If all the sets A_t ($t \in T$) are open and $A = \bigcup_{t \in T} A_t$, then all the sets $X - A_t$ are closed, and consequently, the set $X - A = \bigcap_{t \in T} (X - A_t)$ is closed. This proves that A is open.

3.6. *If A is open and B is closed, then $A-B$ is open and $B-A$ is closed.*

This follows from 3.4, 3.5 and 3.2 because of the identities

$$A - B = A \cap (X - B), \quad B - A = B \cap (X - A).$$

3.7. *Any ball $\mathrm{B}(q, \eta)$ is an open set.*

It suffices to show (see (3)) that for every point $p \in \mathrm{B}(q, \eta)$ there is a number $\delta > 0$ such that $\mathrm{B}(p, \delta) \subset \mathrm{B}(q, \eta)$. The number $\delta = \eta - |p - q|$ has this property. Indeed, if $p_1 \in \mathrm{B}(p, \delta)$, then

$$|p_1 - q| \leqslant |p_1 - p| + |p - q| < \eta, \quad \text{i.e.} \quad p_1 \in \mathrm{B}(q, \eta).$$

By a *neighbourhood* (more precisely, by an *open neighbourhood*) of a point p in the metric space X we shall mean any open set containing p. By 3.7, the ball $\mathrm{B}(q, \eta)$ is a neighbourhood of each of its points.

For any set $A \subset X$, the closed set (see 3.6)

$$\operatorname{cl} A - \operatorname{int} A$$

is said to be the *boundary* of A. If A is open, then $\mathrm{cl}A - A$ is the boundary of A. If A is closed, then $A - \mathrm{int}A$ is the boundary of A.

We shall illustrate the notion of open and closed sets and boundary by the following examples where X is the k-dimensional Euclidean space R^k. Suppose that $a_i < b_i$ for $i = 1, \ldots, k$. The open interval

$$J = \{(x_1, \ldots, x_k):\ a_i < x_i < b_i \text{ for } i = 1, \ldots, k\}$$

is an open set, and the closed interval

$$I = \{(x_1, \ldots, x_k):\ a_i \leqslant x_i \leqslant b_i \text{ for } i = 1, \ldots, k\}$$

is a closed set. Moreover $J = \mathrm{int}I$ and $I = \mathrm{cl}J$. The union or all the $(k-1)$-dimensional faces of I (or J) is the boundary of I and of J.

If $a_i \leqslant b_i$ for $i = 1, \ldots, k$ and $a_j = b_j$ for at least one index j, then the degenerate interval I is a closed set and its interior is empty. The degenerate interval I is the boundary of I.

If A is an open subset of R^k and $p \in A$, then there exists an open interval J such that $p \in J \subset A$. Indeed, there is a number $\eta > 0$ such that $\mathrm{B}(p, \eta) \subset A$. The open interval with center in p and with the length of its edges equal to $2\eta/\sqrt{k}$ is a subset of $\mathrm{B}(p, \eta)$ and, consequently, of A, and it contains p.

EXERCISES

1. For any sequence (A_n) of subsets of a metric space,

$$\mathrm{cl}(A_1 \cup A_2 \cup \ldots) = (\mathrm{cl}A_1 \cup \mathrm{cl}A_2 \cup \ldots) \cup \bigcap_{n=1}^{\infty} \mathrm{cl}(A_n \cup A_{n+1} \cup \ldots).$$

2. If G is open, then for every set A

$$G \cap \mathrm{cl}A \subset \mathrm{cl}(G \cap A), \qquad \mathrm{cl}(G \cap \mathrm{cl}A) = \mathrm{cl}(G \cap A).$$

If F is closed, then for every set A

$$\mathrm{int}(F \cup A) \subset F \cup \mathrm{int}A, \qquad \mathrm{int}(F \cup A) = \mathrm{int}(F \cup \mathrm{int}A).$$

3. For every open set G, $G \subset \mathrm{int}\,\mathrm{cl}G$. For every closed set F, $\mathrm{cl}\,\mathrm{int}F \subset F$.

Give examples of sets in R and R^2 for which the inclusion cannot be replaced by an equality sign.

4. For any sets A_1, A_2, the set

$$G = \{p:\ \mathrm{dist}(p, A_1) < \mathrm{dist}(p, A_2)\}$$

is open and the set

$$F = \{p:\ \mathrm{dist}(p, A_1) \leqslant \mathrm{dist}(p, A_2)\}$$

is closed.

If $\mathrm{cl}A_1 \cap \mathrm{cl}A_2 = 0$, then $\mathrm{cl}A_1 \subset G$ and $\mathrm{cl}A_2 \cap \mathrm{cl}G = 0$.

5. A point p is in the boundary of a set A if and only if there are two sequences (p_n) and (q_n) such that

$$p_n \in A, \quad q_n \in X - A \quad \text{for} \quad n = 1, 2, \ldots \quad \text{and} \quad p_n \to p, \quad q_n \to p.$$

6. For every $\varepsilon > 0$ the set

$$A(\varepsilon) = \{p\colon \operatorname{dist}(p, A) < \varepsilon\}$$

is open. For any sequence $\varepsilon_n \to 0$ of positive numbers,

$$\operatorname{cl} A = \bigcap_{n=1}^{\infty} A(\varepsilon_n).$$

7. Interpret the Euclidean space R^{k+l} as an $(\mathbf{i}, \mathbf{j})$-product of the Euclidean spaces R^k and R^l (see I § 6 (7)-(10)). If G_0 is an open subset of R^{k+l} and $(p, q) \in G_0$, then there are positive numbers ε and δ such that

$$\mathrm{B}(p, \varepsilon) \times \mathrm{B}(q, \delta) \subset G_0.$$

§ 4. Relativization. Any subset Z of a metric space X is also a metric space, the distance between points in Z being the same as in X. For subsets A of Z we can also introduce the notions of the interior and closure and the notions of closed and open sets in the space Z. To assure the uniqueness of notation, we shall denote by $\operatorname{cl} A$ and $\operatorname{int} A$ the closure of A and the interior of A in the metric space X, and by $\operatorname{cl}_Z A$ and $\operatorname{int}_Z A$ the closure and the interior of A in the metric space Z.

It follows from the general definition of closure in § 3 that

$$\operatorname{cl}_Z A = \{q \in Z\colon \operatorname{dist}(q, A) = 0\} = Z \cap \{q \in X\colon \operatorname{dist}(q, A) = 0\}.$$

Hence

(1) $$\operatorname{cl}_Z A = Z \cap \operatorname{cl} A \quad \text{for} \quad A \subset Z.$$

Consequently (see § 3 (5))

$$\begin{aligned}\operatorname{int}_Z A &= Z - \operatorname{cl}_Z(Z - A) = Z - \big(Z \cap \operatorname{cl}(Z - A)\big)\\ &= Z \cap \big(X - \operatorname{cl}(Z - A)\big) = Z \cap \operatorname{int}\big(X - (Z - A)\big),\end{aligned}$$

that is

(2) $$\operatorname{int}_Z A = Z \cap \operatorname{int}\big((X - Z) \cup A\big) \quad \text{for} \quad A \subset Z.$$

4.1. *A set $A \subset Z$ is closed (open) in the metric space Z if and only if it is of the form $A = Z \cap B$ where B is a closed (open) subset of X.*

If A is closed in Z, then $A = \operatorname{cl}_Z A = Z \cap B$ where $B = \operatorname{cl} A$ is closed in X. Conversely, if $A = Z \cap B$ and $B = \operatorname{cl} B$, then $A \subset B$ and consequently $\operatorname{cl} A \subset \operatorname{cl} B = B$. Hence $A \subset Z \cap \operatorname{cl} A \subset Z \cap B = A$ which proves that $A = Z \cap \operatorname{cl} A = \operatorname{cl}_Z A$, i.e. that A is closed in the metric space Z.

If a set $A \cap Z$ is open in Z, then $Z - A$ is closed in Z and, therefore, $Z - A = Z \cap C$ where C is a closed subset of X. The set $B = X - C$ is open in X and $A = Z \cap B$.

Conversely, if $A = Z \cap B$, where B is an open subset in X, then $Z - A = Z \cap (X - B)$. Since $X - B$ is closed in X, the set $Z - A$ is closed in the space Z. This implies that A is open in Z.

It follows from 4.1 that for a set $A \subset Z$ to be a *neighbourhood* of a point $p \in Z$ *in the space* Z it is necessary and sufficient that $A = Z \cap B$ where B is a neighbourhood of p in X.

The next theorem is a direct corollary of 4.1.

4.2. *If Z is an open (closed) subset of X, then a set $A \subset Z$ is open (closed) in Z if and only if it is open (closed) in X.*

EXERCISES

1. Let $X = R^3$ and

$$Z = \{(x_1, x_2, x_3):\ x_1^2 + x_2^2 + x_3^2 = 1\}\,.$$

The set $A = \{(x_1, x_2, x_3) \in Z:\ x_3 > 0\}$ is open in Z, the set $B = \{(x_1, x_2, x_3) \in Z:\ x_3 \geqslant 0\}$ is closed in Z and $B = \mathrm{cl}_Z A$, $A = \mathrm{int}_Z B$.

2. Let A be a circle with center in p_0 ($p_0 \in R^2$, $A \subset R^2$) and let $X = R^2$, $Z = A \cup \{p_0\}$. The sets A and $\{p_0\}$ are open and closed in Z.

3. Let $X = R$ and let $Z = \{1/n:\ n = 1, 2, \dots\}$. Every set $A \subset Z$ is open and closed in Z.

4. Let $X = R$ and let Z be composed of the number 0 and all the numbers $1/n$, $n = 1, 2, \dots$ Each one-point set $\{1/n\}$ is open and closed in Z. The one-point set $\{0\}$ is closed, but not open in Z. The set of all numbers $1/n$, $n = 1, 2, \dots$, is open but not closed in Z.

§ 5. Separable spaces. A sequence (B_n) of open subsets is said to be a *basis* of a metric space X provided every open set is the union of certain sets in this sequence. A metric space X is said to be *separable* if it has a basis.

5.1 (Lindelöf). *In any separable metric space X, if a set A is the union of open sets:*

$$A = \bigcup_{t \in T} A_t\,,$$

then it is the union of countably many of these sets, i.e. there is a countable set $T_0 \subset T$ such that

$$A = \bigcup_{t \in T_0} A_t\,.$$

Let (B_n) be a basis of X. For every point $p \in A$ there exists an index $t(p) \in T$ such that $p \in A_{t(p)}$. Since the set $A_{t(p)}$ is open, it is the union of certain of the sets B_n. Consequently, there is a positive integer, say $m(p)$, such that

$$p \in B_{m(p)} \subset A_{t(p)}\,.$$

The set N of all the integers $m(p)$ ($p \in A$) is countable. For every integer $n \in N$ let p_n be an element in A such that $n = m(p_n)$. Hence $B_n \subset A_{t(p_n)}$. The set T_0 of all the indices $t(p_n)$ where $n \in N$ is countable. Since

$$A \subset \bigcup_{p \in A} B_{m(p)} = \bigcup_{n \in N} B_n \subset \bigcup_{n \in N} A_{t(p_n)} = \bigcup_{t \in T_0} A_t \subset A\,,$$

we infer that $A = \bigcup_{t \in T_0} A_t$.

5.2. *Any subset Z of a separable metric space X is a separable metric space.*

If (B_n) is a basis of X, then the sequence $(Z \cap B_n)$ is a basis of Z. Indeed, if A is an open subset of the space Z, i.e. $A = Z \cap B$ where B is open in X, then there is a set N of positive integers such that $B = \bigcup_{n \in N} B_n$ and, consequently, $A = \bigcup_{n \in N} Z \cap B_n$.

5.3. *The k-dimensional Euclidean space R^k is separable, and the rational open intervals (the empty set included) constitute a basis of R^k.*

Let (B_n) be the sequence formed of the empty sets and of all rational intervals in R^k (such a sequence exists by II 3.1). Let A be any non-empty open subset of R^k. As we stated at the end of § 3, for every point $p \in A$ there is an open interval

$$I = \{(x_1, \ldots, x_k): a_i < x_i < b_i \text{ for } i = 1, \ldots, k\}$$

such that $p \in I \subset A$. We can assume that all the numbers $a_1, \ldots, a_k$, $b_1, \ldots, b_k$ are rational (for if not, replace $a_1, \ldots, a_k$ by a slightly greater rational numbers, and $b_1, \ldots, b_k$ by a slightly smaller rational numbers), i.e. that I is rational. In other words, for every point $p \in A$ there is a positive integer n such that $p \in B_n \subset A$. Let N be the set of all positive integers n such that $B_n \subset A$. By definition, $A \subset \bigcup_{n \in N} B_n \subset A$, i.e. $A = \bigcup_{n \in N} B_n$. If A is empty, then A is also of this form because the empty set appears in (B_n).

EXERCISES

1. A metric space X is separable if and only if there is a countable set A such that $X = \operatorname{cl} A$.

2. Every class of open disjoint subsets of a separable metric space is countable.

3. Let X be the set of all non-negative real numbers with the following definition of the metric: if $x \neq y$, then the distance from x to y is $x + y$; otherwise it is 0. Prove that X is a non-separable metric space.

§ 6. Compact sets. A subset A of a metric space X is said to be *compact* provided every sequence (p_n) of points in A contains a subsequence (p_{m_n}) convergent to a point in A.

6.1. *Every compact set is closed and bounded.*

Suppose A is a compact set.

Let $p_n \in A$ for $n = 1, 2, \ldots, p_n \to p$. The sequence (p_n) contains a subsequence (p_{m_n}) convergent to a point $q \in A$. By 2.1 (c), (p_{m_n}) converges to p. Thus $p = q$, and consequently $p \in A$. We have proved that the limit of every convergent sequence of points in A belongs to A, i.e. that A is closed.

Let q be any point in X. To prove that A is bounded it suffices to show that there is a number $\eta > 0$ such that $A \subset \mathrm{B}(q, \eta)$. Suppose the contrary. Thus for every positive integer n there exists a point $p_n \in A$ such that $p_n \notin \mathrm{B}(q, n)$, i.e. $|p_n - q| \geqslant n$. The sequence (p_n) contains a subsequence (p_{m_n}) convergent to a point p. By 2.2, $|p_{m_n} - q| \to |p - q| < \infty$. On the other hand, $|p_{m_n} - q| \to \infty$ since $|p_{m_n} - q| \geqslant m_n$. Contradiction.

For the Euclidean space R^k the converse theorem is also true. Consequently,

6.2. *A subset A of R^k is compact if and only if it is closed and bounded.*

Suppose that the set $A \subset R^k$ is closed and bounded, i.e. A is a subset of an interval

$$I = \{(x_1, \ldots, x_k) \colon a_i \leqslant x_i \leqslant b_i \text{ for } i = 1, \ldots, k\}\,.$$

Consequently, for any sequence

$$p_n = (x_{n,1}, \ldots, x_{n,k})$$

of points in A, the sequences of the coordinates are bounded:

$$a_1 \leqslant x_{n,1} \leqslant b_1, \quad \ldots, \quad a_k \leqslant x_{n,k} \leqslant b_k \quad \text{for} \quad n = 1, 2, \ldots$$

By the Bolzano-Weierstrass theorem, every bounded sequence of real numbers contains a convergent subsequence. Applying this theorem to the sequence $(x_{n,1})$ we infer that there is an increasing sequence (m_n) of positive integers such that the sequence $(x_{m_n,1})$ converges. Applying this theorem to the sequence $(x_{m_n,2})$ we infer that there exists a subsequence (r_n) of the sequence (m_n) such that the sequence $(x_{r_n,2})$ converges. The sequence $(x_{r_n,1})$ converges also because it is a subsequence of the convergent sequence $(x_{m_n,1})$. Repeating this procedure k times we find an increasing sequence (s_n) of positive integers such that all the sequences

$$(x_{s_n,1}), \ \ldots, \ (x_{s_n,k})$$

converge, i.e. the subsequence (p_{s_n}) of (p_n) converges to a point p. Since $p_{s_n} \in A$ for $n = 1, 2, \ldots$ and since the set A is closed, we infer that $p \in A$. This proves that A is compact.

We say that a class of sets is a *covering* of a set A (or that it *covers* A) if A is a subset of the union of all the sets in the class. If all the sets in the class are open, the covering is said to be *open*.

6.3 (Heine-Borel). *Every countable open covering of a compact set A contains a finite subcovering.*

In other words, *if A is compact, $A_1, A_2, \ldots$ are open and $A \subset A_1 \cup A_2 \cup \ldots$, then there exists an integer n such that $A \subset A_1 \cup \ldots \cup A_n$.*

For suppose such an integer n does not exist, i.e. for every positive integer n there is a point $p_n \in A-(A_1 \cup \ldots \cup A_n)$. The sequence (p_n) contains a subsequence (p_{m_n}) converging to a point $p \in A$. The set $A-(A_1 \cup \ldots \cup A_n)$ is closed by 6.1 and 3.6. Since all the points p_{m_1}, $p_{m_2}, \ldots$, except perhaps a finite number of initial terms, are in this set, we infer that

$$p \in A-(A_1 \cup \ldots \cup A_n) \text{ for every positive integer } n .$$

This is not possible since $A \subset A_1 \cup A_2 \cup \ldots$

6.4. *If A is compact, G is open and $A \subset G$, then there is an $\eta > 0$ such that*

$$\{p\colon \operatorname{dist}(p, A) < \eta\} \subset G .$$

For if not, then for $\eta = 1/n$ $(n = 1, 2, \ldots)$ there would exist points p_n and q_n such that

$$p_n \in A , \quad q_n \in X-G \quad \text{and} \quad |p_n - q_n| < 1/n .$$

The sequence (p_n) contains a subsequence (p_{m_n}) convergent to a point $p \in A$. Since the distance from p_{m_n} to q_{m_n} tends to 0, we infer also that $q_{m_n} \to p$. On the other hand, the points q_{m_n} belong to the closed set $X-G$. Consequently $p \in X-G \subset X-A$ which is impossible since $p \in A$.

6.5. *Every closed subset of a compact set is compact.*

This follows directly from the definition.

EXERCISES

1. If (A_n) is a sequence of non-empty closed subsets, $A_{n+1} \subset A_n$ for $n = 1, 2, \ldots$, and one of these sets is compact, then $A_1 \cap A_2 \cap \ldots \neq 0$.

2. If $\operatorname{cl} A$ is compact, then for every $\varepsilon > 0$ there exists a finite sequence of sets $A_1, \ldots, A_m$ such that

$$A = A_1 \cup \ldots \cup A_m \quad \text{and} \quad \operatorname{dia} A_i < \varepsilon \quad \text{for} \quad i = 1, \ldots, m .$$

3. Let G be an open subset of the plane R^2 and let y_1, y_2 be fixed real numbers. The set of real numbers

$$\Big\{x\colon \bigwedge_y \big((y_1 \leqslant y \leqslant y_2) \Rightarrow ((x, y) \in G)\big)\Big\}$$

is open.

§ 7. The limit of a function at a point. A point p_0 in a metric space X is said to be a *limit point* of a set $A \subset X$ if $p_0 \in \operatorname{cl}(A - \{p_0\})$, i.e. if there exists a sequence (p_n) such that

$$(1) \quad p_n \in A \quad \text{and} \quad p_n \neq p_0 \quad \text{for} \quad n = 1, 2, \ldots, \quad \text{and} \quad p_n \to p_0 .$$

Let p_0 be a limit point of A and let f be a function from A into a metric space Y. A point $q_0 \in Y$ is said to be the *limit of f at p_0* provided that $f(p_n) \to q_0$ for every sequence (p_n) satisfying the conditions (1). We write then

$$q_0 = \lim_{p \to p_0} f(p)$$

or

$$f(p) \to q_0 \quad \text{for} \quad p \to p_0 .$$

7.1. *In order that* $q_0 = \lim_{p \to p_0} f(p)$ *it is necessary and sufficient that, for every* $\varepsilon > 0$, *there exists a* $\delta > 0$ *such that, for every* $p \in A$,

$$0 < |p - p_0| < \delta \quad \textit{implies} \quad |f(p) - q_0| < \varepsilon .$$

Suppose the function f has the property mentioned in 7.1. Let (p_n) satisfy the conditions (1). Then $0 < |p_n - p_0| < \delta$ for $n > n_0$, and consequently $|f(p_n) - q_0| < \varepsilon$ for $n > n_0$. This proves that $f(p_n) \to q_0$.

Suppose now that f does not have the property mentioned in 7.1. Then there is a number $\varepsilon > 0$ such that, for every positive integer n, there is a point $p_n \in A$ such that

$$0 < |p_n - p_0| < 1/n \quad \text{and} \quad |f(p_n) - q_0| \geqslant \varepsilon .$$

The sequence (p_n) has all the properties (1) but $(f(p_n))$ does not converge to q_0. Hence it follows that q_0 is not the limit of f at p_0.

Consider now the case where Y is one of the spaces mentioned in § 1, examples B), D), E).

Let $\Phi = (\varphi_1, \ldots, \varphi_k)$ be a mapping from $A \subset X$ into the Euclidean space R^k, let $q_0 = (y_1, \ldots, y_k) \in R^k$ and let p_0 be a limit point in A. For any sequence (p_n) with properties (1),

$$\Phi(p_n) \to q_0 \text{ if and only if } \varphi_i(p_n) \to y_i \text{ for } i = 1, \ldots, k$$

(see 2.4). In other words,

$$q_0 = \lim_{p \to p_0} \Phi(p) \text{ if and only if } y_i = \lim_{p \to p_0} \varphi_i(p) \text{ for } i = 1, \ldots, k .$$

By the same argument if $\Phi = [\varphi_1, \ldots, \varphi_k]$ is a k-dimensional vector field on A and $\mathfrak{a} = [a_1, \ldots, a_k] \in V_k$, then

$$\mathfrak{a} = \lim_{p \to p_0} \Phi(p) \text{ if and only if } a_i = \lim_{p \to p_0} \varphi_i(p) \text{ for } i = 1, \ldots, k .$$

Similarly, a matrix

$$\mathfrak{A} = [a_{i_1, \ldots, i_m}]_{i_1 \leqslant k_1, \ldots, i_m \leqslant k_m}$$

is the limit of a matrix field

$$\Phi = [\varphi_{i_1, \ldots, i_m}]_{i_1 \leqslant k_1, \ldots, i_m \leqslant k_m}$$

at p_0 if and only if

$$a_{i_1,\ldots,i_m} = \lim_{p \to p_0} \varphi_{i_1,\ldots,i_m}(p)$$

for $i_1 \leqslant k_1, \ldots, i_m \leqslant k_m$.

We summarize these results in the following theorem:

7.2. *A mapping into a Euclidean space (a vector field, a matrix field) has a limit at a point p_0 if and only if all its coordinates have finite limits at p_0. Then the coordinates of the limit of the mapping (vector field, matrix field) are the limits of the coordinates.*

EXERCISES

1. A closed set A is compact if and only if every infinite subset of A has a limit point belonging to A.

2. Let A be a set of real numbers and let x_0 be a fixed real number. Let

$$A^- = \{x \in A\colon x < x_0\}, \qquad A^+ = \{x \in A\colon x > x_0\}.$$

For any function $f\colon A \to Y$ the limit of the function $f|A^-$ at x_0 is called (if it exists) the *left-hand limit* of f at x_0 and is denoted by $f(x_0-)$, and the limit of the function $f|A^+$ at x_0 is called (if it exists) the *right-hand limit* of f at x_0 and is denoted by $f(x_0+)$.

Prove that if f is a *non-decreasing* function (i.e. $f(x_1) \leqslant f(x_2)$ if $x_1 < x_2$, $x_1, x_2 \in A$), then

$$f(x_0-) = \sup\{f(x)\colon x \in A^-\}, \qquad f(x_0+) = \inf\{f(x)\colon x \in A^+\}$$

under the hypothesis that x_0 is a limit point of A^- or A^+, respectively. Under the same hypothesis, if f is *non-increasing* (i.e. $f(x_1) \geqslant f(x_2)$ if $x_1 < x_2$, $x_1, x_2 \in A$), then

$$f(x_0-) = \inf\{f(x)\colon x \in A^-\}, \qquad f(x_0+) = \sup\{f(x)\colon x \in A^+\}.$$

As examples of the right-hand and left-hand limits we mention here the *right-hand* and the *left-hand derivatives*

$$f'(a) = \lim_{\Delta x \to 0+} \frac{f(a+\Delta x) - f(a)}{\Delta x}, \qquad f'(b) = \lim_{\Delta x \to 0-} \frac{f(b+\Delta x) - f(b)}{\Delta x}$$

of a function f defined on an interval $\langle a; b\rangle$.

§ 8. Continuous functions. Let f be a function from a subset A of a metric space X into a metric space Y.

f is said to be *continuous at a point* $p_0 \in A$ provided $f(p_n) \to f(q_0)$ for every sequence $p_n \to p_0$ ($p_n \in A$ for $n = 1, 2, \ldots$).

By definition, f is continuous at a limit point $p_0 \in A$ of the set A if and only if $f(p_0) = \lim_{p \to p_0} f(p)$.

A point $p_0 \in A$ which is not a limit point of A is said to be an *isolated point* of A. If p_0 is an isolated point of A and $p_n \to p_0$, $p_n \in A$ for $n = 1, 2, \ldots$, then $p_n = p_0$ for n sufficiently large and, consequently, $(p_n) = f(p_0)$ for sufficiently large n. This proves that $f(p_n) \to f(p_0)$. Thus any function on A is continuous at every isolated point of A.

Note that a point $p_0 \in A$ is an isolated point of A if and only if the one-point set $\{p_0\}$ is open in the metric space A (see § 4).

8.1. *A function f is continuous at a point p_0 if and only if for every $\varepsilon > 0$ there is a $\delta > 0$ such that, for every $p \in A$,*

$$|p - p_0| < \delta \quad \textit{implies} \quad |f(p) - f(p_0)| < \varepsilon .$$

The proof of 8.1 is the same as that of 7.1. It suffices to replace q_0 by $f(p_0)$ and to omit the conditions $0 < |p - p_0|$ and $p_n \neq p_0$.

A function from A into Y is said to be *continuous* if it is continuous at every point in A.

It follows directly from the definition that if a function f on A is continuous and $B \subset A$, then $f|B$ is also continuous.

A set A is said to be *isolated* provided every point $p_0 \in A$ is an isolated point of A (e.g. every finite set is isolated). It follows from the remark before Theorem 8.1 that every function whose domain is an isolated set is continuous.

Theorem 2.2 says that, for any non-empty subset A of a metric space X, the real function

$$f(q) = \operatorname{dist}(q, A) \qquad (q \in X)$$

is continuous; in particular, the function $f(q) = |q - p|$ (p fixed) is continuous.

8.2. *Each of the following conditions is necessary and sufficient for a function f from A into Y to be continuous:*

(a) *for every closed set $B \subset Y$, the set $f^{-1}(B)$ is closed in the metric space A;*

(b) *for every open set $B \subset Y$ the set $f^{-1}(B)$ is open in the metric space A.*

Suppose f is continuous. Let B be a closed subset of Y and let $p_n \in f^{-1}(B)$, $p_n \to p \in A$. Since $f(p_n) \in B$ for $n = 1, 2, \ldots$ and $f(p_n) \to f(p)$ by the continuity of f, we infer that $f(p) \in B$, i.e. $p \in f^{-1}(B)$. This proves that the set $f^{-1}(B)$ is closed in A, i.e. f has the property (a).

(a) implies (b) since if B is open in Y, then $Y - B$ is closed in Y, $f^{-1}(Y - B)$ is closed in A by (a) and consequently $f^{-1}(B) = A - f^{-1}(Y - B)$ is open in A.

Suppose f has the property (b). Let $p_0 \in A$. For any given $\varepsilon > 0$, the set $f^{-1}\{\mathrm{B}(f(p_0), \varepsilon)\}$ is open in A (see 3.7) and contains p_0. Consequently, there is a $\delta > 0$ such that, for every point $p \in A$,

$$\text{if } |p - p_0| < \delta, \text{ then } p \in f^{-1}\{\mathrm{B}(f(p_0), \varepsilon)\} ,$$

i.e. $|f(p) - f(p_0)| < \varepsilon$. By 8.1, this proves that f is continuous at every point $p_0 \in A$, i.e. f is continuous.

Note that a continuous image of an open set is not always open, and a continuous image of a closed set is not always closed. However,

8.3. *If B is a compact subset of the domain of a continuous function f, then the set $f(B)$ is compact, and therefore closed.*

Let (q_n) be a sequence of points in $f(B)$, i.e. $q_n = f(p_n)$ for a $p_n \in B$, $n = 1, 2, \ldots$ The sequence (p_n) contains a subsequence (p_{m_n}) converging to a point $p \in B$. The subsequence (q_{m_n}) (i.e. $(f(p_{m_n}))$) of (q_n) converges to the point $f(p) \in f(B)$. This proves that $f(B)$ is compact.

8.4. *The inverse f^{-1} of a continuous one-to-one function defined on a compact set A is continuous.*

Let $B = f(A)$ and $g = f^{-1}$: $B \to X$. For every closed set $C \subset X$ the set $g^{-1}(C) = f(A \cap C)$ is compact by 8.3 (see also 6.5) and therefore it is closed in the metric space B. By 8.2 (a) the function g is continuous.

8.5. *Let f: $A \to B$ and g: $B \to C$, the sets A, B, C being subsets of certain metric spaces. If f is continuous at $p_0 \in A$ and g is continuous at $f(p_0) \in B$, then the composition $g \circ f$ is continuous at p_0.*

Consequently, if both f and g are continuous, so is $g \circ f$.

Indeed, if $p_n \to p_0$, then $f(p_n) \to f(p_0)$ and consequently $g(f(p_n)) \to g(f(p_0))$, i.e. $g \circ f(p_n) \to g \circ f(p_0)$.

A function f from a subset A of a metric space X into a metric space Y is said to be *uniformly continuous* provided for every $\varepsilon > 0$ there exists $\delta > 0$ such that for any points $p, q \in A$

$$\text{if } |q-p| < \delta, \text{ then } |f(q)-f(p)| < \varepsilon .$$

To explain the difference between, ordinary continuity and uniform continuity, let us write both definitions by means of the notation of I § 1, p. 17. The function f is continuous if

$$\bigwedge_{\varepsilon>0} \bigwedge_{p\in A} \bigvee_{\delta>0} \bigwedge_{q\in A} \left(|q-p| < \delta \Rightarrow |f(q)-f(p)| < \varepsilon\right).$$

The function f is uniformly continuous if

$$\bigwedge_{\varepsilon>0} \bigvee_{\delta>0} \bigwedge_{p\in A} \bigwedge_{q\in A} \left(|q-p| < \delta \Rightarrow |f(q)-f(p)| < \varepsilon\right).$$

We see that the definitions differ by the transposition of the quantifiers $\bigwedge_{p\in A}$ and $\bigvee_{\delta>0}$. In the definition of continuity the number δ is chosen for ε and p and, consequently, it depends, in general, on both ε and p. In the definition of uniform continuity the number δ depends only on ε and not on p.

Every uniformly continuous function is continuous. The converse statement is not true, in general. However it is true for functions on compact sets:

8.6. *Every continuous function defined on a compact set is uniformly continuous.*

Theorem 8.6 is a particular case of the following general theorem (the case where $A = B$).

8.7. *If B is a compact subset of A, and if f is a continuous function on A, then for every $\varepsilon > 0$ there exists a $\delta > 0$ such that for every $p \in B$ and for every $q \in A$*

$$\textit{if } |q-p| < \delta, \textit{ then } |f(q)-f(p)| < \varepsilon .$$

Suppose the contrary, i.e. suppose that there exists an $\varepsilon > 0$ such that for every $\delta > 0$ there exist points $p \in B$ and $q \in A$ such that $|q-p| < \delta$ and $|f(q)-f(p)| \geqslant \varepsilon$. In particular for $\delta = 1/n$ there are points $p_n \in B$ and $q_n \in A$ such that

$$(1) \qquad |q_n - p_n| < 1/n \quad \text{and} \quad |f(q_n)-f(p_n)| \geqslant \varepsilon .$$

The sequence (p_n) contains a subsequence (p_{m_n}) convergent to a point $p_0 \in B$. It follows from the first of the inequalities (1) that also $q_{m_n} \to p_0$. By the continuity of f, $f(p_{m_n}) \to f(p_0)$ and $f(q_{m_n}) \to f(p_0)$. Hence

$$0 < \varepsilon \leqslant |f(q_{m_n})-f(p_{m_n})| \leqslant |f(q_{m_n})-f(p_0)| + |f(p_{m_n})-f(p_0)| \to 0$$

which is impossible.

Consider now the case where Y is one of the spaces mentioned in § 1, examples B), D), E).

Let $\Phi = (\varphi_1, \ldots, \varphi_k)$ be a mapping from a subset A of a metric space X into the k-dimensional Euclidean space R^k. For any sequence $p_n \to p_0$ $(p_n, p_0 \in A)$

$$\Phi(p_n) \to \Phi(p_0) \text{ if and only if } \varphi_i(p_n) \to \varphi_i(p_0) \text{ for } i = 1, \ldots, k$$

(see 2.4). Hence it follows that Φ is continuous at p_0 if and only if all its coordinates $\varphi_1, \ldots, \varphi_k$ are continuous at p_0. Consequently Φ is continuous if and only if all its coordinates are continuous.

Analogous statements are true for vector fields and matrix fields. We summarize the results in the following theorem.

8.8. *A mapping into a Euclidean space or a vector field or a matrix field is continuous (continuous at p_0) if and only if all its coordinates are continuous (continuous at p_0).*

An analogous theorem is also true for uniform continuity: a mapping into a Euclidean space or a vector field or a matrix field is uniformly continuous if and only if all its coordinates are uniformly continuous.

Now we shall prove two theorems on real functions.

8.9. *Every continuous real function defined on a compact set A is bounded, and there are points $p_1, p_2 \in A$ such that*

$$f(p_1) = \sup f(A), \quad f(p_2) = \inf f(A) .$$

Indeed, it follows from 8.3 that the set $f(A)$ is compact, i.e. bounded and closed (see 6.2). If B is a closed and bounded set of real numbers, then $\sup B \in B$ and $\inf B \in B$. Applying this remark to the set $B = f(A)$ we infer that the points p_1, p_2 mentioned in 8.9 exist.

8.10. *If f, g are continuous real functions on A, then the functions cf, $f+g$, $f-g$, $f\cdot g$, f/g (provided g does not vanish),* $\max(f, g)$ *and* $\min(f, g)$ *are continuous.*

Theorem 8.10 follows directly from the definition of continuity and known properties of convergent sequences of real numbers.

Now we shall give a few examples and make some additional remarks about continuous mappings.

A) The results of algebraic operations performed on continuous vector fields and matrix fields are continuous. More precisely, if φ is a continuous real function and Φ is a continuous matrix (or vector) field, then $\varphi\Phi$ is continuous. If Φ and Ψ are continuous matrix (or vector) fields of the same type, then $\Phi\pm\Psi$ and $\Phi\Psi$ are continuous. If Φ and Ψ are continuous rectangular matrix fields, then $\Phi\bullet\Psi$ is continuous provided the matrix multiplication is feasible. If $\Phi_1, \ldots, \Phi_k$ are continuous vector fields, then $\Phi_1\wedge\ldots\wedge\Phi_k$ and $\Phi_1\vee\ldots\vee\Phi_k$ are continuous, etc.

This follows directly from 8.10 and 8.8 (or can be deduced directly from § 2 (5)-(11)).

B) If $\Phi = [\varphi_{i_1,\ldots,i_r}]$ is a continuous matrix field on a set A, $\Phi(p) \neq 0$ for every $p \in A$, φ is a real function on A and the matrix field $\varphi\Phi$ is continuous, then φ is continuous.

Indeed, let $p_0 \in A$. Since $\Phi(p_0) \neq 0$, at least one coordinate of the matrix is different from zero, say $\varphi_{i_1,\ldots,i_r}(p_0) \neq 0$. Consequently the function $\varphi_{i_1,\ldots,i_r}$ does not vanish in a neighbourhood of p_0. Since the functions $\varphi\varphi_{i_1,\ldots,i_r}$ and $\varphi_{i_1,\ldots,i_r}$ are continuous, their quotient φ is continuous at p_0.

C) The determinant $/\mathfrak{A}/$ is a continuous real function defined on the metric space X of all square matrices of dimension k. Also the inverse matrix $\mathfrak{A}^{-1}$ is a continuous function on the set $A \subset X$ of all matrices whose determinant does not vanish.

The proof is based on the same argument as in A).

D) Let $\mathbf{i} = (i_1, \ldots, i_k)$ be a sequence of positive integers $\leqslant l$. The mapping $\Lambda_{\mathbf{i}}$: $R^l \to R^k$ defined by

$$\Lambda_{\mathbf{i}}(x_1, \ldots, x_l) = (x_{i_1}, \ldots, x_{i_k})$$

is uniformly continuous. In the case where all the integers $i_1, \ldots, i_k$ are different, $\Lambda_{\mathbf{i}}$ is said to be the *projection* (more precisely, the $\mathbf{i}$-*projection*).

E) Linear and multilinear forms and forms of degree m (see II § 6) are continuous. Linear and multilinear mappings from V_k^m into V_l (see II § 7) also are continuous.

F) Let $\mathfrak{A}$ be a k-dimensional m-indexed symmetric matrix and let $p_0 \in R^k$. The real function

$$\varphi(p) = \mathfrak{A}\cdot(p - p_0)^m \tag{2}$$

(i.e. the scalar product of $\mathfrak{A}$ and the mth tensor power of the vector $p - p_0$; see II § 1 (5) and II § 4 (22)) defined on R^k is continuous. Consequently the function $\varphi|A$ is continuous, where

$$A = \{p \in R^k\colon\ |p - p_0| = 1\}\ .$$

The set A is compact by 6.2. By 8.9 there exists a point $p_1 \in A$ such that

$$\inf\{\varphi(p)\colon\ p \in A\} = \varphi(p_1)\ .$$

Suppose now that the integer m is even and the matrix $\mathfrak{A}$ is positive (see II § 6, p. 71). Assuming $\mathfrak{a}_1 = p_1 - p_0$ we infer that $\varphi(p_1) = \mathfrak{A}\cdot\mathfrak{a}_1 > 0$ and consequently

$$\inf\{\mathfrak{A}\cdot\mathfrak{a}^m\colon\ (\mathfrak{a} \in V_k)\ \&\ (|\mathfrak{a}| = 1)\} = \inf\{\varphi(p)\colon\ p \in A\} > 0\ . \tag{3}$$

This inequality will be used in the proof of the following theorem.

8.11. *If a symmetric m-indexed k-dimensional matrix $\mathfrak{A}$ is positive, then there exists an $\varepsilon > 0$ such that every symmetric matrix $\mathfrak{B}$ of the same type, such that $\|\mathfrak{B} - \mathfrak{A}\| < \varepsilon$, is positive.*

The number

$$\varepsilon = \inf\{\mathfrak{A}\cdot\mathfrak{a}^m\colon\ |\mathfrak{a}| = 1\}$$

is positive by (3). If $\|\mathfrak{B} - \mathfrak{A}\| < \varepsilon$, then for every unit vector $\mathfrak{a}$

$$|(\mathfrak{B} - \mathfrak{A})\cdot\mathfrak{a}^m| \leqslant \|\mathfrak{B} - \mathfrak{A}\|\cdot\|\mathfrak{a}^m\| = \|\mathfrak{B} - \mathfrak{A}\|\cdot|\mathfrak{a}|^m < \varepsilon$$

by II § 4 (14) and (19″). Hence for every unit vector $\mathfrak{a}$

$$\mathfrak{B}\cdot\mathfrak{a}^m = \mathfrak{A}\cdot\mathfrak{a}^m + (\mathfrak{B} - \mathfrak{A})\cdot\mathfrak{a}^m > \varepsilon - \varepsilon = 0\ .$$

If $\mathfrak{a}$ is any non-zero vector in V_k, then $\dfrac{\mathfrak{a}}{|\mathfrak{a}|}$ is a unit vector and consequently

$$\mathfrak{B}\cdot\mathfrak{a}^m = |\mathfrak{a}|^m\cdot\left(\mathfrak{B}\cdot\left(\frac{\mathfrak{a}}{|\mathfrak{a}|}\right)^m\right) > 0\ .$$

This proves that $\mathfrak{B}$ is positive.

In the same way one can prove that if $\mathfrak{A}$ is negative, then there is an $\varepsilon > 0$ such that every symmetric matrix $\mathfrak{B}$ satisfying $\|\mathfrak{B} - \mathfrak{A}\| < \varepsilon$ is negative.

EXERCISES

1. The set of all points of discontinuity of a monotonic function f defined on an interval $I \subset R$ is countable.

Hint. Assign to every point x of discontinuity a rational number lying between $f(x-)$ and $f(x+)$.

2. Let X, Y be metric spaces and $f\colon X\to Y$. If every point $p\in X$ has a neighbourhood U such that the function $f|U$ is continuous, then f is continuous. In other words, if X is the union of open sets such that $f|U$ is continuous for each of these sets U, then f is continuous.

3. If $f\colon X\to Y$ and the metric space X is the union of a finite number of closed sets $F_1, \dots, F_n$ such that $f|F_i$ is continuous for $i=1, \dots, n$, then f is continuous.

4. Let f be a continuous function from a subset A of a metric space X into a metric space Y. Let B be the set of all the points $p\in X-A$ such that the limit $\lim\limits_{q\to p} f(q)$ exists. The mapping $g\colon A\cup B\to Y$ defined by

$$g(p)=\begin{cases} f(p) & \text{for} \quad p\in A\,,\\ \lim\limits_{q\to p} f(q) & \text{for} \quad p\in B\end{cases}$$

is continuous.

5. Let $f\colon R^2\to R$ be defined by

$$f(x_1, x_2)=\begin{cases} \dfrac{x_1x_2}{x_1^2+x_2^2} & \text{if} \quad x_1^2+x_2^2>0\,,\\ 0 & \text{if} \quad x_1=0=x_2\,.\end{cases}$$

Prove that for every x_1 the function $f(x_1, \cdot)$ is continuous, for every x_2 the function $f(\cdot, x_2)$ is continuous, but the function f is not continuous. In other words, f is a continuous function of each of the variables x_1, x_2 separately, but it is not a continuous function of the point $p=(x_1, x_2)$.

Hint. Examine the function f on the line $\{(x_1, x_2)\colon x_1=x_2\}$.

6. The set of all integers, the set of all positive integers, the set $\{1/n\colon n=1, 2, \dots\}$ are isolated. The set composed of the numbers 0 and $1/n$, $n=1, 2, \dots,$ is not isolated.

7. Let $X=(0;1)$. The mapping $f(x)=(x, 0)$ of the metric space X into the metric space R^2 is continuous. The set X is open and closed in the space X, but the set $f(X)$ is neither open nor closed in the space R^2.

8. Let $X=(0;1)\cup\{2\}$ and $Y=(0;1\rangle$. The mapping

$$f(x)=\begin{cases} x & \text{if} \quad 0<x<1\,,\\ 1 & \text{if} \quad x=2\end{cases}$$

from X onto Y is continuous and one-to-one. The inverse mapping

$$f^{-1}(x)=\begin{cases} x & \text{if} \quad 0<x<1\,,\\ 2 & \text{if} \quad x=1\end{cases}$$

is not continuous. The set $A=(0;1)$ is closed in X but its image $f(A)$ is not closed in Y. The one-point set $B=\{2\}$ is open in X but its image $f(B)$ is not open in Y.

9. Any continuous additive function $F\colon V_k\to R$ is homogeneous, i.e. linear.

Hint. If F is additive, then $F(a\mathfrak{a})=aF(\mathfrak{a})$ for every rational number a.

§ 9. Convergence of sequences of functions. Let f_n $(n = 1, 2, \ldots)$ and f be functions from a set A into a metric space Y. We say that the sequence (f_n) *converges* to f (or that f is the *limit* of (f_n)) and we write

$$f = \lim_{n\to\infty} f_n \quad \text{or} \quad f_n \to f \tag{1}$$

if $f_n(p) \to f(p)$ for every $p \in A$, i.e. if

$$\bigwedge_{\varepsilon>0} \bigwedge_{p\in A} \bigvee_{n_0} \bigwedge_{n>n_0} |f_n(p) - f(p)| < \varepsilon .$$

We say that the sequence (f_n) *converges uniformly* to f if for every $\varepsilon > 0$ there exists an integer n_0 such that

$$|f_n(p) - f(p)| < \varepsilon \qquad \text{for every } p \in A \text{ and every } n > n_0;$$

in symbols

$$\bigwedge_{\varepsilon>0} \bigvee_{n_0} \bigwedge_{p\in A} \bigwedge_{n>n_0} |f_n(p) - f(p)| < \varepsilon .$$

We also use then the notation (1) but with the addition of the word "uniformly".

The definition of convergence and uniform convergence differ by a transposition of the quantifiers $\bigwedge_{p\in A}$ and $\bigvee_{n_0}$. In the definition of ordinary convergence (also called *pointwise convergence*) the integer n_0 is chosen for a given ε and p; it depends in general on ε and on p. In the definition of uniform convergence the integer n_0 is chosen for a given ε only; it depends only on ε but does not depend on p.

Uniform convergence implies pointwise convergence. The converse statement is not true, in general.

9.1. *A sequence (f_n) of functions from A into a complete metric space Y converges uniformly to a function $f\colon A \to Y$ if and only if it satisfies the following uniform Cauchy condition: for every $\varepsilon > 0$ there is an integer n_0 such that $|f_n(p) - f_m(p)| < \varepsilon$ for every $p \in A$ and all integers $m, n > n_0$.*

Suppose that $f_n \to f$ uniformly. For given $\varepsilon > 0$ there is an integer n_0 such that

$$|f_n(p) - f(p)| < \varepsilon/2$$

for $p \in A$ and $n > n_0$. Consequently

$$|f_m(p) - f(p)| < \varepsilon/2$$

for $p \in A$ and $m > n_0$. Hence it follows that for every $p \in A$ and any $m, n > n_0$

$$|f_n(p) - f_m(p)| \leqslant |f_n(p) - f(p)| + |f_m(p) - f(p)| < \varepsilon ,$$

i.e. the Cauchy condition holds.

Suppose now that (f_n) satisfies the uniform Cauchy conditions. Then, for every point $p \in A$, the sequence $(f_n(p))$ of points of the metric space Y satisfies the Cauchy condition. Since the space Y is complete, the sequence $(f_n(p))$ converges to a point which will be denoted by $f(p)$. For given $\varepsilon > 0$ let n_0 be an integer such that

$$|f_n(p) - f_m(p)| < \varepsilon/2 \quad \text{for all } p \in A \text{ and all } m, n > n_0 .$$

Passing to the limit for $m \to \infty$ we get (see 2.2)

$$|f_n(p) - f(p)| \leqslant \varepsilon/2 < \varepsilon \quad \text{for every } p \in A \text{ and every } n > n_0 .$$

Thus $f_n \to f$ uniformly.

Suppose now that the set A on which all the functions f_n, f are defined is a subset of a metric space X.

9.2. *If $f_n \to f$ uniformly and all the functions f_n are continuous at a point p_0, then the function f is continuous at p_0.*

Consequently, the limit of a uniformly convergent sequence of continuous functions is continuous.

Given any $\varepsilon > 0$, let n_0 be an integer such that

$$|f_n(p) - f(p)| < \varepsilon/3 \quad \text{for every } p \in A \text{ and every } n \geqslant n_0 .$$

In particular,

$$|f_{n_0}(p) - f(p)| < \varepsilon/3 \quad \text{for every } p \in A$$

and

$$|f_{n_0}(p_0) - f(p_0)| < \varepsilon/3 .$$

Since f_{n_0} is continuous at p_0, there is a $\delta > 0$ such that

$$|p - p_0| < \delta \quad \text{implies} \quad |f_{n_0}(p) - f_{n_0}(p_0)| < \varepsilon/3 .$$

It follows from the last three inequalities that $|p - p_0| < \delta$ implies

$$|f(p) - f(p_0)| \leqslant |f(p) - f_{n_0}(p)| + |f_{n_0}(p) - f_{n_0}(p_0)| + |f_{n_0}(p_0) - f(p_0)| < \varepsilon$$

which proves the continuity of f at p_0.

9.3. *If a sequence $f_n \colon A \to Y$ converges uniformly to a continuous function $f \colon A \to Y$, the set A is compact, and g is a continuous function on Y, then $g \circ f_n \to g \circ f$ uniformly.*

By 8.3 the set $f(A)$ is compact. By 8.7, for any given $\varepsilon > 0$ there is a $\delta > 0$ such that

$$\text{if } q \in f(A),\ q_1 \in Y \text{ and } |q_1 - q| < \delta, \text{ then } |g(q_1) - g(q)| < \varepsilon .$$

There is an integer n_0 such that

$$|f_n(p) - f(p)| < \delta \quad \text{for all } n > n_0 \text{ and } p \in A .$$

Consequently

$$|g(f_n(p))-g(f(p))|<\varepsilon \quad \text{for all } n>n_0 \text{ and } p\in A .$$

9.4. *A sequence Φ_n: $X\to R^l$, or a sequence Φ_n: $X\to V_l$, converges uniformly (converges uniformly to a Φ: $X\to R^l$ or a Φ: $X\to V_l$) if and only if the sequences of coordinates of Φ_n converge uniformly (converge uniformly to the corresponding coordinates of Φ).*

This follows directly from the inequalities II § 1 (40)-(41).

An analogous theorem is true for sequences of matrix fields.

EXERCISES

1. A mapping f: $X\to Y$ (where Y is a metric space) is said to be *bounded* if the set $f(X)$ is bounded. Let $\mathbf{B}(X, Y)$ be the set of all bounded mappings f: $X\to Y$. Define the distance between functions $f, g\in\mathbf{B}(X, Y)$ by

$$|f-g| = \sup\{|f(p)-g(p)|:\ p\in X\} .$$

Prove that the distance just defined has properties § 1 (1)-(3), i.e. that $\mathbf{B}(X, Y)$ is a metric space. Prove that $f_n\to f$ in the metric space $\mathbf{B}(X, Y)$ if and only if $f_n\to f$ uniformly.

2. Let $f, f_n\in\mathbf{B}(X, Y)$. Prove that if $f_n\to f$ uniformly, then $\operatorname{dia} f_n(X)\to\operatorname{dia} f(X)$ (in other words, $\operatorname{dia} f(X)$ is a continuous function on the metric space $\mathbf{B}(X, Y)$).

3. Let X, Y be metric spaces, let f_n: $X\to Y$ for $n=1, 2, \ldots$ and let f: $X\to Y$ be continuous.

If $f_n\to f$ uniformly, then

(2) $p_n\to p$ implies $f_n(p_n)\to f(p)$ for every convergent sequence (p_n) of points in X.

Conversely, if X is compact and (2) holds, then $f_n\to f$ uniformly.

§ 10. Homeomorphism. A mapping f of a subset A of a metric space X onto a subset B of a metric space Y is said to be a *homeomorphism* if it is one-to-one and both the functions f and f^{-1} are continuous. Then f^{-1} is also a homeomorphism.

Two sets A, B are said to be *homeomorphic* if there exists a homeomorphism from A onto B.

It follows from 8.4 that every one-to-one continuous function defined on a compact set is a homeomorphism.

If f is a homeomorphism from A onto B and g is a homeomorphism from B onto C, then $g\circ f$ is a homeomorphism from A onto C (see 8.5 and I § 3 (12)).

10.1. *Each of the following conditions is necessary and sufficient for a one-to-one mapping of a metric space X onto a metric space Y to be a homeomorphism:*

(a) *for any sequence (p_n) of points in X,*

$$p_n\to p \text{ in } X \text{ if and only if } f(p_n)\to f(p) \text{ in } Y ;$$

(b) *for every set* $A \subset X$,

$$A \text{ is open in } X \text{ if and only if } f(A) \text{ is open in } Y;$$

(c) *for every set* $A \subset X$,

$$A \text{ is closed in } X \text{ if and only if } f(A) \text{ is closed in } Y;$$

(d) *for every set* $A \subset X$,

$$f(\mathrm{cl}A) = \mathrm{cl}f(A);$$

(e) *for every set* $A \subset X$,

$$f(\mathrm{int}A) = \mathrm{int}f(A).$$

If f is a homeomorphism, then the implication

$$\text{if } p_n \to p, \text{ then } f(p_n) \to f(p)$$

follows from the continuity of f. The converse implication

$$\text{if } f_n(p) \to f(p), \text{ then } p_n \to p$$

follows similarly from the continuity of f^{-1}. Thus (a) holds.

(a) implies (d). In fact, (a) implies that $p \in \mathrm{cl}A$ (i.e. p is the limit of a sequence of points in A) if and only if $f(p) \in \mathrm{cl}f(A)$.

(d) implies (c) since, by (d), $A = \mathrm{cl}A$ if and only if $f(A) = \mathrm{cl}f(A)$.

If (c) holds, then f is a homeomorphism. Indeed, if A is a closed subset of X, then $(f^{-1})^{-1}(A) = f(A)$ is a closed subset of Y, which proves that the mapping $f^{-1}\colon Y \to X$ is continuous. Similarly, if $B \subset Y$ is closed, then the set $A = f^{-1}(B)$ is closed, since $f(A) = B$. This proves that f is continuous.

To complete the proof, observe that (b) is equivalent to (c) by 3.4, and (d) is equivalent to (e) by § 3 (4)-(5′).

Subsets A, B of metric spaces X, Y respectively are said to be *isometric* if there is a mapping f from A onto B such that

$$|f(p_1) - f(p_2)| = |p_1 - p_2| \quad \text{for all points } p_1, p_2 \in A.$$

Any mapping f having this property is said to be an *isometry*.

Every isometry is a homeomorphism since it has the property (a). Thus isometric sets are homeomorphic.

For instance, the k-dimensional Euclidean space R^k is isometric to the space V_k of all k-dimensional vectors, the isometry being given by the formula

$$f(x_1, \ldots, x_k) = [x_1, \ldots, x_k].$$

The space of all matrices of a fixed type $k_1, \ldots, k_r$ (see § 1 E)) is isometric with the $k_1 \cdot \ldots \cdot k_r$-dimensional Euclidean space.

Properties of sets and spaces which are invariant under homeomorphisms (i.e. such that if one of two homeomorphic sets has the property, the other has it, too) are said to be *topological*. Properties invariant under isometries are said to be *metric* properties. Of course every topological property is a metric property. The converse statement is not true, in general.

EXERCISES

1. Prove that the circle $\{(x_1, x_2): x_1^2+x_2^2 = 1\}$ is not homeomorphic with any subset of the set R of real numbers.

2. Prove that the set of all vertices of a pyramid is not isometric to any subset of the plane R^2.

3. Prove that any open disc $B(p_0, \eta) \subset R^2$, any open interval $I \subset R^2$ and any open triangle (i.e. triangle without its boundary) are homeomorphic. Prove that every closed disc $\{p: |p-p_0| \leqslant \eta\} \subset R^2$, any closed interval $I \subset R^2$ and every closed triangle are homeomorphic. No set mentioned in the first sentence is homeomorphic to any set mentioned in the second sentence.

4. Prove that no one-dimensional closed interval is homeomorphic with any two-dimensional closed interval.

§ 11. Connected sets. A subset A of a metric space is said to be *connected* if every real continuous function on A has the following property:

(a) if $f(p_1) < c < f(p_2)$, $p_1, p_2 \in A$, then there is a point $p \in A$ such that $f(p) = c$.

According to this definition, the empty set is connected.

By definition, a subset A of a metric space X is connected if and only if, for any continuous real function f on A, the set $f(A)$ of real numbers is one of the following sets:

$$(1)\quad \left\{\begin{array}{c}\text{the empty set},\\ \text{a one-point set},\\ \text{an (open, closed or half-closed, bounded or unbounded) interval}.\end{array}\right.$$

These and only these sets of real numbers are connected.

11.1. *Every convex subset $A \subset R^k$ (in particular, every open ball, and every open or closed interval) is connected.*

Suppose f is a continuous real function defined on A, $f(p_1) < c < f(p_2)$, $p_1, p_2 \in A$. Since A is convex, we have $p_1+t(p_2-p_1) \in A$ for $0 \leqslant t \leqslant 1$. Consequently

$$g(t) = f\big(p_1+t(p_2-p_1)\big) \qquad (0 \leqslant t \leqslant 1)$$

is a continuous real function on the interval $\langle 0; 1\rangle$. Since $g(0) = f(p_1) < c < f(p_2) = g(1)$, by a known theorem there is a number t_0 in this interval, such that $g(t_0) = c$. Thus $f(p) = c$ for $p = p_1+t_0(p_2-p_1) \in A$.

11.2. *Any continuous image of a connected set is connected.*

Suppose A is a connected subset of a metric space X, and g is a continuous function from A into a metric space Y. Let f be a continuous real function on $g(A)$ and $f(p_1) < c < f(p_2)$ for $p_1, p_2 \in g(A)$. There are points $q_1, q_2 \in A$ such that $g(q_1) = p_1$ and $g(q_2) = p_2$. The real function $f \circ g$ on A is continuous and $f \circ g(q_1) < c < f \circ g(q_2)$. Thus, there is a point $q \in A$ such that $f \circ g(q) = c$, i.e. $f(p) = c$ for $p = g(q) \in g(A)$. This proves that the set $g(A)$ is connected.

The following simple remark will be useful.

11.3. *If the values of a continuous function defined on a connected set are integers, then the function is constant.*

The converse statement is also true: if every integer-valued continuous function defined on A is constant, then A is connected.

11.4. *The union of any number of connected sets having a common point is connected.*

Suppose that $p \in A_t$ for every $t \in T$ and that all the sets A_t are connected. Let f be a continuous real function defined on $A = \bigcup_{t \in T} A_t$. All the sets $f(A_t)$ are of the form (1) and have the common point $f(p)$. Hence it follows that the union of all the sets, i.e. the set $f(A)$, is also of the form (1). This proves that A is connected.

Let A be a subset of a metric space X. For any given point $p \in A$ the union of all connected subsets of A which contain the point p, is connected by 11.4. This union is the greatest connected subset of A which contains the point p. The greatest connected subset containing p is called the *component* of A, containing p.

Let A_1 and A_2 be components of A, containing points p_1 and p_2 respectively. Then

$$\text{either } A_1 = A_2 , \quad \text{or} \quad A_1 \cap A_2 = 0 .$$

Indeed, suppose that $A_1 \cap A_2 \neq 0$. By 11.4 the set $A_1 \cup A_2$ is connected. Since it contains the point p_1, we have $A_1 \cup A_2 \subset A_1$, i.e. $A_2 \subset A_1$. Since it contains the point p_2, we have $A_1 \cup A_2 \subset A_2$, i.e. $A_1 \subset A_2$. This proves that $A_1 = A_2$.

We see that every set A is the union of its disjoint components. A non-empty set is connected if and only if it has only one component.

11.5. *Let f be a homeomorphism of a subset A of a metric space X onto a subset B of a metric space Y. If A_1 is a component of A, then $f(A_1)$ is a component of B.*

Suppose that A_1 is the component containing p. Let B_1 be the component containing the point $f(p)$. Since $f(A_1)$ is connected (see 11.2) and $f(p) \in f(A_1)$, we have $f(A_1) \subset B_1$. Since $f^{-1}(B_1)$ is connected (by Theorem 11.2

applied to the function f^{-1}) and $p \in f^{-1}(B_1)$, we infer that $f^{-1}(B_1) \subset A_1$. It follows from these two inclusions that $f(A_1) = B_1$.

11.6. *The union of at least two non-empty disjoint open sets is not connected.*

Let A be one of the open disjoint sets and let B be the union of all the remaining open sets. Since A and B are open, disjoint, and non-empty, the real function f defined on $A \cup B$ by

$$f(p) = \begin{cases} 1 & \text{if} \quad p \in A\,, \\ 0 & \text{if} \quad p \in B \end{cases}$$

is continuous and has exactly two values: the integers 0 and 1. By 11.3 the set $A \cup B$ is not connected.

By a *polygonal path* in R^k we mean any set

$$K_1 \cup \ldots \cup K_n$$

such that $K_1, \ldots, K_n$ are segments, K_i and K_{i+1} have only one common point, which is their end point, and K_i and K_j are disjoint if $j > i+1$. Applying 11.4 we infer, step by step, that the sets $K_1 \cup K_2$, $K_1 \cup K_2 \cup \cup K_3, \ldots$ are connected. Thus every polygonal path is connected.

11.7. *In order that an open set $G \subset R^k$ be connected it is necessary and sufficient that any two points $q_1, q_2 \in G$ can be joined by a polygonal path lying in G.*

The condition formulated in 11.7 is sufficient since it implies that the whole set G is the union of polygonal paths whose one end point is a given point $p_0 \in G$. By 11.4, G is connected.

The condition is necessary since if it is not satisfied, there exists a point $p_0 \in G$ such that the set A of all points in G, which can be joined to p_0 by a polygonal path lying in G, is a proper subset of G. It is easy to see that the disjoint non-empty sets A and $G-A$ are open. By 11.6, their union G is not connected.

EXERCISES

1. Components of the set of all rational numbers are one-point sets.

2. A subset A of a metric space X is said to be *open-closed* if it is simultaneously open and closed. For instance, the empty set and the whole space X are open-closed.

Prove that the space X is connected if and only if there are no other open-closed subsets.

3. If X is a compact metric space, then the component of X containing a given point p is the intersection of all the open-closed subsets containing the point p.

CHAPTER IV

DIFFERENTIATION OF REAL FUNCTIONS

§ 1. The directional derivative. We recall that if f is a real function defined on an open set G_0 of real numbers, then by the *derivative* $f'(t)$ *of* f *at a point* $t \in G_0$ we mean the number

$$f'(t) = \lim_{\Delta t \to 0} \frac{f(t+\Delta t) - f(t)}{\Delta t}$$

provided that the limit exists and is finite. By the derivative f' of f we mean the real function which assigns the number $f'(t)$ to every point t. The domain of f' is the set of all the points $t \in G_0$ for which $f'(t)$ exists. Instead of "derivative" we shall often say *ordinary derivative* to distinguish this notion from other notions which will be defined in this chapter and whose name will also contain the word "derivative".

Let φ be a real function defined on an open subset G of the k-dimensional Euclidean space R^k. Let $p \in G$ and let $\mathfrak{a}$ be any vector in R^k. By the *derivative of* φ *at* p *in the direction* $\mathfrak{a}$ we mean the number

(1) $$\varphi'_{\mathfrak{a}}(p) = \lim_{\Delta t \to 0} \frac{\varphi(p + \Delta t \cdot \mathfrak{a}) - \varphi(p)}{\Delta t},$$

provided that the limit exists and is finite. The *derivative of* φ *in the direction* $\mathfrak{a}$ is the real function $\varphi'_{\mathfrak{a}}$ which assigns the number (1) to every point p for which the limit (1) exists and is finite. Sometimes we use the words the *directional derivative* instead of "the derivative in the direction $\mathfrak{a}$" if it is not necessary to mention the vector $\mathfrak{a}$. In what follows, we shall be interested only in the case where the whole set G is the domain of $\varphi_{\mathfrak{a}}$ (i.e. if the limit (1) exists and is finite for every $p \in G$) except when the contrary is explicitely stated.

The calculation of the directional derivative $\varphi'_{\mathfrak{a}}(p)$ can be reduced to the calculation of the ordinary derivative of an auxiliary function $\overline{\varphi}$ of one real variable. For this purpose, let

(2) $$\overline{\varphi}(t) = \varphi(p + t\mathfrak{a}) .$$

The function $\overline{\varphi}$ is defined on the set

$$G_0 = \{t \in R\colon\ p + t\mathfrak{a} \in G\}$$

which is open on account of III 8.2 (b). The number 0 is in G_0. Since

$$\frac{\varphi(p+\Delta t\cdot\mathfrak{a})-\varphi(p)}{\Delta t}=\frac{\overline{\varphi}(\Delta t)-\overline{\varphi}(0)}{\Delta t},$$

we have

(3) $$\varphi'_{\mathfrak{a}}(p)=\overline{\varphi}'(0)\,.$$

More generally, for every $t\in G_0$

(4) $$\varphi'_{\mathfrak{a}}(p+t\mathfrak{a})=\overline{\varphi}'(t)\,,$$

since the directional derivative $\varphi'_{\mathfrak{a}}$ at the point $q=p+t\mathfrak{a}$ is equal, by the definition (1), to

$$\varphi'_{\mathfrak{a}}(p+t\mathfrak{a})=\varphi'_{\mathfrak{a}}(q)=\lim_{\Delta t\to 0}\frac{\varphi(q+\Delta t\cdot\mathfrak{a})-\varphi(q)}{\Delta t}=\lim_{\Delta t\to 0}\frac{\overline{\varphi}(t+\Delta t)-\overline{\varphi}(t)}{\Delta t}=\overline{\varphi}'(t)\,.$$

The identity (4) should be interpreted as follows: the left-hand side exists if and only if the right-hand side exists and then they are equal.

In the case where $k=1$ and $\mathfrak{a}$ is the unit vector whose only coordinate is equal to 1, then $\varphi'_{\mathfrak{a}}=\varphi'$.

As an example, we shall calculate the derivative of the function

$$\varphi(x_1,\ldots,x_k)=\sum_{i=1}^{k}x_i^2$$

at the point $p=(1,2,\ldots,k)$ in the direction $\mathfrak{a}=[1,3,5,\ldots,2k-1]$. Since

$$\overline{\varphi}(t)=\sum_{i=1}^{k}(i+(2i-1)t)^2,$$

we have

$$\overline{\varphi}'(t)=\sum_{i=1}^{k}2(2i-1)(i+(2i-1)t)\,.$$

Consequently

$$\varphi'_{\mathfrak{a}}(p)=\overline{\varphi}'(0)=\sum_{i=1}^{k}2(2i-1)i\,.$$

Since $\varphi'_{\mathfrak{a}}$ is equal to the ordinary derivative of the auxiliary function $\overline{\varphi}$, the directional derivative has, roughly speaking, the same properties as the ordinary derivative of a function of a real variable. For instance, we have the following identities which follow directly from the known identities for the auxiliary functions of a real variable:

(5)
$$\begin{gathered}(\varphi+\psi)'_{\mathfrak{a}}=\varphi'_{\mathfrak{a}}+\psi'_{\mathfrak{a}}\,,\\ (\varphi-\psi)'_{\mathfrak{a}}=\varphi'_{\mathfrak{a}}-\psi'_{\mathfrak{a}}\,,\\ (c\varphi)'_{\mathfrak{a}}=c\varphi'_{\mathfrak{a}}\,,\\ (\varphi\cdot\psi)'_{\mathfrak{a}}=\varphi'_{\mathfrak{a}}\cdot\psi+\varphi\cdot\psi'_{\mathfrak{a}}\,,\\ \left(\frac{\varphi}{\psi}\right)'_{\mathfrak{a}}=\frac{\varphi'_{\mathfrak{a}}\cdot\psi-\varphi\cdot\psi'_{\mathfrak{a}}}{\psi^2}\end{gathered}$$

under the hypothesis that the denominator does not vanish. All the identities should be understand as follows: if the directional derivatives on the right-hand side exist, then the directional derivative on the left-hand side exists and the equality holds.

We shall use the identity (4) in the proof of the following *mean-value theorem* for directional derivatives.

1.1. *If φ is a real function on G, the segment S joining the points p and $p+\lambda\mathfrak{a}$ is contained in G, and the directional derivative $\varphi'_{\mathfrak{a}}$ exists on S, then there is a number θ in the interval* $(0;1)$ *such that*

$$\varphi(p+\lambda\mathfrak{a})-\varphi(p)=\lambda\varphi'_{\mathfrak{a}}(p+\theta\lambda\mathfrak{a})\,. \tag{6}$$

The auxiliary function $\overline{\varphi}$ is defined an the closed interval whose end-points are the numbers 0 and λ, and it is differentiable in this interval. By the Lagrange mean-value theorem, there is a number θ in the interval $(0;1)$ such that

$$\overline{\varphi}(\lambda)-\overline{\varphi}(0)=\lambda\overline{\varphi}'(\theta\lambda)\,.$$

Since

$$\overline{\varphi}(\lambda)=\varphi(p+\lambda\mathfrak{a})\,,\quad \overline{\varphi}(0)=\varphi(p)\quad\text{and}\quad \overline{\varphi}'(\theta\lambda)=\varphi'_{\mathfrak{a}}(p+\theta\lambda\mathfrak{a})\,,$$

the last equation is identical to (6).

EXERCISES

1. Calculate the directional derivative of the function

$$\varphi(x_1,x_2,x_3)=e^{x_3/x_1}\sin x_2$$

at the point $p=(3,0,-1)$ in the direction $\mathfrak{a}=[2,-5,7]$.

2. The directional derivative of the function

$$\varphi(x_1,\dots,x_k)=\begin{vmatrix}1 & 1 & \dots & 1\\ x_1 & x_2 & \dots & x_k\\ x_1^2 & x_2^2 & \dots & x_k^2\\ \dots & \dots & \dots & \dots\\ x_1^{k-1} & x_2^{k-1} & \dots & x_k^{k-1}\end{vmatrix}$$

at any point $p=(x_1,\dots,x_k)$ in the direction $\mathfrak{a}=[1,1,\dots,1]$ is equal to 0.

§ 2. Partial derivatives. Let φ be a real function defined on an open subset of R^k and let $\mathfrak{e}_1,\dots,\mathfrak{e}_k$ be the unit vectors of the coordinate axes in R^k, i.e.

$$\begin{aligned}\mathfrak{e}_1&=[1,0,0,\dots,0]\,,\\ \mathfrak{e}_2&=[0,1,0,\dots,0]\,,\\ &\dots\dots\dots\dots\\ \mathfrak{e}_k&=[0,0,\dots,0,1]\,.\end{aligned}$$

The directional derivatives

$$\varphi'_{e_1}, \ldots, \varphi'_{e_k}$$

are called the *partial derivatives of* φ. More precisely, φ'_{e_j} is the *partial derivative of* φ *with respect to the* j*-th coordinate* (or j*-th variable*). To simplify the notation, the partial derivative φ'_{e_j} will be denoted by

$$(1) \qquad \varphi_{|j}\,.$$

Thus

$$(1') \qquad \varphi_{|j}(p) \quad \text{or} \quad \varphi_{|j}(x_1, \ldots, x_k)$$

will denote the value of the partial derivative $\varphi_{|j}$ at a point $p = (x_1, \ldots, x_k)$.

We write a vertical bar before j to indicate that the index on the right-hand side of the vertical bar denotes a partial differentiation. This is especially necessary if the symbol denoting the differentiated function contains lower indices. For instance, $\varphi_{i|j}$ denotes the partial derivative of a function φ_i with respect to the jth coordinate, the symbol $\varphi_{i,j|r}$ denotes the partial derivative of a function $\varphi_{i,j}$ with respect to the rth coordinate, etc.

It follows from the definition of directional derivative in §1 that the partial derivative of a function φ with respect to the jth coordinate at a point $p = (x_1, \ldots, x_k)$ is given by

$$\varphi_{|j}(x_1, \ldots, x_k) = \varphi_{|j}(p) = \varphi'_{e_j}(p) = \lim_{\Delta t \to 0} \frac{\varphi(p + \Delta t \cdot e_j) - \varphi(p)}{\Delta t}$$

$$= \lim_{\Delta t \to 0} \frac{\varphi(x_1, \ldots, x_{j-1}, x_j + \Delta t, x_{j+1}, \ldots, x_k) - \varphi(x_1, \ldots, x_{j-1}, x_j, x_{j+1}, \ldots, x_k)}{\Delta t}.$$

Writing Δx_j instead of Δt we obtain

$$(2) \qquad \varphi_{|j}(x_1, \ldots, x_k)$$

$$= \lim_{\Delta x_j \to 0} \frac{\varphi(x_1, \ldots, x_{j-1}, x_j + \Delta x_j, x_{j+1}, \ldots, x_k) - \varphi(x_1, \ldots, x_{j-1}, x_j, x_{j+1}, \ldots, x_k)}{\Delta x_j}.$$

To understand the meaning of (2), let us consider the expression $\varphi(x_1, \ldots, x_k)$ as a function

$$(3) \qquad f(x_j) = \varphi(x_1, \ldots, x_{j-1}, x_j, x_{j+1}, \ldots, x_k)$$

of one variable x_j only, the $x_1, \ldots, x_{j-1}, x_{j+1}, \ldots, x_k$ being fixed. The right-hand side of (2) is the limit

$$\lim_{\Delta x_j \to 0} \frac{f(x_j + \Delta x_j) - f(x_j)}{\Delta x_j}.$$

We see that the partial derivative $\varphi_{|j}$ of φ is the ordinary derivative of the function f defined by (3):

$$\varphi_{|j}(x_1, \ldots, x_k) = f'(x_j)\,. \tag{4}$$

Identity (4) yields a practical method of calculating the partial derivative $\varphi_{|j}$ in the case where the function φ is defined by an explicit formula as the result of some algebraic, trigonometrical, etc., operations on coordinates $x_1, \ldots, x_k$. For this purpose we treat the expression $\varphi(x_1, \ldots, x_k)$ as a function of one variable x_j only (the remaining variables being treated as fixed parameters) and differentiate this function of one real variable in the ordinary way, according to the known rules for the differentiation of functions of one real variable.

For instance, if $\varphi(x_1, \ldots, x_k) = x_1 x_2^2 \ldots x_k^k$, then

$$\varphi_{|j}(x_1, \ldots, x_k) = j \cdot x_1 x_2^2 \ldots x_{j-1}^{j-1} x_j^{j-1} x_{j+1}^{j+1} \ldots x_k^k\,.$$

Since partial differentiation is ordinary differentiation with respect to one variable, it has, roughly speaking, the same properties that the ordinary differentiation has. For instance,

$$\begin{aligned} (\varphi + \psi)_{|j} &= \varphi_{|j} + \psi_{|j}\,, \\ (\varphi - \psi)_{|j} &= \varphi_{|j} - \psi_{|j}\,, \\ (c\varphi)_{|j} &= c\varphi_{|j}\,, \\ (\varphi \cdot \psi)_{|j} &= \varphi_{|j} \cdot \psi + \varphi \cdot \psi_{|j}\,, \\ \left(\frac{\varphi}{\psi}\right)_{|j} &= \frac{\varphi_{|j} \cdot \psi - \varphi \cdot \psi_{|j}}{\psi^2} \end{aligned} \tag{5}$$

under the hypothesis that the denominator does not vanish. Formulas (5), which are a particular case of those in § 1 (5), should be understood as follows: if the partial derivatives on the right-hand side exist, then the partial derivative on the left-hand side exists and the equality holds.

Note that in the case of $k = 1$, i.e. in the case where φ is a real function of one real variable, the only partial derivative $\varphi_{|1}$ of φ coincides with the ordinary derivative φ'.

The notation traditionally used for partial derivatives differs from (1) and (1′). If the symbols $x_1, \ldots, x_k$ are reserved to denote the coordinates of any point p, then (1) is traditionally denoted by

$$\varphi'_{x_j} \quad \text{or} \quad \varphi_{x_j}\,. \tag{6}$$

Consequently, $\varphi_{|j}(p)$ is denoted by

$$\varphi'_{x_j}(x_1, \ldots, x_k) \quad \text{or} \quad \varphi_{x_j}(x_1, \ldots, x_k)\,. \tag{6′}$$

The traditional notation is sometimes very convenient, especially in numerical calculation, but it is not precise. In expressions (6′) the symbol x_j is used in two different meanings. First it appears as a lower index. Here it should be interpreted only as the symbol of the coordinate with respect to which the differentiation is performed. We may not substitute here for x_j any concrete numerical values, nor any other expressions, etc. Next, x_j appears in the parentheses as one of the coordinates of the point $p = (x_1, \dots, x_k)$ at which the derivative is calculated. Here we may subsitute for $x_1, \dots, x_k$, in particular for x_j, concrete numerical values or other expressions, etc.

Besides (6) we also use the notation

$$(7) \qquad \frac{\partial \varphi}{\partial x_j} \quad \text{or} \quad \frac{\partial}{\partial x_j}\varphi \quad \text{or} \quad \frac{\partial}{\partial x_j}\varphi(x_1, \dots, x_k)\,.$$

This notation is analogous to the notation $\frac{df}{dx}$ for the ordinary derivative of a function f of one real variable. To indicate that the differentiation is partial, we write the letter ∂ instead of d, the latter being reserved only for ordinary derivatives of functions of one real variable.

In the traditional notation we often use, for brevity, certain letters to denote the value of a given function at an arbitrary point $(x_1, \dots, x_k)$. If we assume the convention that z denotes the value of φ at any point $(x_1, \dots, x_k)$:

$$z = \varphi(x_1, \dots, x_k)\,,$$

then the partial derivative of φ with respect to the jth coordinate is traditionally denoted by the following symbols

$$(8) \qquad z'_{x_j}, \quad z_{x_j}, \quad \frac{\partial z}{\partial x_j}, \quad \frac{\partial}{\partial x_j}z\,.$$

This notation is analogous to the notation y' and $\frac{dy}{dx}$ for the ordinary derivative of a function $y = f(x)$ of one real variable.

In the traditional notation (8), the identities (5) can be written as follows

$$\frac{\partial(z \pm u)}{\partial x_j} = \frac{\partial z}{\partial x_j} \pm \frac{\partial u}{\partial x_j},$$

$$\frac{\partial(cz)}{\partial x_j} = c\frac{\partial z}{\partial x_j},$$

$$\frac{\partial(zu)}{\partial x_j} = \frac{\partial z}{\partial x_j}\cdot u + z\cdot\frac{\partial u}{\partial x_j},$$

$$\frac{\partial}{\partial x_j}\left(\frac{z}{u}\right) = \frac{\frac{\partial z}{\partial x_j}\cdot u - z\cdot\frac{\partial u}{\partial x_j}}{u^2}\,.$$

In the traditional notation the directional derivative $\varphi'_{\mathfrak{a}}$ is denoted by the symbol $\frac{\partial \varphi}{\partial \mathfrak{a}}$.

The existence of all the partial derivatives $\varphi_{|j}$ of φ does not imply the existence of the directional derivative $\varphi'_{\mathfrak{a}}$ for every vector $\mathfrak{a}$ (see § 3, exercise 1).

The following theorem is a modification of Theorem 1.1.

2.1. *If a real function φ on an open set $G \subset R^k$ has all the partial derivatives $\varphi_{|j}$ on G $(j=1, \ldots, k)$, $\mathfrak{a} = [a_1, \ldots, a_k] \in V_k$ and a k-dimensional interval $I \subset G$ contains p_0 and $p_0 + \mathfrak{a}$, then there exist numbers $\theta_1, \ldots, \theta_k$ in the interval $(0;1)$ such that*

$$\varphi(p_0+\mathfrak{a}) - \varphi(p_0) = \sum_{j=1}^{k} a_j \varphi_{|j}(p_0 + a_1 \mathfrak{e}_1 + \ldots + a_{j-1}\mathfrak{e}_{j-1} + \theta_j a_j \mathfrak{e}_j) . \tag{9}$$

Denoting

$$p_j = p_0 + a_1 \mathfrak{e}_1 + \ldots + a_j \mathfrak{e}_j \in I \quad (j = 1, \ldots, k)$$

we have

$$\varphi(p_0+\mathfrak{a}) - \varphi(p_0) = \sum_{j=1}^{k} \big(\varphi(p_j) - \varphi(p_{j-1})\big) .$$

The segment joining the points p_{j-1} and $p_j = p_{j-1} + a_j \mathfrak{e}_j$ is a subset of G. By 1.1 there exists a number θ_j in the interval $(0;1)$ such that

$$\varphi(p_j) - \varphi(p_{j-1}) = a_j \varphi_{|j}(p_{j-1} + \theta_j a_j \mathfrak{e}_j) .$$

This yields (9).

EXERCISES

1. Calculate all the partial derivatives of the function

$$\varphi(x_1, \ldots, x_k) = e^{x_1^2 + \ldots + x_k^2}(x_1 + x_2^3 + \ldots + x_k^{2k-1}) .$$

2. What is the sum of all the partial derivatives of the function φ defined in § 1, exercise 2?

§ 3. The derivative. Functions of class C_1. Let φ be, as previously, a real function defined on an open subset G of the k-dimensional Euclidean space. We shall now examine the expression $\varphi'_{\mathfrak{a}}(p)$ as a function of the vector $\mathfrak{a}$.

3.1. *If the directional derivative $\varphi'_{\mathfrak{a}}(p)$ exists, and c is a real number, then $\varphi'_{c\mathfrak{a}}$ also exists and*

$$\varphi'_{c\mathfrak{a}}(p) = c\varphi'_{\mathfrak{a}}(p) .$$

Indeed, if $c \neq 0$, then

$$\varphi'_{c\mathfrak{a}}(p) = \lim_{\Delta t \to 0} \frac{\varphi(p + \Delta t c \cdot \mathfrak{a}) - \varphi(p)}{\Delta t} = c \cdot \lim_{\Delta t \to 0} \frac{\varphi(p + c\Delta t \cdot \mathfrak{a}) - \varphi(p)}{c\Delta t} = c\varphi'_{\mathfrak{a}}(p).$$

If $c = 0$, then $c\mathfrak{a} = 0$ and $\varphi'_{c\mathfrak{a}}(p) = 0$ and, consequently, 3.1 holds.

3.2. *If the directional derivative $\varphi'_{\mathfrak{a}}$ exists in a neighbourhood $\mathrm{B}(p, \eta)$ of a point p, and is continuous at p, and if the directional derivative $\varphi'_{\mathfrak{b}}(p)$ exists, then the directional derivative $\varphi'_{\mathfrak{a}+\mathfrak{b}}(p)$ exists and*

$$\varphi'_{\mathfrak{a}+\mathfrak{b}}(p) = \varphi'_{\mathfrak{a}}(p) + \varphi'_{\mathfrak{b}}(p) . \tag{1}$$

If $|\Delta t| < \eta/(|\mathfrak{a}|+|\mathfrak{b}|)$, then

$$\frac{\varphi(p+\Delta t(\mathfrak{a}+\mathfrak{b}))-\varphi(p)}{\Delta t} = \frac{\varphi((p+\Delta t\cdot\mathfrak{b})+\Delta t\cdot\mathfrak{a})-\varphi(p+\Delta t\cdot\mathfrak{b})}{\Delta t} +$$

$$+\frac{\varphi(p+\Delta t\cdot\mathfrak{b})-\varphi(p)}{\Delta t} = \varphi'_{\mathfrak{a}}(p+\Delta t\cdot\mathfrak{b}+\theta\Delta t\cdot\mathfrak{a})+\frac{\varphi(p+\Delta t\cdot\mathfrak{b})-\varphi(p)}{\Delta t}$$

for a number θ in the interval $(0;1)$ by the mean-value theorem 1.1 (where p should be replaced by $p+\Delta t\cdot\mathfrak{b}$). Consequently, for $\Delta t\to 0$,

$$\varphi'_{\mathfrak{a}+\mathfrak{b}}(p) = \lim_{\Delta t\to 0}\varphi'_{\mathfrak{a}}(p+\Delta t\cdot\mathfrak{b}+\theta\Delta t\cdot\mathfrak{a})+\lim_{\Delta t\to 0}\frac{\varphi(p+\Delta t\cdot\mathfrak{b})-\varphi(p)}{\Delta t} = \varphi'_{\mathfrak{a}}(p)+\varphi'_{\mathfrak{b}}(p)\,.$$

Note that the hypothesis that $\varphi'_{\mathfrak{a}}(p)$ and $\varphi'_{\mathfrak{b}}(p)$ exist implies neither the existence of $\varphi'_{\mathfrak{a}+\mathfrak{b}}(p)$ nor the identity (1) (see exercise 1).

Let $p\in G$. Suppose that the directional derivative $\varphi'_{\mathfrak{a}}(p)$ exists for every vector $\mathfrak{a}$ and (1) holds, i.e. $\varphi'_{\mathfrak{a}}(p)$ is a linear function of $\mathfrak{a}$. It follows from II 6.1 that

$$\varphi'_{\mathfrak{a}}(p) = \mathfrak{w}\mathfrak{a}$$

where $\mathfrak{w}$ is a vector. The vector $\mathfrak{w}$ is called the *derivative of φ at p* and is denoted by the symbol $\varphi'(p)$. By II 6.1, the coordinates of the vector $\varphi'(p)$ are equal to the directional derivatives $\varphi'_{\mathfrak{e}_1}(p), \ldots, \varphi'_{\mathfrak{e}_k}(p)$ where $\mathfrak{e}_1, \ldots, \mathfrak{e}_k$ are unit vectors of the coordinate axes in R^k. In other words,

3.3. *The coordinates of $\varphi'(p)$ are the partial derivatives at p with respect to all the coordinates:*

$$\varphi'(p) = [\varphi_{|1}(p), \ldots, \varphi_{|k}(p)]\,, \tag{2}$$

and

$$\varphi'_{\mathfrak{a}}(p) = \varphi'(p)\cdot\mathfrak{a} = \sum_{j=1}^{k}\varphi_{|j}(p)\cdot a_j \tag{3}$$

for every vector $\mathfrak{a} = [a_1, \ldots, a_k]$.

By the general rules of notation I § 3, the symbol φ' denotes the function (the vector field) which assigns the vector $\varphi'(p)$ to any point p. The function φ' is called the *derivative of φ*. It is defined on the set of all the points $p\in G$ at which the derivative $\varphi'(p)$ exists. In what follows we shall assume that φ' is defined on the whole domain G of φ. It follows from (2) that

$$\varphi' = [\varphi_{|1}, \ldots, \varphi_{|k}]\,. \tag{2'}$$

In the traditional notation, the derivative (2′) is denoted by gradφ and is called the *gradient* of φ.

Instead of the "derivative" we also say the *first derivative* or the *derivative of first order*, to distinguish it from the derivatives of higher orders, which will be defined in § 4 and § 5.

In the case where $k = 1$, i.e. φ is a function of one real variable, the vector $\varphi'(p)$ defined above has only one coordinate which is equal to the ordinary derivative of φ at p. Identifying this vector with its only coordinate, we see that the definition of the derivative given in this section is a generalization of the definition of the ordinary derivative of a function of one real variable.

Observe that according to the notation adopted, the symbol φ_i' denotes the derivative of a function φ_i, i.e. a vector field, but the symbol $\varphi_{|i}$ denotes a partial derivative of a function φ, i.e. a real function.

It is easy to verify the following identities (see § 1 (5) or § 2 (5)):

$$(\varphi \pm \psi)' = \varphi' \pm \psi', \quad (c\varphi)' = c\varphi',$$

(4)
$$(\varphi \cdot \psi)' = \varphi' \cdot \psi + \varphi \cdot \psi', \quad \left(\frac{\varphi}{\psi}\right)' = \frac{\varphi' \cdot \psi - \varphi \cdot \psi'}{\psi^2},$$

provided that the denominator does not vanish. The identities should be understood as follows: if the derivatives on the right-hand side exist at a point, then the derivative on the left-hand side exists at the point and the equality holds.

The next theorem is another formulation (under a little stronger hypothesis) of the mean-value theorem 1.1:

3.4. *If the derivative φ' exists in G and the segment joining p and $p+\mathfrak{a}$ is contained in G, then there exists a number θ in the interval* $(0;1)$ *such that*

(5)
$$\varphi(p+\mathfrak{a}) - \varphi(p) = \mathfrak{a} \cdot \varphi'(p+\theta\mathfrak{a}),$$

i.e., assuming $\mathfrak{a} = [a_1, \ldots, a_k]$,

(5′)
$$\varphi(p+\mathfrak{a}) - \varphi(p) = \sum_{i=1}^{k} a_i \varphi_{|i}(p+\theta\mathfrak{a}).$$

Letting $q = p+\mathfrak{a}$ we can write the identity in the form

(6)
$$\varphi(q) - \varphi(p) = (q-p) \cdot \varphi'(p+\theta(q-p)) \quad (0 < \theta < 1)$$

under the hypothesis that the segment joining p and q is a subset of G.

The existence of the derivative $\varphi'(p)$ of φ at p is a rather weak hypothesis about properties of φ at the point p. For instance, it does not imply the continuity of the function at p (see exercise 3). The composition of functions having derivatives does not always have a derivative (see V § 3, 7). We see that the existence of the derivative in the sense just defined is not a good generalization of the notion of differentiability of a function of one real variable to the case of functions of several variables. A more exact analysis shows that the only pertinent generalization is the notion of differentiability defined below.

A real function φ defined on an open set $G \subset R^k$ is said to be *differentiable at a point* $p \in G$ if the derivative $\varphi'(p)$ exists and

$$\lim_{\mathfrak{a}\to 0} \frac{\varphi(p+\mathfrak{a})-\varphi(p)-\varphi'(p)\mathfrak{a}}{|\mathfrak{a}|} = 0 . \tag{7}$$

In this case the linear form

$$F(\mathfrak{a}) = \varphi'_{\mathfrak{a}}(p) = \varphi'(p)\mathfrak{a} \qquad (\mathfrak{a} \in V_k) \tag{8}$$

(i.e. the directional derivative of φ at p, interpreted as a function of the vector $\mathfrak{a}$) is called the *differential* (or the *complete differential*) of the function φ at the point p.

In other words, a linear form $F\colon V_k \to R$ is said to be the *differential* of φ at p provided that

$$\lim_{\mathfrak{a}\to 0} \frac{\varphi(p+\mathfrak{a})-\varphi(p)-F(\mathfrak{a})}{|\mathfrak{a}|} = 0 . \tag{9}$$

A function φ is said to be *differentiable at a point* p provided that φ has a differential at p, i.e. there exists a linear form F satisfying (9).

To prove that the second definition of differential and differentiability is equivalent to the first definition, it suffices to show that if a linear form F satisfies (9), then $F(\mathfrak{a}) = \varphi'_{\mathfrak{a}}(p)$. Indeed, this implies that $\varphi'_{\mathfrak{a}}(p)$ is a linear function of $\mathfrak{a}$, and therefore $\varphi'(p)$ exists and $F(\mathfrak{a}) = \varphi'(p)\mathfrak{a}$.

For this purpose let us observe that by replacing $\mathfrak{a}$ by $\Delta t\cdot\mathfrak{a}$ in (9) we obtain

$$\frac{\varphi(p+\Delta t\cdot\mathfrak{a})-\varphi(p)-F(\Delta t\cdot\mathfrak{a})}{\Delta t} \to 0 \quad \text{for} \quad \Delta t \to 0$$

for every fixed vector $\mathfrak{a}$. In other words,

$$\frac{\varphi(p+\Delta t\cdot\mathfrak{a})-\varphi(p)}{\Delta t} - F(\mathfrak{a}) \to 0 \quad \text{for} \quad \Delta t \to 0 .$$

This proves that $F(\mathfrak{a}) = \varphi'_{\mathfrak{a}}(p)$.

In the case where $k = 1$ the above definition of differentiability coincides with the ordinary definition of differentiability of a function of one real variable, i.e. with the existence of the ordinary derivative at a given point. In the case where $k > 1$ the condition (7) is not a consequence of the existence of the derivative $\varphi'(p)$ and, therefore, the hypothesis of differentiability is stronger than the hypothesis of the existence of the derivative.

The following theorem yields a simple criterion for the differentiability of a function φ at a point p.

3.5. *If all the partial derivatives* $\varphi_{|j}$ $(j = 1, \ldots, k)$ *exist in a neighbourhood of* p *and are continuous at* p, *then the function* φ *is differentiable at* p.

Defining

$$F(\mathfrak{a}) = \sum_{j=1}^{k} \varphi_{|j}(p)\,\alpha_j \quad \text{for} \quad \mathfrak{a} = [\alpha_1, \ldots, \alpha_k]$$

and denoting

$$\varrho(\mathfrak{a}) = \varphi(p+\mathfrak{a}) - \varphi(p) - F(\mathfrak{a})$$

we obtain by 2.1

$$\varrho(\mathfrak{a}) = \sum_{j=1}^{k} \alpha_j\big(\varphi_{|j}(p + \alpha_1 \mathrm{e}_1 + \ldots + \alpha_{j-1}\mathrm{e}_{j-1} + \theta_j \alpha_j \mathrm{e}_j) - \varphi_{|j}(p)\big)$$

where $\mathrm{e}_1, \ldots, \mathrm{e}_k$ are the unit vectors of the coordinate axes in R^k and $\theta_1, \ldots, \theta_k$ are numbers in the interval $(0;1)$ which depend on $\mathfrak{a}$. Since $\dfrac{|\alpha_j|}{|\mathfrak{a}|} \leqslant 1$, we have

$$\frac{|\varrho(\mathfrak{a})|}{|\mathfrak{a}|} \leqslant \sum_{j=1}^{k} \big|\big(\varphi_{|j}(p + \alpha_1 \mathrm{e}_1 + \ldots + \alpha_{j-1}\mathrm{e}_{j-1} + \theta_j \alpha_j \mathrm{e}_j) - \varphi_{|j}(p)\big)\big| .$$

It follows from the continuity of $\varphi_{|j}$ at p that

$$\lim_{\mathfrak{a}\to 0} \frac{\varrho(\mathfrak{a})}{|\mathfrak{a}|} = 0 ,$$

i.e. (9) holds. This proves that the linear form F is the differential of φ at p and, consequently, that the function φ is differentiable at p.

3.6. *If a function φ is differentiable at p, then it is continuous at p.*

Denoting $\varrho(\mathfrak{a}) = \varphi(p+\mathfrak{a}) - \varphi(p) - \varphi'(p)\mathfrak{a}$ we have

$$\varphi(p+\mathfrak{a}) = \varphi(p) + \varphi'(p)\mathfrak{a} + \varrho(\mathfrak{a})$$

and by (7)

$$\lim_{\mathfrak{a}\to 0} \frac{\varrho(\mathfrak{a})}{|\mathfrak{a}|} = 0 .$$

Consequently

$$\lim_{\mathfrak{a}\to 0} \varrho(\mathfrak{a}) = 0 .$$

This proves that $\lim\limits_{\mathfrak{a}\to 0} \varphi(p+\mathfrak{a}) = \varphi(p)$, i.e., denoting $q = p + \mathfrak{a}$,

$$\lim_{q\to p} \varphi(q) = \varphi(p) .$$

Thus the function φ is continuous at p.

Since, in the case of $k > 1$, the existence of the derivative at a point does not imply continuity at that point, we infer from 3.6 that the hypothesis of differentiability is stronger than the existence of the derivative.

A function $\varphi\colon G \to R$ is said to be *differentiable* if it is differentiable at every point in its domain G.

In this book we shall not develop the theory of differentiation of functions under such weak hypotheses as existence of the derivative or differentiability (the reader can find a few remarks on this subject in exercises). The hypotheses will always be replaced by the stronger condition that the functions under examination are of class C_1.

A real function φ defined on an open set $G \subset R^k$ is said to be *of class* C_1 if its derivative φ' exists in the whole set G and is continuous.

3.7. *The following conditions are equivalent*:

(a) φ *is of class* C_1;

(b) *all the directional derivatives* $\varphi'_{\mathfrak{a}}$ *exist in* G *and are continuous*;

(c) *all the partial derivatives* $\varphi_{|1}, \dots, \varphi_{|k}$ *exist in* G *and are continuous.*

(b) follows from (a) by 3.3. (c) follows trivially from (b). It follows from (c) by 3.5 that the function φ is differentiable and that its derivative is continuous (see (2′)). Thus (c) implies (a).

Incidentally we have proved that

3.8. *Every function of class* C_1 *is differentiable.*

This implies, by 3.6, that

3.9. *Every function of class* C_1 *is continuous.*

EXERCISES

1. Let $p = (0, 0) \in R^2$, $\mathfrak{a} = [1, 0]$ and $\mathfrak{b} = [0, 1]$.

The function

$$\varphi(x_1, x_2) = \begin{cases} \dfrac{x_1 x_2}{x_1^2 + x_2^2} & \text{if} \quad x_1^2 + x_2^2 > 0\,, \\ 0 & \text{if} \quad x_1 = 0 = x_2\,, \end{cases}$$

has derivatives at p in the directions $\mathfrak{a}$ and $\mathfrak{b}$:

$$\varphi'_{\mathfrak{a}}(p) = 0 = \varphi'_{\mathfrak{b}}(p)\,.$$

However φ does not have a derivative at p in the direction $\mathfrak{a} + \mathfrak{b}$.

The function

$$\varphi(x_1, x_2) = \begin{cases} \dfrac{x_1 x_2 (x_1 + x_2)}{x_1^2 + x_2^2} & \text{if} \quad x_1^2 + x_2^2 > 0\,, \\ 0 & \text{if} \quad x_1 = 0 = x_2\,, \end{cases}$$

has derivatives at every point in every direction. Verify that

$$\varphi'_{\mathfrak{a}}(p) = 0 = \varphi'_{\mathfrak{b}}(p) \quad \text{and} \quad \varphi'_{\mathfrak{a}+\mathfrak{b}}(p) = 1\,.$$

Thus (1) does not hold.

2. The function

$$\varphi(x_1, x_2) = \begin{cases} \dfrac{x_1 x_2^2}{x_1^2 + x_2^2} & \text{if} \quad x_1^2 + x_2^2 > 0\,, \\ 0 & \text{if} \quad x_1 = 0 = x_2\,, \end{cases}$$

has a directional derivative at $p = (0, 0)$ in every direction $\mathfrak{a}$ but it is not continuous at this point, nor does it have a derivative there.

3. The function

$$\varphi(x_1, x_2) = \begin{cases} \dfrac{x_1^4 x_2^2}{x_1^8 + x_2^4} & \text{if} \quad x_1^2 + x_2^2 > 0\,, \\ 0 & \text{if} \quad x_1 = 0 = x_2\,, \end{cases}$$

has a derivative at every point but it is not continuous at $(0, 0)$.

4. Let V be the set of all unit vectors in R^k. Let φ be a real function defined on an open set $G \subset R^k$, let p be a fixed point in G and let $F\colon V_k \to R$ be a linear form.

In order that $F(\mathfrak{a}) = \varphi'_{\mathfrak{a}}(p)$ for every vector $\mathfrak{a} \in V_k$ it is necessary and sufficient that

$$\frac{\varphi(p + \Delta t \cdot \mathfrak{a}) - \varphi(p)}{\Delta t} \to F(\mathfrak{a}) \quad \text{as} \quad \Delta t \to 0 \tag{10}$$

for every vector $\mathfrak{a} \in V$, i.e. that

$$\bigwedge_{\varepsilon > 0} \bigwedge_{\mathfrak{a} \in V} \bigvee_{\delta > 0} \left(0 < |\Delta t| < \delta \Rightarrow \left| \frac{\varphi(p + \Delta t \cdot \mathfrak{a}) - \varphi(p)}{\Delta t} - F(\mathfrak{a}) \right| < \varepsilon \right).$$

In order that the linear form F be the differential of φ at p it is necessary and sufficient that convergence (10) be uniform on V, i.e. that

$$\bigwedge_{\varepsilon > 0} \bigvee_{\delta > 0} \bigwedge_{\mathfrak{a} \in V} \left(0 < |\Delta t| < \delta \Rightarrow \left| \frac{\varphi(p + \Delta t \cdot \mathfrak{a}) - \varphi(p)}{\Delta t} - F(\mathfrak{a}) \right| < \varepsilon \right).$$

5. A linear form $F\colon V_k \to R$ is the differential of a function $\varphi\colon G \to R$ at a point $p \in G$ $(G \subset R^k)$ if and only if, for every sequence (Δt_n) of real numbers different from 0 and for every sequence $(\mathfrak{a}_n)$ of vectors in R^k such that $p + \Delta t_n \cdot \mathfrak{a}_n \in G$, the conditions

$$\Delta t_n \to 0\,, \qquad \mathfrak{a}_n \to \mathfrak{a}$$

imply

$$\frac{\varphi(p + \Delta t_n \cdot \mathfrak{a}_n) - \varphi(p)}{\Delta t_n} \to F(\mathfrak{a})\,.$$

Hint. Use III § 9, 3.

6. A linear form $F\colon V_k \to R$ is the differential of a function $\varphi\colon G \to R$ at a point $p \in G$ $(G \subset R^k)$ if and only if

$$\lim_{\Delta t \to 0} \frac{\varphi(p + \Delta t \cdot \Phi(\Delta t)) - \varphi(p)}{\Delta t} = F(\mathfrak{a})$$

for every function Φ with values in V_k, such that

$$\lim_{\Delta t \to 0} \Phi(\Delta t) = \mathfrak{a}$$

and $p + \Delta t \cdot \Phi(\Delta t) \in G$.

7. Let $F\colon V_k \to R$ be a homogeneous (but not necessarily additive) function and let $p \in R^k$ be a fixed point. The formula

$$\varphi(q) = F(q-p) \quad \text{for} \quad q \in R^k$$

defines a real function on R^k. Verify that (9) holds.

This example shows that the condition (9) does not imply the linearity of F.

8. If all the partial derivatives of a real function φ defined on an open convex set are bounded, then φ is uniformly continuous.

9. If all the partial derivatives of a function φ defined on an open connected set $G \subset R^k$ are equal to 0, then φ is constant.

10. The derivative of φ is sometimes denoted by the symbol $\nabla\varphi$. It is then convenient to denote the partial derivative of φ with respect to the ith coordinate by $\nabla_i\varphi$ and to treat ∇ as the symbolic vector

(11) $$\nabla = [\nabla_1, \ldots, \nabla_k]\,.$$

The derivative $\nabla\varphi$ can then be interpreted as the product of the symbolic vector ∇ by the scalar φ, the result of the multiplication being a vector. The product $\nabla\varphi$ should be distinguished from the product

(12) $$\varphi\nabla = [\varphi\nabla_1, \ldots, \varphi\nabla_k]$$

which is also a symbolic vector. More precisely, the symbolic vectors (11) and (12) are *differential operators*, i.e., symbols showing what differential operations should be performed. For instance,

$$\nabla\psi = [\nabla_1\psi, \ldots, \nabla_k\psi]$$

for any ψ, and

$$(\varphi\nabla)\psi = [\varphi\nabla_1\psi, \ldots, \varphi\nabla_k\psi] = [\varphi\psi_{|1}, \ldots, \varphi\psi_{|k}]$$

for any ψ. Note the following associative law:

$$(\varphi\nabla)\psi = \varphi(\nabla\psi)\,.$$

However, the commutative law does not hold, i.e. $\varphi\nabla$ is not identical to $\nabla\varphi$.

Prove that

(13) $$\nabla(\varphi\psi) = (\nabla\varphi)\psi + \varphi(\nabla\psi)\,.$$

This identity is another formulation of the formula for the derivative of the product of functions.

§ 4. Derivatives of second order. Functions of class C_2.

The directional derivative and the partial derivatives defined in § 1 and § 2 are called directional or partial derivatives of *first order* to distinguish them from analogous derivatives of higher orders, which will be defined below.

Let φ be a real function on an open subset G of R^k, and let $\mathfrak{a}$ be a fixed vector. Suppose that the directional derivative $\varphi'_{\mathfrak{a}}$ exists in a neighbourhood of a point $p \in G$. Then we can define the directional derivative of the function $\varphi'_{\mathfrak{a}}$ at the point p in the direction $\mathfrak{b}$, where $\mathfrak{b}$ is any vector. If this directional derivative exists, it is denoted by $\varphi''_{\mathfrak{a},\mathfrak{b}}(p)$. By definition,

(1) $$\varphi''_{\mathfrak{a},\mathfrak{b}}(p) = (\varphi'_{\mathfrak{a}})'_{\mathfrak{b}}(p)\,.$$

The number (1), if it exists, is called the *directional derivative of second order of φ at p*, more precisely, the *derivative of second order of φ at p in the directions $\mathfrak{a}, \mathfrak{b}$*.

According to the general rules of notation from I § 3, $\varphi''_{\mathfrak{a},\mathfrak{b}}$ is the function which assigns the number (1) to every point p at which (1) exists. This function is said to be the *directional derivative of φ of second order*, more precisely, the *derivative of φ of second order in the directions $\mathfrak{a}, \mathfrak{b}$*. In what follows we are interested only in the case where (1) exists for every $p \in G$, i.e. $\varphi''_{\mathfrak{a},\mathfrak{b}}$ is a real function on G.

The directional derivative $\varphi''_{\mathfrak{a},\mathfrak{b}}$ should not be identified, in general, with $\varphi''_{\mathfrak{b},\mathfrak{a}}$. They differ by the alternation of differentiations performed; to calculate $\varphi''_{\mathfrak{a},\mathfrak{b}}$ we first differentiate in the direction $\mathfrak{a}$, and then in the direction $\mathfrak{b}$; to calculate $\varphi''_{\mathfrak{b},\mathfrak{a}}$ we first differentiate in the direction $\mathfrak{b}$, and then in the direction $\mathfrak{a}$. If both $\varphi''_{\mathfrak{a},\mathfrak{b}}$ and $\varphi''_{\mathfrak{b},\mathfrak{a}}$ exist at a point p, they are not equal, in general. However, they are often equal under additional hypotheses. For instance,

4.1. *If both directional derivatives $\varphi''_{\mathfrak{a},\mathfrak{b}}$ and $\varphi''_{\mathfrak{b},\mathfrak{a}}$ exist and are continuous in G, then they are equal*:

(2) $$\varphi''_{\mathfrak{a},\mathfrak{b}} = \varphi''_{\mathfrak{b},\mathfrak{a}} .$$

Let p_0 be any point in G and let $\eta > 0$ be such that $B(p_0, \eta) \subset G$. Let Δt and $\Delta\tau$ be any fixed real numbers such that

$$0 < |\Delta t| < \eta/2|\mathfrak{a}| , \quad 0 < |\Delta\tau| < \eta/2|\mathfrak{b}| .$$

The formulas

$$g(p) = \varphi(p + \Delta t \cdot \mathfrak{a}) - \varphi(p) ,$$
$$h(p) = \varphi(p + \Delta\tau \cdot \mathfrak{b}) - \varphi(p)$$

define real functions g and h on the ball $B(p_0, \eta/2)$ and

$$g'_{\mathfrak{b}}(p) = \varphi'_{\mathfrak{b}}(p + \Delta t \cdot \mathfrak{a}) - \varphi'_{\mathfrak{b}}(p) ,$$
$$h'_{\mathfrak{a}}(p) = \varphi'_{\mathfrak{a}}(p + \Delta\tau \cdot \mathfrak{b}) - \varphi'_{\mathfrak{a}}(p) .$$

Since the segment joining the points p_0 and $p_0 + \Delta\tau \cdot \mathfrak{b}$ is a subset of $B(p_0, \eta/2)$, by the mean-value theorem 1.1

$$\begin{aligned} g(p_0 + \Delta\tau \cdot \mathfrak{b}) - g(p_0) &= \Delta\tau \cdot g'_{\mathfrak{b}}(p_0 + \theta_1 \Delta\tau \cdot \mathfrak{b}) \\ &= \Delta\tau \big(\varphi'_{\mathfrak{b}}(p_0 + \theta_1 \Delta\tau \cdot \mathfrak{b} + \Delta t \cdot \mathfrak{a}) - \varphi'_{\mathfrak{b}}(p_0 + \theta_1 \Delta\tau \cdot \mathfrak{b})\big) , \end{aligned}$$

where $0 < \theta_1 < 1$. By the same argument,

$$\varphi'_{\mathfrak{b}}(p_0 + \theta_1 \Delta\tau \cdot \mathfrak{b} + \Delta t \cdot \mathfrak{a}) - \varphi'_{\mathfrak{b}}(p_0 + \theta_1 \Delta\tau \cdot \mathfrak{b}) = \Delta t \cdot \varphi''_{\mathfrak{b},\mathfrak{a}}(p_0 + \theta_1 \Delta\tau \cdot \mathfrak{b} + \theta_2 \Delta t \cdot \mathfrak{a}) ,$$

where $0 < \theta_2 < 1$. Analogously, we prove that

$$h(p_0+\Delta t\cdot \mathfrak{a})-h(p_0) = \Delta t\cdot h'_{\mathfrak{a}}(p_0+\theta_3\Delta t\cdot \mathfrak{a})$$
$$= \Delta t\big(\varphi'_{\mathfrak{a}}(p_0+\theta_3\Delta t\cdot\mathfrak{a}+\Delta\tau\cdot\mathfrak{b})-\varphi'_{\mathfrak{a}}(p_0+\theta_3\Delta t\cdot\mathfrak{a})\big)\,,$$

where $0 < \theta_3 < 1$, and

$$\varphi'_{\mathfrak{a}}(p_0+\theta_3\Delta t\cdot\mathfrak{a}+\Delta\tau\cdot\mathfrak{b})-\varphi'_{\mathfrak{a}}(p_0+\theta_3\Delta t\cdot\mathfrak{a}) = \Delta\tau\cdot\varphi''_{\mathfrak{a},\mathfrak{b}}(p_0+\theta_3\Delta t\cdot\mathfrak{a}+\theta_4\Delta\tau\cdot\mathfrak{b})\,,$$

where $0 < \theta_4 < 1$.

On the other hand,

$$g(p_0+\Delta\tau\cdot\mathfrak{b})-g(p_0) = h(p_0+\Delta t\cdot\mathfrak{a})-h(p_0)$$

since both sides of the identity are equal to

$$\varphi(p_0+\Delta t\cdot\mathfrak{a}+\Delta\tau\cdot\mathfrak{b})-\varphi(p_0+\Delta t\cdot\mathfrak{a})-\varphi(p_0+\Delta\tau\cdot\mathfrak{b})+\varphi(p_0)\,.$$

It follows from the above identities that

$$\varphi''_{\mathfrak{b},\mathfrak{a}}(p_0+\theta_1\Delta\tau\cdot\mathfrak{b}+\theta_2\Delta t\cdot\mathfrak{a}) = \varphi''_{\mathfrak{a},\mathfrak{b}}(p_0+\theta_3\Delta t\cdot\mathfrak{a}+\theta_4\Delta\tau\cdot\mathfrak{b})\,.$$

If $\Delta t\to 0$ and $\Delta\tau\to 0$, then the left-hand side tends to $\varphi''_{\mathfrak{b},\mathfrak{a}}(p_0)$, and the right-hand side tends to $\varphi''_{\mathfrak{a},\mathfrak{b}}(p_0)$, by the continuity of $\varphi''_{\mathfrak{b},\mathfrak{a}}$ and $\varphi''_{\mathfrak{a},\mathfrak{b}}$ respectively. Thus

$$\varphi''_{\mathfrak{b},\mathfrak{a}}(p_0) = \varphi''_{\mathfrak{a},\mathfrak{b}}(p_0)\,,$$

which proves (2).

The directional derivatives of the second order

(3) $$\varphi''_{\mathfrak{e}_i,\mathfrak{e}_j}$$

in the directions of unit vectors of the coordinate axes are called *partial derivatives of second order of* φ, and are denoted by

(4) $$\varphi_{|i,j}\,.$$

By definition,

(4′) $$\varphi_{|i,j} = (\varphi_{|i})_{|j}\,.$$

Of course, $\varphi_{|i,j}(p)$ denotes the value of (4) at a point p.

The traditional notation for (4) is analogous to the notation for partial derivatives of first order in § 2, (6)-(8).

If $x_1, \ldots, x_k$ denote the coordinates of any point p in G, then (4) is denoted by

(5) $$\varphi''_{x_i,x_j} \quad\text{or}\quad \varphi_{x_i,x_j}\,.$$

Consequently, $\varphi_{|i,j}(p)$ is denoted by

(5′) $$\varphi''_{x_i,x_j}(x_1,\ldots,x_k) \quad\text{or}\quad \varphi_{x_i,x_j}(x_1,\ldots,x_k)\,.$$

The remarks on the two meanings of the symbols $x_1, \ldots, x_k$ from pp. 133-134 are valid. Beside (5), the notation

(6) $$\frac{\partial^2 \varphi}{\partial x_j \partial x_i} \quad \text{or} \quad \frac{\partial^2}{\partial x_j \partial x_i} \varphi(x_1, \ldots, x_k) \quad \text{or} \quad \frac{\partial}{\partial x_j} \frac{\partial}{\partial x_i} \varphi$$

is also commonly used. If the letter z is reserved to denote the values of φ, then the symbols

(7) $$\frac{\partial^2 z}{\partial x_j \partial x_i} \quad \text{and} \quad \frac{\partial}{\partial x_j} \frac{\partial z}{\partial x_i}$$

are also used. In the case where $j = i$, instead of $\frac{\partial^2 \varphi}{\partial x_i \partial x_i}$ and $\frac{\partial^2 z}{\partial x_i \partial x_i}$ we write

(8) $$\frac{\partial^2 \varphi}{\partial x_i^2} \quad \text{or} \quad \frac{\partial^2}{\partial x_i^2} \varphi \quad \text{or} \quad \frac{\partial^2 z}{\partial x_i^2}.$$

In the notation (7), Theorem 4.1 for $\mathfrak{a} = \mathfrak{e}_i$ and $\mathfrak{b} = \mathfrak{e}_j$ can be written:

(9) $$\frac{\partial^2 z}{\partial x_j \partial x_i} = \frac{\partial^2 z}{\partial x_i \partial x_j}$$

under the hypothesis of continuity of both derivatives.

If $i \neq j$, the derivative $\varphi_{|i,j}$ is said to be *mixed*.

The practical method of calculating the partial derivative (4)-(7) of second order is the same as in the case of partial derivatives of first order (see p. 133): first we interpret $\varphi(x_1, \ldots, x_k)$ as a function of one real variable x_i, the remaining variables being treated as constant parameters, and we differentiate this function; then we interpret the derivative as a function of the variable x_j only and we differentiate this function.

Suppose that for a given point $p \in G$ the derivative $\varphi''_{\mathfrak{a},\mathfrak{b}}(p)$ exists for all vectors $\mathfrak{a}, \mathfrak{b}$. It follows from 3.1 that $\varphi''_{\mathfrak{a},\mathfrak{b}}(p)$ is a homogeneous function of $\mathfrak{a}$ and $\mathfrak{b}$. In general, it is not an additive function of $\mathfrak{a}$ and $\mathfrak{b}$.

Suppose that $\varphi''_{\mathfrak{a},\mathfrak{b}}(p)$ is a linear function of each of the vectors $\mathfrak{a}, \mathfrak{b}$, i.e. it is a bilinear form on $V_k \times V_k$. It follows from II 6.2 that there exists a square matrix $\mathfrak{A}$ such that

$$\varphi''_{\mathfrak{a},\mathfrak{b}}(p) = \mathfrak{A} \cdot (\mathfrak{a} \times \mathfrak{b}).$$

The matrix $\mathfrak{A}$ will be called the *second derivative* (or the *derivative of second order*) of φ at p and will be denoted by $\varphi''(p)$. It follows from II 6.2 that the coordinates of $\mathfrak{A}$ are the derivatives $\varphi''_{\mathfrak{e}_i,\mathfrak{e}_j}(p)$. In other words,

4.2. *The coordinates of the derivative $\varphi''(p)$ of second order are the partial derivatives of the second order:*

(10) $$\varphi''(p) = [\varphi_{|i,j}(p)]_{i,j \leqslant k},$$

and

(11) $$\varphi''_{\mathfrak{a},\mathfrak{b}}(p) = \varphi''(p)\cdot(\mathfrak{a}\times\mathfrak{b}) = \sum_{i,j=1}^{k} \varphi_{|i,j}(p)\,\alpha_i\beta_j$$

for any vectors $\mathfrak{a} = [\alpha_1, \ldots, \alpha_k]$ *and* $\mathfrak{b} = [\beta_1, \ldots, \beta_k]$.

According to the general rules of notation from I § 3, the symbol φ'' denotes the function (the matrix field) which assigns the square matrix $\varphi''(p)$ to every point p. The function φ'' is called the *second derivative* (or the *derivative of second order*) of φ. It is defined on the set of all points p for which $\varphi''(p)$ exists. In what follows we are interested only in the case where it exists for every $p \in G$. By (10),

(10′) $$\varphi'' = [\varphi_{|i,j}]_{i,j\leqslant k}\,.$$

We say that φ is *of class* C_2 if φ'' exists on the whole set G and is continuous, i.e. all the real functions $\varphi_{|i,j}$ are continuous.

It follows from 4.1 that

4.3. *If φ is of class C_2, then for every point p the matrix $\varphi''(p)$ is symmetric*:

$$\varphi_{|i,j}(p) = \varphi_{|j,i}(p)\,.$$

Observe that for every fixed positive integer $i \leqslant k$ the vector field

(12) $$[\varphi_{|i,1}, \ldots, \varphi_{|i,k}]$$

is the derivative of the real function $\varphi_{|i}$, i.e. the partial derivative of first order $\varphi_{|i}$. If the function is of class C_2, then all the functions $\varphi_{|i,1}, \ldots$ $\ldots, \varphi_{|i,k}$ are continuous, i.e. the vector field (12) is continuous. In other words, if φ is of class C_2, then all the partial derivatives $\varphi_{|i}$ of first order are of class C_1. It follows from 3.9 that the $\varphi_{|i}$ are continuous. Hence, by 3.7, φ is of class C_1. Thus we have proved that

4.4. *Every function of class C_2 is of class C_1.*

Hence it follows, by 3.9, that every function of class C_2 is continuous.

The following theorem is analogous to 3.7.

4.5. *The following conditions are equivalent*:

(a) *φ is of class C_2*;

(b) *all the directional derivatives $\varphi''_{\mathfrak{a},\mathfrak{b}}$ of second order exist and are continuous*;

(c) *all the partial derivatives $\varphi_{|i,j}$ of second order exist and are continuous.*

(b) results from (a) by (11). (c) is a trivial consequence of (b). We shall show that (c) implies (a).

Suppose that (c) holds. Since $\varphi_{|i,1}, \ldots, \varphi_{|i,k}$ are partial derivatives of $\varphi_{|i}$, it follows from 3.7 that $\varphi_{|i}$ is of class C_1. By 3.3 (where φ should be replaced by $\varphi_{|i}$),

$$(\varphi_{|i})'_{\mathfrak{b}} = \sum_{j=1}^{k} \varphi_{|i,j}\beta_j \quad \text{for any vector } \mathfrak{b} = [\beta_1, \ldots, \beta_k].$$

By 3.9, $\varphi_{|i}$ are continuous, i.e. φ is of class C_1 by 3.7. Hence (see 3.3)

$$\varphi'_{\mathfrak{a}} = \sum_{i=1}^{k} \varphi_{|i}\alpha_i \quad \text{for every vector } \mathfrak{a} = [\alpha_1, \ldots, \alpha_k].$$

The last two identities imply that $\varphi''_{\mathfrak{a},\mathfrak{b}}$ exists everywhere in G, viz.

$$\varphi''_{\mathfrak{a},\mathfrak{b}} = (\varphi'_{\mathfrak{a}})'_{\mathfrak{b}} = \Big(\sum_{i=1}^{k} \varphi_{|i}\alpha_i\Big)'_{\mathfrak{b}} = \sum_{i=1}^{k} (\varphi_{|i})'_{\mathfrak{b}}\alpha_i = \sum_{i=1}^{k}\Big(\sum_{j=1}^{k} \varphi_{|i,j}\beta_j\Big)\alpha_i = \sum_{i,j=1}^{k} \varphi_{|i,j}\alpha_i\beta_j.$$

We infer that $\varphi''_{\mathfrak{a},\mathfrak{b}}(p)$ is a bilinear form in $\mathfrak{a}$, $\mathfrak{b}$, i.e. the derivative $\varphi''(p)$ exists for every point $p \in G$. The continuity of φ'' follows directly from (c). The function φ is therefore of class C_2.

EXERCISES

1. Prove that $(\varphi\cdot\psi)'' = \varphi''\cdot\psi + \varphi'\times\psi' + \psi'\times\varphi' + \varphi\cdot\psi''$.

2. Prove that, using the notation of § 3, exercise 10,

$$\varphi'' = (\nabla\times\underline{\nabla})\varphi = \nabla\times(\nabla\varphi).$$

On the other hand,

$$(\nabla\nabla)\varphi = \nabla(\nabla\varphi) = \sum_{i=1}^{k} \varphi_{|i,i} = \sum_{i=1}^{k} \frac{\partial^2\varphi}{\partial x_i^2}. \tag{13}$$

The expression (13) is usually denoted by $\Delta\varphi$. Symbolically, Δ is the scalar product of ∇:

$$\Delta = \nabla\nabla = \sum_{i=1}^{k} \frac{\partial^2}{\partial x_i^2}. \tag{14}$$

Prove that

$$\Delta(\varphi\psi) = (\Delta\varphi)\psi + 2\nabla\varphi\nabla\psi + \varphi\Delta\psi. \tag{15}$$

3. Let

$$\varphi(x_1, x_2) = \frac{x_1x_2(x_1^2 - x_2^2)}{x_1^2 + x_2^2} \quad \text{if} \quad x_1^2 + x_2^2 > 0, \quad \text{and} \quad \varphi(0, 0) = 0.$$

Calculate $\varphi''_{\mathfrak{a},\mathfrak{b}}(0, 0)$. Prove that the identity $\varphi''_{\mathfrak{a},\mathfrak{b}}(0, 0)$ that $\varphi''_{\mathfrak{b},\mathfrak{a}}(0, 0)$ does not hold, in general.

§ 5. Derivatives of higher orders. Continuing the procedure described at the beginning of § 4 we define the directional derivative of the third order $\varphi'''_{\mathfrak{a},\mathfrak{b},\mathfrak{c}}$ as the derivative in the direction $\mathfrak{c}$ of the directional derivative $\varphi''_{\mathfrak{a},\mathfrak{b}}$; in symbols

$$\varphi'''_{\mathfrak{a},\mathfrak{b},\mathfrak{c}} = (\varphi''_{\mathfrak{a},\mathfrak{b}})'_{\mathfrak{c}}.$$

More generally, we define by induction the *directional derivative* $\varphi^{(m)}_{\mathfrak{a}_1,\ldots,\mathfrak{a}_m}$ *of order* $m>1$ as the derivative in the direction $\mathfrak{a}_m$ of the directional derivative $\varphi^{(m-1)}_{\mathfrak{a}_1,\ldots,\mathfrak{a}_{m-1}}$; in symbols,

$$\varphi^{(m)}_{\mathfrak{a}_1,\ldots,\mathfrak{a}_m} = (\varphi^{(m-1)}_{\mathfrak{a}_1,\ldots,\mathfrak{a}_{m-1}})'_{\mathfrak{a}_m}\,. \tag{1}$$

Of course, we can introduce the notion of the directional derivative $\varphi^{(m)}_{\mathfrak{a}_1,\ldots,\mathfrak{a}_m}(p)$ of order m at a point p only in the case where the directional derivative $\varphi^{(m-1)}_{\mathfrak{a}_1,\ldots,\mathfrak{a}_{m-1}}$ exists in a neighbourhood of p. In what follows we shall be interested only in the case where all the directional derivatives in question exist in the whole open set $G\subset R^k$ on which the function φ is defined.

The directional derivative

$$\varphi^{(m)}_{\mathfrak{e}_{j_1},\ldots,\mathfrak{e}_{j_m}}, \tag{2}$$

where $\mathfrak{e}_1,\ldots,\mathfrak{e}_k$ are unit vectors of the coordinate axes and $j_1,\ldots,j_m$ is any sequence of positive integers $\leqslant k$, is called the *m-th partial derivative* (or the *partial derivative of order m*) *of* φ with respect to the $(j_1,\ldots,j_m)$-coordinates.

The partial derivative (2) will be denoted, for brevity, by

$$\varphi_{|j_1,\ldots,j_m}\,. \tag{3}$$

By definition,

$$\varphi_{|j_1,\ldots,j_m} = (\varphi_{|j_1,\ldots,j_{m-1}})_{|j_m} \qquad (m>1)\,. \tag{3'}$$

As was stated in § 2, the vertical bar denotes that all the indices on the right side of the vertical bar refer to a partial differentiation. The indication which indices refer to a partial differentiation is especially necessary in the case where the symbol denoting the function under consideration includes lower indices. For instance, $\varphi_{i|j,r,s}$ denotes a partial derivative of third order of a function φ_i. The symbol $\varphi_{i,j|r,s}$ denotes a partial derivative of second order of a function $\varphi_{i,j}$. The symbol $\varphi_{i,j,r|s}$ denotes a partial derivative of first order of a function $\varphi_{i,j,r}$, etc.

The traditional notation for partial derivatives of order m is analogous to that in § 2 (6)-(8) and § 4 (5)-(8). If $x_1,\ldots,x_k$ denote the coordinates of any point p in G, then instead of (3) we write

$$\varphi^{(m)}_{x_{j_1},\ldots,x_{j_m}} \quad \text{or} \quad \varphi_{x_{j_1},\ldots,x_{j_m}}\,. \tag{4}$$

Consequently, instead of $\varphi_{|j_1,\ldots,j_m}(p)$ we write

$$\varphi^{(m)}_{x_{j_1},\ldots,x_{j_m}}(x_1,\ldots,x_k) \quad \text{or} \quad \varphi_{x_{j_1},\ldots,x_{j_m}}(x_1,\ldots,x_k) \tag{4'}$$

to denote the value of (3) at a point $p=(x_1,\ldots,x_k)$. The remarks on pp. 133-134 about the two meanings of the symbols $x_1,\ldots,x_k$ are valid.

Beside (4), we also write

$$(5)\qquad \frac{\partial^m \varphi}{\partial x_{j_m} \dots \partial x_{j_1}} \quad \text{or} \quad \frac{\partial^m}{\partial x_{j_m} \dots \partial x_{j_1}} \varphi(x_1, \dots, x_k) \quad \text{or} \quad \frac{\partial}{\partial x_{j_m}} \dots \frac{\partial}{\partial x_{j_1}} \varphi .$$

If the letter z is reserved to denote the value of φ at any point, we write

$$(6)\qquad \frac{\partial^m z}{\partial x_{j_m} \dots \partial x_{j_1}} \quad \text{or} \quad \frac{\partial}{\partial x_{j_m}} \dots \frac{\partial}{\partial x_{j_1}} z$$

instead of (5). In the case where the differentiation with respect to a coordinate is repeated, we use certain obvious abbreviations. For instance we write

$$\frac{\partial^3 \varphi}{\partial x_1^3}$$

instead of $\dfrac{\partial^3 \varphi}{\partial x_1 \partial x_1 \partial x_1}$ (i.e. $\varphi_{|1,1,1}$). The symbol

$$\frac{\partial^4}{\partial x_1^4} \frac{\partial^2}{\partial x_2^2} \varphi \quad \text{or} \quad \frac{\partial^6 \varphi}{\partial x_1^4 \partial x_2^2}$$

indicates that we differentiate twice with respect to x_2 and then four times with respect to x_1, etc.

Let φ be a real function on an open set $G \subset R^k$, as always. Suppose that the directional derivative $\varphi^{(m)}_{\mathfrak{a}_1,\dots,\mathfrak{a}_m}(p)$ at a given point $p \in G$ exists for all vectors $\mathfrak{a}_1, \dots, \mathfrak{a}_m$ and that it is a linear function of each of the vectors $\mathfrak{a}_1, \dots, \mathfrak{a}_m$, i.e. that it is an m-linear form on V_k^m. It follows from II 6.3 that there exists a k-dimensional m-indexed matrix $\mathfrak{A}$ such that

$$\varphi^{(m)}_{\mathfrak{a}_1,\dots,\mathfrak{a}_m}(p) = \mathfrak{A} \cdot (\mathfrak{a}_1 \times \dots \times \mathfrak{a}_m) \quad \text{for all } \mathfrak{a}_1, \dots, \mathfrak{a}_m .$$

The matrix $\mathfrak{A}$ will be called the *m-th derivative* (or the *derivative of order m*) of φ at p and will be denoted by $\varphi^{(m)}(p)$. By II 6.3 the directional derivatives (2) at p are the coordinates of $\mathfrak{A}$. In other words,

5.1. *The partial derivatives of order m of φ at p are the coordinates of $\varphi^{(m)}(p)$:*

$$(7)\qquad \varphi^{(m)}(p) = [\varphi_{|j_1,\dots,j_m}(p)]_{j_1,\dots,j_m \leqslant k} ,$$

and

$$(8)\qquad \varphi^{(m)}_{\mathfrak{a}_1,\dots,\mathfrak{a}_m}(p) = \varphi^{(m)}(p) \cdot (\mathfrak{a}_1 \times \dots \times \mathfrak{a}_m) = \sum_{j_1,\dots,j_m=1}^{k} \varphi_{|j_1,\dots,j_m}(p)\, a_{1,j_1} \dots a_{m,j_m}$$

for any vectors

$$(9)\qquad \begin{aligned} \mathfrak{a}_1 &= [a_{1,1}, \dots, a_{1,k}] , \\ &\dots\dots\dots\dots \\ \mathfrak{a}_m &= [a_{m,1}, \dots, a_{m,k}] . \end{aligned}$$

According to the general rules of notation in I § 3, the symbol $\varphi^{(m)}$ will denote the function (the m-indexed matrix field) which assigns the matrix $\varphi^{(m)}(p)$ to any point p. The function $\varphi^{(m)}$ is called the *m-th derivative* (or the *derivative of order m*) of φ. It is defined on the set of all points p at which $\varphi^{(m)}(p)$ exists. In what follows we shall assume that it is defined on the whole domain G of φ. By (7),

$$\varphi^{(m)} = [\varphi_{|j_1,\ldots,j_m}]_{j_1,\ldots,j_m \leqslant k}\,. \tag{7'}$$

We say that φ is of *class* C_m provided $\varphi^{(m)}$ exists everywhere in G and is continuous (i.e. all the real functions $\varphi_{|j_1,\ldots,j_m}$ are continuous — see III 8.8). We say that φ is *of class* C_∞ if it is of class C_m for every positive integer m.

The following theorem is a generalization of 3.7 and 4.5.

5.2. *The following conditions are equivalent*:

(a) *φ is of class C_m in G;*

(b) *for any vectors $\mathfrak{a}_1, \ldots, \mathfrak{a}_m$ the directional derivative $\varphi^{(m)}_{\mathfrak{a}_1,\ldots,\mathfrak{a}_m}$ exists and is continuous in G;*

(c) *all the partial derivatives $\varphi_{|j_1,\ldots,j_m}$ of order m exist and are continuous in G.*

(b) results from (a) by (8). (c) is a particular case of (b). The proof that (c) implies (a) is by induction with respect to m. The case of $m = 1$ was proved in 3.7. Suppose that the implication is true for $m-1$ $(m > 1)$. For any fixed positive integers $j_1, \ldots, j_{m-1}$, the functions

$$\varphi_{|j_1,\ldots,j_{m-1},1}, \ldots, \varphi_{|j_1,\ldots,j_{m-1},k}$$

are the partial derivatives of first order of the real function $\varphi_{|j_1,\ldots,j_{m-1}}$. By 3.7 (where φ should be replaced by $\varphi_{|j_1,\ldots,j_{m-1}}$) the function $\varphi_{|j_1,\ldots,j_{m-1}}$ is of class C_1 and, consequently,

$$(\varphi_{|j_1,\ldots,j_m})'_{\mathfrak{a}_m} = \sum_{j_m=1}^{k} \varphi_{|j_1,\ldots,j_m}\, a_{m,j_m}$$

for any vector $\mathfrak{a}_m$ (see (9)). By 3.9, the functions $\varphi_{|j_1,\ldots,j_{m-1}}$ are continuous. This implies, by the induction hypothesis, that φ is of class C_{m-1} and therefore for any vectors $\mathfrak{a}_1, \ldots, \mathfrak{a}_{m-1}$ (see (9)),

$$\varphi^{(m-1)}_{\mathfrak{a}_1,\ldots,\mathfrak{a}_{m-1}} = \sum_{j_1,\ldots,j_{m-1}=1}^{k} \varphi_{|j_1,\ldots,j_{m-1}}\, a_{1,j_1} \cdots a_{m-1,j_{m-1}}\,.$$

Consequently,

$$\varphi^{(m)}_{\mathfrak{a}_1,\ldots,\mathfrak{a}_m} = (\varphi^{(m-1)}_{\mathfrak{a}_1,\ldots,\mathfrak{a}_{m-1}})'_{\mathfrak{a}_m} = \sum_{j_1,\ldots,j_m=1}^{k} \varphi_{|j_1,\ldots,j_m}\, a_{1,j_1} \cdots a_{m,j_m}$$

which proves that, for every point $p \in G$, $\varphi^{(m)}_{\mathfrak{a}_1,\ldots,\mathfrak{a}_m}(p)$ is an m-linear form in $\mathfrak{a}_1, \ldots, \mathfrak{a}_m$. Thus $\varphi^{(m)}(p)$ exists for every $p \in G$. By (c) and (7′), $\varphi^{(m)}$ is continuous, i.e. φ is of class C_m.

5.3. *Every function of class* C_m *is of class* $C_{m-1}, \ldots, C_1$ *and is continuous.*

Since φ has the property (c) from 5.2, similarly as in the proof of the implication (c) $\Rightarrow$ (a) in 5.2 we infer that φ is of class C_{m-1}. Consequently, by the part of 5.3 just proved (where m should be now replaced by $m-1$), we infer that φ is of class C_{m-2}. Repeating this argument, we infer, step by step, that φ is of class $C_{m-3}, \ldots, C_1$. By 3.9, φ is continuous.

The following theorem is a generalization of 4.3.

5.4. *If* φ *is of class* C_m, *then for every point* p *the matrix* $\varphi^{(m)}(p)$ *is symmetric, i.e.*

(10)
$$\varphi_{|j_{r_1},\ldots,j_{r_m}} = \varphi_{|j_1,\ldots,j_m}$$

for any permutation $r_1, \ldots, r_m$ *of the integers* $1, \ldots, m$.

First observe that (10) holds if $j_{r_1}, \ldots, j_{r_m}$ was obtained from $j_1, \ldots, j_m$ by a transposition of two consecutive indices, i.e. that

$$\varphi_{|j_1,\ldots,j_{s-1},j_{s+1},j_s,j_{s+2},\ldots,j_m} = \varphi_{|j_1,\ldots,j_{s-1},j_s,j_{s+1},j_{s+2},\ldots,j_m}.$$

Indeed replacing φ by $\varphi_{|j_1,\ldots,j_{s-1}}$ in 4.3 we infer that

$$\varphi_{|j_1,\ldots,j_{s-1},j_{s+1},j_s} = \varphi_{|j_1,\ldots,j_{s-1},j_s,j_{s+1}}.$$

Differentiating this identity with respect to the $j_{s+2}, \ldots, j_m$-th coordinates we get the identity required.

The general identity (10) follows from the case just proved because every permutation is the result of a finite number of transpositions of consecutive terms.

Theorem 5.4 can be sharpened as follows: if both partial derivatives in (10) exist in G and are continuous, then the equation holds.

Suppose that φ is a function of class C_m on an open set $G \subset R^k$ and that

(11)
$$\varphi_{|j_1,\ldots,j_m}$$

is its partial derivative of order m. Thus

$$j_1, \ldots, j_m$$

is a sequence of positive integers $\leqslant k$. Suppose the integer i ($i = 1, \ldots, k$) appears in this sequence m_i-times. In other words, to obtain the partial derivative (11) we must differentiate m_1-times with respect to the first coordinate, m_2-times with respect to the second coordinate, and so on. It follows from 5.4 that the order of the differentiations is of no consequence; only the multiplicities $m_1, \ldots, m_k$ of differentiation with respect to each of the coordinates play an essential part. Therefore the partial derivative (11) is sometimes denoted by the symbol $\varphi^{(m_1,\dots,m_k)}$ or by the the shorter symbol $\varphi^{(\mathbf{m})}$ where $\mathbf{m}$ is the sequence $(m_1, \dots, m_k)$.

The corresponding traditional notation for the partial derivative (11) is

$$(12)\qquad \frac{\partial^{m_1}}{\partial x_1^{m_1}}\cdot\frac{\partial^{m_2}}{\partial x_2^{m_2}}\cdots\frac{\partial^{m_k}}{\partial x_k^{m_k}}\varphi \quad\text{or}\quad \frac{\partial^{m}\varphi}{\partial x_1^{m_1}\dots\partial x_k^{m_k}}, \quad\text{etc}.$$

5.5. *Suppose $0 < n < m$. A function φ is of class C_m if and only if all its partial derivatives $\varphi_{|j_1,\dots,j_n}$ of order n are of class C_{m-n}.*

This follows immediately from the equivalence of (a) and (c) in 5.2 and from the following generalization of (4)

$$\varphi_{|j_1,\dots,j_m} = (\varphi_{|j_1,\dots,j_n})_{|j_{n+1},\dots,j_m}.$$

5.6. *The sum, difference, product and quotient of functions of class C_m is of class C_m.*

For the sum and difference this follows directly from the identity:

$$(\varphi\pm\psi)^{(m)} = \varphi^{(m)}\pm\psi^{(m)}.$$

For the product the proof is by induction with respect to m. For $m = 1$ it follows from the formula

$$(\varphi\cdot\psi)_{|j} = \varphi_{|j}\cdot\psi+\varphi\cdot\psi_{|j}$$

(see § 2 (5)) which guarantees the continuity of all the partial derivatives $(\varphi\psi)_{|j}$. Suppose that the theorem is true for $m-1$ $(m > 1)$. It follows from the quoted formula and the induction hypothesis that all the partial derivatives $(\varphi\psi)_{|j}$ of first order are of class C_{m-1}. By 5.5, φ is of class C_m.

The proof for the quotient is similar to that for the product.

Observe that if φ is of class C_m and c is a real number, then $c\varphi$ is of class C_m and

$$(c\varphi)^{(m)} = c\varphi^{(m)}.$$

Suppose φ is of class C_m. According to II § 4 (22), for every vector $\mathfrak{a} = [a_1, \dots, a_k]$,

$$(13)\quad \varphi^{(m)}(p)\cdot\mathfrak{a}^m = \varphi^{(m)}(p)\cdot\underbrace{(\mathfrak{a}\times\mathfrak{a}\times\dots\times\mathfrak{a})}_{m\text{-times}} = \sum_{j_1,\dots,j_m=1}^{k}\varphi_{|j_1,\dots,j_m}(p)\cdot a_{j_1}\dots a_{j_m}.$$

In other words, the scalar product of the matrix $\varphi^{(m)}(p)$ and the mth tensor power of $\mathfrak{a}$ is the directional derivative of order m, the differentiation always being performed in the direction $\mathfrak{a}$. The expression $\varphi^{(m)}(p)\cdot\mathfrak{a}^m$ is a form of degree m when considered as a function of $\mathfrak{a}$.

In the traditional notation we have

$$(13') \qquad \varphi^{(m)} \cdot \mathfrak{a}^m = \sum_{j_1, \dots, j_m = 1}^{k} a_{j_1} \dots a_{j_m} \cdot \frac{\partial^m \varphi}{\partial x_{j_1} \dots \partial x_{j_m}} .$$

In the case $k = 2$ $(\mathfrak{a} = [a_1, a_2])$ we verify easily that

$$(14) \qquad \varphi^{(m)} \cdot \mathfrak{a}^m = \sum_{i=0}^{m} \binom{m}{i} a_1^{m-i} a_2 \cdot \frac{\partial^m \varphi}{\partial x_1^{m-i} \partial x_2^i} ,$$

which can be symbolically written

$$(14') \qquad \varphi^{(m)} \cdot \mathfrak{a}^m = \left(a_1 \cdot \frac{\partial}{\partial x_1} + a_2 \cdot \frac{\partial}{\partial x_2} \right)^m \varphi .$$

Indeed, applying the binomial formula for the symbolic power

$$\left(a_1 \frac{\partial}{\partial x_1} + a_2 \frac{\partial}{\partial x_2} \right)^m$$

and multiplying the result formally by φ from the right-hand side, we get the right-hand side of (14).

The general formula (13') can also be formally written in the similar form

$$(15) \qquad \varphi^{(m)} \cdot \mathfrak{a}^m = \left(a_1 \frac{\partial}{\partial x_1} + \dots + a_k \frac{\partial}{\partial x_k} \right)^m \varphi .$$

As an illustration of the notions introduced in this section we shall calculate directional derivatives, partial derivatives and derivatives of all orders of a function φ defined below.

Let F be a symmetric m-linear form on V_k^m. By definition (see II § 6, p. 70)

$$F(\mathfrak{a}_1, \dots, \mathfrak{a}_m) = \mathfrak{A} \cdot (\mathfrak{a}_1 \times \dots \times \mathfrak{a}_m)$$

where

$$\mathfrak{A} = [a_{j_1, \dots, j_m}]_{j_1, \dots, j_m \leqslant k}$$

is a symmetric matrix. Let $p_0 = (x_{0,1}, \dots, x_{0,k})$ be a fixed point in R^k and let

$$\varphi(p) = F(p - p_0, \dots, p - p_0) = \mathfrak{A} \cdot (p - p_0)^m \qquad \text{for} \qquad p \in R^k .$$

In other symbols,

$$\varphi(x_1, \dots, x_k) = \sum_{j_1, \dots, j_m = 1}^{k} a_{j_1, \dots, j_m} (x_{j_1} - x_{0, j_1}) \cdot \dots \cdot (x_{j_m} - x_{0, j_m}) .$$

Since F is m-linear, we have for any vector $\mathfrak{a}$ and real number Δt

$$\begin{aligned} \varphi(p + \Delta t \cdot \mathfrak{a}) &= F\big((p - p_0) + \Delta t \cdot \mathfrak{a}, \dots, (p - p_0) + \Delta t \cdot \mathfrak{a}\big) \\ &= a_0 + a_1 \cdot \Delta t + a_2 \cdot (\Delta t)^2 + \dots + a_m \cdot (\Delta t)^m , \end{aligned}$$

where

$$a_0 = F(p-p_0, p-p_0, \dots, p-p_0) = \varphi(p),$$

and

$$\begin{aligned} a_1 &= F(\mathfrak{a}, p-p_0, p-p_0, \dots, p-p_0)+F(p-p_0, \mathfrak{a}, p-p_0, \dots, p-p_0)+ \\ &\quad +F(p-p_0, p-p_0, \mathfrak{a}, \dots, p-p_0)+\dots+F(p-p_0, p-p_0, \dots, p-p_0, \mathfrak{a}) \\ &= m\cdot F(\mathfrak{a}, p-p_0, p-p_0, \dots, p-p_0) \end{aligned}$$

by the symmetry of F. Similar formulas for the numbers $a_2, \dots, a_m$ which do not depend on Δt are not quoted because we shall not need to use them.

It follows from the above formulas that

$$\lim_{\Delta t\to 0} \frac{\varphi(p+\Delta t\cdot\mathfrak{a})-\varphi(p)}{\Delta t} = a_1,$$

i.e.

$$\begin{aligned} \varphi'_{\mathfrak{a}}(p) &= m\cdot F\big(\mathfrak{a}, p-p_0, p-p_0, \dots, p-p_0\big) = m\cdot\mathfrak{A}\cdot\big(\mathfrak{a}\times(p-p_0)^{m-1}\big) \\ &= m\cdot\big(\mathfrak{A}\cdot(p-p_0)^{m-1}\big)\cdot\mathfrak{a}. \end{aligned}$$

Consequently

$$\varphi'(p) = m\cdot\mathfrak{A}\cdot(p-p_0)^{m-1}.$$

To calculate the directional derivative $\varphi''_{\mathfrak{a},\mathfrak{b}}$ let us first observe that, by the m-linearlity of F,

$$\begin{aligned} \varphi'_{\mathfrak{a}}(p+\Delta t\cdot\mathfrak{b}) &= m\cdot F\big(\mathfrak{a}, (p-p_0)+\Delta t\cdot\mathfrak{b}, \dots, (p-p_0)+\Delta t\cdot\mathfrak{b}\big) \\ &= b_0+b_1\cdot\Delta t+b_2\cdot(\Delta t)^2+\dots+b_{m-1}\cdot(\Delta t)^{m-1}, \end{aligned}$$

where

$$b_0 = mF(\mathfrak{a}, p-p_0, \dots, p-p_0) = \varphi'_{\mathfrak{a}}(p),$$

$$\begin{aligned} b_1 &= m\cdot\big(F(\mathfrak{a}, \mathfrak{b}, p-p_0, p-p_0, \dots, p-p_0)+F(\mathfrak{a}, p-p_0, \mathfrak{b}, p-p_0, \dots, p-p_0)+ \\ &\qquad +\dots+F(\mathfrak{a}, p-p_0, \dots, p-p_0, \mathfrak{b})\big) \\ &= m(m-1)F(\mathfrak{a}, \mathfrak{b}, p-p_0, \dots, p-p_0), \end{aligned}$$

and the numbers $b_2, \dots, b_{m-1}$ do not depend on Δt. Consequently

$$\lim_{\Delta t\to 0} \frac{\varphi'_{\mathfrak{a}}(p+\Delta t\cdot\mathfrak{b})-\varphi'_{\mathfrak{a}}(p)}{\Delta t} = b_1,$$

i.e.

$$\begin{aligned} \varphi''_{\mathfrak{a},\mathfrak{b}}(p) &= m(m-1)\cdot F(\mathfrak{a}, \mathfrak{b}, p-p_0, \dots, p-p_0) \\ &= m(m-1)\cdot\mathfrak{A}\cdot\big(\mathfrak{a}\times\mathfrak{b}\times(p-p_0)^{m-2}\big) = m(m-1)\cdot\big(\mathfrak{A}\cdot(p-p_0)^{m-2}\big)\cdot(\mathfrak{a}\times\mathfrak{b}). \end{aligned}$$

Hence it follows that

$$\varphi''(p) = m(m-1)\cdot\mathfrak{A}\cdot(p-p_0)^{m-2}.$$

Continuing the above procedure, we prove that for every positive integer $r \leqslant m$

$$\begin{aligned}\varphi^{(r)}_{\mathfrak{a}_1,\ldots,\mathfrak{a}_r}(p) &= m(m-1)\ldots(m-r+1)\cdot\mathfrak{A}\cdot\big(\mathfrak{a}_1\times\ldots\times\mathfrak{a}_r\times(p-p_0)^{m-r}\big)\\ &= m(m-1)\ldots(m-r+1)\cdot\big(\mathfrak{A}\cdot(p-p_0)^{m-r}\big)\cdot(\mathfrak{a}_1\times\ldots\times\mathfrak{a}_r)\,,\end{aligned}$$

and consequently,

$$\varphi^{(r)}(p) = m(m-1)\ldots(m-r+1)\cdot\mathfrak{A}\cdot(p-p_0)^{m-r}.$$

The first formula yields the following formula for the partial derivatives

$$\begin{aligned}&\varphi_{|j_1,\ldots,j_r}(x_1,\,\ldots,\,x_k)\\ &= m(m-1)\ldots(m-r+1)\sum_{j_{r+1},\ldots,j_m=1}^{k} a_{j_1,\ldots,j_r,j_{r+1},\ldots,j_m}(x_{j_{r+1}}-x_{0,j_{r+1}})\ldots(x_{j_m}-x_{0,j_m})\,.\end{aligned}$$

In particular for $r = m$ we have

$$\varphi^{(m)}_{\mathfrak{a}_1,\ldots,\mathfrak{a}_m}(p) = m!\,\mathfrak{A}\cdot(\mathfrak{a}_1\times\ldots\times\mathfrak{a}_m)\,,$$

$$\varphi^{(m)}(p) = m!\,\mathfrak{A}\,,$$

$$\varphi_{|j_1,\ldots,j_m}(p) = m!\,a_{j_1,\ldots,j_m}\,.$$

Since all the directional derivatives of order m are constant, we have for $r > m$

$$\varphi^{(r)}_{\mathfrak{a}_1,\ldots,\mathfrak{a}_r}(p) = 0$$

and consequently,

$$\varphi^{(r)}(p) = 0 \qquad \text{and} \qquad \varphi_{|j_1,\ldots,j_r}(p) = 0\,.$$

The function $\varphi(p) = \mathfrak{A}\cdot(p-p_0)^m$ is the k-dimensional analogue of the function $f(x) = a(x-x_0)^m$ of one real variable x. We see that, from the point of view of differentiation, the function φ has properties similar to the function f. In particular the formula for $\varphi^{(r)}$ has the same form as the analogous formula

$$f^{(r)}(x) = m(m-1)\ldots(m-r+1)\cdot a\cdot(x-x_0)^{m-r} \qquad (r \leqslant m)\,.$$

As an application of the formulas obtained we shall prove Theorem II 6.4 announced on p. 70: if a matrix

$$\mathfrak{A} = [a_{j_1,\ldots,j_m}]_{j_1,\ldots,j_m\leqslant k}$$

is symmetric and $\mathfrak{A}\cdot\mathfrak{a}^m = 0$ for every vector $\mathfrak{a}\in V_k$, then $\mathfrak{A} = 0$, i.e. $\mathfrak{A}$ is a zero matrix.

Indeed, under the above hypothesis on $\mathfrak{A}$, the function φ defined above is identically equal to 0. Hence it follows that all its derivatives are equal to 0. In particular,

$$\mathfrak{A} = \frac{1}{m!}\varphi^m = 0 .$$

EXERCISES

1. Prove that

$$\varphi^{(m)}\cdot \mathfrak{a}^m = (\mathfrak{a}\nabla)^m \varphi$$

for any function φ of class C_m (for notation see § 3, exercise 10).

2. Calculate the partial derivatives of third order of the function

$$\varphi(x_1, x_2) = e^{\sin x_1}\cdot \sin e^{x_2} .$$

3. Prove that the second derivative of the function

$$\varphi(x_1, x_2, x_3) = e^{x_1^2+x_2^2+x_3^2}$$

at the point (0, 0, 0) is a positive matrix (see II § 6, p. 71).

4. The derivative of order $2m$ at the point $(0, \ldots, 0)$ of the function

$$\varphi(x_1, \ldots, x_k) = e^{x_1^{2m}+x_2^{2m}+\ldots+x_k^{2m}}$$

is positive.

§ 6. The Taylor formula. Let φ be a real function defined on an open subset G of the Euclidean space R^k. Let $p \in G$ and $\mathfrak{a} \in V_k$. Consider the auxiliary function (see § 1 (2))

$$\bar{\varphi}(t) = \varphi(p + t\mathfrak{a})$$

defined in the open set

$$G_0 = \{t\colon\ p + t\mathfrak{a} \in G\} \subset R .$$

We shall show that the existence of the directional derivative $\varphi^{(m)}_{\mathfrak{a},\ldots,\mathfrak{a}}$ in G implies the existence of the derivative $\bar{\varphi}^{(m)}$ in G_0 and the equation

(1) $$\bar{\varphi}^{(m)}(t) = \varphi^{(m)}_{\mathfrak{a},\ldots,\mathfrak{a}}(p + t\mathfrak{a}) \quad \text{for} \quad t \in G_0 .$$

The proof of (1) is by induction. The case $m = 1$ was proved in § 1 (4). Suppose that (1) holds for $m-1$ $(m > 1)$, i.e.

$$\bar{\varphi}^{(m-1)}(t) = \psi(p + t\mathfrak{a}) ,$$

where, for brevity,

$$\psi(q) = \varphi^{(m-1)}_{\mathfrak{a},\ldots,\mathfrak{a}}(q) \quad \text{for} \quad q \in G .$$

By hypothesis, $\psi'_{\mathfrak{a}}(q)$ exists for every $q \in G$, viz.

$$\psi'^{(q)}_{\mathfrak{a}}(q) = \varphi^{(m)}_{\mathfrak{a},\ldots,\mathfrak{a}}(q) .$$

For $q = p + t\mathfrak{a}$ we have

$$\bar{\varphi}^{(m)}(t) = \lim_{\Delta t \to 0} \frac{\bar{\varphi}^{(m-1)}(t+\Delta t) - \bar{\varphi}^{(m-1)}(t)}{\Delta t} = \lim_{\Delta t \to 0} \frac{\psi(q+\Delta t \cdot \mathfrak{a}) - \psi(q)}{\Delta t}$$
$$= \psi'_{\mathfrak{a}}(q) = \varphi^{(m)}_{\mathfrak{a},\ldots,\mathfrak{a}}(p+t\mathfrak{a}) .$$

Identity (1) will be used in the proof of the following *Taylor formula* for the function φ.

6.1. *If φ is of class C_m in G and the segment joining the points p and $p+\mathfrak{a}$ is contained in G, then there is a number θ in the interval $(0;1)$, such that*

(2) $$\varphi(p+\mathfrak{a}) = \varphi(p) + \frac{\varphi'(p)\cdot\mathfrak{a}}{1!} + \frac{\varphi''(p)\cdot\mathfrak{a}^2}{2!} + \ldots + \frac{\varphi^{(m-1)}(p)\cdot\mathfrak{a}^{m-1}}{(m-1)!} + \frac{\varphi^{(m)}(p+\theta\mathfrak{a})\cdot\mathfrak{a}^m}{m!} .$$

The hypothesis that the segment joining p and $p+\mathfrak{a}$ is a subset of G implies that the interval $\langle 0;1\rangle$ is a subset of G_0. Applying the Taylor formula for functions of one real variable to the function $\bar{\varphi}$, we get

$$\bar{\varphi}(1) = \bar{\varphi}(0) + \frac{\bar{\varphi}'(0)}{1!} + \frac{\bar{\varphi}''(0)}{2!} + \ldots + \frac{\bar{\varphi}^{(m-1)}(0)}{(m-1)!} + \frac{\bar{\varphi}^{(m)}(\theta)}{m!} ,$$

where θ is a number in $(0;1)$. By the definition of $\bar{\varphi}$,

$$\bar{\varphi}(1) = \varphi(p+\mathfrak{a}) \quad \text{and} \quad \bar{\varphi}(0) = \varphi(p) .$$

By (1) and 5.1 (8),

$$\bar{\varphi}'(0) = \varphi'(p)\cdot\mathfrak{a} ,$$

$$\bar{\varphi}''(0) = \varphi''(p)\cdot\mathfrak{a}^2, \quad \ldots, \quad \bar{\varphi}^{(m-1)}(0) = \varphi^{(m-1)}(p)\cdot\mathfrak{a}^{m-1}$$

and

$$\bar{\varphi}^{(m)}(\theta) = \varphi^{(m)}(p+\theta\mathfrak{a})\cdot\mathfrak{a}^m .$$

This implies (2).

Assuming

$$p = (x_1, \ldots, x_k) , \quad \mathfrak{a} = [a_1, \ldots, a_k] ,$$

we can write (2) as follows (see 5.1):

(2') $$\varphi(x_1+a_1, \ldots, x_k+a_k)$$
$$= \varphi(x_1, \ldots, x_k) + \sum_{r=1}^{m-1} \frac{1}{r!} \sum_{j_1,\ldots,j_r=1}^{k} \varphi_{|j_1,\ldots,j_r}(x_1, \ldots, x_k)\, a_{j_1} \ldots a_{j_r} +$$
$$+ \frac{1}{m!} \sum_{j_1,\ldots,j_m=1}^{k} \varphi_{|j_1,\ldots,j_m}(x_1+\theta a_1, \ldots, x_k+\theta a_k)\, a_{j_1} \ldots a_{j_m} .$$

In the traditional notation the Taylor formula (2) is written as follows (see § 5 (15)):

$$(2'')\quad \varphi(p+\mathfrak{a}) = \varphi(p) + \frac{1}{1!}\left(a_1\frac{\partial}{\partial x_1}+\ldots+a_k\frac{\partial}{\partial x_k}\right)\varphi(p) + \\ +\frac{1}{2!}\left(a_1\frac{\partial}{\partial x_1}+\ldots+a_k\frac{\partial}{\partial x_k}\right)^2\varphi(p)+\ldots+\frac{1}{(m-1)!}\left(a_1\frac{\partial}{\partial x_1}+\ldots+a_k\frac{\partial}{\partial x_k}\right)^{m-1}\varphi(p) + \\ +\frac{1}{m!}\left(a_1\frac{\partial}{\partial x_1}+\ldots+a_k\frac{\partial}{\partial x_k}\right)^{m}\varphi(p+\theta\mathfrak{a}).$$

In particular, for $k=2$ the Taylor formula can be written (see § 5 (14))

$$(2''')\quad \varphi(x_1+a_1, x_2+a_2) = \varphi(x_1, x_2) + \sum_{r=1}^{m-1}\frac{1}{r!}\sum_{i=0}^{r}\binom{r}{i} a_1^{r-i}a_2^{i}\frac{\partial^r}{\partial x_1^{r-i}\partial x_2^{i}}\varphi(x_1, x_2) + \\ +\frac{1}{m!}\sum_{i=0}^{n}\binom{m}{i} a_1^{m-i}a_2^{i}\frac{\partial^r}{\partial x_1^{m-i}\partial x_2^{i}}\varphi(x_1+\theta a_1, x_2+\theta a_2).$$

Observe that, if we replace $\mathfrak{a}$ by $q-p$ in (2), we can formulate Theorem 6.1 as follows: If φ is of class C_m in G and the segment joining p and q is a subset of G, then there is a θ in $(0;1)$ such that

$$(3)\quad \varphi(q) = \varphi(p) + \frac{\varphi'(p)\cdot(q-p)}{1!} + \frac{\varphi''(p)\cdot(q-p)^2}{2!} + \ldots + \\ + \frac{\varphi^{(m-1)}(p)\cdot(q-p)^{m-1}}{(m-1)!} + \frac{\varphi^{(m)}\big(p+\theta(q-p)\big)\cdot(q-p)^m}{m!}.$$

The following slightly weaker formulation of 6.1 is often used in practice.

6.2. *If φ is of class C_m in G and $p\in G$ is fixed, then*

$$(4)\quad \varphi(q) = \varphi(p) + \frac{\varphi'(p)\cdot(q-p)}{1!} + \frac{\varphi''(p)\cdot(q-p)^2}{2!} + \ldots + \\ + \frac{\varphi^{(m-1)}(p)\cdot(q-p)^{m-1}}{(m-1)!} + \frac{\varphi^{(m)}(p)\cdot(q-p)^m}{m!} + \varrho(q) \quad \textit{for} \quad q\in G,$$

where ϱ is a continuous function on G such that

$$(5)\quad \lim_{q\to p}\frac{\varrho(q)}{|q-p|^m} = 0.$$

Namely, the function

$$\varrho(q) = \varphi(q) - \left(\varphi(p) + \frac{\varphi'(p)\cdot(q-p)}{1!} + \ldots + \frac{\varphi^{(m)}(p)\cdot(q-p)^m}{m!}\right)$$

has the properties required. In fact, by definition ϱ is continuous and (4) holds. Let η be a positive number such that $B(p,\eta)\subset G$. Therefore, if $q \in \mathrm{B}(p, \eta)$, then the segment joining p and q is a subset of G. Hence it follows from (3) that

$$\varrho(q) = \frac{1}{m!}\left(\varphi^{(m)}\big(p+\theta(q-p)\big) - \varphi^{(m)}(p)\right)\cdot(q-p)^m \quad \text{for} \quad q \in \mathrm{B}(p, \eta),$$

where θ is a number in $(0;1)$ depending on q. By II § 4 (14) and (19'')

$$|\varrho(q)| \leqslant \frac{1}{m!}\|\varphi^{(m)}\big(p+\theta(q-p)\big)-\varphi^{(m)}(p)\|\cdot|q-p|$$

and consequently,

$$0 \leqslant \frac{|\varrho(q)|}{|q-p|^m} \leqslant \frac{1}{m!}\|\varphi^{(m)}\big(p+\theta(q-p)\big)-\varphi^{(m)}(p)\|.$$

Since φ if of class C_m, $\varphi^{(m)}\big(p+\theta(q-p)\big)$ tends to $\varphi^{(m)}(p)$ as $q \to p$, i.e.

$$\left\|\varphi^{(m)}\big(p+\theta(q-p)\big)-\varphi^{(m)}(p)\right\| \to 0.$$

This proves (5).

Sometimes Theorem 6.2 is formulated as follows:

6.3. *If φ is a function of class C_m in an open set $G \subset R^k$ and $p \in G$ is fixed, then there is an m-indexed k-dimensional matrix field $\mathfrak{r}$ on G such that*

$$\varphi(q) = \varphi(p) + \frac{\varphi'(p)\cdot(q-p)}{1!} + \ldots + \frac{\varphi^{(m)}(p)\cdot(q-p)^m}{m!} + \mathfrak{r}(q)\cdot(q-p)^m \quad \textit{for } q \in G \tag{4'}$$

and

$$\lim_{q\to p} \mathfrak{r}(q) = \mathfrak{r}(p) = 0. \tag{5'}$$

Let $\eta > 0$ be such that $\mathrm{B}(p, \eta) \subset G$. If $q \in G - \mathrm{B}(p, \eta)$, let $\mathfrak{r}(q)$ be any matrix such that (4') holds. If $q \in \mathrm{B}(p, \eta)$, then the segment joining q and p is a subset of G and therefore there is a number θ in $(0;1)$ such that (3) holds. Define

$$\mathfrak{r}(q) = \frac{1}{m!}\left(\varphi^{(m)}\big(p+\theta(q-p)\big) - \varphi^{(m)}(p)\right).$$

(4') follows from (3), and (5') follows from the continuity of $\varphi^{(m)}$.

Now we introduce, by induction, the notion of a differential of any order r $(r = 1, 2, \ldots)$ of φ at a point $p \in G$, the differential of order r being a form of degree r on V_k. By the *differential of first order* at p we shall mean the differential of φ at p, defined in § 3, p. 138. Suppose that the differentials $F_1, \ldots, F_{m-1}$ of φ at p of orders $1, \ldots, m-1$ are defined. Then a form F_m of degree m is said to be a *differential of order m of φ at p* if

$$\lim_{q\to p} \frac{\varphi(q) - \left(\varphi(p) + \frac{1}{1!}F_1(q-p) + \ldots + \frac{1}{(m-1)!}F_{m-1}(q-p) + \frac{1}{m!}F_m(q-p)\right)}{|q-p|^m} = 0,$$

i.e. if

(8) $$\varphi(q) = \varphi(p) + \frac{1}{1!}F_1(q-p) + \ldots + \frac{1}{(m-1)!}F_{m-1}(q-p) + \frac{1}{m!}F_m(q-p) + \varrho(q) \quad \text{for} \quad q \in G$$

and

(9) $$\lim_{q\to p} \frac{\varrho(q)}{|q-p|^m} = 0 .$$

It follows from 6.2 that if φ is of class C_m, then φ has differentials of all the orders $1, \ldots, m$ at any point $p \in G$, namely

$$F_1(\mathfrak{a}) = \varphi'(p)\cdot\mathfrak{a} , \quad \ldots, \quad F_m(\mathfrak{a}) = \varphi^{(m)}(p)\cdot\mathfrak{a}^m$$

are the differentials of φ at p of orders $1, \ldots, m$ respectively.

EXERCISES

1. Write the Taylor formula for the function

$$\varphi(x_1, \ldots, x_k) = e^{x_1+\ldots+x_k} .$$

2. Write the Taylor formula at the point $p = (0, \ldots, 0)$ for the function

$$\varphi(x_1, \ldots, x_k) = \log(1+x_1+\ldots+x_k) .$$

3. Write the Taylor formula at the point $p = (0, 0)$ for the function

$$\varphi(x_1, x_2) = \sin x_1 \cdot e^{2x_2}$$

and $m = 3$.

4. If $\mathfrak{A}$ is a symmetric m-indexed k-dimensional matrix and $\lim_{\mathfrak{a}\to 0} \frac{\mathfrak{A}\cdot\mathfrak{a}^m}{|\mathfrak{a}|^m} = 0$, then $\mathfrak{A} = 0$.

Thus, if the differential of order m of a function φ at a point p exists, it is unique.

§ 7. Extrema of a function. Let φ be a real function defined on a subset G of a metric space and let $p \in G$. We say that φ has a *local maximum* at the point p if there exists a neighbourhood U of p in the metric space G (see III § 4) such that

(1) $$\varphi(q) \leqslant \varphi(p) \quad \text{for every } q \in U .$$

If there exists a neighbourhood U of p in G such that

(1′) $$\varphi(q) < \varphi(p) \quad \text{for every } q \in U, q \neq p ,$$

we say that φ has a *proper local maximum* at p.

Similarly, we say that φ has a *local minimum* at p if there is a neighbourhood U of p in the metric space G such that

(2) $$\varphi(q) \geqslant \varphi(p) \quad \text{for every } q \in U .$$

If there exists a neighbourhood U of p in G, such that

(2′) $$\varphi(q) > \varphi(p) \quad \text{for every } q \in U, \quad q \neq p\,,$$

we say that φ has a *proper local minimum* at p.

We say that φ has an *extremum* at p (or that p is an *extremal point* of φ) if φ has a local maximum or a local minimum at p.

In this section G will be an open subset of the Euclidean space R^k. We may always assume that the naighbourhood U mentioned in the above definitions is a ball, $U = \mathrm{B}(p, \eta)$.

7.1. *If p is an extremal point of φ and if the derivative $\varphi'(p)$ exists, then $\varphi'(p)$ is the zero vector*:

(3) $$\varphi'(p) = 0\,.$$

Let $\mathfrak{a}$ be any vector in R^k, and let $\bar{\varphi}$ be the auxiliary function

$$\bar{\varphi}(t) = \varphi(p + t\mathfrak{a})$$

as in § 1 and § 6. If (1) holds, then $\bar{\varphi}(t) \leqslant \bar{\varphi}(0)$ for $|t| < \eta/|\mathfrak{a}|$. If (2) holds, then $\bar{\varphi}(t) \geqslant \bar{\varphi}(0)$ for $|t| < \eta/|\mathfrak{a}|$. Thus if p is an extremal point of φ, then 0 is an extremal point of $\bar{\varphi}$. By a known theorem from differential calculus of functions of one real variable, $\bar{\varphi}'(0) = 0$. Since $\varphi'(p)\cdot\mathfrak{a} = \varphi'_{\mathfrak{a}}(p) = \bar{\varphi}'(0)$, we infer that $\varphi'(p)\cdot\mathfrak{a} = 0$ for every vector $\mathfrak{a}$. This proves (3).

A point p is said to be a *critical* point of φ if (3) holds. Condition (3) is necessary but not sufficient for the function φ to have an extremum at p. In order to determine whether φ has an extremum at a critical point p we must examine derivatives of higher orders. This is the subject of Theorem 7.2 below.

We recall that, according to a general definition from II § 6, p. 71, if φ is of class C_m and p is fixed, then the differential of order m at p

(4) $$F(\mathfrak{a}) = \varphi^{(m)}(p)\cdot\mathfrak{a}^m$$

is positive if $F(\mathfrak{a}) > 0$ for every vector $\mathfrak{a} \neq 0$;

is negative if $F(\mathfrak{a}) < 0$ for every vector $\mathfrak{a} \neq 0$;

has no sign if there are vectors $\mathfrak{a}_1, \mathfrak{a}_2$ such that

$$F(\mathfrak{a}_1) > 0 \quad \text{and} \quad F(\mathfrak{a}_2) < 0\,.$$

If m is odd and $\varphi^{(m)}(p) \neq 0$, then F has no sign. If m is even, then the derivative $\varphi^{(m)}$ at p

is positive if F is positive;

is negative if F is negative;

has no sign if F has no sign.

7.2. *Suppose that for a given point* $p \in G$

(5) $$\varphi'(p) = 0\,, \quad \varphi''(p) = 0\,, \quad \ldots, \quad \varphi^{(m-1)}(p) = 0 \qquad (m>1)$$

but

(6) $$\varphi^{(m)}(p) \neq 0\,,$$

φ *being a function of class* C_m.

If the differential (4) *is positive, then* φ *has a proper local minimum at* p. *If the differential* (4) *is negative, then* φ *has a proper local maximum at* p. *If the differential* (4) *has no sign, then the function* φ *has no extremum at* p.

In other words, *if* m *is odd, then* φ *has no extremum at* p. *If* m *is even, then*:

if $\varphi^{(m)}(p)$ *is positive, then* φ *has a proper local minimum at* p;

if $\varphi^{(m)}(p)$ *is negative, then* φ *has a proper local maximum at* p;

if $\varphi^{(m)}(p)$ *has no sign, then* φ *has no extremum at* p

Let $\eta_0 > 0$ be such that $B(p, \eta_0) \subset G$. By the Taylor formula § 6 (3) and the hypothesis (5), for every $q \in B(p, \eta_0)$ there exists a number θ in $(0; 1)$ such that

(7) $$\varphi(q) = \varphi(p) + \frac{1}{m!}\varphi^{(m)}\big(p+\theta(q-p)\big)\cdot(q-p)^m\,.$$

Suppose that F is positive, i.e. m is even and the matrix $\varphi^{(m)}(p)$ is positive. By III 8.11, there exists a number $\varepsilon > 0$ such that every symmetric matrix $\mathfrak{B}$ satisfying the inequality

$$\|\mathfrak{B} - \varphi^{(m)}(p)\| < \varepsilon$$

is positive. Since $\varphi^{(m)}$ is continuous, there exists an $\eta > 0$ $(\eta < \eta_0)$ such that

$$|q-p| < \eta \quad \text{implies} \quad \|\varphi^{(m)}(q) - \varphi^{(m)}(p)\| < \varepsilon\,.$$

If $q \in B(p, \eta)$ and $0 < \theta < 1$, then $\big|\big(p+\theta(q-p)\big)-p\big| < \eta$ and consequently

$$\big\|\varphi^{(m)}\big(p+\theta(q-p)\big) - \varphi^{(m)}(p)\big\| < \varepsilon\,.$$

Thus the matrix $\varphi^{(m)}\big(p+\theta(q-p)\big)$ is positive, i.e.

$$\varphi^{(m)}\big(p+\theta(q-p)\big)\cdot\mathfrak{a}^m > 0$$

for every non-zero vector $\mathfrak{a}$. Assuming $\mathfrak{a} = q-p$, we infer that for $q \in B(p, \eta)$ $(q \neq p)$ the last term in (7) is positive, and consequently,

$$\varphi(q) > \varphi(p) \quad \text{for} \quad q \in B(p, \eta), \quad q \neq p\,.$$

This proves that φ has a proper local minimum at p.

In the same way we prove that if F is negative, i.e. m is even and $\varphi^{(m)}(p)$ is negative, then φ has a proper local maximum at p.

Suppose now that F has no sign, i.e. there exist vectors $\mathfrak{a}_1$ and $\mathfrak{a}_2$ such that

$$\varphi^{(m)}(p)\cdot\mathfrak{a}_1^m > 0 \quad \text{and} \quad \varphi^{(m)}(p)\cdot\mathfrak{a}_2^m < 0 .$$

To prove that φ has no extremum at p it suffices to show that for every $\eta > 0$ $(\eta < \eta_0)$ there are points $p_1, p_2 \in \mathrm{B}(p, \eta)$ such that

$$\varphi(p_1) > \varphi(p) \quad \text{and} \quad \varphi(p_2) < \varphi(p) .$$

Since for every fixed vector $\mathfrak{a}$ the expression $\varphi^{(m)}(q)\cdot\mathfrak{a}^m$ is a continuous function of q, there is a number $\delta > 0$ $(\delta < \eta)$ such that

$$\varphi^{(m)}(q)\cdot\mathfrak{a}_1^m > 0 \quad \text{and} \quad \varphi^{(m)}(q)\cdot\mathfrak{a}_2^m < 0 \quad \text{if} \quad |q-p| \leqslant \delta .$$

If $|q-p| \leqslant \delta$ and $0 < \theta < 1$, then $\|(p+\theta(q-p))-p| < \delta$ and consequently

$$\varphi^{(m)}\big(p+\theta(q-p)\big)\cdot\mathfrak{a}_1^m > 0 \quad \text{and} \quad \varphi^{(m)}\big(p+\theta(q-p)\big)\cdot\mathfrak{a}_2^m < 0 .$$

Let $p_1 = p + t_1\mathfrak{a}_1$ and $p_2 = p + t_2\mathfrak{a}_2$, where $t_1 = \dfrac{\delta}{|\mathfrak{a}_1|} > 0$ and $t_2 = \dfrac{\delta}{|\mathfrak{a}_2|} > 0$. By definition, $|p_1-p| = t_1\mathfrak{a}_1$, $|p_2-p| = t_2\mathfrak{a}_2$, and consequently $|p_1-p| = \delta < \eta$, $|p_2-p| = \delta < \eta$. Assuming $q = p_1$ in (7) we see that the last term in (7) is positive since it is equal to the number

$$\frac{1}{m!}t_1^m\varphi^{(m)}\big(p+\theta(p_1-p)\big)\cdot\mathfrak{a}_1^m .$$

Thus

$$\varphi(p_1) > \varphi(p) .$$

Letting $q = p_2$ in (7) we conclude similarly that the last term in (7) is negative and consequently

$$\varphi(p_2) < \varphi(p) .$$

This completes the proof of 7.2.

Theorems 7.1 and 7.2 yield the following practical method of finding extremal points of a function φ of sufficiently high class C_m. First we look for critical points, i.e. for all the solutions $p = (x_1, \ldots, x_k)$ of the following system of equations:

$$(8) \qquad \begin{aligned} &\varphi_{|1}(x_1, \ldots, x_k) = 0 , \\ &\cdots\cdots\cdots\cdots \\ &\varphi_{|k}(x_1, \ldots, x_k) = 0 . \end{aligned}$$

The solutions $p = (x_1, \ldots, x_k)$ of (8) are points which can be extremal points of φ (other points are not extremal). To verify whether a point p satisfying (8) is really extremal, we have to apply Theorem 7.2 to this point. The first derivative of φ at p is equal to zero. We calculate consecutively the derivatives of higher orders, until we find the smallest integer m such that $\varphi^{(m)}(p) \neq 0$. If m is odd, φ has no extremum at p. If m is even, we must check the sign of $\varphi^{(m)}(p)$. If $\varphi^{(m)}(p)$ is positive, negative or has no sign, our problem is completely solved by 7.2: φ has a minimum, a maximum or no extremum at p, respectively. However, Theorem 7.2 does not always give a definite answer. If the matrix $\varphi^{(m)}(p)$ is non-negative but not positive (or non-positive but not negative), Theorem 7.2 does not give any answer. Actually, it is possible that φ has an extremum at such a point but it is also possible that φ has no extremum.

In practice we mainly consider the case where $m = 2$. For $m > 2$ it is very difficult to decide whether a given m-indexed matrix is positive or negative, etc.

The case where $k = 2$ is especially simple. To discuss it, let us recall the following theorem from algebra.

Let

$$\mathfrak{A} = \begin{bmatrix} a_{1,1} & a_{1,2} \\ a_{2,1} & a_{2,2} \end{bmatrix} \qquad (a_{1,2} = a_{2,1})$$

be a symmetric matrix and let

$$F(\mathfrak{a}) = \mathfrak{A}\cdot\mathfrak{a}^2 = a_{1,1}\alpha_1^2 + 2a_{1,2}\alpha_1\alpha_2 + a_{2,2}\alpha_2^2 \quad \text{for} \quad \mathfrak{a} = [\alpha_1, \alpha_2]$$

be the corresponding form of degree 2. If the determinant

$$/\mathfrak{A}/ = \begin{vmatrix} a_{1,1} & a_{1,2} \\ a_{2,1} & a_{2,2} \end{vmatrix} = a_{1,1}a_{2,2} - a_{1,2}^2$$

is negative, then F has no sign. If the determinant $/\mathfrak{A}/$ is positive, then the numbers $a_{1,1}$, $a_{2,2}$ have the same sign; if they are positive, then F is positive, and if they are negative, then F is negative.

If the determinant $/\mathfrak{A}/$ is equal to zero, then either both $a_{1,1}$ and $a_{2,2}$ are non-negative or both are non-positive. In the first case F is non-negative. In the second case F is non-positive.

To prove the above remarks it is enough to discuss the binomial $f(x) = a_{1,1}x^2 + 2a_{1,2}x + a_{2,2}$ and to substitute $x = \alpha_1/\alpha_2$.

Suppose now that φ is a function of class C_2 defined on an open subset G of the plane R^2 and $p \in G$, $\varphi'(p) = 0$, i.e.

$$\varphi_{|1}(p) = 0\,, \quad \varphi_{|2}(p) = 0\,,$$

but the second derivative

$$\varphi''(p) = \begin{bmatrix} \varphi_{|1,1}(p) & \varphi_{|1,2}(p) \\ \varphi_{|2,1}(p) & \varphi_{|2,2}(p) \end{bmatrix} \qquad (\varphi_{|1,2}(p) = \varphi_{|2,1}(p))$$

is not the zero matrix. Let

$$F(\mathfrak{a}) = \varphi''(p)\cdot\mathfrak{a} = \varphi_{|1,1}(p)\alpha_1^2 + 2\varphi_{|1,2}(p)\alpha_1\alpha_2 + \varphi_{|2,2}(p)\alpha_2^2$$

for $\mathfrak{a} = [\alpha_1, \alpha_2]$. If the determinant

$$/\varphi''(p)/ = \begin{vmatrix} \varphi_{|1,1}(p) & \varphi_{|1,2}(p) \\ \varphi_{|2,1}(p) & \varphi_{|2,2}(p) \end{vmatrix} = \varphi_{|1,1}(p)\varphi_{|2,2}(p) - (\varphi_{|1,2}(p))^2$$

is negative, then F has no sign and therefore p is not an extremal point of φ. If the determinant $/\varphi''(p)/$ is positive, then neither of the partial derivatives $\varphi_{|1,1}(p), \varphi_{|2,2}(p)$ vanishes and both have the same sign. If they are positive, then F is positive and φ has a proper local minimum at p. If they are negative, then F is negative and φ has a proper local maximum at p.

If $/\varphi''(p)/ = 0$, then F is either non-negative or non-positive. However, Theorem 7.2 does not give any answer in this case.

EXERCISES

1. Determine all extremal points of the function

$$\varphi(x_1, x_2) = \sin x_1 \cdot \sin x_2 .$$

2. Find the least upper bound and the greatest lower bound of the function

$$\varphi(x_1, x_2) = (x_1 + x_2)e^{-(x_1^2 + x_2^2)}$$

on the whole plane R^2.

3. Prove the inequality

$$xe^{x(1+y^2)^2} \geqslant -e^{-1}$$

for all real x, y.

4. Let K be the class of all triangles T for which the sum of the lengths of the edges of T is equal to a given number $a > 0$. For what triangles $T \in K$ does the area of T have the greatest value?

5. Find the largest and the smallest value of the function

$$\varphi(x_1, x_2, x_3) = (x_1 + x_2 + x_3)e^{-(x_1+2x_2+3x_3)}$$

on the set $\{(x_1, x_2, x_3): (x_1 \geqslant 0) \& (x_2 \geqslant 0) \& (x_3 \geqslant 0)\}$.

6. Let φ be a function of class C_m, defined on an open set $G \subset R^k$. If $m > 1$ is an even integer,

$$\varphi'(p_0) = 0, \quad \varphi''(p_0) = 0, \quad \ldots, \quad \varphi^{(m-1)}(p_0) = 0$$

but $\varphi^{(m)}(p)$ is non-negative (non-positive) for every point p in a neighbourhood U of p_0, then φ has a local minimum (maximum) at p_0.

7. Let φ be a real function defined on a separable metric space X. The set of al points p at which φ has a proper local maximum (minimum) is countable.

Hint. Take an element of a basis of X as the set U in (1').

CHAPTER V

DIFFERENTIABLE MAPPINGS

§ 1. The derivative of mapping, vector field and matrix field. Let G be an open subset of the k-dimensional Euclidean space R^k. Let Φ be either a mapping from G into the l-dimensional Euclidean space R^l:

(1) $$\Phi = (\varphi_1, \dots, \varphi_l),$$

or an l-dimensional vector field on G:

(2) $$\Phi = [\varphi_1, \dots, \varphi_l],$$

or, more generally, a matrix field on G:

(3) $$\Phi = [\varphi_{i_1,\dots,i_r}]_{i_1\leqslant k_1,\dots,i_r\leqslant k_r}.$$

Let $p \in G$ and $\mathfrak{a} \in V_k$. Imitating the definition and notation from IV § 1, we shall call the limit

(4) $$\Phi'_{\mathfrak{a}}(p) = \lim_{\Delta t\to 0} \frac{\Phi(p+\Delta t\cdot\mathfrak{a})-\Phi(p)}{\Delta t}$$

(if it exists) the *derivative of Φ at p in the direction* $\mathfrak{a}$. In the case of the mapping (1), the expression

$$\frac{\Phi(p+\Delta t\cdot\mathfrak{a})-\Phi(p)}{\Delta t}$$

is an l-dimensional vector and, consequently, $\Phi'_{\mathfrak{a}}(p)$ is an l-dimensional vector. In the case of the vector field (2), $\Phi'_{\mathfrak{a}}(p)$ is also an l-dimensional vector. In the case of the matrix field (3), $\Phi'_{\mathfrak{a}}(p)$ is a matrix of the type $k_1, \dots, k_r$.

Since

$$\frac{\Phi(p+\Delta t\cdot\mathfrak{a})-\Phi(p)}{\Delta t} = \left[\frac{\varphi_1(p+\Delta t\cdot\mathfrak{a})-\varphi_1(p)}{\Delta t}, \dots, \frac{\varphi_l(p+\Delta t\cdot\mathfrak{a})-\varphi_l(p)}{\Delta t}\right]$$

in the cases (1) and (2), and

$$\frac{\Phi(p+\Delta t\cdot\mathfrak{a})-\Phi(p)}{\Delta t} = \left[\frac{\varphi_{i_1,\dots,i_r}(p+\Delta t\cdot\mathfrak{a})-\varphi_{i_1,\dots,i_r}(p)}{\Delta t}\right]_{i_1\leqslant k_1,\dots,i_r\leqslant k_r}$$

in the case (3), it follows from III 7.3 that

1.1. *The derivative $\Phi'_{\mathfrak{a}}(p)$ of Φ at p in the direction $\mathfrak{a}$ exists if and only if the derivatives at p in the direction $\mathfrak{a}$ of all the coordinates of Φ exist. Then*

$$\Phi'_{\mathfrak{a}}(p) = [(\varphi_1)'_{\mathfrak{a}}(p), \ldots, (\varphi_l)'_{\mathfrak{a}}(p)] \tag{5}$$

in the cases (1) *and* (2), *and*

$$\Phi'_{\mathfrak{a}}(p) = [(\varphi_{i_1,\ldots,i_r})'_{\mathfrak{a}}(p)]_{i_1 \leqslant k_1,\ldots,i_r \leqslant k_r} \tag{6}$$

in the case (3).

The *derivative of Φ in direction* $\mathfrak{a}$ (or, for short, the *directional derivative of* Φ) is the function $\Phi'_{\mathfrak{a}}$ that assigns the vector or matrix $\Phi'_{\mathfrak{a}}(p)$ to every point $p \in G$ at which the limit (4) exists. We shall be interested only in the case where the whole set G is the domain of $\Phi'_{\mathfrak{a}}$. It follows from (5) and (6) that

$$\Phi'_{\mathfrak{a}} = [(\varphi_1)'_{\mathfrak{a}}, \ldots, (\varphi_l)'_{\mathfrak{a}}] \tag{5'}$$

in the case (1) and (2), and

$$\Phi'_{\mathfrak{a}} = [(\varphi_{i_1,\ldots,i_r})'_{\mathfrak{a}}]_{i_1 \leqslant k_1,\ldots,i_r \leqslant k_r} \tag{6'}$$

in the case (3). We see that $\Phi'_{\mathfrak{a}}$ is an l-dimensional vector field in the cases (1) and (2), and in the case (3) it is a matrix field of the same type as Φ.

Imitating the definitions and notation from IV § 2, we shall call the directional derivatives

$$\Phi'_{\mathfrak{e}_1}, \ldots, \Phi'_{\mathfrak{e}_k},$$

where $\mathfrak{e}_1, \ldots, \mathfrak{e}_k$ are the unit vectors of the coordinate axes in R^k, the *partial derivatives* of Φ and we shall denote them by the symbols

$$\Phi_{|1}, \ldots, \Phi_{|k},$$

respectively. More precisely, $\Phi_{|j}$ is the *partial derivative of first order with respect to the j-th coordinate*. By definition,

$$\Phi_{|j} = [\varphi_{1|j}, \ldots, \varphi_{l|j}], \qquad \Phi_{|j}(p) = [\varphi_{1|j}(p), \ldots, \varphi_{l|j}(p)] \tag{7}$$

in the cases (1) and (2), and

$$\Phi_{|j} = [\varphi_{i_1,\ldots,i_r|j}]_{i_1 \leqslant k_1,\ldots,i_r \leqslant k_r}, \qquad \Phi_{|j}(p) = [\varphi_{i_1,\ldots,i_r|j}(p)]_{i_1 \leqslant k_1,\ldots,i_r \leqslant k_r} \tag{8}$$

in the case (3).

The traditional notation for partial derivatives of functions of the form (1), (2) or (3) is the same as the traditional notation for partial derivatives of real functions. If $x_1, \ldots, x_k$ denote the coordinates of any point $p \in G$, then instead of $\Phi_{|j}$ we write

$$\Phi'_{x_j}, \quad \Phi_{x_j}, \quad \frac{\partial \Phi}{\partial x_j}, \quad \frac{\partial}{\partial x_j}\Phi, \quad \frac{\partial}{\partial x_j}\Phi(x_1, \ldots, x_k),$$

and instead of $\Phi_{|j}(p)$ we write

$$\Phi'_{x_j}(x_1, \ldots, x_k) \quad \text{or} \quad \Phi_{x_j}(x_1, \ldots, x_k),$$

etc. The remark on pp. 133-134 about the two different interpretations of the variables $x_1, \ldots, x_k$ in those symbols is valid.

It follows from (5) and (6) and the known properties of directional derivatives of real functions that

$$\Phi'_{c\mathfrak{a}}(p) = c\Phi'_{\mathfrak{a}}(p) .$$

However the equation

$$\Phi'_{\mathfrak{a}_1+\mathfrak{a}_2}(p) = \Phi'_{\mathfrak{a}_1}(p)+\Phi'_{\mathfrak{a}_2}(p) \tag{9}$$

does not always hold. If the directional derivative $\Phi'_{\mathfrak{a}}(p)$ at p exists for every vector $\mathfrak{a}$ and the identity (9) holds, i.e. if $\Phi'_{\mathfrak{a}}(p)$ is a linear function of $\mathfrak{a}$, then by II 7.1 and II 7.3

$$\Phi'_{\mathfrak{a}}(p) = \mathfrak{A}\cdot\mathfrak{a} ,$$

where $\mathfrak{A}$ is a rectangular matrix of the type l, k in the cases (1) and (2), and $\mathfrak{A}$ is a matrix of the type $k_1, \ldots, k_r, k$ in the case (3). The matrix $\mathfrak{A}$ is then called the *derivative of Φ at p* and is denoted by the symbol $\Phi'(p)$. It follows from II § 7 (pp. 74-76) and, in particular, from II 7.1 and II 7.3 that

1.2. *Φ has a derivative $\Phi'(p)$ at p if and only if all the coordinates of Φ have derivatives at p. Then*

$$\Phi'_{\mathfrak{a}}(p) = \Phi'(p)\mathfrak{a} . \tag{10}$$

Moreover,

$$\Phi'(p) = [\varphi_{i|j}(p)]_{i\leqslant l,\, j\leqslant k} \tag{11}$$

in the cases (1) *and* (2), *and*

$$\Phi'(p) = [\varphi_{i_1,\ldots,i_r|j}(p)]_{i_1\leqslant k_1,\ldots,\, i_r\leqslant k_r,\, j\leqslant k} \tag{12}$$

in the case (3).

According to the general rules of notation from I § 3, the symbol Φ' will denote the function (the matrix field) which assigns the matrix $\Phi'(p)$ to every point p. The function Φ' is called the *derivative of Φ*. It is defined on the set of all points p at which $\Phi'(p)$ exists. We shall be interested only in the case where the whole set G is the domain of Φ'. It follows from (11) and (12) that

$$\Phi' = [\varphi_{i|j}]_{i\leqslant l,\, j\leqslant k} = \begin{bmatrix} \varphi_{1|1} & \cdots & \varphi_{1|k} \\ \cdot & \cdot\;\cdot\;\cdot & \cdot \\ \varphi_{l|1} & \cdots & \varphi_{l|k} \end{bmatrix} \tag{11'}$$

in the cases (1) and (2), and

$$\Phi' = [\varphi_{i_1,\ldots,i_r|j}]_{i_1\leqslant k_1,\ldots,\, i_r\leqslant k_r,\, j\leqslant k} \tag{12'}$$

in the case (3).

Similarly to the case of real functions of several variables, the existence of the derivative of a function (1), (2) or (3) at a point p is a very weak hypothesis on properties of the function at the point p, and it is not a good generalization of the notion of differentiability of a real function of one real variable. The only pertinent generalization of the notion can be introduced in the same way as in IV § 3, p. 138.

A function Φ of a form (1), (2) or (3) is said to be *differentiable at a point* p if the derivative $\Phi'(p)$ exists and

$$\lim_{\mathfrak{a}\to 0} \frac{\Phi(p+\mathfrak{a})-\Phi(p)-\Phi'(p)\mathfrak{a}}{|\mathfrak{a}|} = 0 .$$

In this case, the linear function

$$F(\mathfrak{a}) = \Phi'_{\mathfrak{a}}(p) = \Phi'(p)\mathfrak{a} \qquad (\mathfrak{a} \in V_k)$$

(i.e. the directional derivative at p, considered as a function of the vector $\mathfrak{a}$) is called the *differential* (or the *complete differential*) of Φ at the point p. A function Φ is said to be *differentiable* if it is differentiable at every point in its domain G.

It follows from the definition (see also III 7.2) that Φ is differentiable (differentiable at p) if and only if all the coordinates of Φ are differentiable (differentiable at p) in the sense defined in IV § 3, p. 138.

In this book we shall not develop the theory of differentiability of functions (1), (2) or (3) under such weak hypotheses as the existence of the derivative or differentiability. The hypotheses will be replaced by a stronger condition that all functions under consideration are of class C_1.

We say that a function Φ of the form (1), (2) or (3) is *of class* C_1 if the derivative Φ' exists on the whole domain G of Φ and is continuous. It follows from (11′) and (12′) that Φ is of class C_1 if and only if all the coordinates of Φ are of class C_1 (see III 8.8). Since every real function of class C_1 is differentiable, we infer that if a mapping (1) or a vector field (2) or a matrix field (3) is of class C_1, then it is differentiable.

In the case where $k = 1$, i.e. when the functions (1), (2), (3) are defined in an open set G of real numbers, the above definition of the derivative can be simplified. In formulas (11), (12) the index j is always equal to 1; therefore we can omit it (see the remarks on p. 56). Consequently by the *derivative at a point* p of a mapping (1) of an open set G of real numbers into R^l we mean the vector

(13) $$\Phi'(p) = [\varphi_1'(p), \ldots, \varphi_l'(p)] ,$$

(where φ_i' is the ordinary derivative of φ_i) and by the *derivative* of (1) we mean the vector field

(14) $$\Phi' = [\varphi_1', \ldots, \varphi_l'] .$$

By the *derivative at p* of a vector field (2) defined on an open set G of real numbers we mean the vector

(15) $$\Phi'(p) = [\varphi_1'(p), \ldots, \varphi_l'(p)],$$

and by the *derivative* of (2) we mean the vector field

(16) $$\Phi' = [\varphi_1', \ldots, \varphi_l'].$$

By the *derivative at p* of a matrix field (3) defined on an open set G of real numbers we mean the matrix

(17) $$\Phi'(p) = [\varphi'_{i_1,\ldots,i_r}(p)]_{i_1\leqslant k_1,\ldots,i_r\leqslant k_r}$$

of the same type as $\Phi(p)$, and by the *derivative* of (3) we mean the matrix field

(18) $$\Phi' = [\varphi'_{i_1,\ldots,i_r}]_{i_1\leqslant k_1,\ldots,i_r\leqslant k_r}.$$

In each of the three cases (1), (2), (3) under consideration,

$$\Phi'(t) = \lim_{\Delta t\to 0} \frac{\Phi(t+\Delta t)-\Phi(t)}{\Delta t}.$$

In the case of the mapping (1) where $k = 1$, the set $\Phi(G)$ is, from an intuitive point of view, a curve in the l-dimensional space R^l (the exact definition of a curve will be formulated in § 6). The vector $\dfrac{\Phi(t+\Delta t)-\Phi(t)}{\Delta t}$ is parallel to the secant line through the points $\Phi(t+\Delta t)$, $\Phi(t)$. For $\Delta t\to 0$, the point $\Phi(t+\Delta t)$ tends to $\Phi(t)$, and the secant line tends, from the intuitive point of view, to the line through the point $\Phi(t)$, parallel to the vector $\Phi'(t)$, provided that this vector exists and is not the zero vector. The intuition suggests calling this line the *line tangent to the curve at the point* $\Phi(t)$. We see that the derivative $\Phi'(t)$ has a simple geometrical interpretation: it is the vector parallel to the line tangent at $\Phi(t)$ to the curve $C = \{\Phi(t)\colon t \in G\} = \Phi(G)$.

Now let us consider the case where Φ is of the form (1) with $l = 1$. Then Φ is a real function, $\Phi = (\varphi_1) = (\varphi)$, defined on an open subset G of the space R^k ($k \geqslant 1$). The index i in the matrix $\Phi'(p) = [\varphi_{i|j}]$ is always equal to 1. We can omit this index and interpret $\Phi'(p)$ as the vector

$$[\varphi_{|1}(p), \ldots, \varphi_{|k}(p)].$$

We see that, in this case, the derivative Φ' in the sense defined in this section is identical with the derivative φ' of the function φ in the sense defined in IV § 3.

Let us return to the general case. The following identities follow from (11)-(12) and from IV § 1 (5) or IV § 2 (5):

(19) $$(c\Phi)' = c\Phi',$$

(20) $$(\Phi+\Psi)' = \Phi'+\Psi', \quad (\Phi-\Psi)' = \Phi'-\Psi'.$$

We suppose here that Φ and Ψ are l-dimensional vector fields or—more generally—matrix fields of the same type. The first of formulas (20) also holds when Φ is a mapping (1) and Ψ is an l-dimensional vector field. The second of formulas (20) also holds when Φ and Ψ are mappings of the form (1).

If Φ is an l-dimensional vector field (2) and $\mathfrak{A} = [a_{i,j}]_{i\leqslant\lambda, j\leqslant l}$ is a rectangular matrix with constant coordinates, then

$$(21) \qquad (\mathfrak{A}\cdot\Phi)' = \mathfrak{A}\bullet\Phi',$$

where the operations $\cdot$ and $\bullet$ are those defined in II § 5 (4) and (10).

If Φ and Ψ are l-dimensional vector fields defined on an open set of real numbers (the case of $k = 1$), and φ is a real function on this set, then

$$(22) \qquad (\Phi\Psi)' = \Phi'\Psi+\Phi\Psi',$$

$$(23) \qquad (\varphi\Phi)' = \varphi'\Phi+\varphi\Phi'.$$

It follows from (7), (8) and IV § 2 (5) that the following identities analogous to (19) and (20) hold for partial derivatives:

$$(19') \qquad (c\Phi)_{|j} = c\Phi_{|j},$$

$$(20') \qquad (\Phi+\Psi)_{|j} = \Phi_{|j}+\Psi_{|j}.$$

The hypotheses about Φ and Ψ are the same as in the case of (19) and (20).

If Φ and $\mathfrak{A}$ satisfy the same conditions as in (21), then

$$(21') \qquad (\mathfrak{A}\cdot\Phi)_{|j} = \mathfrak{A}\cdot\Phi_{|j}.$$

If Φ and Ψ are l-dimensional vector fields on an open set $G\subset R^k$ ($k\geqslant 1$), and φ is a real function on G, then

$$(22') \qquad (\Phi\Psi)_{|j} = \Phi_{|j}\Psi+\Phi\Psi_{|j},$$

$$(23') \qquad (\varphi\Phi)_{|j} = \varphi_{|j}\Phi+\varphi\Phi_{|j}.$$

All the identities (19)-(23′) should be understood as follows: if the derivatives on the right-hand side of an identity exist at a point p, then the derivative on the left-hand side also exists, and the identity holds.

Observe that the remark about partial derivatives of real functions in IV § 2 (pp. 133-134) is also valid for partial derivatives of functions of the form (1), (2) or (3): If $\Phi(p)$ is interpreted as a function of the jth coordinate x_j of the point $p = (x_1, \ldots, x_k)$, the remaining coordinates being fixed, then the derivative (see (14), (16) or (18) respectively) of the function of one real variable x_j coincides with the partial derivative $\Phi_{|j}$.

The symbol $\Phi_{|j}$ will be generalized as follows. Suppose that Φ is a function of the form (1), (2) or (3), defined on an open set $G \subset R^k$. Let

$$\mathbf{j} = (j_1, \ldots, j_\varkappa) \quad (1 \leqslant \varkappa \leqslant k) \tag{24}$$

be a sequence of different positive integers $\leqslant k$. Consider the function $\Phi(x_1, \ldots, x_k)$ as a function of coordinates $x_{j_1}, \ldots, x_{j_\varkappa}$ only, the remaining coordinates being fixed. More precisely, we consider a new function Ψ (of the form (1), (2) or (3) respectively) defined by

$$\Psi(x_{j_1}, \ldots, x_{j_\varkappa}) = \Phi(x_1, \ldots, x_k)$$

on the open set

$$G_0 = \{(x_{j_1}, \ldots, x_{j_\varkappa}) \colon (x_1, \ldots, x_k) \in G\} \subset R^\varkappa .$$

We suppose that G_0 is not empty. The derivative Ψ' of Ψ will be called the *partial derivative of Φ with respect to the* $\mathbf{j}$*-th coordinates* and will be denoted by the symbol $\Phi_{|\mathbf{j}}$. By definition,

$$\Phi_{|\mathbf{j}} = [\varphi_{i|j_s}]_{i \leqslant l, s \leqslant \varkappa} = \begin{bmatrix} \varphi_{1|j_1} & \cdots & \varphi_{1|j_\varkappa} \\ \cdot & \cdot\ \cdot\ \cdot & \cdot \\ \varphi_{l|j_1} & \cdots & \varphi_{l|j_\varkappa} \end{bmatrix} \tag{25}$$

in the cases (1) and (2), and

$$\Phi_{|\mathbf{j}} = [\varphi_{i_1, \ldots, i_r|j_s}]_{i_1 \leqslant k_1, \ldots, i_r \leqslant k_r, s \leqslant \varkappa} \tag{26}$$

in the case (3).

If the sequence $\mathbf{j}$ consists of only one integer j, then $\Phi_{|\mathbf{j}}$ coincides with $\Phi_{|j}$. If $\mathbf{j} = (1, \ldots, k)$, then $\Phi_{|\mathbf{j}} = \Phi'$.

If $\Phi_{|\mathbf{j}}$ is a continuous function on G, we say that Φ is *of class* C_1 *with respect to the* $\mathbf{j}$*-th coordinates*.

The partial derivatives $\Phi_{|\mathbf{j}}$ also satisfy identities (19′)-(21′) under the same hypotheses. If $\varkappa > 1$, the sign $\cdot$ on the right side of (21′) should be replaced by $\bullet$ (see the analogous formula (21)).

The notation adopted in this section implies the identities (27)-(30) which follow. Suppose that $\varphi_1, \ldots, \varphi_l$ are real functions defined on an open set $G \subset R^k$ and having derivatives on the whole set G. Then

$$(\varphi_1, \ldots, \varphi_l)' = \begin{bmatrix} \varphi_{1|1} & \cdots & \varphi_{1|k} \\ \cdot & \cdot\ \cdot\ \cdot & \cdot \\ \varphi_{l|1} & \cdots & \varphi_{l|k} \end{bmatrix}. \tag{27}$$

For any sequence (24) of different positive integers $\leqslant k$,

$$(\varphi_1, \ldots, \varphi_l)_{|\mathbf{j}} = (\varphi_1, \ldots, \varphi_l)_{|(j_1, \ldots, j_\varkappa)} = \begin{bmatrix} \varphi_{1|j_1} & \cdots & \varphi_{1|j_\varkappa} \\ \cdot & \cdot\ \cdot\ \cdot & \cdot \\ \varphi_{l|j_1} & \cdots & \varphi_{l|j_\varkappa} \end{bmatrix}. \tag{28}$$

In the case where $\varkappa = l \leqslant k$, we have, applying the notation for determinants from II § 5 and denoting $\Phi = (\varphi_1, \ldots, \varphi_l)$ for short,

(29) $$/(\varphi_1, \ldots, \varphi_l)_{|\mathrm{j}}/ = /(\varphi_1, \ldots, \varphi_l)_{|(j_1, \ldots, j_l)}/ = /\Phi_{|\mathrm{j}}/$$
$$= /\Phi_{|(j_1, \ldots, j_l)}/ = \begin{vmatrix} \varphi_{1|j_1} & \cdots & \varphi_{1|j_l} \\ \cdot & \cdot \cdot \cdot & \cdot \\ \varphi_{l|j_1} & \cdots & \varphi_{l|j_l} \end{vmatrix}.$$

In the particular case where $k = l = \varkappa$, $\mathrm{j} = (1, 2, \ldots, k)$ and $\Phi = (\varphi_1, \ldots, \varphi_k)$: $G \to R^k$ $(G \subset R^k)$, we have

(30) $$/\Phi'/ = /(\varphi_1, \ldots, \varphi_k)'/ = \begin{vmatrix} \varphi_{1|1} & \cdots & \varphi_{1|k} \\ \cdot & \cdot \cdot \cdot & \cdot \\ \varphi_{k|1} & \cdots & \varphi_{k|k} \end{vmatrix}.$$

Determinants of the form (29), (30) are called *jacobians*. In particular, the real function (30) is called the *jacobian of the mapping* Φ: $G \to R^k$ $(G \subset R^k)$. As we shall see later, the jacobian of a mapping Φ has many of the properties of the ordinary derivative of a real function of one real variable. To emphasize this fact, the jacobian $/\Phi'/$ of Φ will also be called the *scalar derivative* of the mapping Φ. In the case where $k = 1$ and $\Phi = (\varphi)$, the scalar derivative $/\Phi'/$ just defined coincides with the ordinary derivative φ' of the real function φ of one real variable.

The notation (28), (29) will also be applied in the case where some of the integers $j_1, \ldots, j_\varkappa$ in the sequence $\mathrm{j} = (j_1, \ldots, j_\varkappa)$ are equal. Observe that then the jacobian (29) is equal to 0 since the determinant on the right-hand side has two equal columns. Also in the case where two of the functions $\varphi_1, \ldots, \varphi_l$ are equal, the jacobians (29) and (30) are equal to 0 because the determinants on the right-hand sides have two identical rows.

If we do not postulate that $k = l$, we cannot introduce the notion of the scalar derivative of the mapping (1). However, we can introduce the following useful substitute for this notion: by the *modulus of the derivative* of a mapping (1) we mean the non-negative real function $|\Phi'|$ (see II § 5 (6)-(6′)) whose value at any point $p \in G$ is equal to the modulus $|\Phi'(p)|$ of the rectangular matrix $\Phi'(p)$. By definition,

(31) $$|\Phi'| = \sqrt{\sum_{i_1 < i_2 < \ldots < i_k} \begin{vmatrix} \varphi_{i_1|1} & \cdots & \varphi_{i_1|k} \\ \cdot & \cdot \cdot \cdot & \cdot \\ \varphi_{i_k|1} & \cdots & \varphi_{i_k|k} \end{vmatrix}^2} \quad \text{if} \quad l \geqslant k,$$

(32) $$|\Phi'| = \sqrt{\sum_{j_1 < j_2 < \ldots < j_l} \begin{vmatrix} \varphi_{1|j_1} & \cdots & \varphi_{1|j_l} \\ \cdot & \cdot \cdot \cdot & \cdot \\ \varphi_{l|j_1} & \cdots & \varphi_{l|j_l} \end{vmatrix}^2} \quad \text{if} \quad l \leqslant k.$$

Observe that

$$|\Phi'| = |/\Phi'/| \quad \text{if} \quad k = l,$$

i.e., in that case, the modulus of the derivative is equal to the absolute value of the scalar derivative.

Applying the notation (29)-(30) we can write definitions (31) and (32) as follows for any mapping $\Phi = (\varphi_1, \ldots, \varphi_l)$: $G \to R^k$ $(G \subset R^k)$:

If $l \geqslant k$, then

$$|\Phi'| = \sqrt{\sum |(\varphi_{i_1}, \ldots, \varphi_{i_k})'|^2}, \tag{31'}$$

the summation being extended over all sequences $(i_1, \ldots, i_k)$ of positive integers $\leqslant l$ such that $i_1 < i_2 < \ldots < i_k$.

If $l \leqslant k$, then

$$|\Phi'| = \sqrt{\sum |\Phi_{|\mathbf{j}}|^2} = \sqrt{\sum |(\varphi_1, \ldots, \varphi_l)_{|(j_1, \ldots, j_l)}|^2}, \tag{32'}$$

the summation being extended over all sequences $\mathbf{j} = (j_1, \ldots, j_l)$ of positive integers $\leqslant k$ such that $j_1 < j_2 < \ldots < j_l$.

It follows from II § 5 (20) and (21) that the modulus of the derivative of Φ can be defined by means of Gram determinants. Namely

$$|\Phi'| = \sqrt{\begin{vmatrix} \Phi_{|1}\Phi_{|1} & \ldots & \Phi_{|1}\Phi_{|k} \\ \ldots & \ldots & \ldots \\ \Phi_{|k}\Phi_{|1} & \ldots & \Phi_{|k}\Phi_{|k} \end{vmatrix}} \quad \text{for} \quad l \geqslant k, \tag{31''}$$

$$|\Phi'| = \sqrt{\begin{vmatrix} \varphi_1'\varphi_1' & \ldots & \varphi_1'\varphi_l' \\ \ldots & \ldots & \ldots \\ \varphi_l'\varphi_1' & \ldots & \varphi_l'\varphi_l' \end{vmatrix}} \quad \text{for} \quad l \leqslant k. \tag{32''}$$

The definitions (31)-(32) will also be adopted in the case where $\Phi = [\varphi_1, \ldots, \varphi_l]$ is an l-dimensional vector field on an open set $G \subset R^k$. The non-negative real function $|\Phi'|$ is then called the *modulus of the derivative* of the vector field Φ.

The traditional notation for jacobians is as follows. If $\varphi_1, \ldots, \varphi_r$ are real functions defined on an open set $G \subset R^k$ and $j_1, \ldots, j_r$ are positive integers $\leqslant k$, then the jacobian

$$\begin{vmatrix} \varphi_{1|j_1} & \ldots & \varphi_{1|j_r} \\ \ldots & \ldots & \ldots \\ \varphi_{r|j_1} & \ldots & \varphi_{r|j_r} \end{vmatrix}, \tag{33}$$

i.e. the determinant

$$\begin{vmatrix} \dfrac{\partial \varphi_1}{\partial x_{j_1}} & \ldots & \dfrac{\partial \varphi_1}{\partial x_{j_r}} \\ \ldots & \ldots & \ldots \\ \dfrac{\partial \varphi_r}{\partial x_{j_1}} & \ldots & \dfrac{\partial \varphi_r}{\partial x_{j_r}} \end{vmatrix} \tag{33'}$$

(where $x_1, \ldots, x_k$ are the coordinates of any point $p \in G$) is denoted by the symbol

$$\frac{\partial(\varphi_1, \ldots, \varphi_r)}{\partial(x_{j_1}, \ldots, x_{j_r})}. \tag{34}$$

If $z_1, \ldots, z_r$ denote the values of $\varphi_1, \ldots, \varphi_r$ at any point $p = (x_1, \ldots, x_r)$:

$$z_1 = \varphi_1(x_1, \ldots, x_r),$$

$$\cdots\cdots\cdots\cdots$$

$$z_r = \varphi_r(x_1, \ldots, x_r),$$

then the notation

$$\frac{\partial(z_1, \ldots, z_r)}{\partial(x_{j_1}, \ldots, x_{j_r})} \tag{34'}$$

is also used.

EXERCISES

1. Calculate the partial derivatives, the derivative and the modulus of the derivative of the mapping $\Phi: R^2 \to R^4$ defined by

$$\Phi(x_1, x_2) = (e^{x_1 x_2}, \cos x_1, \log(1+x_1^2+x_2^2), x_1 + \operatorname{arctg} x_2).$$

2. Calculate the derivative and the modulus of the derivative of the mapping $\Phi: R \to R^3$ defined by

$$\Phi(x) = (\cos x, \sin x, cx).$$

The curve $\{\Phi(x): x \in R\}$ is called a *helix*.

3. Let $\Phi: R^k \to R^{k+1}$ be the mapping defined as follows. Let $p_0 = (0, \ldots, 0, 1) \in R^{k+1}$ and

$$S = \{(x_1, \ldots, x_{k+1}): x_1^2 + \ldots + x_{k+1}^2 = 1\}.$$

For any point $p = (x_1, \ldots, x_k) \in R^k$, let L_p be the straight line through p and p_0. Let $\Phi(p)$ be the point of $L_p \cap S$ which is different from p_0.

Find formulas for the coordinates of Φ, the derivative and the modulus of the derivative of Φ.

The mapping $\Phi^{-1}: S - \{p_0\} \to R^k$ is called the *stereographic projection*.

4. Find the derivative of the matrix field on R^k:

$$\Phi(x_1, \ldots, x_k) = [x_{i_1} + x_{i_2} + \ldots + x_{i_r}]_{i_1, \ldots, i_r \leqslant k}.$$

5. Let Ψ be an isometry from R^l onto itself and let Φ be a mapping of an open set $G \subset R^k$ into R^l, $k \leqslant l$. Prove that

$$|(\Psi \circ \Phi)'| = |\Phi'|.$$

Hint. Use (31″).

6. Prove the following generalization of (22): if Φ and Ψ are l-dimensional vector fields on an open set $G \subset R^k$, then

$$(\Phi\Psi)' = \Psi \cdot \Phi' + \Phi \cdot \Psi',$$

provided that the derivatives on the right-hand side exist.

7. Prove the following generalization of (23): if $\Phi: G \to V_l$ is an l-dimensional vector field on an open set $G \subset R^k$, and φ is a real function on G, then

$$(\varphi\Phi)' = \Phi \times \varphi' + \varphi\Phi',$$

provided that the derivatives on the right-hand side exist.

8. For any matrix fields Φ and Ψ,

$$(\Phi\times\Psi)_{|j} = \Phi_{|j}\times\Psi+\Phi\times\Psi_{|j}\,.$$

More generally, for any matrix fields $\Phi_1, \ldots, \Phi_k$,

$$(\Phi_1\times\ldots\times\Phi_k)_{|j} = \sum_{r=1}^{k}\Phi_1\times\ldots\times\Phi_{r-1}\times\Phi_{r|j}\times\Phi_{r+1}\times\ldots\times\Phi_k\,.$$

The identities should be interpreted as follows: if the derivatives on the right-hand side exist at a point p, then the derivative on the left-hand side exists at that point and the equation holds.

9. Generalize exercises IV § 3, 4-6 to the case of functions of form (1), (2) or (3).

§ 2. Derivatives of higher orders. The derivatives, partial derivatives and directional derivatives defined in § 1 are called derivatives of *first order* (or *first derivatives*) to distinguish them from derivatives of higher orders which will be defined below.

Similarly as in § 1, let G be an open subset of R^k and let Φ be either a mapping from G into R^l:

(1) $$\Phi = (\varphi_1, \ldots, \varphi_l)\,,$$

or an l-dimensional vector field on G:

(2) $$\Phi = [\varphi_1, \ldots, \varphi_l]\,,$$

or, more generally, a matrix field on G:

(3) $$\Phi = [\varphi_{i_1,\ldots,i_r}]_{i_1\leqslant k_1,\ldots,i_r\leqslant k_r}\,.$$

The *directional derivative of order* m (or m-*th directional derivative*) of the function Φ in the direction of vectors $\mathfrak{a}_1, \ldots, \mathfrak{a}_m$ is defined by induction as follows:

$$\Phi^{(m)}_{\mathfrak{a}_1,\ldots,\mathfrak{a}_{m-1},\mathfrak{a}_m} = (\Phi^{(m-1)}_{\mathfrak{a}_1,\ldots,\mathfrak{a}_{m-1}})'_{\mathfrak{a}_m}\,.$$

This definition and notation is analogous to that for real functions in IV § 5. It is easy to verify that the directional derivative exists if and only if the corresponding directional derivatives of the coordinates of Φ exist. Then

(4) $$\Phi^{(m)}_{\mathfrak{a}_1,\ldots,\mathfrak{a}_m} = [(\varphi_1)^{(m)}_{\mathfrak{a}_1,\ldots,\mathfrak{a}_m}, \ldots, (\varphi_l)^{(m)}_{\mathfrak{a}_1,\ldots,\mathfrak{a}_m}]$$

in the cases (1) and (2), and

(5) $$\Phi^{(m)}_{\mathfrak{a}_1,\ldots,\mathfrak{a}_m} = [(\varphi_{i_1,\ldots,i_r})^{(m)}_{\mathfrak{a}_1,\ldots,\mathfrak{a}_m}]_{i_1\leqslant k_1,\ldots,i_r\leqslant k_r}$$

in the case (3).

Let $\mathfrak{e}_1, \ldots, \mathfrak{e}_k$ be the unit vectors of the coordinate axes in R^k. Instead of $\Phi^{(m)}_{\mathfrak{e}_{j_1},\ldots,\mathfrak{e}_{j_m}}$ we shall write

(6) $$\Phi_{|j_1,\ldots,j_m}$$

and we shall call the derivative (6) the *partial derivative of Φ of order m* (or the *m-th partial derivative of Φ*) with respect to $(j_1, \ldots, j_m)$-coordinates. This definition and notation is analogous to that in IV § 5 for real functions. It follows from (4) and (5) that

$$\Phi_{|j_1,\ldots,j_m} = [\varphi_{1|j_1,\ldots,j_m}, \ldots, \varphi_{l|j_1,\ldots,j_m}] \tag{7}$$

in the cases (1) and (2), and

$$\Phi_{|j_1,\ldots,j_m} = [\varphi_{i_1,\ldots,i_r|j_1,\ldots,j_m}]_{i_1\leqslant k_1,\ldots,i_r\leqslant k_r} \tag{8}$$

in the case (3).

Suppose that $\Phi^{(m)}_{\mathfrak{a}_1,\ldots,\mathfrak{a}_m}(p)$ is a linear function of each of the vectors $\mathfrak{a}_1, \ldots, \mathfrak{a}_m$ (i.e. it is an m-linear function on V_k^m). It follows from II 7.4 that

$$\Phi^{(m)}_{\mathfrak{a}_1,\ldots,\mathfrak{a}_m}(p) = \mathfrak{A}\cdot(\mathfrak{a}_1 \times \ldots \times \mathfrak{a}_m)$$

where $\mathfrak{A}$ is a matrix. The matrix $\mathfrak{A}$ will be called the *derivative of order m* (or the *m-th derivative*) *of Φ at p* and will be denoted by the symbol $\Phi^{(m)}(p)$. It follows from II 7.4 that

2.1. *Φ has a derivative $\Phi^{(m)}(p)$ at a point p if and only if all the coordinates of Φ have derivatives of order m at p. Then*

$$\Phi^{(m)}_{\mathfrak{a}_1,\ldots,\mathfrak{a}_m}(p) = \Phi^{(m)}(p)\cdot(\mathfrak{a}_1 \times \ldots \times \mathfrak{a}_m)\,. \tag{9}$$

Moreover

$$\Phi^{(m)}(p) = [\varphi_{i|j_1,\ldots,j_m}(p)]_{i\leqslant l,\, j_1,\ldots,j_m\leqslant k} \tag{10}$$

in the cases (1) *and* (2), *and*

$$\Phi^{(m)}(p) = [\varphi_{i_1,\ldots,i_r|j_1,\ldots,j_m}(p)]_{i_1\leqslant k_1,\ldots,i_r\leqslant k_r,\, j_1,\ldots,j_m\leqslant k} \tag{11}$$

in the case (3).

According to the general rules of notation from I § 3, the symbol $\Phi^{(m)}$ will denote the mapping (matrix field) which assigns the matrix $\Phi^{(m)}(p)$ to each point p. The function $\Phi^{(m)}$ is called the *derivative of order m* (or the *m-th derivative*) of Φ. It is defined on the set of all $p \in G$ at which $\Phi^{(m)}(p)$ exists. We shall be interested only in the case where the whole set G is the domain of $\Phi^{(m)}$. The above definition of the mth derivative is analogous to that for real functions in IV § 5. It follows from (10) and (11) that

$$\Phi^{(m)} = [\varphi_{i|j_1,\ldots,j_m}]_{i\leqslant l,\, j_1,\ldots,j_m\leqslant k} \tag{10'}$$

in the cases (1) and (2), and

$$\Phi^{(m)} = [\varphi_{i_1,\ldots,i_r|j_1,\ldots,j_m}]_{i_1\leqslant k_1,\ldots,i_r\leqslant k_r,\, j_1,\ldots,j_m\leqslant k} \tag{11'}$$

in the case (3).

We say that Φ is *of class* C_m if the derivative $\Phi^{(m)}$ exists on the whole set G and is continuous. It follows from (10′) and (11′) that Φ is of class C_m if and only if all the coordinates of Φ are of class C_m. Hence it follows that if Φ is of class C_m, then it is also of class $C_{m-1}, \ldots, C_1$ and it is continuous (see IV 5.3). This implies that Φ is of class C_m if and only if Φ' is of class C_{m-1} (or if $\Phi^{(m-1)}$ is of class C_1). Observe also that if Φ is of class C_m and $1 < n < m$, then

$$\Phi^{(m)} = (\Phi^{(n)})^{(m-n)} .$$

We say that Φ is *of class* C_∞ if it is of class C_m for every positive integer m, i.e. if all its coordinates are of class C_∞.

It follows from IV 5.4 and formulas (9)-(11) that if Φ is of class C_m, then the matrix $\Phi^{(m)}(p)$ is symmetric in last m indices.

In the case where $k = 1$, i.e., when the function (1), (2) or (3) under consideration is defined in an open set G of real numbers, the definition of the mth derivative can be simplified. In formulas (10)-(11′) the indices $j_1, \ldots, j_m$ are always equal to 1, therefore they can be omitted (see p. 56). Consequently, by the *m-th derivative* of a mapping (1) of an open set G of real numbers into R^l we mean the vector field

$$\Phi^{(m)} = [\varphi_1^{(m)}, \ldots, \varphi_l^{(m)}] . \tag{12}$$

By the *m-th derivative* of a vector field (2) on an open set G of real numbers we mean the vector field

$$\Phi^{(m)} = [\varphi_1^{(m)}, \ldots, \varphi_l^{(m)}] . \tag{13}$$

By the *m-th derivative* of a matrix field (3) on an open set G of real numbers we mean the matrix field

$$\Phi^{(m)} = [\varphi_{i_1,\ldots,i_r}^{(m)}]_{i_1 \leqslant k_1,\ldots,i_r \leqslant k_r} . \tag{14}$$

The symbol $\varphi^{(m)}$ denotes here the ordinary mth derivative of a real function φ of one real variable.

Let us now consider the case where (1) is a mapping of an open set $G \subset R^k$ $(k \geqslant 1)$ into the set R of real numbers, i.e. $l = 1$. Then $\Phi = (\varphi)$ where φ is a real function on G. It follows from (10′) that

$$\Phi^{(m)} = [\varphi_{|j_1,\ldots,j_m}]_{j_1,\ldots,j_m \leqslant k} .$$

Thus the mth derivative $\Phi^{(m)}$ coincides with the mth derivative $\varphi^{(m)}$ in the sense defined in IV § 5. Similarly the partial derivatives $\Phi_{|j_1,\ldots,j_m}$ defined in this section coincide with the partial derivatives $\varphi_{|j_1,\ldots,j_m}$ defined in IV § 5. The same is true for directional derivatives. We see that all the notions introduced in this section are a natural generalization of those from IV § 5.

The following theorem is a generalization of Theorem IV 6.3.

2.2. *Let Φ be a function of the form* (1), (2) *or* (3), *of class* C_m, *defined on an open set* $G \subset R^k$. *For every fixed point* $p \in G$ *there exists a matrix field* $\mathfrak{R}$ *on* G, *of the same type as* $\Phi^{(m)}$, *such that*

$$(15) \qquad \Phi(q) = \Phi(p) + \frac{\Phi'(p)\cdot(q-p)}{1!} + \frac{\Phi''(p)\cdot(q-p)^2}{2!} + \ldots + \frac{\Phi^{(m)}(p)\cdot(q-p)^m}{m!} + \mathfrak{R}(q)\cdot(q-p)^m \quad \textit{for} \quad q \in G\,,$$

and

$$(16) \qquad \lim_{q\to p} \mathfrak{R}(q) = \mathfrak{R}(p) = 0\,.$$

We shall prove Theorem 2.2 in the case where Φ is of the form (1). The proofs in the cases (2) and (3) are analogous.

By hypothesis, $\Phi = (\varphi_1, \ldots, \varphi_l)$ where $\varphi_1, \ldots, \varphi_l$ are real functions on G of class C_m. By Theorem IV 6.3 for the function φ_i $(i = 1, \ldots, l)$, there exists a matrix field

$$\mathfrak{r}_i = [\varrho_{i,j_1,\ldots,j_m}]_{j_1,\ldots,j_m\leqslant k}$$

such that

$$(17) \qquad \varphi_i(q) = \varphi_i(p) + \frac{\varphi_i'(p)\cdot(q-p)}{1!} + \frac{\varphi_i''(p)\cdot(q-p)^2}{2!} + \ldots + \frac{\varphi_i^{(m)}(p)\cdot(q-p)^m}{m!} + \mathfrak{r}_i(q)\cdot(q-p)^m$$

and

$$(18) \qquad \lim_{q\to p} \mathfrak{r}_i(q) = \mathfrak{r}(p) = 0\,,$$

that is,

$$\lim_{q\to p} \varrho_{i,j_1,\ldots,j_m}(q) = \varrho_{i,j_1,\ldots,j_m}(p) = 0\,.$$

It suffices to assume

$$\mathfrak{R} = [\varrho_{i,j_1,\ldots,j_m}]_{i\leqslant l,j_1,\ldots,j_m\leqslant k}\,.$$

Indeed, (15) is identical with all the identities (17) for $i = 1, \ldots, l$, and (16) is identical with (18) for $i = 1, \ldots, l$.

Theorem 2.2 is a substitute for the Taylor formula in IV § 6. In the case $m = 1$, Theorem 2.2 yields the following substitute for the mean-value theorem for real functions (see IV 1.1 and IV 3.4): if Φ is of class C_1 and of the form (1), (2) or (3) on $G \subset R^k$, then

$$(19) \qquad \Phi(q) - \Phi(p) = \big(\Phi'(p) + \mathfrak{R}(q)\big)\cdot(q-p)$$

where $\mathfrak{R}$ is a matrix field of the same type as Φ', and

$$(20) \qquad \lim_{q\to p} \mathfrak{R}(q) = \mathfrak{R}(p) = 0\,.$$

EXERCISES

1. Generalize Theorem IV 6.2 to the case of functions of the form (1), (2) or (3).

2. Show that the Taylor formula IV 6.1 is not true for mappings $\Phi: R \to R^2$ even for $m = 1$ (i.e. that the mean-value theorems IV 1.1 and IV 3.4 cannot be generalized to such mappings).

3. Find the second derivatives of the mappings defined in § 1, exercises 1-3.

§ 3. Differentiation of composite functions. First we shall prove the following theorem which will be later generalized.

3.1. *If φ is a real function of class C_1 defined on an open set $G \subset R^k$ and*

$$\Lambda = (\lambda_1, \ldots, \lambda_k)$$

is a mapping of class C_1 from an open set $G_0 \subset R$ into G, then the composite function $\psi = \varphi \circ \Lambda: G_0 \to R$ is of class C_1 and

$$\psi'(t) = (\varphi \circ \Lambda)'(t) = \sum_{j=1}^{k} \varphi_{|j}\big(\Lambda(t)\big)\lambda_j'(t) = \varphi'\big(\Lambda(t)\big)\cdot \Lambda'(t) \tag{1}$$

for $t \in G_0$, i.e., the derivative of the composite function $\varphi \circ \Lambda$ at a point t is equal to the scalar product of the derivative of φ at the point $\Lambda(t)$ and of the derivative of Λ at t. In other words,

$$(\varphi \circ \Lambda)' = \sum_{j=1}^{k} (\varphi_{|j} \circ \Lambda)\lambda_j', \tag{2}$$

that is,

$$(\varphi \circ \Lambda)' = (\varphi' \circ \Lambda)\cdot \Lambda'. \tag{3}$$

If φ and Λ are of class C_m, then $\varphi \circ \Lambda$ is also of class C_m.

Formulas (1)-(3) establish a connection between differentiation and composition of functions. On the left-hand side of (3) we first form the composition of φ and Λ and then we differentiate the composition. On the right-hand side of (3) we first differentiate the function φ, then we form the composition of φ' and Λ and multiply it by Λ'. Analogous remarks hold for formulas (1) and (2), and also for other formulas for differentiation of composite mappings proved below. Observe that formula (3) has the same form as the formula for differentiation of a composition of real functions of one real variable—see I § 3 (13) (the same remark holds for formulas (6), (12) and (18) proved below).

In the traditional notation the formula for the derivative of a composition can be written as follows:

$$\frac{d}{dt}\varphi(\Lambda(t)) = \psi'(t) = \sum_{j=1}^{k} \varphi_{x_j}\big(\lambda_1(t), \ldots, \lambda_k(t)\big)\lambda_j'(t)\,. \tag{4}$$

If z is used to denote the values of the functions φ and ψ, and if $x_1, \ldots, x_k$ are used to denote the coordinates of any point in R^k and to denote the values of the functions $\lambda_1, \ldots, \lambda_k$ respectively, then formula (4) can also be written in the form

$$\frac{dz}{dt} = \sum_{j=1}^{k} \frac{\partial z}{\partial x_j} \cdot \frac{dx_j}{dt}. \tag{5}$$

Of course, this formula should be understood as follows. The derivatives $\frac{\partial z}{\partial x_j}$ on the right-hand side are the partial derivatives of z considered as a function of the variables $x_1, \ldots, x_k$, the differentation being performed before the substitution $x_1 = \lambda_1(t), \ldots, x_k = \lambda_k(t)$. On the left-hand side the derivative $\frac{dz}{dt}$ is the derivative of z considered as a function of t, i.e. it is the ordinary derivative of the function $z = \varphi(\lambda_1(t), \ldots, \lambda_k(t))$. Thus here we first form the composition and then we perform the differentiation. Of course we have to guess that the partial derivatives $\frac{\partial z}{\partial x_j}$ on the right-hand side are taken at the point $(x_1, \ldots, x_k) \in G$ corresponding to the value of t under consideration, i.e. at the point whose coordinates are

$$x_1 = \lambda_1(t), \ldots, x_k = \lambda_k(t).$$

To prove 3.1 let us observe that, for small $|\Delta t|$, the segment joining the points $\Lambda(t)$ and $\Lambda(t+\Delta t)$ is a subset of G. Assuming that $p = \Lambda(t)$ and that

$$a = \Lambda(t+\Delta t) - \Lambda(t) = [\lambda_1(t+\Delta t) - \lambda_1(t), \ldots, \lambda_k(t+\Delta t) - \lambda_k(t)]$$

in the mean-value theorem IV 3.4, we infer that for small $|\Delta t|$

$$\varphi(\Lambda(t+\Delta t)) - \varphi(\Lambda(t)) = \sum_{j=1}^{k} \varphi_{|j}\big(\Lambda(t) + \theta(\Lambda(t+\Delta t) - \Lambda(t))\big)\big(\lambda_j(t+\Delta t) - \lambda_j(t)\big),$$

where θ is a number from the interval $(0;1)$ depending on Δt. Hence

$$\frac{\psi(t+\Delta t) - \psi(t)}{\Delta t} = \sum_{j=1}^{k} \varphi_{|j}\big(\Lambda(t) + \theta(\Lambda(t+\Delta t) - \Lambda(t))\big) \frac{\lambda_j(t+\Delta t) - \lambda_j(t)}{\Delta t}.$$

The partial derivatives $\varphi_{|j}$ being continuous, we obtain (1) for $\Delta t \to 0$. Formulas (2) and (3) are another formulation of (1). The continuity of ψ' follows directly from (1) - (3).

The last part of 3.1 follows from the part just proved. The proof is by induction. The case $m = 1$ has already been proved. Suppose that the composition of any function and any mapping of class C_{m-1} is of class C_{m-1} and suppose that φ and Λ are of class C_m. It follows from (3), Theorem IV 5.6 and the induction hypothesis that ψ' is of class C_{m-1}. Thus the function $\psi = \varphi \circ \Lambda$ is of class C_m.

3.2. *Let Φ be either a mapping of class* C_1

$$\Phi = (\varphi_1, \ldots, \varphi_l)$$

from an open set $G \subset R^k$ *into* R^l, *or an* l*-dimensional vector field*

$$\Phi = [\varphi_1, \ldots, \varphi_l]$$

on an open set $G \subset R^k$. *Let* $\Lambda = (\lambda_1, \ldots, \lambda_k)$ *be a mapping of class* C_1 *from an open set* $G_0 \subset R$ *into* G. *Then the composition* $\Phi \circ \Lambda$ *is of class* C_1 *and*

$$(\Phi \circ \Lambda)' = (\Phi' \circ \Lambda) \cdot \Lambda', \tag{6}$$

that is,

$$(\Phi \circ \Lambda)'(t) = \Phi'(\Lambda(t)) \cdot \Lambda'(t) \quad \text{for} \quad t \in G_0, \tag{7}$$

i.e. the derivative of the composition $\Phi \circ \Lambda$ *at a point* t *is equal to the product of the rectangular matrix* $\Phi'(\Lambda(t))$ (*the derivative of* Φ *at the point* $\Lambda(t)$) *and the vector* $\Lambda'(t)$. *In other words,*

$$(\varphi_i \circ \Lambda)' = \sum_{j=1}^{k} (\varphi_{i|j} \circ \Lambda) \lambda_j' \quad \text{for} \quad i = 1, \ldots, l. \tag{8}$$

If Φ *and* Λ *are of class* C_m, *then* $\Phi \circ \Lambda$ *is also of class* C_m.

Identity (8) is a direct consequence of (2). Formulas (6) and (7) are other formulations of (8).

The last part of 3.2 follows directly from the analogous part of 3.1.

If $x_1, \ldots, x_k$ denote the coordinates of any point in G and, simultaneously, the values of functions $\lambda_1, \ldots, \lambda_k$, then formulas (6)-(8) can be written in the traditional notation as follows

$$\frac{d}{dt} \Phi(\Lambda(t)) = \sum_{j=1}^{k} \Phi_{x_j}(\Lambda(t)) \lambda_j'(t), \tag{9}$$

or

$$\frac{d}{dt} \Phi(\Lambda(t)) = \sum_{j=1}^{k} \frac{\partial \Phi}{\partial x_j} \cdot \frac{dx_j}{dt}. \tag{10}$$

The interpretation of formula (10) is the same as that of (5).

3.3. *If* φ *is a real function of class* C_1 *on an open set* $G \subset R^k$ *and* $\Lambda = (\lambda_1, \ldots, \lambda_k)$ *is a mapping of class* C_1 *from an open set* $G' \subset R^\varkappa$ *into* G, *then the composite function* $\psi = \varphi \circ \Lambda \colon G' \to R$ *is of class* C_1 *and*

$$\psi_{|r} = \sum_{j=1}^{k} (\varphi_{|j} \circ \Lambda) \lambda_{j|r} = (\varphi' \circ \Lambda) \cdot \Lambda_{|r} \quad \text{for} \quad r = 1, \ldots, \varkappa, \tag{11}$$

i.e. the partial derivative of $\psi = \varphi \circ \Lambda$ *at a point* $p \in G'$ *with respect to the* r*-th coordinate is the scalar product of two vectors: the derivative* φ' *of* φ *at* *the point $\Lambda(p)$ and the partial derivative of the mapping Λ at p with respect to the r-th coordinate. Thus*

$$(12)\qquad (\varphi \circ \Lambda)' = (\varphi' \circ \Lambda)\cdot \Lambda',$$

that is,

$$(13)\qquad (\varphi \circ \Lambda)'(p) = \varphi'\big(\Lambda(p)\big)\cdot \Lambda'(p) \quad \text{for} \quad p \in G',$$

i.e. the derivative of the composite function $\varphi \circ \Lambda$ is the product of the vector $\varphi'\big(\Lambda(p)\big)$ (the derivative of φ at the point $\Lambda(p)$) and the rectangular matrix $\Lambda'(p)$.

If φ and Λ are of class C_m, then $\varphi \circ \Lambda$ is also of class C_m.

Theorem 3.3 follows from 3.1. Indeed, interpreting φ, ψ and Λ as functions of one real variable t_r ($t_1, \ldots, t_{r-1}, t_{r+1}, \ldots, t_\varkappa$ being fixed) on the open set

$$G_0 = \{t_r\colon (t_1, \ldots, t_\varkappa) \in G'\} \subset R,$$

we see that (11) is a particular case of (2) or (3). Formulas (12) and (13) are other formulations of (11).

The proof of the last part of 3.3 is the same as the proof of the analogous part of 3.1.

If z denotes values of φ and ψ at any point, $x_1, \ldots, x_k$ denote the coordinates of any point in G and, simultaneously, the values of the functions $\lambda_1, \ldots, \lambda_k$, and if $t_1, \ldots, t_\varkappa$ denote the coordinates of any point in G', then formulas (11)-(13) can be written in the traditional notation as follows:

$$(14)\qquad \psi_{t_r}(t_1, \ldots, t_\varkappa) = \sum_{j=1}^{k} \varphi_{x_j}\big(\lambda_1(t_1, \ldots, t_\varkappa), \ldots, \lambda_k(t_1, \ldots, t_\varkappa)\big)\cdot (\lambda_j)_{t_r}(t_1, \ldots, t_\varkappa),$$

or in the form

$$(15)\qquad \frac{\partial z}{\partial t_r} = \sum_{j=1}^{k} \frac{\partial z}{\partial x_j}\cdot\frac{\partial x_j}{\partial t_r}.$$

Of course, the partial derivatives $\partial z/\partial x_j$ (i.e. the partial derivatives of the function $z = \varphi(x_1, \ldots, x_k)$) should be taken at the point whose coordinates are

$$x_1 = \lambda_1(t_1, \ldots, t_\varkappa),\ \ldots,\ x_k = \lambda_k(t_1, \ldots, t_\varkappa).$$

The derivative on the left-hand side is the partial derivative of the function

$$z = \psi(t_1, \ldots, t_\varkappa) = \varphi\big(\lambda_1(t_1, \ldots, t_\varkappa), \ldots, \lambda_k(t_1, \ldots, t_\varkappa)\big).$$

The symbol $\partial x_j/\partial t_r$ denotes the partial derivative of the function $x_j = \lambda_j(t_1, \ldots, t_\varkappa)$ with respect to t_r.

3.4. *Let Φ be either a mapping of class C_1*

$$\Phi = (\varphi_1, \ldots, \varphi_l)$$

from an open set $G \subset R^k$ *into* R^l, *or an* l*-dimensional vector field of class* C_1

$$\Phi = [\varphi_1, \ldots, \varphi_l]$$

on an open set $G \subset R^k$. *Let* $\Lambda = (\lambda_1, \ldots, \lambda_k)$ *be a mapping of class* C_1 *from an open set* $G' \subset R^\varkappa$ *into* G. *Then the composition* $\Phi \circ \Lambda$ *is of class* C_1 *and*

$$(16) \qquad (\varphi_i \circ \Lambda)_{|r} = \sum_{j=1}^{k} (\varphi_{i|j} \circ \Lambda)\lambda_{j|r} \quad \text{for} \quad i = 1, \ldots, l \text{ and } r = 1, \ldots, \varkappa,$$

that is,

$$(17) \qquad (\Phi \circ \Lambda)_{|r} = (\Phi' \circ \Lambda) \cdot \Lambda_{|r} \quad \text{for} \quad r = 1, \ldots, \varkappa,$$

i.e. the partial derivative of the composition $\Phi \circ \Lambda$ *at a point* $p \in G'$ *with respect to the* r*-th coordinate is the product of the rectangular matrix* $\Phi'(\Lambda(p))$ *and the vector* $\Lambda_{|r}(p)$. *Thus*

$$(18) \qquad (\Phi \circ \Lambda)' = (\Phi' \circ \Lambda) \bullet \Lambda',$$

that is,

$$(19) \qquad (\Phi \circ \Lambda)'(p) = \Phi'(\Lambda(p)) \bullet \Lambda'(p) \quad \text{for} \quad p \in G',$$

i.e. the derivative of the composition $\Phi \circ \Lambda$ *at a point* $p \in G'$ *is the matrix product of the rectangular matrix* $\Phi'(\Lambda(p))$ *and the rectangular matrix* $\Lambda'(p)$.

If Φ *and* Λ *are of class* C_m, *then* $\Phi \circ \Lambda$ *is also of class* C_m.

Identity (16) follows directly from (11). Identities (17), (18), (19) are other formulations of (16). The last part of 3.4 follows directly from the last part of 3.3.

If $x_1, \ldots, x_k$ denote the coordinates of any point in G and, simultaneously, values of the functions $\lambda_1, \ldots, \lambda_k$, and if $t_1, \ldots, t_\varkappa$ denote the coordinates of any point in G', then formulas (16)-(19) can be written in the traditional notation as follows:

$$(20) \qquad \frac{\partial}{\partial t_r}\Phi(\Lambda(t_1, \ldots, t_\varkappa)) = \sum_{j=1}^{k} \frac{\partial \Phi}{\partial x_j} \cdot \frac{\partial x_j}{\partial t_r}.$$

The interpretation of formulas (20) is the same as that of (15). The formulation of the identity analogous to (14) is left to the reader.

The following theorem follows from (18) and from II 5.1 and the known theorem on multiplication of determinants. The hypotheses about Φ and Λ are the same as in the first part of 3.4.

3.5. *If either* $l = k \leqslant \varkappa$ *or* $l \geqslant k = \varkappa$, *then the modulus of the derivative of the composition* $\Phi \circ \Lambda$ *is equal to the product of the modulus of the derivative of the mapping (or vector field)* Φ *and the modulus of the derivative of the modulus of the mapping* Λ; *more precisely,*

$$(21) \qquad |(\Phi \circ \Lambda)'| = (|\Phi'| \circ \Lambda) \cdot |\Lambda'|,$$

that is,

$$|(\Phi \circ \Lambda)'(p)| = |\Phi'(\Lambda(p))| \cdot |\Lambda'(p)| . \tag{22}$$

If $l = k = \varkappa$, then the scalar derivative of the composition $\Phi \circ \Lambda$ is equal to the product of the scalar derivative of the mapping (or vector field) Φ and the scalar derivative of the mapping Λ; more precisely,

$$/(\Phi \circ \Lambda)'/ = (/\Phi'/ \circ \Lambda) \cdot /\Lambda'/ , \tag{23}$$

that is,

$$/(\Phi \circ \Lambda)'(p)/ = /\Phi'(\Lambda(p))/ \cdot /\Lambda'(p)/ . \tag{24}$$

In the traditional notation for jacobians, the identity (23) can be written as follows. Let

$$\begin{aligned} z_1 &= \varphi_1(x_1, \ldots, x_k) , \\ &\cdots\cdots\cdots\cdots \\ z_k &= \varphi_k(x_1, \ldots, x_k) \end{aligned} \tag{25}$$

and

$$\begin{aligned} x_1 &= \lambda_1(t_1, \ldots, t_k) , \\ &\cdots\cdots\cdots\cdots \\ x_k &= \lambda_k(t_1, \ldots, t_k) \end{aligned} \tag{26}$$

be two systems of real functions of class C_1. The jacobian of the compositions

$$\begin{aligned} z_1 &= \psi_1(t_1, \ldots, t_k) = \varphi_1(\lambda_1(t_1, \ldots, t_k), \ldots, \lambda_k(t_1, \ldots, t_k)) , \\ &\cdots\cdots\cdots\cdots\cdots\cdots\cdots\cdots\cdots\cdots \\ z_k &= \psi_k(t_1, \ldots, t_k) = \varphi_k(\lambda_1(t_1, \ldots, t_k), \ldots, \lambda_k(t_1, \ldots, t_k)) \end{aligned} \tag{27}$$

is the product of the jacobians of (25) and (26):

$$\frac{\partial(z_1, \ldots, z_k)}{\partial(t_1, \ldots, t_k)} = \frac{\partial(z_1, \ldots, z_k)}{\partial(x_1, \ldots, x_k)} \cdot \frac{\partial(x_1, \ldots, x_k)}{\partial(t_1, \ldots, t_k)} . \tag{28}$$

EXERCISES

1. If Φ is a matrix field of class C_1 on an open set $G \subset R^k$, and Λ is a mapping of class C_1 from an open set $G' \subset R$ into G, then the composition $\Phi \circ \Lambda$ is of class C_1 and

$$(\Phi \circ \Lambda)_{|r} = (\Phi' \circ \Lambda) \cdot \Lambda_{|r} \quad \text{for} \quad r = 1, \ldots, \varkappa .$$

If Φ and Λ are of class C_m, then $\Phi \circ \Lambda$ is also of class C_m.

2. Let φ and Λ satisfy the hypotheses of 3.1. If φ and Λ are of class C_2, then

$$(\varphi \circ \Lambda)'' = \sum_{i,j=1}^{k} (\varphi_{|i,j} \circ \Lambda) \cdot \lambda_i' \lambda_j' + \sum_{j=1}^{k} (\varphi_{|j} \circ \Lambda) \cdot \lambda_j'' ,$$

that is,

$$(\varphi \circ \Lambda)'' = (\varphi'' \circ \Lambda) \cdot (\Lambda' \times \Lambda') + (\varphi' \circ \Lambda) \cdot \Lambda'' .$$

In the traditional notation,

$$\frac{d^2}{dt^2}\varphi(\Lambda(t)) = \sum_{i,j=1}^{k} \frac{\partial^2\varphi}{\partial x_i \partial x_j}\cdot\frac{d\lambda_i}{dt}\cdot\frac{d\lambda_j}{dt} + \sum_{j=1}^{k}\frac{\partial\varphi}{\partial x_j}\cdot\frac{d^2\lambda_j}{dt^2};$$

the partial derivatives of φ on the right-hand side should be taken at the point $\Lambda(t)$.

Find analogous formulas for the third derivative of $\varphi\circ\Lambda$ under the hypothesis that φ and Λ are of class C_3.

3. Let Φ and Λ satisfy the hypotheses of 3.2. If Φ and Λ are of class C_2, then

$$(\Phi\circ\Lambda)'' = (\Phi''\circ\Lambda)\cdot(\Lambda'\times\Lambda') + (\Phi'\circ\Lambda)\cdot\Lambda'',$$

that is,

$$(\varphi_i\circ\Lambda)'' = \sum_{r,s=1}^{k}(\varphi_{i|r,s}\circ\Lambda)\cdot\lambda_r'\lambda_s' + \sum_{r=1}^{k}(\varphi_{i|r}\circ\Lambda)\cdot\lambda_r''.$$

In the traditional notation,

$$\frac{d^2}{dt^2}\Phi(\Lambda(t)) = \sum_{r,s=1}^{k}\frac{\partial^2\Phi}{\partial x_r\partial x_s}\cdot\frac{d\lambda_r}{dt}\cdot\frac{d\lambda_s}{dt} + \sum_{r=1}^{k}\frac{\partial\Phi}{\partial x_r}\cdot\frac{d^2\lambda_r}{dt^2}$$

$$= \sum_{r,s=1}^{k}\Phi_{x_r,x_s}(\Lambda(t))\cdot\lambda_r'(t)\cdot\lambda_s'(t) + \sum_{r=1}^{k}\Phi_{x_r}(\Lambda(t))\cdot\lambda_r''(t).$$

4. Let Φ and Λ satisfy the hypotheses of 3.4. If Φ and Λ are of class C_2, then

$$(\Phi\circ\Lambda)_{|i,j} = (\Phi''\circ\Lambda)\cdot(\Lambda_{|i}\times\Lambda_{|j}) + (\Phi'\circ\Lambda)\cdot\Lambda_{|i,j}.$$

5. Prove that formula (1) holds if φ is differentiable at $\Lambda(t)$ and Λ is differentiable at t.

Hint. Use IV § 3, 6.

6. Prove the following generalization of 3.3: If φ is differentiable at $\Lambda(p)$ and Λ is differentiable (has a derivative) at p, then the composition $\varphi\circ\Lambda$ is differentiable (has a derivative) at p and (13) holds.

Formulate and prove an analogous generalization of 3.4.

7. Prove that a function $\varphi\colon G\to R$ ($G\subset R^k$) is differentiable at a point $p\in G$ if and only if (1) holds for every mapping from an open set of real numbers into G, such that $\Lambda(t_0)=p$ for a $t_0\in G_0$, and $\Lambda'(t_0)$ exists.

8. Let φ be a real function of class C_1 on an open set $G\subset R^{1+l}$, and let $\lambda_1, \ldots, \lambda_l$ be real functions of class C_1 on an open set $G_0\subset R$, such that the composite function

$$\psi(x) = \varphi(x, \lambda_1(x), \ldots, \lambda_l(x)) \tag{29}$$

is defined on the whole set G_0. Prove that ψ is of class C_1 and

$$\psi'(x) = \varphi_{|1}(x, \lambda_1(x), \ldots, \lambda_l(x)) +$$
$$+\varphi_{|2}(x, \lambda_1(x), \ldots, \lambda_l(x))\lambda_1'(x) + \ldots + \varphi_{|1+l}(x, \lambda_1(x), \ldots, \lambda_l(x))\lambda_l'(x)$$

for every $x\in G_0$.

Denoting the coordinates of any point in G by the symbols $x, y_1, \ldots, y_l$ we can formulate this theorem in traditional notation as follows: If $\varphi(x, y_1, \ldots, y_l)$ and

$\lambda_1(x), \ldots, \lambda_l(x)$ are functions of class C_1, then the composite function (29) is of class C_1, and

$$\psi'(x) = \varphi_x(x, \lambda_1(x), \ldots, \lambda_l(x)) + \sum_{j=1}^{k} \varphi_{y_j}(x, \lambda_1(x), \ldots, \lambda_l(x))\lambda_j'(x).$$

Denoting the values of the functions $\lambda_1, \ldots, \lambda_l$ at any point by the same symbols $y_1, \ldots, y_l$, we can write this identity as follows

$$\frac{d\psi}{dx} = \frac{\partial\varphi}{\partial x} + \sum_{j=1}^{l} \frac{\partial\varphi}{\partial y_j} \cdot \frac{dy_j}{dx}.$$

Hint. Replace in 3.1 k by $1+l$ and assume $\Lambda(x) = (x, \lambda_1(x), \ldots, \lambda_l(x))$.

9. If the functions $\varphi, \lambda_1, \ldots, \lambda_l$ in exercise 8 are of class C_2, then the composite function (29) is of class C_2 and

$$\psi''(x) = \varphi_{|1,1}(x, \lambda_1(x), \ldots, \lambda_l(x)) + 2\sum_{j=1}^{l} \varphi_{|1,1+j}(x, \lambda_1(x), \ldots, \lambda_l(x))\lambda_j'(x) +$$
$$+ \sum_{i=1}^{l}\sum_{j=1}^{l} \varphi_{|1+i,1+j}(x, \lambda_1(x), \ldots, \lambda_l(x))\lambda_i'(x)\lambda_j'(x) +$$
$$+ \sum_{j=1}^{l} \varphi_{|1+j}(x, \lambda_1(x), \ldots, \lambda_l(x))\lambda_j''(x).$$

Using the traditional notation as in exercise 8, we have

$$\psi''(x) = \varphi_{x,x}(x, \lambda_1(x), \ldots, \lambda_l(x)) + 2\sum_{j=1}^{l} \varphi_{x,y_j}(x, \lambda_1(x), \ldots, \lambda_l(x))\lambda_j'(x) +$$
$$+ \sum_{i=1}^{l}\sum_{j=1}^{l} \varphi_{y_i,y_j}(x, \lambda_1(x), \ldots, \lambda_l(x))\lambda_i'(x)\lambda_j'(x) +$$
$$+ \sum_{j=1}^{l} \varphi_{y_j}(x, \lambda_1(x), \ldots, \lambda_l(x))\lambda_j''(x),$$

that is,

$$\frac{d^2\psi}{dx^2} = \frac{\partial^2\varphi}{\partial x^2} + 2\sum_{j=1}^{l} \frac{\partial^2\varphi}{\partial x\partial y_j} \cdot \frac{dy_j}{dx} + \sum_{i=1}^{l}\sum_{j=1}^{l} \frac{\partial^2\varphi}{\partial y_i\partial y_j} \cdot \frac{dy_i}{dx} \cdot \frac{dy_j}{dx} + \sum_{j=1}^{l} \frac{\partial\varphi}{\partial y_j} \cdot \frac{d^2y_j}{dx^2}.$$

If the functions $\varphi, \lambda_1, \ldots, \lambda_l$ are of class C_m, then the composite function (29) is of class C_m.

10. Prove the following generalization of exercise 8.

Let φ be a real function of class C_1 on an open set $G \subset R^{k+l}$ and let $\lambda_1, \ldots, \lambda_l$ be real functions of class C_1 on an open set $G_0 \subset R^k$, such that the composite function

$$\psi(x_1, \ldots, x_k) = \varphi(x_1, \ldots, x_k, \lambda_1(x_1, \ldots, x_k), \ldots, \lambda_l(x_1, \ldots, x_k)) \tag{30}$$

is defined on the whole set G_0. The function ψ is of class C_1 and

$$\psi_{|i}(x_1, \ldots, x_k) = \varphi_{|i}(x_1, \ldots, x_k, \lambda_1(x_1, \ldots, x_k), \ldots, \lambda_l(x_1, \ldots, x_k)) +$$
$$+ \sum_{j=1}^{l} \varphi_{|k+j}(x_1, \ldots, x_k, \lambda_1(x_1, \ldots, x_k), \ldots, \lambda_l(x_1, \ldots, x_k))\lambda_{j|i}(x_1, \ldots, x_k).$$

Denoting the coordinates of any point in G by $x_1, \ldots, x_k, y_1, \ldots, y_l$ and, simultaneously, the coordinates of any point in G_0 by $x_1, \ldots, x_k$, we can formulate this theorem in the traditional notation as follows: If $\varphi(x_1, \ldots, x_k, y_1, \ldots, y_l)$ and $\lambda_1(x_1, \ldots, x_k), \ldots, \lambda_l(x_1, \ldots, x_k)$ are functions of class C_1, then the composite function (30) is of class C_1 and

$$\psi_{x_i}(x_1, \ldots, x_k) = \varphi_{x_i}(x_1, \ldots, x_k, \lambda_1(x_1, \ldots, x_k), \ldots, \lambda_l(x_1, \ldots, x_k)) +$$
$$+ \sum_{j=1}^{l} \varphi_{y_j}(x_1, \ldots, x_k, \lambda_1(x_1, \ldots, x_k), \ldots, \lambda_l(x_1, \ldots, x_k)) \lambda_{j,x_i}(x_1, \ldots, x_k),$$

that is,

$$\frac{\partial \psi}{\partial x_i} = \frac{\partial \varphi}{\partial x_i} + \sum_{j=1}^{l} \frac{\partial \varphi}{\partial y_j} \cdot \frac{\partial \lambda_j}{\partial x_i}.$$

11. Verify that the composite function (30) in exercise 10 is of class C_2 if $\varphi, \lambda_1, \ldots, \lambda_l$ are of class C_2. Write formulas for the partial derivatives of the second order of ψ.

Verify that, more generally, if $\varphi, \lambda_1, \ldots, \lambda_l$ are of class C_m, then the composite function (30) is of class C_m.

12. Let $p \in R^k$ be a fixed point and let $\varphi\colon B(p, \eta) \to R$ be of class C_∞. There exist functions φ_i, ψ_i of class C_∞ such that

$$\varphi(q) = \varphi(p) + \sum_{i=1}^{k} \varphi_i(q)\psi_i(q) \quad \text{for} \quad q \in B(p, \eta),$$

$$\varphi_i(p) = \varphi_{|i}(p) \quad \text{and} \quad \psi_i(p) = 0, \quad i = 1, \ldots, k.$$

Hint. Observe that

$$\varphi(q) = \varphi(p) + \int_0^1 \frac{d}{dt}\varphi(p + (q-p)t)\,dt.$$

Use 3.1.

13. For any fixed point $p \in R^k$, let $\mathcal{F}_\infty(p)$ be the set of all real functions φ such that

(a) φ is defined in a neighbourhood of p,

(b) φ is of class C_∞.

The mapping $\mathfrak{w}\colon \mathcal{F}_\infty(p) \to R$ defined by

$$\mathfrak{w}(\varphi) = \varphi'_a(p), \tag{31}$$

where a is a fixed vector in R^k, has the properties

$$\mathfrak{w}(\varphi + \psi) = \mathfrak{w}(\varphi) + \mathfrak{w}(\psi), \quad \mathfrak{w}(c\varphi) = c\mathfrak{w}(\varphi), \quad \mathfrak{w}(\varphi\psi) = \mathfrak{w}(\varphi)\psi(p) + \varphi(p)\mathfrak{w}(\psi). \tag{32}$$

Prove that, conversely, every mapping $\mathfrak{w}\colon \mathcal{F}_\infty(p) \to R$ satisfying the conditions (32) is of the form (31), and that the correspondence between $\mathfrak{w}$ and a given by (31) is one-to-one.

Hint. Verify that $\mathfrak{w}(c) = 0$ for every constant function c. Use exercise 12.

§ 4. Systems of equations. The subject of this section can be formulated as follows. Let $\omega_1, \ldots, \omega_l$ be given continuous real functions defined on an open subset G_0 of the $(k+l)$-dimensional Euclidean space. Let

$$\mathfrak{j} = (j_1, \ldots, j_l) \tag{1}$$

be a given sequence of different positive integers $\leqslant k+l$, and let

$$\mathbf{i} = (i_1, \ldots, i_k) \tag{2}$$

be a sequence formed of all the positive integers $\leqslant k+l$ which do not appear in the sequence (1). We shall be interested in the problem, under what conditions we can solve, in a continuous manner, the system of equations

$$\begin{aligned} &\omega_1(x_1, \ldots, x_{k+l}) = 0\,, \\ &\ldots\ldots\ldots\ldots\ldots \\ &\omega_l(x_1, \ldots, x_{k+l}) = 0 \end{aligned} \tag{3}$$

with respect to $x_{j_1}, \ldots, x_{j_l}$, i.e. when do there exist continuous real functions $\psi_1, \ldots, \psi_l$ defined on an open subset $G \subset R^k$ such that if we perform the substitution

$$x_{j_1} = \psi_1(x_{i_1}, \ldots, x_{i_k}),\ \ldots,\ x_{j_l} = \psi_l(x_{i_1}, \ldots, x_{i_k})\,,$$

the left-hand sides of the equations (3) will be equal to 0 for every point $(x_{i_1}, \ldots, x_{i_k}) \in G$.

To simplify the notation we shall interpret the space R^{k+l} as the $(\mathbf{i}, \mathbf{j})$-product of the spaces R^k and R^l (see I § 6), i.e. any point

$$(x_1, \ldots, x_{k+l}) \in R^{k+l} \tag{4}$$

will be denoted by

$$(p, q) \tag{5}$$

where

$$p = (x_{i_1}, \ldots, x_{i_k}) \quad \text{and} \quad q = (x_{j_1}, \ldots, x_{j_l})\,. \tag{6}$$

Consequently the value of any function f at a point (4) will also be denoted by $f(p, q)$ where p and q are defined by (6).

In this section the letter p (with indices) will always denote a point in R^k, and the letter q will denote a point in R^l.

Denoting

$$\Omega = [\omega_1, \ldots, \omega_l] \tag{7}$$

we can formulate the main problem of this section as follows. Let (7) be a given continuous l-dimensional vector field on an open set $G_0 \subset R^{k+l}$. Under what conditions does a continuous mapping $\Psi = (\psi_1, \ldots, \psi_l)$ from an open set $G \subset R^k$ into R^l exist, such that

$$\Omega(p, \Psi(p)) = 0 \tag{8}$$

for every $p \in G$?

It is easy to give an example of a continuous vector field Ω such that the equation $\Omega(p, q) = 0$ never holds (for instance, if the coordinates $\omega_1, \ldots, \omega_l$ of Ω are everywhere different from 0). Then, of course, no solution Ψ of (8) exists. Thus the existence of a point $(p_0, q_0) \in G_0$ such that $\Omega(p_0, q_0) = 0$ is a necessary condition for the existence of a solution Ψ of (8). We shall show that this condition is also sufficient provided that Ω satisfies some simple differentiability conditions. More precisely, then there exists a continuous mapping Ψ of a neighbourhood G of p_0 such that (8) holds and $\Psi(p_0) = q_0$. This will be proved in Theorem 4.3 below.

First we shall examine the case where the vector field (7) is of the form

$$\Omega(p, q) = q - \Lambda(p, q),$$

Λ being a mapping from G_0 into R^l.

We shall say that a mapping Λ of a set $A \subset R^{k+l}$ into R^l satisfies the *Lipschitz condition with a constant* $M \geqslant 0$ *with respect to the* **j**-*th coordinates* if

$$|\Lambda(p, q_1) - \Lambda(p, q_2)| \leqslant M \cdot |q_1 - q_2| \tag{9}$$

for any points $(p, q_1) \in A$, $(p, q_2) \in A$.

In the theorem that follows (p_0, q_0) is a fixed point in R^{k+l}.

4.1. *Let Λ be a continuous mapping from an open set*

$$A = \mathrm{B}(p_0, \delta) \times \mathrm{B}(q_0, \eta) \subset R^{k+l}$$

into R^l. If Λ satisfies the Lipschitz condition with a constant $M < 1$ with respect to the **j**-*th coordinates and the number*

$$K = \sup\{|\Lambda(p, q_0) - q_0| : p \in \mathrm{B}(p_0, \delta)\}$$

satisfies the inequality

$$\frac{K}{1-M} < \eta, \tag{10}$$

then there exists a continuous mapping Ψ: $\mathrm{B}(p_0, \delta) \to \mathrm{B}(q_0, \eta)$ such that

$$\Psi(p) = \Lambda(p, \Psi(p)) \quad \textit{for every } p \in \mathrm{B}(p_0, \delta).$$

Moreover, for every $p \in \mathrm{B}(p_0, \delta)$, the point $q = \Psi(p)$ is the only point q in $\mathrm{B}(q_0, \eta)$ such that

$$q = \Lambda(p, q).$$

We recall that $\mathrm{B}(p_0, \delta)$ and $\mathrm{B}(q_0, \eta)$ are open balls in the spaces R^k and R^l respectively (see II § 1, p. 45).

We shall define by induction a sequence of mappings

$$\Psi_n: \mathrm{B}(p_0, \delta) \to \mathrm{B}(q_0, \eta) \quad (n = 0, 1, 2, \ldots) \tag{11}$$

as follows:

$$\Psi_0(p) = q_0,$$

$$\Psi_{n+1}(p) = \Lambda(p, \Psi_n(p)).$$

It follows from (9) that for $n > 1$

$$|\Psi_{n+1}(p) - \Psi_n(p)| \leqslant M \cdot |\Psi_n(p) - \Psi_{n-1}(p)| .$$

Hence

$$|\Psi_{n+1}(p) - \Psi_n(p)| \leqslant M^n \cdot |\Psi_1(p) - \Psi_0(p)| = M^n \cdot |\Lambda(p, q_0) - q_0| .$$

Consequently

$$(12) \qquad |\Psi_{n+1}(p) - \Psi_n(p)| \leqslant M^n \cdot K .$$

Thus

$$(13) \qquad \begin{aligned} &|\Psi_{n+1}(p) - q_0| \\ &\leqslant |\Psi_{n+1}(p) - \Psi_n(p)| + |\Psi_n(p) - \Psi_{n-1}(p)| + \ldots + |\Psi_1(p) - \Psi_0(p)| \\ &\leqslant K \cdot (M^n + M^{n-1} + \ldots + 1) < K/(1-M) < \eta . \end{aligned}$$

This proves that (11) really holds. It follows directly from the definition that all the mappings Ψ_n are continuous. The sequence (Ψ_n) satisfies the uniform Cauchy condition since for $m > n$

$$\begin{aligned} |\Psi_m(p) - \Psi_n(p)| &< |\Psi_m(p) - \Psi_{m-1}(p)| + \ldots + |\Psi_{n+1}(p) - \Psi_n(p)| \\ &\leqslant K \cdot (M^{m-1} + \ldots + M^n) < K \cdot M^n/(1-M) \end{aligned}$$

by (12), and $M^n \to 0$ as $n \to \infty$. By III 9.1 and 9.2, the sequence (Ψ_n) converges to a continuous mapping $\Psi\colon \mathrm{B}(p_0, \delta) \to R^l$. By (13),

$$|\Psi(p) - q_0| = \lim_{n\to\infty} |\Psi_{n+1}(p) - q_0| \leqslant K/(1-M) < \eta .$$

Thus $\Psi\colon \mathrm{B}(p_0, \delta) \to \mathrm{B}(q_0, \eta)$. It follows from the inductive definition of (Ψ_n) and from the continuity of Λ that

$$\Psi(p) = \lim_{n\to\infty} \Psi_{n+1}(p) = \lim_{n\to\infty} \Lambda(p, \Psi_n(p)) = \Lambda(p, \lim_{n\to\infty} \Psi_n(p)) = \Lambda(p, \Psi(p)) .$$

To prove the final part of 4.1 let us suppose that simultaneously

$$q_1 = \Lambda(p, q_1) \quad \text{and} \quad q_2 = \Lambda(p, q_2) .$$

By (9),

$$|q_1 - q_2| = |\Lambda(p, q_1) - \Lambda(p, q_2)| \leqslant M \cdot |q_1 - q_2| .$$

Since $M < 1$, this inequality can only hold if $|q_1 - q_2| = 0$, i.e. $q_1 = q_2$. We see that for every $p \in \mathrm{B}(p_0, \delta)$ there exists at most one point q satisfying the equation $q = \Lambda(p, q)$. In the first part of the proof we showed that the point $q = \Psi(p)$ satisfies the equation. Thus it is the only point satisfying the equation.

4.2. *Let* $\Lambda = (\lambda_1, \ldots, \lambda_l)$ *be a continuous mapping of an open subset* $G_0 \subset R^{k+l}$ *into* R^l, *of class* C_1 *with respect to the* **j**-*th variables, and let* $(p_0, q_0) \in G_0$. *If*

$$q_0 = \Lambda(p_0, q_0), \qquad \Lambda_{|\mathbf{j}}(p_0, q_0) = 0, \tag{14}$$

then there are numbers $\delta > 0$ *and* $\eta > 0$ *such that*

(a) $A = B(p_0, \delta) \times B(q_0, \eta) \subset G_0$;

(b) *there exists a continuous mapping* $\Psi\colon B(p_0, \delta) \to B(q_0, \eta)$ *such that*

$$\Psi(p) = \Lambda\big(p, \Psi(p)\big) \quad \textit{for every } p \in B(p_0, \delta);$$

(c) *for every point* $p \in B(p_0, \delta)$ *the point* $q = \Psi(p)$ *is the only point* q *in* $B(q_0, \eta)$ *satisfying the equation*

$$q = \Lambda(p, q). \tag{15}$$

By hypothesis, all the partial derivatives $\lambda_{i|j_r}$ $(i = 1, \ldots, l,\ r = 1, \ldots, l)$ are continuous and vanish at the point (p_0, q_0) (see the second of the conditions (14)). Consequently there exist numbers $\delta_0 > 0$ and $\eta > 0$ such that

$$|\lambda_{i|j_r}(p, q)| < \frac{1}{2l} \quad \text{for} \quad p \in B(p_0, \delta_0) \text{ and } q \in B(q_0, \eta),\ i, r = 1, \ldots, l.$$

The set G_0 being open, we may assume, moreover, that

$$B(p_0, \delta_0) \times B(q_0, \eta) \subset G_0$$

(for if not, we can replace these numbers by smaller ones). It follows from the continuity of Λ and from the first of conditions (14) that there is a positive number $\delta \leqslant \delta_0$ such that the number

$$K = \sup\{|\Lambda(p, q_0) - q_0|\colon p \in B(p_0, \delta)\}$$

satisfies the inequality

$$2K < \eta. \tag{16}$$

By the definition of δ, the inclusion (a) holds.

Now we shall prove that $\Lambda|A$ satisfies the Lipschitz condition with the constant $M = \frac{1}{2}$ with respect to the **j**-th coordinates, A being the set defined in (a). Considering the function $\lambda_i(p, q)$ as a function of q, the point p being fixed, and applying the mean-value theorem III 3.4 we get for any $(p, q_1), (p, q_2)$ in A

$$|\lambda_i(p, q_1) - \lambda_i(p, q_2)| = \big|(q_1 - q_2)\cdot \lambda_{i|\mathbf{j}}\big(p, q_2 + \theta(q_1 - q_2)\big)\big| \leqslant \frac{1}{2\sqrt{l}}|q_1 - q_2|.$$

Consequently

$$|\Lambda(p, q_1) - \Lambda(p, q_2)| = \sqrt{\sum_{i=1}^{l} \big(\lambda_i(p, q_1) - \lambda_i(p, q_2)\big)^2} \leqslant \tfrac{1}{2}|q_1 - q_2|.$$

By (16), the numbers δ and η, the set A and the mapping Λ restricted to A satisfy all the hypothesis of 4.1. Thus (b) and (c) follow directly from 4.1.

We recall that if $\Omega = [\omega_1, \ldots, \omega_l]$, then

$$|\Omega_{|\mathbf{j}}| = \begin{vmatrix} \omega_{1|j_1} & \ldots & \omega_{1|j_l} \\ \cdot & \cdot & \cdot \\ \omega_{l|j_1} & \ldots & \omega_{l|j_l} \end{vmatrix}$$

(see § 1 (25) and (29)).

4.3. *Let Ω be a continuous l-dimensional vector field, of class C_1 with respect to the* **j***-th coordinates, defined on an open set $G_0 \subset R^{k+l}$, and let $(p_0, q_0) \in G_0$. If*

(17) $$\Omega(p_0, q_0) = 0 \quad \textit{and} \quad |\Omega_{|\mathbf{j}}(p_0, q_0)| \neq 0,$$

then there exist numbers $\delta > 0$ and $\eta > 0$ such that

(a) $B(p_0, \delta) \times B(q_0, \eta) \subset G_0$;

(b) *there exists a continuous mapping Ψ: $B(p_0, \delta) \to B(q_0, \eta)$ such that*

(18) $$\Omega(p, \Psi(p)) = 0 \quad \textit{for every } p \in B(p_0, \delta),$$

(19) $$\Psi(p_0) = q_0;$$

(c) *for every $p \in B(p_0, \delta)$, the point $q = \Psi(p)$ is the only point q in $B(q_0, \eta)$ satisfying the equation*

$$\Omega(p, q) = 0.$$

Assuming $\Omega = [\omega_1, \ldots, \omega_l]$ and $\Psi = (\psi_1, \ldots, \psi_l)$, we can formulate the main part (b) of 4.3 as follows:

If

(α) *$\omega_1, \ldots, \omega_l$ are continuous real functions on an open set $G_0 \subset R^{k+l}$,*

(β) *a point*

(20) $$(x_{0,1}, \ldots, x_{0,k+l})$$

satisfies all the equations (3),

(γ) *the partial derivatives*

$$\frac{\partial \omega_i}{\partial x_{j_r}} \quad (i = 1, \ldots, l, \ r = 1, \ldots, l)$$

are continuous on G_0, and

(δ) *the jacobian*

$$\frac{\partial(\omega_1, \ldots, \omega_l)}{\partial(x_{j_1}, \ldots, x_{j_l})} = \begin{vmatrix} \dfrac{\partial \omega_1}{\partial x_{j_1}} & \ldots & \dfrac{\partial \omega_1}{\partial x_{j_l}} \\ \cdot & \cdot & \cdot \\ \dfrac{\partial \omega_l}{\partial x_{j_1}} & \ldots & \dfrac{\partial \omega_l}{\partial x_{j_l}} \end{vmatrix}$$

does not vanish at the point (20),

then in a neighbourhood of the point $(x_{0,i_1}, \ldots, x_{0,i_k}) \in R^k$ *there exist continuous real functions*

$$x_{j_1} = \psi_1(x_{i_1}, \ldots, x_{i_k}),$$
$$\ldots\ldots\ldots\ldots$$
$$x_{j_l} = \psi_l(x_{i_1}, \ldots, x_{i_k}),$$

that are solutions of the equations (3) *satisfying the following initial conditions*

$$\psi_1(x_{0,i_1}, \ldots, x_{0,i_k}) = x_{0,j_1},$$
$$\ldots\ldots\ldots\ldots\ldots$$
$$\psi_l(x_{0,i_1}, \ldots, x_{0,i_k}) = x_{0,j_l}.$$

We shall also formulate a few particular cases which often appear in applications. We assume that the *functions* ω, λ *and their derivatives* which appear below *are continuous* on an open set G_0.

A) The case where $k = 1 = l$. For any function $\omega(x, y)$, if

$$\omega(x_0, y_0) = 0 \quad \text{and} \quad \omega_y(x_0, y_0) \neq 0,$$

then there exists a continuous function $\psi(x)$ defined on a neighbourhood of x_0, such that

$$\omega(x, \psi(x)) = 0 \quad \text{and} \quad \psi(x_0) = y_0.$$

B) The case where $k = 2$ and $l = 1$. For any function $\omega(x, y, z)$, if

$$\omega(x_0, y_0, z_0) = 0 \quad \text{and} \quad \omega_z(x_0, y_0, z_0) \neq 0,$$

then there exists a continuous function $\psi(x, y)$ defined on a neighbourhood of the point (x_0, y_0), such that

$$\omega(x, y, \psi(x, y)) = 0 \quad \text{and} \quad \psi(x_0, y_0) = z_0.$$

C) The case where $k = 1$ and $l = 2$. For any functions $\lambda(x, y, z)$ and $\omega(x, y, z)$, if

$$\lambda(x_0, y_0, z_0) = 0, \quad \omega(x_0, y_0, z_0) = 0 \quad \text{and} \quad \begin{vmatrix} \lambda_y(x_0, y_0, z_0) & \lambda_z(x_0, y_0, z_0) \\ \omega_y(x_0, y_0, z_0) & \omega_z(x_0, y_0, z_0) \end{vmatrix} \neq 0,$$

then there exist continuous functions $\varphi(x)$ and $\psi(x)$ defined on a neighbourhood of x_0, such that

$$\lambda(x, \varphi(x), \psi(x)) = 0, \quad \omega(x, \varphi(x), \psi(x)) = 0, \quad \text{and} \quad \varphi(x_0) = y_0, \quad \psi(x_0) = z_0.$$

D) The case where $k = 2 = l$. For any functions $\lambda(x, y, z, w)$, $\omega(x, y, z, w)$, if

$$\lambda(x_0, y_0, z_0, w_0) = 0, \quad \omega(x_0, y_0, z_0, w_0) = 0$$

and

$$\begin{vmatrix} \lambda_z(x_0, y_0, z_0, w_0) & \lambda_w(x_0, y_0, z_0, w_0) \\ \omega_z(z_0, y_0, z_0, w_0) & \omega_w(x_0, y_0, z_0, w_0) \end{vmatrix} \neq 0,$$

then there exist continuous functions $\varphi(x, y)$ and $\psi(x, y)$ defined on a neighbourhood of the point (x_0, y_0), such that

$$\lambda(x, y, \varphi(x, y), \psi(x, y)) = 0, \quad \omega(x, y, \varphi(x, y), \psi(x, y)) = 0$$

and

$$\varphi(x_0, y_0) = z_0, \quad \psi(x_0, y_0) = w_0.$$

To prove Theorem 4.3, let

$$\mathfrak{A} = \Omega_{|j}(p_0, q_0)$$

for short. By the second of the hypotheses (17), the determinant $/\mathfrak{A}/$ of the square matrix $\mathfrak{A}$ does not vanish, i.e. the inverse matrix $\mathfrak{A}^{-1}$ exists. The mapping

$$\Lambda(p, q) = q - \mathfrak{A}^{-1} \cdot \Omega(p, q)$$

of G_0 into R^l satisfies all the hypotheses of 4.2. Indeed, Λ is continuous and

$$\Lambda_{|j}(p, q) = \mathfrak{J} - \mathfrak{A}^{-1} \cdot \Omega_{|j}(p, q)$$

(see § 1 (21-(21′)) where $\mathfrak{J}$ denotes the l-dimensional unit matrix. Thus $\Lambda_{|j}$ is continuous. Hypotheses (14) are also satisfied since by (17)

$$\Lambda(p_0, q_0) = q_0 - \mathfrak{A}^{-1} \cdot \Omega(p_0, q_0) = q_0 - \mathfrak{A}^{-1} \cdot 0 = q_0$$

and

$$\Lambda_{|j}(p_0, q_0) = \mathfrak{J} - \mathfrak{A}^{-1} \cdot \mathfrak{A} = \mathfrak{J} - \mathfrak{J} = 0 .$$

Thus there exist numbers $\delta > 0$ and $\eta > 0$ and a mapping Ψ satisfying conditions (a), (b), (c) of 4.2. On the other hand, condition (a) in 4.2 is identical with condition (a) in 4.3. Moreover, (18) and condition (c) in 4.3 are equivalent to (b) and (c) in 4.2, respectively. This follows from the fact that the equation

$$q = \Lambda(p, q)$$

holds if and only if the equation

$$\Omega(p, q) = 0$$

holds. Indeed, the first equation holds if and only if $\mathfrak{A}^{-1} \cdot \Omega(p, q) = 0$. Multiplying this equation by $\mathfrak{A}$ from the left we infer that it holds if and only if $\Omega(p, q) = 0$.

We have proved that the numbers δ, η and the mapping Ψ from 4.2 satisfy conditions (a), (18) and (c) in 4.3.

Since, for $p = p_0$, both points q_0 and $\Psi(p_0)$ are solutions of the equation $\Omega(p_0, q) = 0$ (see the first of conditions (17), and (18)), it follows from (c) that $q_0 = \Psi(p_0)$, i.e. (19) is true. This completes the proof of 4.3.

4.4. *Let $\Omega = [\omega_1, \ldots, \omega_l]$ be an l-dimensional vector field on an open set $G_0 \subset R^{k+l}$ such that*

(21) $$/\Omega_{|j}(p, q)/ \neq 0 \quad \textit{for every point } (p, q) \in G_0 ,$$

and let $\Psi = (\psi_1, \ldots, \psi_l)$ be a continuous mapping of an open set $G \subset R^k$ into R^l such that

$$\Omega(p, \Psi(p)) = 0 \quad \textit{for every } p \in G .$$

If Ω is of class C_1, then Ψ is also of class C_1 and

$$\Psi'(p) = -\big(\Omega_{|j}(p, \Psi(p))\big)^{-1} \cdot \Omega_{|i}(p, \Psi(p)) \quad \textit{for} \quad p \in G, \tag{22}$$

that is,

$$\psi_{r|s} = -\frac{\begin{vmatrix} \omega_{1|j_1} & \cdots & \omega_{1|j_{r-1}} & \omega_{1|i_s} & \omega_{1|j_{r+1}} & \cdots & \omega_{1|j_l} \\ \cdot & \cdot & \cdot & \cdot & \cdot & \cdot & \cdot \\ \omega_{l|j_1} & \cdots & \omega_{l|j_{r-1}} & \omega_{l|i_s} & \omega_{l|j_{r+1}} & \cdots & \omega_{l|j_l} \end{vmatrix}}{\begin{vmatrix} \omega_{1|j_1} & \cdots & \omega_{1|j_l} \\ \cdot & \cdot & \cdot \\ \omega_{l|j_1} & \cdots & \omega_{l|j_l} \end{vmatrix}} \tag{23}$$

for $r = 1, \ldots, l$ and $s = 1, \ldots, k$. All the partial derivatives of the right-hand side should be taken at the point $(p, \Psi(p))$.

If Ω is of class C_m, then Ψ is also of class C_m.

In the traditional notation formula (23) can be written as follows:

$$\frac{\partial x_{j_r}}{\partial x_{i_s}} = -\frac{\dfrac{\partial(\omega_1, \ldots, \omega_l)}{\partial(x_{j_1}, \ldots, x_{j_{r-1}}, x_{i_s}, x_{j_{r+1}}, \ldots, x_{j_l})}}{\dfrac{\partial(\omega_1, \ldots, \omega_l)}{\partial(x_{j_1}, \ldots, x_{j_l})}} \tag{24}$$

where $\partial x_{j_r}/\partial x_{i_s}$ denotes the partial derivative of the function

$$x_{j_r} = \psi_r(x_{i_1}, \ldots, x_{i_k}), \qquad r = 1, \ldots, l$$

with respect to the variable x_{i_s}, $s = 1, \ldots, k$.

We shall also rewrite formula (24) in the particular cases A)-D) discussed on p. 194. We assume that the *functions* ω, λ appearing below *are of class* C_1.

A) If $y = \psi(x)$ is a continuous solution of the equation $\omega(x, y) = 0$, then

$$\psi'(x) = -\frac{\omega_x(x, \psi(x))}{\omega_y(x, \psi(x))},$$

that is,

$$\frac{dy}{dx} = -\frac{\dfrac{\partial\omega}{\partial x}}{\dfrac{\partial\omega}{\partial y}}.$$

B) If $z = \psi(x, y)$ is a continuous solution of the equation $\omega(x, y, z) = 0$, then

$$\psi_x(x, y) = -\frac{\omega_x(x, y, \psi(x, y))}{\omega_z(x, y, \psi(x, y))}, \quad \psi_y(x, y) = -\frac{\omega_y(x, y, \psi(x, y))}{\omega_z(x, y, \psi(x, y))},$$

that is,

$$\frac{\partial z}{\partial x} = -\frac{\dfrac{\partial\omega}{\partial x}}{\dfrac{\partial\omega}{\partial z}}, \quad \frac{\partial z}{\partial y} = -\frac{\dfrac{\partial\omega}{\partial y}}{\dfrac{\partial\omega}{\partial z}}.$$

C) If continuous functions $y = \varphi(x)$, $z = \psi(x)$ constitute a solution of the system of equations

$$\lambda(x, y, z) = 0, \quad \omega(x, y, z) = 0,$$

then

$$\varphi'(x) = -\frac{\begin{vmatrix} \lambda_x(x, \varphi(x), \psi(x)) & \lambda_z(x, \varphi(x), \psi(x)) \\ \omega_x(x, \varphi(x), \psi(x)) & \omega_z(x, \varphi(x), \psi(x)) \end{vmatrix}}{\begin{vmatrix} \lambda_y(x, \varphi(x), \psi(x)) & \lambda_z(x, \varphi(x), \psi(x)) \\ \omega_y(x, \varphi(x), \psi(x)) & \omega_z(x, \varphi(x), \psi(x)) \end{vmatrix}},$$

$$\psi'(x) = -\frac{\begin{vmatrix} \lambda_y(x, \varphi(x), \psi(x)) & \lambda_x(x, \varphi(x), \psi(x)) \\ \omega_y(x, \varphi(x), \psi(x)) & \omega_x(x, \varphi(x), \psi(x)) \end{vmatrix}}{\begin{vmatrix} \lambda_y(x, \varphi(x), \psi(x)) & \lambda_z(x, \varphi(x), \psi(x)) \\ \omega_y(x, \varphi(x), \psi(x)) & \omega_z(x, \varphi(x), \psi(x)) \end{vmatrix}},$$

that is,

$$\frac{dy}{dx} = -\frac{\begin{vmatrix} \dfrac{\partial\lambda}{\partial x} & \dfrac{\partial\lambda}{\partial z} \\ \dfrac{\partial\omega}{\partial x} & \dfrac{\partial\omega}{\partial z} \end{vmatrix}}{\begin{vmatrix} \dfrac{\partial\lambda}{\partial y} & \dfrac{\partial\lambda}{\partial z} \\ \dfrac{\partial\omega}{\partial y} & \dfrac{\partial\omega}{\partial z} \end{vmatrix}}, \qquad \frac{dz}{dx} = -\frac{\begin{vmatrix} \dfrac{\partial\lambda}{\partial y} & \dfrac{\partial\lambda}{\partial x} \\ \dfrac{\partial\omega}{\partial y} & \dfrac{\partial\omega}{\partial x} \end{vmatrix}}{\begin{vmatrix} \dfrac{\partial\lambda}{\partial y} & \dfrac{\partial\lambda}{\partial z} \\ \dfrac{\partial\omega}{\partial y} & \dfrac{\partial\omega}{\partial z} \end{vmatrix}}.$$

D) If continuous functions $z = \varphi(x, y)$, $w = \psi(x, y)$ constitute a solution of the system of equations

$$\lambda(x, y, z, w) = 0, \quad \omega(x, y, z, w) = 0,$$

then

$$\varphi_x(x, y) = -\frac{\begin{vmatrix} \lambda_x(x, y, \varphi(x, y), \psi(x, y)) & \lambda_w(x, y, \varphi(x, y), \psi(x, y)) \\ \omega_x(x, y, \varphi(x, y), \psi(x, y)) & \omega_w(x, y, \varphi(x, y), \psi(x, y)) \end{vmatrix}}{\begin{vmatrix} \lambda_z(x, y, \varphi(x, y), \psi(x, y)) & \lambda_w(x, y, \varphi(x, y), \psi(x, y)) \\ \omega_z(x, y, \varphi(x, y), \psi(x, y)) & \omega_w(x, y, \varphi(x, y), \psi(x, y)) \end{vmatrix}},$$

$$\varphi_y(x, y) = -\frac{\begin{vmatrix} \lambda_y(x, y, \varphi(x, y), \psi(x, y)) & \lambda_w(x, y, \varphi(x, y), \psi(x, y)) \\ \omega_y(x, y, \varphi(x, y), \psi(x, y)) & \omega_w(x, y, \varphi(x, y), \psi(x, y)) \end{vmatrix}}{\begin{vmatrix} \lambda_z(x, y, \varphi(x, y), \psi(x, y)) & \lambda_w(x, y, \varphi(x, y), \psi(x, y)) \\ \omega_z(x, y, \varphi(x, y), \psi(x, y)) & \omega_w(x, y, \varphi(x, y), \psi(x, y)) \end{vmatrix}}$$

$$\psi_x(x, y) = -\frac{\begin{vmatrix} \lambda_z(x, y, \varphi(x, y), \psi(x, y)) & \lambda_x(x, y, \varphi(x, y), \psi(x, y)) \\ \omega_z(x, y, \varphi(x, y), \psi(x, y)) & \omega_x(x, y, \varphi(x, y), \psi(x, y)) \end{vmatrix}}{\begin{vmatrix} \lambda_z(x, y, \varphi(x, y), \psi(x, y)) & \lambda_w(x, y, \varphi(x, y), \psi(x, y)) \\ \omega_z(x, y, \varphi(x, y), \psi(x, y)) & \omega_w(x, y, \varphi(x, y), \psi(x, y)) \end{vmatrix}},$$

$$\psi_y(x, y) = -\frac{\begin{vmatrix} \lambda_z(x, y, \varphi(x, y), \psi(x, y)) & \lambda_y(x, y, \varphi(x, y), \psi(x, y)) \\ \omega_z(x, y, \varphi(x, y), \psi(x, y)) & \omega_y(x, y, \varphi(x, y), \psi(x, y)) \end{vmatrix}}{\begin{vmatrix} \lambda_z(x, y, \varphi(x, y), \psi(x, y)) & \lambda_w(x, y, \varphi(x, y), \psi(x, y)) \\ \omega_z(x, y, \varphi(x, y), \psi(x, y)) & \omega_w(x, y, \varphi(x, y), \psi(x, y)) \end{vmatrix}},$$

that is,

$$\frac{\partial z}{\partial x} = -\frac{\begin{vmatrix} \frac{\partial\lambda}{\partial x} & \frac{\partial\lambda}{\partial w} \\ \frac{\partial\omega}{\partial x} & \frac{\partial\omega}{\partial w} \end{vmatrix}}{\begin{vmatrix} \frac{\partial\lambda}{\partial z} & \frac{\partial\lambda}{\partial w} \\ \frac{\partial\omega}{\partial z} & \frac{\partial\omega}{\partial w} \end{vmatrix}}, \qquad \frac{\partial z}{\partial y} = -\frac{\begin{vmatrix} \frac{\partial\lambda}{\partial y} & \frac{\partial\lambda}{\partial w} \\ \frac{\partial\omega}{\partial y} & \frac{\partial\omega}{\partial w} \end{vmatrix}}{\begin{vmatrix} \frac{\partial\lambda}{\partial z} & \frac{\partial\lambda}{\partial w} \\ \frac{\partial\omega}{\partial z} & \frac{\partial\omega}{\partial w} \end{vmatrix}},$$

$$\frac{\partial w}{\partial x} = -\frac{\begin{vmatrix} \frac{\partial\lambda}{\partial z} & \frac{\partial\lambda}{\partial x} \\ \frac{\partial\omega}{\partial z} & \frac{\partial\omega}{\partial x} \end{vmatrix}}{\begin{vmatrix} \frac{\partial\lambda}{\partial z} & \frac{\partial\lambda}{\partial w} \\ \frac{\partial\omega}{\partial z} & \frac{\partial\omega}{\partial w} \end{vmatrix}}, \qquad \frac{\partial w}{\partial y} = -\frac{\begin{vmatrix} \frac{\partial\lambda}{\partial z} & \frac{\partial\lambda}{\partial y} \\ \frac{\partial\omega}{\partial z} & \frac{\partial\omega}{\partial y} \end{vmatrix}}{\begin{vmatrix} \frac{\partial\lambda}{\partial z} & \frac{\partial\lambda}{\partial w} \\ \frac{\partial\omega}{\partial z} & \frac{\partial\omega}{\partial w} \end{vmatrix}}.$$

To prove 4.4 let us recall that $\mathbf{i}$ and $\mathbf{j}$ are sequences (1) and (2). By the general definition in § 1 (25)

$$\Omega_{|\mathbf{i}} = [\omega_{i|i_r}]_{i\leqslant l, r\leqslant k}, \qquad \Omega_{|\mathbf{j}} = [\omega_{i|j_r}]_{i\leqslant l, r\leqslant l}.$$

Let (p, q) be a fixed point in G_0. By § 2 (19)-(20), there exists a matrix field

$$\mathfrak{R} = [\varrho_{i,j}]_{i\leqslant l, j\leqslant k+l}$$

such that

$$\Omega(\overline{p}, \overline{q}) - \Omega(p, q) = \big(\Omega'(p, q) + \mathfrak{R}(\overline{p}, \overline{q})\big)\cdot\big((\overline{p}, \overline{q}) - (p, q)\big) \quad \text{for} \quad (\overline{p}, \overline{q}) \in G_0$$

and

$$\mathfrak{R}(\overline{p}, \overline{q}) \to 0 \quad \text{for} \quad (\overline{p}, \overline{q}) \to (p, q).$$

Let

$$\mathfrak{A} = [\varrho_{i,i_r}]_{i\leqslant l, r\leqslant k}, \qquad \mathfrak{B} = [\varrho_{i,j_r}]_{i\leqslant l, r\leqslant l}.$$

It is easy to verify that

$$\mathfrak{R}(\overline{p}, \overline{q})\cdot\big((\overline{p}, \overline{q}) - (p, q)\big) = \mathfrak{A}(\overline{p}, \overline{q})\cdot(\overline{p} - p) + \mathfrak{B}(\overline{p}, \overline{q})\cdot(\overline{q} - q).$$

Similarly

$$\Omega'(p, q)\cdot\big((\overline{p}, \overline{q}) - (p, q)\big) = \Omega_{|\mathbf{i}}(p, q)\cdot(\overline{p} - p) + \Omega_{|\mathbf{j}}(p, q)\cdot(\overline{q} - q).$$

Consequently

$$(25) \quad \Omega(\overline{p}, \overline{q}) - \Omega(p, q) = \big(\Omega_{|\mathbf{i}}(p, q) + \mathfrak{A}(\overline{p}, \overline{q})\big)\cdot(\overline{p} - p) + \big(\Omega_{|\mathbf{j}}(p, q) + \mathfrak{B}(\overline{p}, \overline{q})\big)\cdot(\overline{q} - q),$$

where $\mathfrak{A}$ and $\mathfrak{B}$ are matrix fields of types l, k and l, l respectively, such that

$$\mathfrak{A}(\bar{p}, \bar{q}) \to 0 \quad \text{as} \quad (\bar{p}, \bar{q}) \to (p, q)$$

and

$$\mathfrak{B}(\bar{p}, \bar{q}) \to 0 \quad \text{as} \quad (\bar{p}, \bar{q}) \to (p, q) .$$

Now let p be a fixed point in G, let $\mathfrak{a}$ be a fixed vector in V_k, and let

$$q = \Psi(p) , \quad \bar{p} = p + \Delta t \cdot \mathfrak{a} , \quad \bar{q} = \Psi(\bar{p}) = \Psi(p + \Delta t \cdot \mathfrak{a}) .$$

Since Ψ is continuous, we have

$$(\bar{p}, \bar{q}) \to (p, q) \quad \text{as} \quad \Delta t \to 0 .$$

It follows from (25) that, for small $|\Delta t|$,

$$\begin{aligned} &\Omega\big(p+\Delta t\cdot\mathfrak{a}, \Psi(p+\Delta t\cdot\mathfrak{a})\big) - \Omega\big(p, \Psi(p)\big) \\ &= \big(\Omega_{|\mathfrak{i}}(p, \Psi(p)) + \mathfrak{A}(\Delta t)\big)\cdot \Delta t \cdot \mathfrak{a} + \big(\Omega_{|\mathfrak{j}}(p, \Psi(p)) + \mathfrak{B}(\Delta t)\big)\cdot\big(\Psi(p+\Delta t\cdot\mathfrak{a}) - \Psi(p)\big) \end{aligned}$$

where $\mathfrak{A}(\Delta t)$ and $\mathfrak{B}(\Delta t)$ are matrices (depending on Δt) of types l, k and l, l respectively, such that

$$\mathfrak{A}(\Delta t) \to 0 \quad \text{and} \quad \mathfrak{B}(\Delta t) \to 0 \quad \text{as} \quad \Delta t \to 0 .$$

On the other hand,

$$\Omega\big(p, \Psi(p)\big) = 0 \quad \text{and} \quad \Omega\big(p+\Delta t\cdot\mathfrak{a}, \Psi(p+\Delta t\cdot\mathfrak{a})\big) = 0$$

since Ψ is a solution of the equation $\Omega(p, q) = 0$. Consequently,

$$(26) \quad \big(\Omega_{|\mathfrak{i}}(p, \Psi(p)) + \mathfrak{A}(\Delta t)\big)\cdot\mathfrak{a} + \\ + \big(\Omega_{|\mathfrak{j}}(p, \Psi(p)) + \mathfrak{B}(\Delta t)\big)\cdot \frac{\Psi(p+\Delta t\cdot\mathfrak{a}) - \Psi(p)}{\Delta t} = 0 .$$

Since

$$/\Omega_{|\mathfrak{j}}(p, \Psi(p)) + \mathfrak{B}(\Delta t)/ \to /\Omega_{|\mathfrak{j}}(p, \Psi(p))/ \neq 0 \quad \text{for} \quad \Delta t \to 0$$

(see (21)), we infer that the inverse matrix

$$\big(\Omega_{|\mathfrak{j}}(p, \Psi(p)) + \mathfrak{B}(\Delta t)\big)^{-1}$$

exists for $|\Delta t|$ sufficiently small. Multiplying (26) by the inverse matrix we obtain

$$\begin{aligned} &\frac{\Psi(p+\Delta t\cdot\mathfrak{a}) - \Psi(p)}{\Delta t} \\ &\quad = -\Big(\big(\Omega_{|\mathfrak{j}}(p, \Psi(p)) + \mathfrak{B}(\Delta t)\big)^{-1} \cdot \big(\Omega_{|\mathfrak{i}}(p, \Psi(p)) + \mathfrak{A}(\Delta t)\big)\Big)\cdot\mathfrak{a} . \end{aligned}$$

Consequently, for $\Delta t \to 0$ we obtain

$$\Psi'_{\mathfrak{a}}(p) = -\Big(\big(\Omega_{|\mathfrak{j}}(p, \Psi(p))\big)^{-1} \cdot \Omega_{|\mathfrak{i}}(p, \Psi(p))\Big)\cdot\mathfrak{a} .$$

This proves that the directional derivative $\Psi'_{\mathfrak{a}}(p)$ exists and is a linear function of the vector $\mathfrak{a}$, i.e. the derivative $\Psi'(p)$ exists. Comparing the identity $\Psi'_{\mathfrak{a}}(p) = \Psi'(p)\cdot\mathfrak{a}$ (see IV 3.3 (3)) with the formula for $\Psi'_{\mathfrak{a}}(p)$ just obtained, we get identity (22). The continuity of Ψ' follows directly from (22). This proves the first part of 4.4.

The proof of the last part of 4.4 is by induction with respect to m. The case of $m=1$ has just been proved. Suppose the statement is true for $m-1$. If Ω is of class C_m, then $\Omega_{|\mathrm{i}}$ and $\Omega_{|\mathrm{j}}$ are of class C_{m-1}. By the induction hypothesis, Ψ is of class C_{m-1}. Thus the right-hand side of (22) is of class C_{m-1} (see the last part of 3.4), i.e. the derivative Ψ' is of class C_{m-1}. This proves that Ψ is of class C_m.

Theorem 4.4 is a particular case (where $s=l$) of the following more general theorem. We recall that if $\mathfrak{A}$ is a rectangular matrix, then $\mathfrak{A}^*$ is the transpose of $\mathfrak{A}$ (see II § 5, p. 61).

4.5. *Let $\Omega = [\omega_1, \dots, \omega_s]$ be an s-dimensional vector field on an open set $G_0 \subset R^{k+l}$ such that*

$$(27) \qquad |\Omega_{|\mathrm{j}}(p, q)| \neq 0 \quad \text{for every point } (p, q) \in G_0,$$

let $s \geqslant l$ and let $\Psi = (\psi_1, \dots, \psi_l)$ be a continuous mapping of an open set $G \subset R^k$ into R^l such that $(p, \Psi(p)) \in G_0$ and

$$(28) \qquad \Omega(p, \Psi(p)) = 0 \quad \text{for every } p \in G.$$

If Ω is of class C_1, then Ψ is also of class C_1 and

$$(29) \quad \Psi'(p) = -\left(\Omega_{|\mathrm{j}}(p, \Psi(p))^* \cdot \Omega_{|\mathrm{j}}(p, \Psi(p))\right)^{-1} \cdot \Omega_{|\mathrm{j}}(p, \Psi(p))^* \cdot \Omega_{|\mathrm{i}}(p, \Psi(p)).$$

If Ω is of class C_m, then Ψ is also of class C_m.

The difference between 4.5 and 4.4 is that now we admit the case where the number s of coordinates of Ω is greater than the number l of coordinates of q. Then the matrix $\Omega_{|\mathrm{j}}(p, q)$ is rectangular but not a square matrix, therefore it has no inverse and formula (22) has no sense for $s > l$. If $s = l$, then hypothesis (21) is identical with (27), and formula (22) is identical with (29) since $\Omega_{|\mathrm{j}}(p, q)$ is then a square matrix,

$$\left(\Omega_{|\mathrm{j}}(p, \Psi(p))^* \cdot \Omega_{|\mathrm{j}}(p, \Psi(p))\right)^{-1} = \Omega_{|\mathrm{j}}(p, \Psi(p))^{-1} \cdot \left(\Omega_{|\mathrm{j}}(p, \Psi(p))^*\right)^{-1}$$

(see II § 5 (15)) and consequently

$$\left(\Omega_{|\mathrm{j}}(p, \Psi(p))^* \cdot \Omega_{|\mathrm{j}}(p, \Psi(p))\right)^{-1} \cdot \Omega_{|\mathrm{j}}(p, \Psi(p))^* = \Omega_{|\mathrm{j}}(p, \Psi(p))^{-1}.$$

To prove 4.5, let p_0 be a fixed point in G. Since $|\Omega_{|\mathrm{j}}(p_0, \Psi(p_0))| \neq 0$ by (27), there exists a sequence $r_1, \dots, r_l$ of positive integers $\leqslant s$ such that the jacobian

$$\begin{vmatrix} \omega_{r_1|j_1} & \cdots & \omega_{r_1|j_l} \\ \cdot & \cdot & \cdot \\ \omega_{r_l|j_1} & \cdots & \omega_{r_l|j_l} \end{vmatrix}$$

does not vanish at the point $(p_0, \Psi(p_0))$. Denoting

$$\overline{\Omega} = [\omega_{r_1}, \dots, \omega_{r_l}]$$

we can write the last statement as follows:

$$/\overline{\Omega}_{|j}(p_0, \Psi(p_0))/ \neq 0 .$$

By definition and (28),

$$\overline{\Omega}(p, \Psi(p)) = 0 \quad \text{for every } p \in G .$$

Replacing in 4.4 the open set G_0 by the open set

$$G_0' = \{(p, q) \in G_0 \colon /\overline{\Omega}_{|j}(p, q)/ \neq 0\} ,$$

and the open set G by the open set

$$G' = \{p \in G \colon (p, \Psi(p)) \in G_0'\}$$

containing the point p_0, and replacing Ω by $\overline{\Omega}$ and Ψ by $\Psi|G'$, we infer from 4.4 that if Ω is of class C_m, then $\Psi|G'$ is of class C_m. In other words, we have proved that if Ω is of class C_m, then Ψ is of class C_m on a neighbourhood G' of any given point $p_0 \in G$. This proves that Ψ is of class C_m.

To prove (29) let us observe that if Ω is of class C_1, then the mapping $\Phi\colon G \to G_0$ defined by

$$\Phi(p) = (p, \Psi(p))$$

is also of class C_1 by the part of 4.5 just proved. Consequently, by 3.4,

$$(\Omega \circ \Phi)'(p) = \Omega'(\Phi(p)) \cdot \Phi'(p) .$$

By a simple calculation we verify that the expression on the right-hand side is equal to

$$\Omega_{|i}(p, \Psi(p)) + \Omega_{|j}(p, \Psi(p)) \cdot \Psi'(p) . \tag{30}$$

On the other hand, $\Omega \circ \Phi = 0$ by (28) and, consequently, $(\Omega \circ \Phi)'(p) = 0$ for every $p \in G$. Thus (30) is equal to 0, i.e.

$$\Omega_{|j}(p, \Psi(p)) \cdot \Psi'(p) = -\Omega_{|i}(p, \Psi(p)) .$$

This proves (29) by II 5.2 (22) where $\mathfrak{A} = \Omega_{|j}(p, \Psi(p))$, $\mathfrak{B} = \Psi'(p)$ and $\mathfrak{C} = -\Omega_{|i}(p, \Psi(p))$.

EXERCISES

1. Prove the following formula for the second derivative of a continuous solution $y = \psi(x)$ of an equation $\omega(x, y) = 0$ (ω is supposed to be of class C_2):

$$\psi''(x) = \frac{2\omega_{x,y}(x, \psi(x))\,\omega_x(x, \psi(x))\,\omega_y(x, \psi(x))}{\omega_y(x, \psi(x))^3} \\ \qquad \frac{-\,\omega_{x,x}(x, \psi(x))\,\omega_y(x, \psi(x))^2 - \omega_{y,y}(x, \psi(x))\,\omega_x(x, \psi(x))^2}{\omega_y(x, \psi(x))^3} .$$

2. Calculate the second partial derivatives of the function ψ in the example B) on p. 196 under the hypothesis that ω is of class C_2.

3. If the functions $x = f(y, z)$, $y = g(z, x)$, $z = h(x, y)$ are solutions of an equation $\omega(x, y, z) = 0$ satisfying the following initial condition

$$x_0 = f(y_0, z_0), \quad y_0 = g(z_0, x_0), \quad z_0 = h(x_0, y_0)$$

and $\omega(x_0, y_0, z_0) = 0$, $\omega_x(x_0, y_0, z_0) \neq 0$, $\omega_y'(x_0, y_0, z_0) \neq 0$, $\omega_z(x_0, y_0, z_0) \neq 0$, then

$$f_z(y_0, z_0) \cdot g_x(z_0, x_0) \cdot h_y(x_0, y_0) = -1$$

(we assume that ω is of class C_1).

4. Calculate the derivative $\psi'(0)$ of a continuous solution $y = \psi(x)$ of the equation

$$e^{2x\cos y} + e^{y\cos 2x} - 2 = 0$$

satisfying the initial condition $\psi(0) = 0$. Calculate also $\psi''(0)$.

5. Calculate all the partial derivatives of the first order at the point $(1, 1)$ of a continuos solution $z = \psi(x, y)$ of the equation

$$\log(x + y + z - 2) \cdot e^{x+y} - 2x + y + z = 0$$

satisfying the initial condition $\psi(1, 1) = 1$.

6. Calculate the derivatives $\varphi'(1)$, $\psi'(1)$ of a continuous solution $y = \varphi(x)$, $z = \psi(x)$ of the system of equations

$$2xe^{2x+3y-z} - z\cos y = 0,$$

$$\log(z - x) + \sin y + 2x + y - z = 0,$$

satisfying the initial condition $\varphi(1) = 0$, $\psi(1) = 2$.

7. Let $\omega(x, y) = x^2 + y^2 - 1$. We have $\omega(1, 0) = 0$ and $\omega_y(1, 0) = 0$. There exists no continuous function $y = \psi(x)$ defined on a neighbourhood of the point $x_0 = 1$ and satisfying the equation $\omega(x, \psi(x)) = 0$.

8. Let $\omega(x, y) = (x^2 + y^2 + 2y)(x^2 + y^2 - 2y)$. Verify that $\omega(0, 0) = 0$, $\omega_y(0, 0) = 0$ and that there exist four continuous functions (and even of class C_1) defined on the interval $(-1; 1)$ and satisfying the equation $\omega(x, \psi(x)) = 0$ and the initial condition $\psi(0) = 0$.

9. Let $\omega(x, y) = (x^2 + y^2 - 1)^2$. Verify that $\omega(0, 1) = 0$, $\omega_y(0, 1) = 0$ and that, on every interval $(a; b)$ where $-1 < a < 0 < b < 1$, there exists exactly one continuous function ψ such that $\omega(x, \psi(x)) = 0$ and $\psi(0) = 1$.

10. Theorem 4.1 remains true if the space R^k is replaced by a metric space X, the space R^l is replaced by a complete metric space Y, and the space R^{k+l} is replaced by the product $X \times Y$ with the metric defined in III § 1, 3. Formulate and prove this modification of 4.1.

§ 5. Regular mappings and vector fields. Diffeomorphisms. Let G be an open subset of the k-dimensional Euclidean space R^k, and let Φ be either a mapping

$$\Phi = (\varphi_1, \dots, \varphi_l)$$

from G into R^l, or an l-dimensional vector field

$$\Phi = [\varphi_1, \dots, \varphi_l]$$

on G. We say that Φ is *regular* if it is of class C_1 and

(1) $$|\Phi'(p)| \neq 0 \quad \text{for every } p \in G\,.$$

In the case where $l \geqslant k$ condition (1) states that for every point $p \in G$ at least one of the jacobians

$$\begin{vmatrix} \varphi_{i_1|1}(p) & \dots & \varphi_{i_1|k}(p) \\ \cdot & \cdots & \cdot \\ \varphi_{i_k|1}(p) & \dots & \varphi_{i_k|k}(p) \end{vmatrix} \qquad (1 \leqslant i_1 < i_2 < \dots < i_k \leqslant l)$$

is not equal to 0. Since the vectors

(2) $$\Phi_{|1}(p),\ \dots,\ \Phi_{|k}(p)$$

are the columns of the rectangular matrix $\Phi'(p)$, condition (1) holds if and only if, for every point $p \in G$, the vectors (2) are linearly independent (see II § 5 (7)). By II § 5 (8), condition (1) holds if and only if for every point $p \in G$ and for every non-zero vector $\mathfrak{a} = [a_1, \dots, a_k]$,

(2′) $$\Phi'(p)\mathfrak{a} = a_1\Phi_{|1}(p) + \dots + a_k\Phi_{|k}(p) \neq 0\,.$$

It is easy to verify that condition (2′) holds if and only if the differential $\Phi'(p)\mathfrak{a}$ of Φ at p is a one-to-one function of the vector $\mathfrak{a}$. Thus a mapping $\Phi\colon G \to R^l$ of class C_1 is regular if and only if, for every point $p \in G$, the differential of Φ at p is a one-to-one function.

In the case where $l \leqslant k$ condition (1) states that for every point $p \in G$ at least one of the jacobians

$$\begin{vmatrix} \varphi_{1|j_1}(p) & \dots & \varphi_{1|j_l}(p) \\ \cdot & \cdots & \cdot \\ \varphi_{l|j_1}(p) & \dots & \varphi_{l|j_l}(p) \end{vmatrix} \qquad (1 \leqslant j_1 < j_2 < \dots < j_l \leqslant k)$$

is not equal to zero. Since the vectors

(3) $$\varphi_1'(p),\ \dots,\ \varphi_l'(p)\,,$$

i.e. the derivatives of the real functions $\varphi_1, \dots, \varphi_l$ at p, are rows of the rectangular matrix $\Phi'(p)$, condition (1) holds if and only if for every point $p \in G$ the vectors (3) are linearly independent (see II § 5 (7′)). By II § 5 (8′), condition (1) holds if and only if for every point $p \in G$ and for every non-zero vector $\mathfrak{a} = [a_1, \dots, a_l]$,

(3′) $$\mathfrak{a}\Phi'(p) = a_1\varphi_1'(p) + \dots + a_l\varphi_l'(p) \neq 0\,.$$

In the case where $l = k$ condition (1) states that

(1′) $$/\Phi'(p)/ = \begin{vmatrix} \varphi_{1|1}(p) & \dots & \varphi_{1|k}(p) \\ \cdot & \cdots & \cdot \\ \varphi_{k|1}(p) & \dots & \varphi_{k|k}(p) \end{vmatrix} \neq 0 \quad \text{for every } p \in G\,,$$

i.e. that the scalar derivative $/\Phi'(p)/$ does not vanish in G.

A mapping Φ of an open set $G \subset R^k$ into R^l is said to be a *diffeomorphism* if it is regular and if it is a homeomorphism. We shall prove later that if $l < k$, then there exists no diffeomorphism from $G \subset R^k$ into R^l (see Theorem 6.7). Therefore in this section we always assume that $l \geqslant k$ in all theorems on diffeomorphisms.

First we shall examine the case where $l = k$.

5.1. *If Φ is a regular mapping of an open set $G \subset R^k$ into R^k, then*

(a) *the set $\Phi(G)$ is open,*

(b) *every point $q_0 \in G$ has a neighbourhood $U \subset G$ such that $\Phi|U$ is a diffeomorphism of U onto a neighbourhood V of the point $p_0 = \Phi(q_0)$.*

Theorem 5.1 will be deduced from Theorem 4.3. For this purpose let us assume in the notation adopted in § 4 that

$$l = k\,, \quad \mathbf{i} = (1, \ldots, k) \quad \text{and} \quad \mathbf{j} = (k+1, \ldots, 2k)\,.$$

Consequently, according to the notation in § 4, if

$$p = (x_1, \ldots, x_k)\,, \qquad q = (x_{k+1}, \ldots, x_{2k})\,,$$

then

$$(p, q) = (x_1, \ldots, x_{2k})\,.$$

We shall apply Theorem 4.3 to the vector field

$$\Omega(p, q) = \Phi(q) - p$$

defined in the open set

$$G_0 = \{(p, q)\colon (p \in R^k) \,\&\, (q \in G)\} \subset R^{2k}.$$

It follows directly from the definition that $\Omega_{|\mathbf{j}}(p, q) = \Phi'(q)$. Hence

$$/\Omega_{|\mathbf{j}}(p, q)/ = /\Phi'(q)/ \neq 0 \qquad \text{for every point } (p, q) \in G_0\,.$$

In particular, $/\Omega_{|\mathbf{j}}(p_0, q_0)/ \neq 0$, where p_0 and q_0 are the points mentioned in (b), i.e. $q_0 \in G$ and $p_0 = \Phi(q_0)$. Since $\Omega(p_0, q_0) = 0$, there exist numbers $\delta > 0$ and $\eta > 0$ and a mapping Ψ satisfying conditions (a), (b), (c) of 4.3. By (a), $\mathrm{B}(q_0, \eta) \subset G$.

The sets $V = \mathrm{B}(p_0, \delta)$ and $U = \Phi^{-1}(V) \cap \mathrm{B}(q_0, \eta)$ are open, and $p_0 \in V$, $q_0 \in U$. Thus V and U are neighbourhoods of p_0 and q_0 respectively. It follows from 4.3 (b) that

$$\Phi\big(\Psi(p)\big) = p \qquad \text{for every } p \in V.$$

By 4.3 (c), for every $p \in V$ the point $q = \Psi(p)$ is the only point in U such that

$$\Phi(q) = p\,.$$

Since $U \subset \Phi^{-1}(V)$, we infer that

$$\Psi(V) = U\,, \quad \Phi|U \text{ is one-to-one} \quad \text{and} \quad \Psi = (\Phi|U)^{-1}.$$

Consequently $\Phi(U) = V$. Since Ψ is continuous, $\Phi|U$ is a homeomorphism, and therefore a diffeomorphism from U onto V. This proves part (b) of 5.1.

Part (a) follows from part (b) just proved. Indeed, if $p_0 \in \Phi(G)$, then $p_0 = \Phi(q_0)$ for some $q_0 \in G$. By (b), Φ maps a neighbourhood U of q_0 onto a neighbourhood V of p_0. Since $p_0 \in V \subset \Phi(G)$, p_0 is an interior point of $\Phi(G)$. Since every point p_0 in $\Phi(G)$ is an interior point, the set $\Phi(G)$ is open.

5.2. *If Φ is a one-to-one regular mapping of an open set $G \subset R^k$ into R^k, then the set $\Phi(G)$ is open, both the mappings Φ and Φ^{-1} are diffeomorphisms, and*

$$(\Phi^{-1})' = (\Phi')^{-1} \circ \Phi^{-1}, \tag{4}$$

$$|(\Phi^{-1})'| = |\Phi'|^{-1} \circ \Phi^{-1}, \tag{5}$$

that is,

$$(\Phi^{-1})'(p) = \big(\Phi'(\Phi^{-1}(p))\big)^{-1}, \tag{4'}$$

$$|(\Phi^{-1})'(p)| = \frac{1}{|\Phi'(\Phi^{-1}(p))|} \tag{5'}$$

for every point $p \in \Phi(G)$.

If Φ is of class C_m, then Φ^{-1} is also of class C_m.

Formulas (4) and (5) can also be written in the following form which is a little less precise but easier to remember:

$$(\Phi^{-1})'(p) = \big(\Phi'(q)\big)^{-1}, \tag{4''}$$

$$|(\Phi^{-1})'(p)| = |\Phi'(q)|^{-1}. \tag{5''}$$

Here we must remember that $p = \Phi(q)$, i.e. $q = \Phi^{-1}(p)$. Roughly speaking, the identities express the fact that the operation $^{-1}$ is commutative with differentiation.

If we adopt the notation

$$\Phi = (\varphi_1, \ldots, \varphi_k), \quad \Phi^{-1} = (\psi_1, \ldots, \psi_k),$$
$$p = (u_1, \ldots, u_k), \quad q = (v_1, \ldots, v_k),$$

then the equation $p = \Phi(q)$ states that

$$\begin{aligned} u_1 &= \varphi_1(v_1, \ldots, v_k), \\ &\ldots\ldots\ldots\ldots \\ u_k &= \varphi_k(v_1, \ldots, v_k), \end{aligned} \tag{6}$$

and the equation $q = \Phi^{-1}(p)$ states that

$$\begin{aligned} v_1 &= \psi_1(u_1, \ldots, u_k), \\ &\ldots\ldots\ldots\ldots \\ v_k &= \psi_k(u_1, \ldots, u_k). \end{aligned} \tag{6'}$$

Moreover, equations (6′) are obtained by solving the system (6), and conversely. Using the traditional notation for jacobians, we can write identity (5) in the form

$$\frac{\partial(\psi_1, \ldots, \psi_k)}{\partial(u_1, \ldots, u_k)} = \frac{1}{\dfrac{\partial(\varphi_1, \ldots, \varphi_k)}{\partial(v_1, \ldots, v_k)}}, \tag{7}$$

or in the form

$$\frac{\partial(v_1, \ldots, v_k)}{\partial(u_1, \ldots, u_k)} = \frac{1}{\dfrac{\partial(u_1, \ldots, u_k)}{\partial(v_1, \ldots, v_k)}}, \tag{7'}$$

which is analogous to the known formula $\frac{dv}{du} = \frac{1}{du/dv}$ for the derivative of the inverse of a real function of one real variable. We have to remember that $u_1, \ldots, u_k$ and $v_1, \ldots, v_k$ in identities (7) and (7′) are connected by equations (6), i.e. (6′).

It follows from 5.1 (a) that the set $G' = \Phi(G)$ is open.

We shall prove that Φ is a diffeomorphism. For this purpose it suffices to show that the mapping Φ^{-1} is continuous at any point $p_0 \in G'$. Let $q_0 \in G$ be such that $p_0 = \Phi(q_0)$. By 5.1 (b) there exist neighbourhoods U and V of points q_0 and p_0 respectively, such that $\Phi|U$ is a homeomorphism from U onto V. Since $\Phi^{-1}|V = (\Phi|U)^{-1}$, the mapping $\Phi^{-1}|V$ is continuous. This proves that Φ^{-1} is continuous at p_0.

Now we shall assume the same notation as in the proof of 5.1. The vector field $\Omega(p, q) = \Phi(q) - p$ is of class C_1. Moreover,

$$\Omega_{|i} = -\mathfrak{I}$$

where $\mathfrak{I}$ is the unit matrix with k rows and k columns, and

$$\Omega_{|j}(p, q) = \Phi'(q)$$

which implies that

$$/\Omega_{|j}(p, q)/ = /\Phi'(q)/ \neq 0 .$$

The continuous mapping $\Psi = \Phi^{-1}$ satisfies the equation

$$\Omega(p, \Psi(p)) = 0 \quad \text{for all } p \in G'.$$

By 4.4, Ψ is of class C_1 and

$$(\Phi^{-1})'(p) = \Psi'(p) = \big(\Omega_{|j}(p, \Psi(p))\big)^{-1} = \big(\Phi'(\Psi(p))\big)^{-1}$$

for every $p \in G'$. This proves (4). Identity (5) is a direct consequence of (4). It follows from (5) that

$$/\Psi'(p)/ = /\Phi'(\Psi(p))/^{-1} \neq 0 .$$

Hence Ψ is regular. Since it is a homeomorphism, it is therefore a diffeomorphism.

If Φ is of class C_m, then Ω is also of class C_m. It follows from the final part of 4.4 that Ψ is of class C_m.

It can be proved that, for every homeomorphism Φ of an open set $G \subset R^k$ into R^k, the set $\Phi(G)$ is open in R^k. The proof of this theorem is much more difficult than the proof of the corresponding parts in 5.1 and 5.2.

5.3. *In order that a one-to-one mapping Φ from an open set $G \subset R^k$ onto an open set $G' \subset R^k$ be a diffeomorphism it is necessary and sufficient that both the mappings Φ and Φ^{-1} be of class C_1.*

The necessity follows from 5.2. To prove the sufficiency, it is enough to show that $|\Phi'(q)| \neq 0$ for every $q \in G$. Let $\Psi = \Phi^{-1}$. The composite mapping $\Psi \circ \Phi$ is the identity mapping of G onto itself. The derivative of the identity mapping is the unit matrix, and its scalar derivative is equal to 1. If Φ and Ψ are of class C_1, then by 3.5 (24)

$$|\Psi'(\Phi(q))| \cdot |\Phi'(q)| = 1 .$$

Hence $|\Phi'(q)| \neq 0$.

The following theorem is called the *theorem on three mappings*. We recall that $\Phi'(q)^*$ denotes the transpose of the matrix $\Phi'(q)$ (see p. 61).

5.4. *Let Φ be a diffeomorphism of an open set $G \subset R^k$ into R^l $(l \geqslant k)$, let Ψ be a mapping of an open set $G' \subset R^\varkappa$ into the set $\Phi(G)$, and let $\Lambda = \Phi^{-1} \circ \Psi$.*

(a) *If Ψ is of class C_1, then Λ is also of class C_1 and, for every $p \in G'$,*

$$\Lambda'(p) = \big(\Phi'(\Lambda(p))^* \cdot \Phi'(\Lambda(p))\big)^{-1} \cdot \Phi'(\Lambda(p))^* \cdot \Psi'(p) . \tag{8}$$

If Φ and Ψ are of class C_m, then Λ is also of class C_m.

(b) *If Ψ is regular and $\varkappa \leqslant k$, then Λ is regular.*

(c) *If Ψ is a diffeomorphism and $\varkappa \leqslant k$, then Λ is a diffeomorphism.*

Observe that identity (8) can also be written as follows

$$\Lambda'(p) = \big(\Phi'(q)^* \cdot \Phi'(q)\big)^{-1} \cdot \Phi'(q)^* \cdot \Psi'(p) \quad \text{where} \quad q = \Lambda(p) , \tag{8'}$$

or

$$\Lambda' = \Big(\big((\Phi'^* \cdot \Phi')^{-1} \cdot \Phi'^*\big) \circ \Lambda\Big) \cdot \Psi' . \tag{8''}$$

Part (a) of the theorem follows from 4.5 where

$$\Omega(p, q) = \Phi(q) - \Psi(p) \quad \text{for} \quad p \in G' \text{ and } q \in G ,$$

$$\mathbf{i} = (1, \ldots, \varkappa) , \quad \mathbf{j} = (\varkappa+1, \ldots, \varkappa+k)$$

(the letters k, l, s in 4.5 should now be replaced by $\varkappa, k, l$ respectively, and Ψ should be replaced by Λ). By definition, Λ is continuous and

$$\Omega\big(p, \Lambda(p)\big) = 0 \quad \text{for} \quad p \in G' .$$

Thus Λ is of class C_m if Φ and Ψ are of class C_m. Since $\Omega_{|i}(p,q) = -\Psi'(p)$ and $\Omega_{|j}(p,q) = \Phi'(q)$, (8) is a direct consequence of § 4 (29). Formula (8) can also be obtained directly. By definition $\Psi = \Phi \circ \Lambda$ and, consequently, $\Psi'(p) = \Phi'(\Lambda(p)) \cdot \Lambda'(p)$. This and II 5.2 (22) imply (8). This implies also that $|\Lambda'(p)| \neq 0$ if $|\Psi'(p)| \neq 0$ (see II 5.1), i.e. (b) is true. Since the composition of two homeomorphisms is a homeomorphism, (b) implies (c).

5.5. *If $\varkappa \leqslant k \leqslant l$, Λ is a regular mapping of an open set $G' \subset R^{\varkappa}$ into an open set $G \subset R^k$, and Φ is a regular mapping from G into R^l, then the composition $\Psi = \Phi \circ \Lambda$ is regular. If Φ and Λ are diffeomorphisms, then Ψ is a diffeomorphism.*

By 3.4, Ψ is of class C_1 and, for every $p \in G'$, the matrix $\Psi'(p)$ is the matrix product of the rectangular matrix $\Phi'(\Lambda(p))$ and the rectangular matrix $\Lambda'(p)$. Since $|\Phi'(q)| \neq 0$ and $|\Lambda'(p)| \neq 0$ for all points $q \in G$ and $p \in G'$, we have $|\Psi'(p)| \neq 0$ by II 5.1.

This proves the first part of 5.5. The second part follows from the first because the composition of two homeomorphisms is a homeomorphism.

5.6. *If $\Phi = (\varphi_1, \ldots, \varphi_l)$ is a regular mapping from an open set $G_0 \subset R^k$ into R^l and $k \leqslant l$, then every point $p_0 \in G_0$ has a neighbourhood U such that $\Phi|U$ is a diffeomorphism.*

This generalization of 5.1 (b) can be deduced from 5.1 (b) as follows.

Since $|\Phi'(p_0)| \neq 0$, there exists a sequence $\mathbf{i} = (i_1, \ldots, i_k)$ of positive integers $\leqslant l$ such that the jacobian of the mapping $\Psi = (\varphi_{i_1}, \ldots, \varphi_{i_k})$ at the point p_0 is not equal to 0. The set $G = \{p \in G_0 \colon /\Psi'(p)/ \neq 0\}$ is open and contains the point p_0, and the mapping $\Psi|G$ is regular. By 5.1 (b) there exists a neighbourhood $U \subset G$ of the point p_0 such that $\Psi|U$ is a diffeomorphism. Since $\Psi|U = \Lambda \circ (\Phi|U)$ where $\Lambda \colon \Phi(U) \to R^k$ is the $\mathbf{i}$-projection reduced to $\Phi(U)$, i.e.

$$\Lambda(x_1, \ldots, x_l) = (x_{i_1}, \ldots, x_{i_k}) \quad \text{for} \quad (x_1, \ldots, x_l) \in \Phi(U),$$

the mapping $\Phi|U$ is one-to-one and $(\Phi|U)^{-1} = (\Psi|U)^{-1} \circ \Lambda$. Thus $(\Phi|U)^{-1}$ is continuous, i.e. $\Phi|U$ is a homeomorphism and, therefore, a diffeomorphism.

EXERCISES

1. Let

$$\Phi(u, v) = (u + v\cos u, e^{2u}(v+1)).$$

There exists a neighbourhood U of the point $(0, 0)$ such that $\Phi|U$ is a diffeomorphism. Calculate the derivative and the scalar derivative of $(\Phi|U)^{-1}$ at the point $(0, 1)$.

2. Let Φ, S and p_0 have the same meaning as in § 1, exercise 3. Let $\Psi \colon R^k \to S - \{p_0\}$ be a mapping of class C_1 such that $\psi(0) = (1, 0, \ldots, 0)$. Then $\Psi'(0) = [0, a_1, \ldots, a_k]$. Let $\Lambda = \Phi^{-1} \circ \Psi$. Calculate $\Lambda'(0)$ using identity (8).

3. Let $p_0 \in R^k$ and $\varrho > 0$. By the *inversion with center* p_0 *and radius* ϱ we mean the mapping Φ of the set $R^k - \{p_0\}$ onto itself, defined by

$$\Phi(p) = p_0 + \frac{\varrho^2}{|p-p_0|^2}(p-p_0).$$

Calculate the derivative and the scalar derivative of Φ. Prove that Φ is a diffeomorphism and that it maps the set $\{p\colon 0 < |p-p_0| \leqslant \varrho\}$ onto the set $\{p\colon |p-p_0| \geqslant \varrho\}$, and conversely.

4. The mapping

$$\Phi(p) = p_0 + \frac{p-p_0}{\sqrt{1+|p-p_0|^2}}$$

is a diffeomorphism of R^k onto $\mathrm{B}(p_0, 1) \subset R^k$. Calculate the derivative and the scalar derivative of Φ.

§ 6. Hypersurfaces. The notion of hypersurface which is the subject of this section is a natural multi-dimensional generalization of the notion of surface (without the boundary contour). A characteristic property of any surface H is that every point in H has a neighbourhood in H that is homeomorphic with an open subset of the plane. This property is the starting point for the definition of the k-dimensional hyperplane formulated below. We shall be interested only in k-dimensional analogues of smooth surfaces, i.e. of surfaces which have, at every point p, a tangent plane depending in a continuous manner on p. We shall call them regular hypersurfaces. It will be convenient not to assume that surfaces and hypersurfaces are connected sets. In other words, we shall admit surfaces and hypersurfaces composed of a number of disjoint pieces. After this intuitive explanation of what the subject of our study is, we formulate the corresponding precise definitions as follows.

A non-empty subset H of the l-dimensional Euclidean space R^l is said to be a *k-dimensional hypersurface* $(1 \leqslant k \leqslant l)$ if every point $q_0 \in H$ has a neighbourhood U in H such that U is homeomorphic to an open subset of the k-dimensional Euclidean space R^k. It should be emphasized that the neighbourhood U mentioned in the definition is a neighbourhood of q_0 in the metric space H (not in the space R^l)—see III § 4, p. 110.

In other words, a non-empty set $H \subset R^l$ is a k-dimensional hypersurface if, for every point $q_0 \in H$, there exists a homeomorphism Φ from an open set $G \subset R^k$ onto a neighbourhood U of q_0 in H.

One-dimensional hypersurfaces are called *curves*. Two-dimensional hypersurfaces are called *surfaces*.

Sometimes it is convenient to introduce the notion of zero-dimensional hypersurface. By a *zero-dimensional hypersurface* we shall mean any isolated non-empty subset of R^l (see III § 8, p. 116), i.e. every set H with the property: for every $q_0 \in H$, the one-point set $\{q_0\}$ is a neighbourhood of q_0 in H. In particular, every finite non-empty set $H \subset R^l$ is a zero-dimensional hypersurface. Observe that the general definition of k-dimensional hypersurface can also be extended to the case $k = 0$ by assuming the convention that the zero-dimensional Euclidean space R^0 is the set $\{0\}$ composed of the number zero only (then $\{0\}$ is also the only non-empty open subset of R^0).

It follows directly from the definition of a k-dimensional hypersurface H that every non-empty set $H_0 \subset H$, open in H, is also a k-dimensional hypersurface.

We shall be interested only in the case where the homeomorphism in the definition of hypersurface is regular, i.e. it is a diffeomorphism. We shall say that a k-dimensional hypersurface H is *regular* (of class C_m) if for every point $q_0 \in H$ there exists a diffeomorphism Φ (of class C_m) from an open set $G \subset R^k$ onto a neighbourhood U of q_0 in H ($1 \leqslant m \leqslant \infty$).

Every zero-dimensional hypersurface is considered to be regular (of class C_m for every m). The notion of zero-dimensional hypersurface will be useful only in Chapter IX. Now we shall restrict ourselves to k-dimensional hypersurfaces with positive k.

It follows from 5.1 (a) that the only l-dimensional regular hypersurfaces in R^l are non-empty open subsets of R^l (as a diffeomorphism Φ we can use the identity mapping of the set onto itself).

It follows directly from the definition that if H is a regular k-dimensional hypersurface of class C_m and a set $H_0 \subset H$ is non-empty and open in H, then H_0 is also a regular k-dimensional hypersurface of class C_m.

The final part of Theorem 6.1 below states that the converse statement is also true.

6.1. *If $H \subset R^l$ is a k-dimensional regular hypersurface and Φ_0 is a diffeomorphism from an open set $G_0 \subset R^k$ into H, then the set $\Phi_0(G_0)$ is open in H.*

Consequently, every k-dimensional regular hypersurface $H' \subset H$ is an open subset of H.

To prove the first part of 6.1 it suffices to show that, for every $p_0 \in G_0$, there exists a set U, open in H, such that $\Phi_0(p_0) \in U \subset \Phi_0(G_0)$.

There exists a diffeomorphism Φ from an open set $G \subset R^k$ into H such that $\Phi(G)$ is a neighbourhood of the point $\Phi_0(p_0)$ in H. The set $G' = \Phi_0^{-1}(\Phi(G))$ is open in R^k by III 8.2 (b) and III 4.2. Moreover, $p_0 \in G'$. The mappings Φ and $\Psi = \Phi_0|G'$ satisfy the hypotheses of 5.4 (c). Thus the mapping $\Lambda = \Phi^{-1} \circ \Psi$ is a diffeomorphism from G' into G. By 5.1, the set $\Lambda(G')$ is open in R^k. Therefore the set $U = \Phi(\Lambda(G')) = \Psi(G') = \Phi_0(G')$ is open in the set $\Phi(G)$ (see III 10.1) which is open in H. This proves that U is open in H. Since $\Phi_0(p_0) \in U \subset \Phi_0(G_0)$, the first part of 6.1 is proved.

The second part follows from the first since H' is the union of sets of the form $\Phi_0(G_0)$.

To construct simple examples of k-dimensional regular hypersurfaces in R^l $(0<k<l)$, let us interpret the space R^l as the $(\mathbf{i}, \mathbf{j})$-product of the spaces R^k and R^{l-k} where

$$\mathbf{i} = (i_1, \ldots, i_k), \qquad \mathbf{j} = (j_1, \ldots, j_{l-k})$$

are two fixed sequences of positive integers such that $i_1, \ldots, i_k, j_1, \ldots, j_{l-k}$ is a permutation of $1, \ldots, l$. In other words, a point $(x_1, \ldots, x_l) \in R^l$ will be identified with the point $(p, q) \in R^k \times R^{l-k}$ where

$$p = (x_{i_1}, \ldots, x_{i_k}), \qquad q = (x_{j_1}, \ldots, x_{j_{l-k}}).$$

According to III § 8 D), let $\Lambda_\mathbf{i}$ and $\Lambda_\mathbf{j}$ be the $\mathbf{i}$-projection and the $\mathbf{j}$-projection in R^l. By definition,

$$\Lambda_\mathbf{i}(x_1, \ldots, x_l) = (x_{i_1}, \ldots, x_{i_k}), \qquad \Lambda_\mathbf{j}(x_1, \ldots, x_l) = (x_{j_1}, \ldots, x_{j_{l-k}}),$$

that is,

$$\Lambda_\mathbf{i}(p, q) = p \quad \text{and} \quad \Lambda_\mathbf{j}(p, q) = q \quad \text{for} \quad p \in R^k \text{ and } q \in R^{l-k}.$$

Let G be a subset of R^k. For any mapping $\Psi\colon G \to R^{l-k}$ the formula

$$\Phi(p) = \big(p, \Psi(p)\big) \quad \text{for} \quad p \in G \tag{1}$$

defines a mapping $\Phi\colon G \to R^l$. A mapping $\Phi\colon G \to R^l$ is of the form (1) if and only if the composition $\Lambda_\mathbf{i} \circ \Phi$ is the identity mapping on G, i.e.

$$\Lambda_\mathbf{i} \circ \Phi(p) = p \quad \text{for every } p \in G.$$

Then $\Psi = \Lambda_\mathbf{j} \circ \Phi$.

Let Φ be of the form (1). If $p_1 \neq p_2$, then $\Phi(p_1) \neq \Phi(p_2)$. Thus Φ is a one-to-one mapping from G onto the set $H = \Phi(G)$ i.e. onto the graph

$$H = \big\{\big(p, \Psi(p)\big)\colon p \in G\big\} \subset R^l \tag{2}$$

of the mapping Ψ. The mapping Φ is continuous if and only if Ψ is continuous. Then Φ is a homeomorphism since the inverse mapping $\Phi^{-1} = \Lambda_\mathbf{i}|H$ is continuous (see III § 8 D)).

Suppose now that the set $G \subset R^k$ is open. The mapping (1) is of class C_m if and only if Ψ is of class C_m. If Ψ is of class C_1, then $|\Phi'| \geqslant 1$ since the jacobian formed of all partial derivatives of the $\mathbf{i}$th coordinates of Φ is equal to 1. Thus if Ψ is of class C_m, then Φ is a diffeomorphism of class C_m and the set $U = \Phi(G) = H$ is a neighbourhood in H of every point in H. This proves that if Ψ is of class C_m, then the graph H of Ψ is a regular k-dimensional hypersurface of class C_m.

Regular hypersurfaces H of form (2) where Ψ is a mapping of class C_m will be called *graphs of class* C_m. They are the simplest examples of regular hypersurfaces.

Every regular k-dimensional hypersurface which is a graph has the property that the whole hypersurface is a diffeomorphic image of an open set $G \subset R^k$ (viz. the diffeomorphism (1) is a mapping onto the whole set (2)). Not all hypersurfaces have this property. In particular, not all hypersurfaces are graphs. However, every hypersurface is locally a graph. More precisely,

6.2. *If $H \subset R^l$ is a regular k-dimensional hypersurface of class C_m, $0 < k < l$, then every point $q_0 \in H$ has a neighbourhood U in H such that the hypersurface U is a graph of class C_m.*

By hypothesis, there exists a diffeomorphism $\Phi_0 = (\varphi_1, \ldots, \varphi_l)$ from an open set $G_0 \subset R^k$ into H, such that the set $U = \Phi_0(G_0)$ is a neighbourhood of $q_0 = \Phi_0(p_0)$ $(p_0 \in G_0)$ in the space H. Since $|\Phi_0'(p_0)| \neq 0$, there exists a sequence $\mathbf{i} = (i_1, \ldots, i_k)$ of different positive integers $\leqslant l$, such that the jacobian

$$J = \begin{vmatrix} \varphi_{i_1|1}(p_0) & \ldots & \varphi_{i_1|k}(p_0) \\ \cdot\;\cdot\;\cdot & \cdot\;\cdot\;\cdot & \cdot\;\cdot\;\cdot \\ \varphi_{i_k|1}(p_0) & \ldots & \varphi_{i_k|k}(p_0) \end{vmatrix}$$

is not equal to zero. Let $\mathbf{j} = (j_1, \ldots, j_{l-k})$ be a sequence formed from the remaining positive integers $\leqslant l$. Interpret R^l as the $(\mathbf{i}, \mathbf{j})$-product of R^k and R^{l-k}, as before. Let

$$\Lambda = \Lambda_{\mathbf{i}} \circ \Phi = (\varphi_{i_1}, \ldots, \varphi_{i_k}): \ G_0 \to R^k$$

be the composition of the $\mathbf{i}$-projection $\Lambda_{\mathbf{i}}$ and Φ. By definition, $|\Lambda'(p_0)| = J \neq 0$. Since Λ' is continuous, we have $|\Lambda'(p)| \neq 0$ for all points p in a neighbourhood of p_0. Consequently, by 5.1 (b), there exists a neighbourhood V of p_0 such that $\Lambda|V$ is a diffeomorphism. We may assume for simplicity that $V = G_0$, i.e. that Λ is a diffeomorphism (for if not, replace everywhere G_0 by V, and Φ_0 by $\Phi_0|V$). The set $G = \Lambda(G_0) \subset R^k$ is open by 5.1 (a). The diffeomorphism $\Phi = \Phi_0 \circ \Lambda^{-1}$ of class C_m from G onto the neighbourhood $U = \Phi_0(G_0) = \Phi(G)$ of q_0 in H is of the form (1) since the mapping

$$\Lambda_{\mathbf{i}} \circ \Phi = (\Lambda_{\mathbf{i}} \circ \Phi_0) \circ \Lambda^{-1} = \Lambda \circ \Lambda^{-1}$$

is the identity mapping on G. Thus U is the graph of the mapping $\Psi = \Lambda_{\mathbf{j}} \circ \Phi$ of class C_m.

The converse statement to 6.2 is also true: if every point of a non-empty set $H \subset R^l$ has a neighbourhood in H which is a graph of class C_m, then H is a regular hypersurface of class C_m.

As an example, we shall show that the $(l-1)$-*dimensional sphere*

$$S = \{(x_1, \ldots, x_l): \ x_1^2 + \ldots + x_l^2 = \varrho^2\}$$

is a regular $(l-1)$-dimensional hypersurface of class C_∞. In fact, S is the union of the sets

$$S_i^+ = S \cap \{(x_1, \dots, x_l): x_i > 0\}, \qquad S_i^- = S \cap \{(x_1, \dots, x_l): x_i < 0\}$$

where $i = 1, \dots, k$. Each of the sets is open in S because it is the intersection of S and an open set in R^l. Each of the sets is a graph of class C_∞ since

$$S_i^+ = \{(x_1, \dots, x_{i-1}, \sqrt{\varrho^2 - x_1^2 - \dots - x_{i-1}^2 - x_{i+1}^2 - \dots - x_l^2}, x_{i+1}, \dots, x_l):$$
$$(x_1, \dots, x_{i-1}, x_{i+1}, \dots, x_l) \in G\},$$

where G is the open ball in R^{l-1} with center $(0, \dots, 0)$ and radius ϱ, and similarly for S_i^- (replace $\sqrt{\ }$ by $-\sqrt{\ }$).

By the same method we prove that, more generally, every $(l-1)$-*dimensional sphere*

$$\{p \in R^l: |p - p_0| = \varrho\}$$

with center p_0 and radius ϱ is an $(l-1)$-dimensional regular hypersurface of class C_∞.

Observe that the $(l-1)$-dimensional sphere S is not of the form $S = \Phi(G)$, where G is an open set in R^{l-1} and Φ is a diffeomorphism from G onto S. For suppose it is of this form. Then the open non-empty set $G = \Phi^{-1}(S)$ would be compact since S is compact and Φ^{-1} is continuous (see III 6.2). This is impossible (see III 11.1 and III § 11, exercise 2).

The set

$$S_k = \{(x_1, \dots, x_{k+1}, 0, \dots, 0) \in R^l: x_1^2 + \dots + x_{k+1}^2 = \varrho^2\},$$

i.e. the intersection of the sphere S with the $(k+1)$-dimensional hyperplane composed of points whose coordinates with indices $k+2, \dots, l$ are equal to zero, is an example of a k-dimensional regular hypersurface in R^l of class C_∞, $0 < k < l$. To prove it, let us define the sets S_i^+, S_i^- (where $i = 1, \dots, k+1$) as previously, but replacing S by S_k. The sets are open in S_k and S_k is their union. Each of the sets is a graph of class C_∞. Indeed,

$$S_i^+ = \{(x_1, \dots, x_{i-1}, \sqrt{\varrho^2 - x_1^2 - \dots - x_{i-1}^2 - x_{i+1}^2 - \dots - x_{k+1}^2},$$
$$x_{i+1}, \dots, x_{k+1}, 0, \dots, 0): (x_1, \dots, x_{i-1}, x_{i+1}, \dots, x_{k+1}) \in G\},$$

where G is the open ball in R^k with center $(0, \dots, 0)$ and radius ϱ, and similarly for S_i^-.

An interesting example of a regular surface of class C_∞ in R^3 is given by the set

$$T = \{\Phi(p): p \in R^2\}$$

where

$$\Phi(u, v) = \big((\varrho_1 + \varrho_2 \cos v)\cos u, (\varrho_1 + \varrho_2 \cos v)\sin u, \varrho_2 \sin v\big), \qquad \varrho_1 > \varrho_2 > 0 .$$

The surface T shown in fig. 1 is called a *torus*.

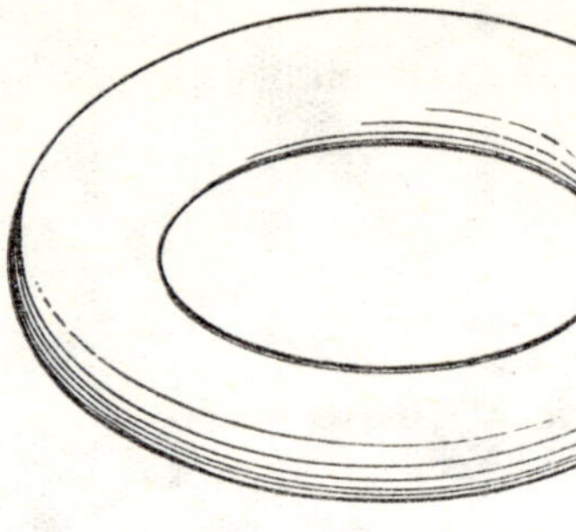

Fig. 1

It is easy to calculate that $|\Phi'(u, v)| = \varrho_2(\varrho_1 + \varrho_2 \cos v) > 0$. The mapping Φ is not one-to-one; however it is regular, of class C_∞ and every point $p = (u_0, v_0) \in R^2$ has a neighbourhood G such that $\Phi|G$ is a diffeomorphism of G onto a set open in T. For instance, the set

$$G = \{(u, v): (|u - u_0| < \pi) \,\&\, (|v - v_0| < \pi)\}$$

has this property. This proves that T is a regular surface of class C_∞.

6.3. *Let $H \subset R^l$ be a k-dimensional regular hypersurface and let H' be a $\varkappa$-dimensional regular hypersurface, $H' \subset H$, $\varkappa \leqslant k$. For every diffeomorphism Φ from an open set $G \subset R^k$ into H, the set $\Phi^{-1}(H')$ is either empty or it is a $\varkappa$-dimensional regular hypersurface in R^k.*

Let $q_0 \in \Phi^{-1}(H')$. Since $\Phi(q_0) \in H'$, there exists a diffeomorphism $\overline{\Psi}$ from an open set $G_0' \subset R^\varkappa$ into H', such that $\overline{\Psi}(G_0')$ is a neighbourhood of the point $\Phi(q_0)$ in the space H'. By III 8.2 (b) the set $G' = \overline{\Psi}^{-1}(\Phi(G))$ is open in $R^\varkappa$. The mapping $\Psi = \overline{\Psi}|G'$ is a diffeomorphism from G' onto the neighbourhood $\Psi(G') = \overline{\Psi}(G_0) \cap \Phi(G)$ of $\Phi(q_0)$ in the space H'. It follows from 5.4 (c) that the mapping $\Lambda = \Phi^{-1} \circ \Psi$ is a diffeomorphism from G' into $\Phi^{-1}(H')$. Since the set $\Psi(G')$ is a neighbourhood of $\Phi(q_0)$ in H', the set $\Lambda(G') = \Phi^{-1}(\Psi(G'))$ is a neighbourhood of q_0 in the set $\Phi^{-1}(H')$. The point $q_0 \in \Phi^{-1}(H')$ being arbitrary, we infer that $\Phi^{-1}(H')$ is a $\varkappa$-dimensional regular hypersurface.

We shall say that a vector $\mathfrak{a} \in V_l$ is *tangent* at a point q_0 to a (non-empty) set $Z \subset R^l$ if there is a mapping Ψ of class C_1 of an open set G_0 of real numbers into Z, such that

$$\Psi(t_0) = q_0 \quad \text{and} \quad \Psi'(t_0) = \mathfrak{a}$$

for a number $t_0 \in G_0$. Observe that we can always assume that $t_0 = 0$ (replace Ψ by the mapping $\overline{\Psi}(t) = \Psi(t + t_0)$).

It follows from the definition that the zero vector is tangent to every non-empty set Z at any point $q_0 \in Z$ (assume $\Psi(t) = q_0$ for every $t \in R$). Assuming $\Psi(t) = q_0 + t\mathfrak{a}$ for small t, we see that every vector $\mathfrak{a} \in V_l$ is tangent to every non-empty open set $Z \subset R^l$ at any point $q_0 \in Z$.

6.4. *Let H be a k-dimensional regular hypersurface in R^l and let Φ be a diffeomorphism from an open set in R^k onto a neighbourhood U of a point $q_0 = \Phi(p_0)$ in H. A vector $\mathfrak{a} \in V_l$ is tangent at q_0 to H if and only if it is a linear combination of the partial derivatives of Φ at p_0:*

(3) $$\mathfrak{a} = c_1 \Phi_{|1}(p_0) + \ldots + c_k \Phi_{|k}(p_0)\,,$$

that is, if and only if

(4) $$\mathfrak{a} = \Phi'(p_0)\mathfrak{c} \quad \textit{for a vector } \mathfrak{c} = [c_1, \ldots, c_k]\,.$$

The vectors

(5) $$\Phi_{|1}(p_0),\ \ldots,\ \Phi_{|k}(p_0)$$

are linearly independent.

The linear independence of the vectors (5) follows directly from $|\Phi'(p_0)| \neq 0$ (see § 5 (2), (2')).

If $\mathfrak{a}$ is of the form (4), then assuming $\Psi(t) = \Phi(p_0 + t\mathfrak{c})$ we have

$$\Psi(t) \in H \quad \text{for } t \text{ in a neighbourhood of } 0\,,$$

$$\Psi(0) = q_0 \quad \text{and} \quad \Psi'(0) = \mathfrak{a}$$

(see 3.2). This proves that $\mathfrak{a}$ is tangent at q_0 to H.

Conversely, if $\mathfrak{a}$ is tangent at q_0 to H, then there exists a mapping Ψ of class C_1, of an open set G' of real numbers into H, such that $\Psi(t_0) = q_0$ and $\Psi'(t_0) = \mathfrak{a}$ for some $t_0 \in G'$. We may suppose that $\Psi(G') \subset U$ (for if not, replace G' by $\Psi^{-1}(U)$ and Ψ by $\Psi|\Psi^{-1}(U)$). By 5.1 (a) the mapping

$$\Lambda = \Phi^{-1} \circ \Psi = (\lambda_1, \ldots, \lambda_k)$$

is of class C_1. By definition, $\Lambda(t_0) = p_0$ and $\Psi = \Phi \circ \Lambda$. By 3.2,

$$\mathfrak{a} = \Psi'(t_0) = \Phi_{|1}(p_0)\lambda_1'(t_0) + \ldots + \Phi_{|k}(p_0)\lambda_k'(t_0)\,.$$

Thus $\mathfrak{a}$ is of the form (3) where $c_i = \lambda_i'(t_0)$, $i = 1, \ldots, k$.

Let H be a k-dimensional hypersurface in R^l and let $q_0 \in H$, $0 < k < l$.

The k-dimensional hyperplane K through the point q_0, such that every vector $\mathfrak{a}$ tangent at q_0 to H is parallel to K, is called the *hyperplane tangent at q_0 to H* (in the case of $k = 1$ we call it the *line tangent at q_0 to H*; in the case of $k = 2$ we call it the *plane tangent at q_0 to H*). Every k-vector parallel to the hyperplane tangent at q_0 to H is said to be *tangent at q_0 to H*.

The $(l-k)$-dimensional hyperplane through q_0, which is orthogonal to the hyperplane tangent at q_0 to H, is called the *hyperplane orthogonal* (or *normal*) *at* q_0 *to* H (in the case of $k = l-1$ it is called the *line orthogonal to* H *at* q_0). Every $(l-k)$-vector parallel to the $(l-k)$-dimensional hyperplane orthogonal at q_0 to H is said to be *orthogonal* (or *normal*) *at* q_0 *to* H.

The following theorem follows directly from 6.4 and II § 10 (25)-(26).

6.5. *If* $\Phi = (\varphi_1, \ldots, \varphi_l)$ *is a diffeomorphism from an open subset of* R^k *onto a neighbourhood of a point* $q_0 = \Phi(p_0)$ *of a* k*-dimensional regular hypersurface* H *in* R^l $(0 < k < l)$, *then the hyperplane tangent at* q_0 *to* H *is the hyperplane through* q_0, *parallel to the vectors*

$$\Phi_{|1}(p_0), \ldots, \Phi_{|k}(p_0),$$

i.e., the set of all points

$$q = q_0 + c_1\Phi_{|1}(p_0) + \ldots + c_k\Phi_{|k}(p_0)$$

where $c_1, \ldots, c_k$ *are any real numbers.*

$\Phi_{|1}(p_0)\wedge\ldots\wedge\Phi_{|k}(p_0)$ *is a* k*-vector tangent at* q_0 *to* H, *and*

$$\frac{\Phi_{|1}(p_0)\wedge\ldots\wedge\Phi_{|k}(p_0)}{|\Phi'(p_0)|} \tag{6}$$

is a unit k*-vector tangent at* q_0 *to* H.

$\Phi_{|1}(p_0)\vee\ldots\vee\Phi_{|k}(p_0)$ *is an* $(l-k)$*-vector orthogonal at* q_0 *to* H, *and*

$$\frac{\Phi_{|1}(p_0)\vee\ldots\vee\Phi_{|k}(p_0)}{|\Phi'(p_0)|} \tag{7}$$

is a unit $(l-k)$*-vector orthogonal at* q_0 *to* H.

The set of all points $q \in R^l$ *satisfying the system of equations*

$$\Phi_{|1}(p)\cdot(q-q_0) = 0, \ \ldots, \ \Phi_{|k}(p_0)\cdot(q-q_0) = 0$$

is the hyperplane orthogonal at q_0 *to* H.

We recall that

$$\Phi_{|j} = [\varphi_{1|j}, \ldots, \varphi_{l|j}] \quad (j = 1, \ldots, k).$$

Consequently, by II § 10 (3)-(3′),

$$\Phi_{|1}\wedge\ldots\wedge\Phi_{|k} = [J_{i_1,\ldots,i_k}]_{i_1,\ldots,i_k\leqslant l} \tag{8}$$

where

$$J_{i_1,\ldots,i_k} = \begin{vmatrix} \varphi_{i_1|1} & \cdots & \varphi_{i_k|1} \\ \cdot & \cdots & \cdot \\ \varphi_{i_1|k} & \cdots & \varphi_{i_k|k} \end{vmatrix}. \tag{9}$$

By the convention on p. 173, the jacobian (9) is also denoted by $/(\varphi_{i_1}, \ldots, \varphi_{i_k})'/$. Hence

$$\Phi_{|1}\wedge\ldots\wedge\Phi_{|k} = [/(\varphi_{i_1}, \ldots, \varphi_{i_k})'/]_{i_1,\ldots,i_k\leqslant l}. \tag{10}$$

Using the traditional notation, we can write formula (10) as follows:

$$(10') \qquad \frac{\partial \Phi}{\partial x_1} \wedge \ldots \wedge \frac{\partial \Phi}{\partial x_k} = \left[\frac{\partial(\varphi_{i_1}, \ldots, \varphi_{i_k})}{\partial(x_1, \ldots, x_k)}\right]_{i_1, \ldots, i_k \leqslant l}.$$

We mention a few of the most important particular cases of 6.5.

A) Let $\Phi\colon G \to R^l$ be a diffeomorphism on an open set of real numbers. The line

$$\{\Phi(t_0) + t\Phi'(t_0)\colon\ t \in R\}$$

is tangent at the point $q_0 = \Phi(t_0)$ to the curve

$$H = \{\Phi(t)\colon\ t \in G\}.$$

In other words,

$$(11) \qquad q = \Phi(t_0) + t\Phi'(t_0)$$

is the parametric equation of the tangent line.

B) Let $\Phi\colon G \to R^3$ be a diffeomorphism onto a neighbourhood of a point $q_0 = \Phi(p_0)$ of a surface $H \subset R^3$ ($p_0 \in G \subset R^2$). The vector product $\Phi_{|1}(p_0) \vee \Phi_{|2}(p_0)$ of the partial derivatives of Φ is a vector orthogonal at q_0 to H. Consequently the equation of the plane tangent at q_0 to H has the form

$$(12) \qquad \big(q - \Phi(p_0)\big)\cdot\big(\Phi_{|1}(p_0) \vee \Phi_{|2}(p_0)\big) = 0,$$

i.e. the tangent plane is the set of all points $q \in R^3$ satisfying equation (12).

Denoting $q = (x, y, z)$, $\Phi = (\varphi, \psi, \zeta)$, we can write equation (12) as follows:

$$(12') \qquad \begin{vmatrix} x - \varphi(p_0) & y - \psi(p_0) & z - \zeta(p_0) \\ \varphi_{|1}(p_0) & \psi_{|1}(p_0) & \zeta_{|1}(p_0) \\ \varphi_{|2}(p_0) & \psi_{|2}(p_0) & \zeta_{|2}(p_0) \end{vmatrix} = 0,$$

that is, the tangent plane is the set of all points (x, y, z) satisfying equation (12′).

If u, v denote the coordinates of any point $p \in G$, and $p_0 = (u_0, v_0)$, equation (12) can be written:

$$(12'') \qquad \big(q - \Phi(u_0, v_0)\big)\cdot\big(\Phi_u(u_0, v_0) \vee \Phi_v(u_0, v_0)\big) = 0,$$

or

$$(12''') \qquad \begin{vmatrix} x - \varphi(u_0, v_0) & y - \psi(u_0, v_0) & z - \zeta(u_0, v_0) \\ \varphi_u(u_0, v_0) & \psi_u(u_0, v_0) & \zeta_u(u_0, v_0) \\ \varphi_v(u_0, v_0) & \psi_v(u_0, v_0) & \zeta_v(u_0, v_0) \end{vmatrix} = 0.$$

C) The formulas given in B) can be generalized as follows.

Let $\Phi = (\varphi_1, \ldots, \varphi_l)\colon G \to R^l$ be a diffeomorphism onto a neighbourhood of a point $q_0 = \Phi(p_0)$ in an $(l-1)$-dimensional regular hypersurface $H \subset R^l$ ($p_0 \in G \subset R^{l-1}$). The vector product

$$\Phi_{|1}(p_0) \vee \ldots \vee \Phi_{|l-1}(p_0)$$

(see II § 10, p. 95) is a vector orthogonal at q_0 to H. Thus the equation of the $(l-1)$-dimensional hyperplane tangent at q_0 to H has the form

$$(q - \Phi(p_0)) \cdot (\Phi_{|1}(p_0) \vee \ldots \vee \Phi_{|l-1}(p_0)) = 0 , \tag{13}$$

i.e. the tangent hyperplane is the set of all points q satisfying equation (13).

Denoting $q = (x_1, \ldots, x_l)$, we can write equation (13) in the following form (see II § 10 (23)):

$$\begin{vmatrix} x - \varphi_1(p_0) & \ldots & x_l - \varphi_l(p_0) \\ \varphi_{1|1}(p_0) & \ldots & \varphi_{l|1}(p_0) \\ \cdot & \cdot & \cdot \\ \varphi_{1|l-1}(p_0) & \ldots & \varphi_{l|l-1}(p_0) \end{vmatrix} = 0 . \tag{13'}$$

A k-dimensional surfaces in R^l is often defined as the set of all points $q = (x_1, \ldots, x_l)$ whose coordinates satisfy a system of $(l-k)$ equations:

$$\omega_1(x_1, \ldots, x_l) = 0 ,$$
$$\ldots\ldots\ldots$$
$$\omega_{l-k}(x_1, \ldots, x_l) = 0 ,$$

i.e. as the set of all points q satisfying the equation $\Omega(q) = 0$ where $\Omega = [\omega_1, \ldots, \omega_{l-k}]$ is an $(l-k)$-dimensional vector field. Some additional hypotheses about Ω, i.e. about the real functions $\omega_1, \ldots, \omega_{l-k}$, are necessary of course. They are given in the following theorem.

6.6. *If* $\Omega - [\omega_1, \ldots, \omega_{l-k}]$ *is a regular* $(l-k)$*-dimensional vector field of class* C_m *on an open set* $G_0 \subset R^l$ $(0 < k < l)$, *then the set*

$$H = \{q\colon \Omega(q) = 0\}$$

is either empty or is a regular k*-dimensional hypersurface of class* C_m.

A vector $\mathfrak{a} \in V_l$ *is tangent to* H *at a point* $q_0 \in H$ *if and only if*

$$\omega_1'(q_0) \cdot \mathfrak{a} = 0, \ \ldots, \ \omega_{l-k}'(q_0) \cdot \mathfrak{a} = 0 , \tag{14}$$

i.e. if it is orthogonal to all the derivatives

$$\omega_1'(q_0), \ \ldots, \ \omega_{l-k}'(q_0) \tag{15}$$

of the functions $\omega_1, \ldots, \omega_{l-k}$. *The vectors* (15) *are linearly independent.*

A vector $\mathfrak{b}$ *is orthogonal at* q_0 *to* H *if and only if it is a linear combination of the vectors* (15):

$$\mathfrak{b} = c_1 \omega_1'(q_0) + \ldots + c_{l-k} \omega_{l-k}'(q_0) ,$$

i.e. if it is of the form

$$\mathfrak{b} = \mathfrak{c}\cdot\Omega'(q_0) \quad \textit{for a vector } \mathfrak{c} = [c_1, \ldots, c_{l-k}] .$$

The hyperplane tangent at q_0 to H is the set of all points q satisfying all the equations

(16) $$\omega_1'(q_0)\cdot(q-q_0) = 0 , \ \ldots, \ \omega_{l-k}'(q_0)\cdot(q-q_0) = 0 .$$

The set of all points q of the form

(17) $$q = q_0 + c_1\omega_1'(q_0) + \ldots + c_{l-k}\omega_{l-k}'(q_0) ,$$

where $c_1, \ldots, c_{l-k}$ are any real numbers, is the hyperplane orthogonal at q_0 to H.

$\omega_1'(q_0)\wedge\ldots\wedge\omega_{l-k}'(q_0)$ is an $(l-k)$-vector orthogonal to H at q_0, and

(18) $$\frac{\omega_1'(q_0)\wedge\ldots\wedge\omega_{l-k}'(q_0)}{|\Omega'(q_0)|}$$

is a unit $(l-k)$-vector orthogonal to H at q_0.

$\omega_1'(q_0)\vee\ldots\vee\omega_{l-k}'(q_0)$ is a k-vector tangent to H at q_0, and

(19) $$\frac{\omega_1'(q_0)\vee\ldots\vee\omega_{l-k}'(q_0)}{|\Omega'(q_0)|}$$

is a unit k-vector tangent to H at q_0.

Let q_0 be a point in H. Let $s = l-k$, for short. By hypothesis, Ω is a regular s-dimensional vector field on $G_0 \subset R^{k+s}$. Since $|\Omega'(q_0)| \neq 0$, there exists a sequence $\mathbf{j} = (j_1, \ldots, j_s)$ of positive integers $\leqslant k+s$, such that $|\Omega_{|\mathbf{j}}(q_0)| \neq 0$. Let $\mathbf{i} = (i_1, \ldots, i_k)$ be a sequence formed of all the remaining positive integers $\leqslant k+s$. We shall consider the space R^{k+s} as the $(\mathbf{i}, \mathbf{j})$-product of R^k and R^s, i.e. a point $q = (x_1, \ldots, x_{k+s}) \in R^{k+s}$ will be identified with the point

$$(p, \overline{p}) \in R^k \times R^s ,$$

where

$$p = (x_{i_1}, \ldots, x_{i_k}) \in R^k , \qquad \overline{p} = (x_{j_1}, \ldots, x_{j_s}) \in R^s .$$

In particular, let

$$q_0 = (p_0, \overline{p}_0) .$$

It follows from 4.3 (where l should be replaced by s) that there exist a continuous mapping Ψ of a ball $G = \mathrm{B}(p_0, \delta) \subset R^k$ into a ball $V = \mathrm{B}(\overline{p}_0, \eta) \subset R^s$ such that $\Psi(p_0) = \overline{p}_0$ and

(20) $$\Omega\big(p, \Psi(p)\big) = 0 \quad \text{for every } p \in G;$$

(21) $$\text{if } p \in G,\ \overline{p} \in V \text{ and } \Omega(p, \overline{p}) = 0, \text{ then } \overline{p} = \Psi(p) .$$

It follows from 4.4 that Ψ is of class C_m. The mapping

$$\Phi(p) = \big(p, \Psi(p)\big) \quad \text{for} \quad p \in G$$

is a diffeomorphism of class C_m (see p. 211). The set

$$U_0 = G \times V = \{(p, \bar{p}): (p \in G) \& (\bar{p} \in V)\}$$

is open in R^l. It follows from the inclusion $\Psi(G) \subset V$ and from (20) that $\Phi(G) \subset H \cap U_0$. It follows from (21) that $H \cap U_0 \subset \Phi(G)$. Thus $\Phi(G) = H \cap U_0$ which proves that the set $U = \Phi(G)$ is open in H. Since $q_0 = \Phi(p_0)$, U is a neighbourhood of q_0 in H. Since q_0 is arbitrary, this proves that H is a k-dimensional hypersurface of class C_m.

It follows from (20) that $\Omega(\Phi(p)) = 0$, i.e.

$$\omega_i \circ \Phi(p) = \omega_i(\Phi(p)) = 0 \qquad \text{for every point } p \in G \qquad (i = 1, \ldots, l-k).$$

Differentiating the composition $\omega_i \circ \Phi$ we get, by 3.3,

$$\omega_i'(q_0) \cdot \Phi_{|j}(p_0) = 0 \qquad \text{for } i = 1, \ldots, l-k \text{ and } j = 1, \ldots, k$$

since $\Phi(p_0) = q_0$. We have proved that each of the linearly independent vectors $\Phi_{|1}(p_0), \ldots, \Phi_{|k}(p_0)$ is orthogonal to all the vectors (15). The vectors (15) are also linearly independent since $|\Omega'(q_0)| \neq 0$ (see § 5 (1), (3)). By 6.5, the hyperplane tangent at q_0 to H is the k-dimensional hyperplane through q_0, parallel to all the vectors $\Phi_{|1}(p_0), \ldots, \Phi_{|k}(p_0)$, i.e. orthogonal to all the vectors (15). In other words, the tangent hyperplane coincides with the set of points satisfying the equations (16).

The remaining part of 6.6 follows directly from the part just proved, and from II § 10.

We shall discuss a few of the most important particular cases of 6.6.

D) Let ω and λ be real functions of class C_1, defined on an open set $G_0 \subset R^3$, such that the vector product of their derivatives is never equal to the zero vector:

$$\omega'(q) \vee \lambda'(q) \neq 0 \qquad \text{for every } q \in G_0 . \tag{22}$$

Then the set H of all points $q = (x, y, z)$ such that

$$\omega(x, y, z) = 0 \qquad \text{and} \qquad \lambda(x, y, z) = 0$$

is a regular curve provided that it is not empty. Indeed, (22) states that the vector field $\Omega = [\omega, \lambda]$ is regular. By definition, $H = \{q: \Omega(q) = 0\}$. For every point $q_0 = (x_0, y_0, z_0) \in H$, the vector $\omega'(q_0) \vee \lambda'(q_0)$ is tangent to H at q_0. The set of all points $q = (x, y, z) \in R^3$ satisfying both the equations

$$\omega'(q_0) \cdot (q - q_0) = 0 , \qquad \lambda'(q_0) \cdot (q - q_0) = 0 , \tag{23}$$

i.e., the equations

$$\begin{aligned} &\omega_{|1}(x_0, y_0, z_0)(x - x_0) + \omega_{|2}(x_0, y_0, z_0)(y - y_0) + \omega_{|3}(x_0, y_0, z_0)(z - z_0) = 0 , \\ &\lambda_{|1}(x_0, y_0, z_0)(x - x_0) + \lambda_{|2}(x_0, y_0, z_0)(y - y_0) + \lambda_{|3}(x_0, y_0, z_0)(z - z_0) = 0 , \end{aligned} \tag{23'}$$

is the line tangent to H at q_0. The set of all points $q = (x, y, z)$ satisfying the equation

(24) $$(q - q_0)\cdot(\omega'(q_0) \vee \lambda'(q_0)) = 0 \,,$$

i.e., the equation

(24′) $$\begin{vmatrix} x - x_0 & y - y_0 & z - z_0 \\ \omega_{|1}(x_0, y_0, z_0) & \omega_{|2}(x_0, y_0, z_0) & \omega_{|3}(x_0, y_0, z_0) \\ \lambda_{|1}(x_0, y_0, z_0) & \lambda_{|2}(x_0, y_0, z_0) & \lambda_{|3}(x_0, y_0, z_0) \end{vmatrix} = 0 \,,$$

is the plane orthogonal to H at $q_0 = (x_0, y_0, z_0) \in H$.

In the traditional notation equations (23′) have the form

(23″) $$\begin{aligned} &\omega_x(x_0, y_0, z_0)(x - x_0) + \omega_y(x_0, y_0, z_0)(y - y_0) + \omega_z(x_0, y_0, z_0)(z - z_0) = 0 \,, \\ &\lambda_x(x_0, y_0, z_0)(x - x_0) + \lambda_y(x_0, y_0, z_0)(y - y_0) + \lambda_z(x_0, y_0, z_0)(z - z_0) = 0 \,, \end{aligned}$$

and equation (24′) has the form

(24″) $$\begin{vmatrix} x - x_0 & y - y_0 & z - z_0 \\ \omega_x(x_0, y_0, z_0) & \omega_y(x_0, y_0, z_0) & \omega_z(x_0, y_0, z_0) \\ \lambda_x(x_0, y_0, z_0) & \lambda_y(x_0, y_0, z_0) & \lambda_z(x_0, y_0, z_0) \end{vmatrix} = 0 \,.$$

E) Let ω be a real function of class C_1, defined on an open set $G_0 \subset R^3$, such that $\omega'(q) \neq 0$ for every $q \in G_0$, i.e. such that for every point $q \in G_0$ at least one of the first partial derivatives of ω does not vanish. The set H of all points $q = (x, y, z)$ satisfying the equation

$$\omega(x, y, z) = 0$$

is a regular surface provided that it is not empty.

For every point $q_0 = (x_0, y_0, z_0) \in H$, the vector $\omega'(q_0)$ is orthogonal to H at q_0. The set of all points $q = (x, y, z) \in R^3$ satisfying the equation

(25) $$\omega'(q_0)(q - q_0) = 0 \,,$$

i.e., the equation

(25′) $$\omega_{|1}(x_0, y_0, z_0)(x - x_0) + \omega_{|2}(x_0, y_0, z_0)(y - y_0) + \omega_{|3}(x_0, y_0, z_0)(z - z_0) = 0 \,,$$

is the plane tangent to H at q_0. The line

$$\{q_0 + t\omega'(q_0) \colon \ t \in R\}$$

is orthogonal to H at q_0. In other words,

(26) $$q = q_0 + t\omega'(q_0)$$

is the parametric equation of the line orthogonal to H at q_0.

In the traditional notation the equation (25′) of the plane tangent to H at $q_0 = (x_0, y_0, z_0)$ has the form

(25″) $$\omega_x(x_0, y_0, z_0)(x - x_0) + \omega_y(x_0, y_0, z_0)(y - y_0) + \omega_z(x_0, y_0, z_0)(z - z_0) = 0 \,.$$

F) Let ω be a real function of class C_1, defined on an open set $G_0 \subset R^2$, such that $\omega'(q) \neq 0$ for every $q \in G_0$, i.e. such that at every $q \in G_0$ at least one of the first partial derivatives of ω does not vanish. The set H of all points $q = (x, y)$ satisfying the equation

$$\omega(x, y) = 0$$

is a regular curve provided that it is not empty.

For every point $q_0 = (x_0, y_0) \in H$ the vector $\omega'(q_0)$ is orthogonal to H at q_0. The set of all points $q = (x, y) \in R^2$ satisfying the equation

$$\omega'(q_0)(q - q_0) = 0 , \tag{27}$$

i.e. the equation

$$\omega_{|1}(x_0, y_0)(x - x_0) + \omega_{|2}(x_0, y_0)(y - y_0) = 0 , \tag{27'}$$

is the line tangent to H at q_0. The set of all points $q = (x, y)$ satisfying the equation

$$\begin{vmatrix} x - x_0 & y - y_0 \\ \omega_{|1}(x_0, y_0) & \omega_{|2}(x_0, y_0) \end{vmatrix} = 0 \tag{28}$$

is the line orthogonal to H at q_0.

In the traditional notation equations (27′) and (28) have respectively the following form:

$$\omega_x(x_0, y_0)(x - x_0) + \omega_y(x_0, y_0)(y - y_0) = 0 , \tag{27''}$$

$$\begin{vmatrix} x - x_0 & y - y_0 \\ \omega_x(x_0, y_0) & \omega_y(x_0, y_0) \end{vmatrix} = 0 . \tag{28'}$$

G) The formulas given in E) and F) can be generalized as follows.

Let ω be a real function of class C_1, defined on an open set $G_0 \subset R^l$, such that $\omega'(q) \neq 0$ for every $q \in G_0$. The set H of all points $q = (x_1, \ldots, x_l) \in R^l$ satisfying the equation

$$\omega(x_1, \ldots, x_l) = 0$$

is a regular $(l-1)$-dimensional hypersurface provided that it is not empty.

For every point $q_0 = (x_{0,1}, \ldots, x_{0,l}) \in H$ the vector $\omega'(q_0)$ is orthogonal to H at q_0. The set of all points $q = (x_1, \ldots, x_l)$ satisfying the equation

$$\omega'(q_0)(q - q_0) = 0 , \tag{29}$$

i.e., the equation

$$\omega_{|1}(x_{0,1}, \ldots, x_{0,l})(x_1 - x_{0,1}) + \ldots + \omega_{|l}(x_{0,1}, \ldots, x_{0,l})(x_l - x_{0,l}) = 0 , \tag{29'}$$

is the $(l-1)$-dimensional hyperplane tangent to H at q_0. The line

$$\{q_0 + t\omega'(q_0):\ t \in R\}$$

is orthogonal to H at q_0.

The following theorem announced on p. 204 is a simple corollary of 6.6.

6.7. *There exists no diffeomorphism from an open non-empty subset of a Euclidean space into a Euclidean space of a smaller dimension.*

Suppose that $\Phi: G_0 \to R^k$ is a regular mapping, $G_0 \subset R^l$, $l > k$. Let $q_0 \in G_0$. The vector field

$$\Omega(q) = \Phi(q) - \Phi(q_0) \quad \text{for} \quad q \in G_0$$

is regular since $\Omega' = \Phi'$. By 6.6 (where k should be replaced by $l-k$) the set

$$H = \{q \in G_0:\ \Omega(q) = 0\} = \{q \in G_0:\ \Phi(q) = \Phi(q_0)\}$$

is an $(l-k)$-dimensional hypersurface (it is not empty because $q_0 \in H$). Therefore it contains more than one point, i.e. there exists a point $q_1 \in G_0$ such that $q_1 \neq q_0$ and $\Phi(q_1) = \Phi(q_0)$. This proves that Φ is not one-to-one. Consequently Φ is not a diffeomorphism.

Theorem 6.7 remains true if the word "diffeomorphism" is replaced by "homeomorphism". The proof is then much more difficult.

EXERCISES

1. The following subsets of R^3 are regular surfaces (of class C_∞): the ellipsoid, the hyperboloid of one sheet, the hyperboloid of two sheets, the elliptic paraboloid, the hyperbolic paraboloid, the elliptic or hyperbolic or parabolic cylinder. However, the elliptic cone is not a surface in the sense defined in this section (it becomes a regular surface of class C_∞ after removing its vertex).

2. Every connected curve is homeomorphic either with the line R or with a circle on the plane.

3. The following subset of R^4

$$T_0 = \{(\varrho_1 \cos u,\ \varrho_1 \sin u,\ \varrho_2 \cos v,\ \varrho_2 \sin v):\ u, v \in R\} \qquad (\varrho_1, \varrho_2 > 0)$$

is a regular surface (of class C_∞) homeomorphic with the torus T defined on p. 214. The surface T_0 is also called a *torus*.

4. By the *open Möbius strip* we mean every set (surface) homeomorphic with the set

$$M = \{\Phi(u, v):\ (|u| < \varrho_0) \& (|v| \leqslant \pi)\} \subset R^3$$

where

$$\Phi(u, v) = \big((\varrho + u\cos\tfrac{1}{2}v)\cos v,\ (\varrho + u\cos\tfrac{1}{2}v)\sin v,\ u\sin\tfrac{1}{2}v\big) \quad \text{and} \quad \varrho_0 < \varrho\,.$$

To understand the construction of M, let us imagine the plane H through the first and the third coordinate axes, and a segment P (without the end-points) in H such that P lies on the first coordinate axis, has length $2\varrho_0$ and center at the point $(\varrho, 0, 0)$. Now let us imagine that the plane H (together with the segment P) rotates about the third coordinate axis and, simultaneously, the segment P rotates, in the plane H, about its center; however, the rotation of P about its center is at half the speed of the rotation of H. Consequently, when H rotates by 2π, the segment P rotates by π, i.e. it comes back to the initial position but with end-points reversed. The set M is the set of all points in R^3 swept out by P (see figure 2).

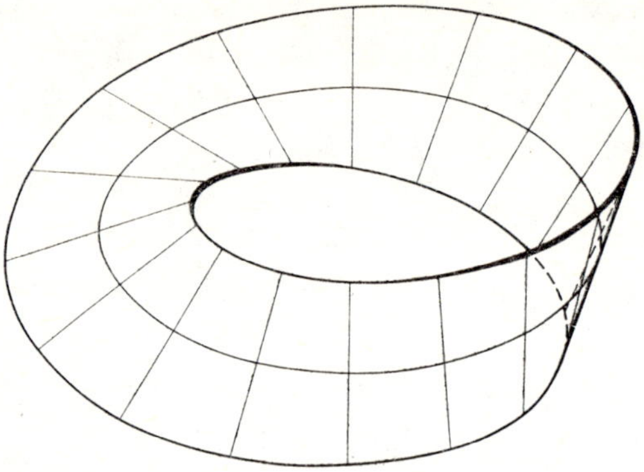

Fig. 2

An open Möbius strip can also be visualized as follows. Take a strip made of a stretched material (see figure 3; the upper and lower edges are not included to the strip). Turn one end of the strip by 180° and glue it to the other end. The result of the operation is a Möbius strip.

Fig. 3.

Prove that the set M is a regular surface of class C_∞.

Prove that the set

$$M_0 = \{\Phi_0(u, v)\colon (|u| < \varrho_0) \& (|v| \leqslant \pi)\} \subset R^4$$

where

$$\Phi_0(u, v) = (\varrho \cos v, \varrho \sin v, u \cos \tfrac{1}{2} v, u \sin \tfrac{1}{2} v)$$

is also a regular surface of class C_∞, homeomorphic with M, i.e. an open Möbius strip.

5. Write the equations of the tangent planes at any point of the torus T from p. 214, of the torus T_0 from exercise 3 and of the Möbius strips M and M_0 from exercise 4. Write the equations of the lines orthogonal to T and M and of the planes orthogonal to T_0 and M_0 at any point.

6. If H is a k_1-dimensional hypersurface in R^{l_1}, and H_2 is a k_2-dimensional surface in R^{l_2}, then $H_1 \times H_2$ is a $(k_1 + k_2)$-dimensional hypersurface in the space $R^{l_1+l_2}$ interpreted as a product of R^{l_1} and R^{l_2}. If H_1 and H_2 are regular of class C_m, so is $H_1 \times H_2$.

7. Write the equation of the four-dimensional hyperplanes tangent to the product $T \times M \subset R^6$ and $T_0 \times M_0 \subset R^8$, where T, T_0, M, M_0 have the same meaning as in exercise 5.

8. Prove that the set

$$H = \{(x, y):\ e^{x\cos y}\cos y + e^{x\sin y}\sin y + x + y - 2 = 0\}$$

is a regular curve of class C_∞ on the plane R^2. Write the equation of the line tangent to H at $(0, 0)$.

9. Prove that the set

$$H = \{(x, y, z):\ e^{2x+y+z} + e^{3x-y} + \log(1+y+z) - 2 = 0\} \subset R^3$$

is a regular surface of class C_∞. Write the equation of the tangent line and the orthogonal line to H, at the point $(0, 0, 0)$.

10. Prove that the set of all points $p = (x, y, u, v) \in R^4$ satisfying both the equations

$$e^{x+y+u} + e^{x+y+v} + u = 2\,, \qquad e^{x+u} + e^{x+y-u} + u - v = 2$$

is a regular surface of class C_∞. Write the equations of the tangent plane and the orthogonal plane to H, at the point $(0, 0, 0, 0)$.

11. Let H be a k-dimensional regular hypersuface of class C_m $(1 \leqslant m \leqslant \infty)$ in R^l. A real function φ defined on an open subset A of the space H is said to be *of class* C_m if, for any diffeomorphism Φ: $G \to H$ of class C_m $(G \subset R^k)$, the composition $\varphi \circ \Phi$ is of class C_m.

Let $\mathcal{K}$ be a set of diffeomorphisms of class C_m, from open subsets of R^k into H, such that H is the union of the ranges of all the diffeomorphisms in $\mathcal{K}$. Prove that if $\varphi \circ \Phi$ is of class C_m for every diffeomorphism Φ in $\mathcal{K}$, then φ is of class C_m.

12. Let H be a k-dimensional regular hypersurface in R^l, let $q \in H$, and let $\mathfrak{a}$ be a vector tangent to H at q. For every function φ of class C_1, defined on a neighbourhood U of q in H, by the *directional derivative of* φ *at* q *in the direction* $\mathfrak{a}$ we mean the number

$$\varphi'_{\mathfrak{a}}(q) = \bar{\varphi}'(0)$$

where $\bar{\varphi} = \varphi \circ \Lambda$ and Λ is a mapping of class C_1, of an interval $(-a\,; a) \subset R$ into H, such that $\Lambda(0) = q$ and $\Lambda'(0) = \mathfrak{a}$.

Verify that $\varphi'_{\mathfrak{a}}(q)$ does not depend on the choice of the auxiliary mapping Λ, i.e. that it depends only on φ, q and $\mathfrak{a}$. Prove that

$$(\varphi+\psi)'_{\mathfrak{a}}(q) = \varphi'_{\mathfrak{a}}(q) + \psi'_{\mathfrak{a}}(q)\,, \qquad (c\varphi)'_{\mathfrak{a}}(q) = c\varphi'_{\mathfrak{a}}(q)\,, \qquad (\varphi\psi)'_{\mathfrak{a}}(q) = \varphi'_{\mathfrak{a}}(q)\psi(q) + \varphi(q)\psi'_{\mathfrak{a}}(q)\,.$$

13. Let H be a k-dimensional regular hypersurface of class C_∞. Let q be a fixed point in H and let C denote the class of all real functions φ of class C_∞ defined on a neighbourhood U_φ of q in H (U_φ depends, in general, on φ). Prove that every function $\mathfrak{w}$: $C \to R$ such that

$$\mathfrak{w}(\varphi+\psi) = \mathfrak{w}(\varphi) + \mathfrak{w}(\psi)\,, \qquad \mathfrak{w}(c\varphi) = c\mathfrak{w}(\varphi)\,, \qquad \mathfrak{w}(\varphi\psi) = \mathfrak{w}(\varphi)\psi(q) + \varphi(q)\mathfrak{w}(\psi)$$

is of the form

$$\mathfrak{w}(\varphi) = \varphi'_{\mathfrak{a}}(q)\,, \tag{30}$$

where $\mathfrak{a}$ is a vector tangent to H at q. Prove that the mapping which assigns, to every vector $\mathfrak{a}$ tangent to H at q, the function $\mathfrak{w}$ defined by (30), is one-to-one.

Hint. Use § 3, exercise 13.

14. Write formulas (13) and (13′) using the traditional notation.

15. Every component of a k-dimensional surface in R^l is a k-dimensional surface.

16. A vector $\mathfrak{a}$ is tangent to a set Z at a point q if and only if it is tangent at q to a curve lying in Z.

§ 7. Extremal and critical points on hypersurfaces. Let φ be a real function defined on a k-dimensional hypersurface $H \subset R^l$. According to the general definition from IV § 7, pp. 160-161, the function φ has an extremum (i.e. a local maximum or a local minimum) at a point $q_0 \in H$ if there exists a neighbourhood U_0 of q_0 in H, such that either

$$(1) \qquad \varphi(q) \leqslant \varphi(q_0) \quad \text{for every } q \in U_0 \quad \text{(local maximum)}$$

or

$$(2) \qquad \varphi(q) \geqslant \varphi(q_0) \quad \text{for every } q \in U_0 \quad \text{(local minimum)}.$$

Suppose that Φ is a diffeomorphism of an open set $G \subset R^k$ onto a neighbourhood U of q_0 and $q_0 = \Phi(p_0)$, $p_0 \in G$. It is easy to show that Φ has an extremum at q_0 if and only if the function $\varphi \circ \Phi$ has an extremum at p_0.

Indeed, if (1) holds, then the set $V = \Phi^{-1}(U_0)$ is a neighbourhood of p_0 and

$$(1') \qquad \varphi \circ \Phi(p) \leqslant \varphi \circ \Phi(p_0) \quad \text{for every } p \in V,$$

i.e. $\varphi \circ \Phi$ has a local maximum at p_0. Conversely, if $\varphi \circ \Phi$ has a local maximum at p_0, i.e. there exists a neighbourhood V of p_0 such that (1′) holds, then the set $U_0 = \Phi(V)$ is a neighbourhood of q_0 in H and (1) holds, i.e. φ has a local maximum at q_0. The same is true for the case of local minimum, (1′) should be now replaced by

$$(2') \qquad \varphi \circ \Phi(p) \geqslant \varphi \circ \Phi(p_0) \quad \text{for every } p \in V.$$

We see that the problem of whether φ has a local maximum or minimum at q_0 can be reduced to the problem of whether the composite mapping $\varphi \circ \Phi$ defined on an open subset of R^k has a local maximum or minimum at p_0. The last problem can be solved e.g. by means of the method given in IV § 7. We see, in particular, that the following condition

$$(3) \qquad (\varphi \circ \Phi)'(p_0) = 0$$

is necessary for φ to have an extremum at the point q_0, of course under the hypothesis that the derivative in question exists.

Generalizing the definition from p. 161, we shall say that a point $q_0 \in H$ is a *critical* point of a function φ: $H \to R$ if there exists a diffeomorphism Φ of class C_1 from an open set $G \subset R^k$ onto a neighbourhood of q_0 in H, such that the function $\varphi \circ \Phi$ is of class C_1 in a neighbourhood V of the point $p_0 = \Phi^{-1}(q_0)$ and (3) holds. Then for every diffeomorphism Φ_1 from an open set $G_1 \subset R^k$ onto a neighbourhood of q_0 in H the composition $\varphi \circ \Phi_1$ is of class C_1 in a neighbourhood V_1 of the point $p_1 = \Phi_1^{-1}(q_0)$ and (3) holds, i.e.

$$(\varphi \circ \Phi_1)'(p_1) = 0 .$$

Indeed, the open set $V_1 = \Phi_1^{-1}(\Phi(V))$ is a neighbourhood of p_1, the mappings $\Psi = \Phi_1|V_1$ and $\Lambda = \Phi^{-1} \circ \Psi$ are diffeomorphism (see 5.4), $p_0 = \Lambda(p_1)$ and

$$(\varphi \circ \Phi_1)|V_1 = (\varphi \circ \Phi) \circ \Lambda .$$

This proves that the function $\varphi \circ \Phi_1$ is of class C_1 on the neighbourhood V_1 of p_1, and

$$(\varphi \circ \Phi_1)'(p_1) = (\varphi \circ \Phi)'(p_0) \cdot \Lambda'(p_1) = 0$$

by 3.3 and the property (3).

Often the hypersurface H in question is defined as the set

$$H = \{q\colon \Omega(q) = 0\}$$

where Ω is a regular $(l-k)$-dimensional vector field on an open set $G_0 \subset R^l$. Then, as a rule, no diffeomorphism onto a neighbourhood of any point in H is explicitly given, and the problem of finding one is rather difficult. Therefore, it is more convenient to find, by means of another simpler method, all critical points. Only these points can be extremal and only they should be investigated in more detail. Other points are not extremal.

Theorem 7.1 below yields such a method called the *method of Lagrange multipliers.*

7.1. *Let* $\Omega = [\omega_1, \ldots, \omega_{l-k}]$ *be a regular vector field defined on an open set* $G_0 \subset R^l$ $(0 < k < l)$, *let* $H = \{q\colon \Omega(q) = 0\}$ *and let* φ *be a real function of class* C_1 *defined on* G_0. *A point* $q_0 \in H$ *is a critical point of the function* $\varphi|H$ *if and only if the vector* $\varphi'(q_0)$ *is orthogonal to* H *at* q_0, *i.e. there exist real numbers* $c_1, \ldots, c_{l-k}$ *such that*

(4) $$\varphi'(q_0) = c_1 \omega_1'(q_0) + \ldots + c_{l-k}\omega_{l-k}'(q_0) ,$$

that is,

(4′) $$\varphi'(q_0) = \mathfrak{c}\Omega'(q_0) \quad \textit{for a vector } \mathfrak{c} = [c_1, \ldots, c_{l-k}] .$$

The practical value of the first part of 7.1 is as follows. To find all critical points of $\varphi|H$ it suffices to solve the following system of equations:

$$\Omega(q) = 0 , \quad \varphi'(q) = c_1\omega_1'(q) + \ldots + c_{l-k}\omega_{l-k}'(q) ,$$

i.e. to find all the solutions $x_1, \ldots, x_l, c_1, \ldots, c_{l-k}$ of the following system of $l+(l-k)$ of scalar equations:

$$\omega_1(x_1, \ldots, x_l) = 0 \,,$$
$$\ldots\ldots\ldots\ldots$$
$$\omega_{l-k}(x_1, \ldots, x_l) = 0 \,,$$
$$\varphi_{|1}(x_1, \ldots, x_l) = \sum_{j=1}^{l-k} c_j \omega_{j|1}(x_1, \ldots, x_l),$$
$$\ldots\ldots\ldots\ldots\ldots\ldots\ldots\ldots$$
$$\varphi_{|l}(x_1, \ldots, x_l) = \sum_{j=1}^{l-k} c_j \omega_{j|l}(x_1, \ldots, x_l) \,.$$

If $x_1, \ldots, x_l, c_1, \ldots, c_{l-k}$ satisfy this system of equations, then the point $q_0 = (x_1, \ldots, x_l)$ may be an extremal point of $\varphi|H$. If the coordinates $x_1, \ldots, x_l$ of q_0 do not satisfy this system of equations for certain numbers $c_1, \ldots, c_{l-k}$, then q_0 is not an extremal point of $\varphi|H$.

Let Φ be a diffeomorphism onto a neighbourhood U of q_0 in H, $q_0 = \Phi(p_0)$. The function $(\varphi|H) \circ \Phi$, i.e. the function $\varphi \circ \Phi$, is of class C_1. It follows from 3.3 that

$$(\varphi \circ \Phi)_{|r}(p_0) = \varphi'(q_0) \cdot \Phi_{|r}(p_0) \,.$$

(3) holds if and only if $(\varphi \circ \Phi)_{|r}(p_0) = 0$ for $r = 1, \ldots, k$, i.e. if

$$\varphi'(q_0) \cdot \Phi_{|r}(p_0) = 0 \qquad \text{for} \qquad r = 1, \ldots, k \,.$$

In other words, if and only if the vector $\varphi'(q_0)$ is orthogonal to the hyperplane tangent to H at q_0 (see 6.4 and 6.5), i.e., is orthogonal to H at q_0. It follows from 6.6 that $\varphi'(q_0)$ has this property if and only if it is of the form (4).

EXERCISES

1. Find the greatest and the smallest value of the function $\varphi(x, y, z) = xyz$ on the curve $H = \{(x, y, z)\colon (x+y+z = 5) \& (xy+yz+zx = 8)\}$.

2. Find the smallest and the greatest value of the function $\varphi(x, y, z) = x - 2y + 2z$ on the sphere with radius 1 and center (0, 0, 0).

3. Find all extremal points of the function $\varphi(x, y, z) = x^2 + 2y^2 + 3z^2$ on the circle $H = \{(x, y, z)\colon (x^2+y^2+z^2 = 1) \& (x+2y+3z = 0)\}$.

CHAPTER VI

MEASURE THEORY

§ 1. Introduction. In elementary geometry we assign, to certain sets of points on the line, in the plane or in three-dimensional space, certain non-negative numbers called the length, area or volume of these sets.

For instance the number $b-a$ is called the length of the interval $\{x\colon a \leqslant x \leqslant b\}$. Aside from intervals we do not examine any other subset of the straight line R in elementary geometry.

In the plane R^2 we assign to the rectangle

$$\{(x_1, x_2)\colon\ a_i \leqslant x_i \leqslant b_i \ \text{for}\ i = 1, 2\}$$

the number $(b_1-a_1)(b_2-a_2)$ called the area of the rectangle. We generalize the definition of the area for certain other sets like polygons, discs, etc. We introduce also the notion of length for certain curves.

In the case of the three-dimensional space R^3 we also introduce the notion of length for some curves, the notion of area for certain surfaces, and the notion of volume for certain sets which are considered to be three-dimensional, from the intuitive point of view. In particular, by the volume of the three-dimensional interval

$$\{(x_1, x_2, x_3)\colon\ a_i \leqslant x_i \leqslant b_i \ \text{for}\ i = 1, 2, 3\}$$

we mean the number $(b_1-a_1)(b_2-a_2)(b_3-a_3)$.

The mathematical notion which is a common generalization of the notions of length, area and volume, is called *measure*. We say "the one-dimensional measure" instead of "the length", "the two-dimensional measure" instead of "the area", and "the three-dimensional measure" instead of "the volume".

The class of sets to which we assign measures in elementary geometry is rather small. For instance, the length of the set of all rational numbers in the unit interval $(0;1)$ is not defined at all, and the intuition does not suggest here a definition. Moreover the definition of measures of various sets in elementary geometry are based on different ideas; for instance, the definition of the area of a plane polygon is different from the definition of the area of a disc, etc. The question arises whether it is possible to formulate a general definition of r-dimensional measure for a rather large class of subsets of the k-dimensional Euclidean space R^k $(r = 1, 2, \ldots, k)$. The notion should be a common generalization of the notions of length, area and volume from elementary geometry.

The problem has a positive solution. In this chapter we shall give a definition of the k-dimensional measure in the k-dimensional Euclidean space; this measure is called the *Lebesgue measure* in R^k. In Chapter IX we shall discuss the case of *r-dimensional measures* in R^k.

All the measures mentioned above have two common properties:

(a) The finite or countable set-theoretical operations performed on measurable sets (i.e. sets, for which the measure in question is defined) yield measurable sets.

(b) The measure is a non-negative countably additive set function.

By the countable additivity we mean the property that the measure of the union of a sequence of disjoint measurable sets is equal to the sum of the measures of these sets.

A more precise formulation of properties (a) and (b) will be given later.

A deeper analysis of the problem shows that (a) and (b) are the most fundamental properties of measure. Many other properties of measure can be deduced from (a) and (b). For this reason, as well as for other reasons which cannot be given at this moment, it is convenient to introduce a very general notion of measure as a set function satisfying conditions (a) and (b). The Lebesgue measure and the more general notion of the r-dimensional measure in R^k are particular cases of this general notion.

The general notion of measure will be the subject of § 3. The Lebesgue measure will be defined and examined later, in § 6 and § 7.

§ 2. Algebras of sets. A non-void class $\mathcal{M}$ of subsets of a set X is called an *algebra of sets* (more precisely, an *algebra of subsets of X*) if the following conditions are satisfied:

(a) if $A \in \mathcal{M}$, then $X - A \in \mathcal{M}$;

(b) if $A, B \in \mathcal{M}$, then $A \cup B \in \mathcal{M}$;

(c) if $A, B \in \mathcal{M}$, then $A \cap B \in \mathcal{M}$.

2.1. *For a non-void class $\mathcal{M}$ of subsets of X to be an algebra of sets it suffices that condition* (a) *and one of conditions* (b), (c) *hold.*

Indeed, suppose that (a) and (b) hold. We shall prove that (c) holds, too. For this purpose we recall the de Morgan law

$$A \cap B = X - \big((X - A) \cup (X - B)\big).$$

If $A, B \in \mathcal{M}$, then $X - A \in \mathcal{M}$ and $X - B \in \mathcal{M}$ by (a). Hence $(X - A) \cup (X - B) \in \mathcal{M}$ by (b). Using (a) again we infer that $X - \big((X - A) \cup (X - B)\big) \in \mathcal{M}$, i.e. $A \cap B \in \mathcal{M}$.

Replacing the above de Morgan law by the following one

$$A \cup B = X - ((X - A) \cap (X - B))$$

we prove, in the same way, that (a) and (c) imply (b).

2.2. *For every algebra $\mathcal{M}$ of subsets of X:*

(1) $$0 \in \mathcal{M} \quad \textit{and} \quad X \in \mathcal{M},$$

i.e. the empty set and the whole space X belong to $\mathcal{M}$. Moreover,

(2) $$\textit{if } A, B \in \mathcal{M}, \textit{ then } A - B \in \mathcal{M}.$$

By definition, $\mathcal{M}$ is a non-empty class of sets, i.e. there is a set $A \in \mathcal{M}$ ($A \subset X$). By (a), $X - A \in \mathcal{M}$. Hence, by (c) and (b),

$$0 = A \cap (X - A) \in \mathcal{M} \quad \text{and} \quad X = A \cup (X - A) \in \mathcal{M}.$$

If $A, B \in \mathcal{M}$, then also $X - B \in \mathcal{M}$ by (a), and consequently by (c)

$$A - B = A \cap (X - B) \in \mathcal{M}.$$

It follows from (b) and (c) that the union and the intersection of any finite number of sets in an algebra of sets $\mathcal{M}$ belong to $\mathcal{M}$. The simple proof by induction is omitted.

The above properties of an algebra of sets $\mathcal{M}$, together with (a) and (2), can be formulated, in brief, as follows: all the finite set-theoretical operations on sets in $\mathcal{M}$ yield sets in $\mathcal{M}$.

A non-empty class $\mathcal{M}$ of subsets of a set X is said to be a *countably additive algebra of sets* (more precisely, a *countably additive algebra of subsets of X*) if condition (a) and the following conditions are satisfied:

(b′) if $A_n \in \mathcal{M}$ for $n = 1, 2, \ldots$, then $A_1 \cup A_2 \cup \ldots \in \mathcal{M}$;

(c′) if $A_n \in \mathcal{M}$ for $n = 1, 2, \ldots$, then $A_1 \cap A_2 \cap \ldots \in \mathcal{M}$.

We often say *σ-algebra* instead of "countably additive algebra".

2.3. *For a non-void class $\mathcal{M}$ of subsets of X to be a σ-algebra it suffices that condition* (a) *and one of conditions* (b′), (c′) *hold.*

Indeed, suppose that (a) and (b′) hold. We shall prove that (c′) holds, too. For this purpose we recall the following de Morgan law:

$$\bigcap_{n=1}^{\infty} A_n = X - \bigcup_{n=1}^{\infty} (X - A_n).$$

If $A_n \in \mathcal{M}$ for $n = 1, 2, \ldots$, then $X - A_n \in \mathcal{M}$ by (a) and, consequently, $\bigcup_{n=1}^{\infty} (X - A_n) \in \mathcal{M}$ by (b′). Using (a) again we infer that $X - \bigcup_{n=1}^{\infty} (X - A_n) \in \mathcal{M}$, i.e. $\bigcap_{n=1}^{\infty} A_n \in \mathcal{M}$.

Replacing the above de Morgan law by the following one

$$\bigcup_{n=1}^{\infty} A_n = X - \bigcap_{n=1}^{\infty} (X - A_n)$$

we prove, in the same way, that (a) and (c′) imply (b′).

2.4. ***Every σ-algebra of sets is an algebra of sets.***

Conditions (b′) and (c′) imply (b) and (c) respectively. Indeed, if $A, B \in \mathcal{M}$, then assuming

$$A_1 = A \quad \text{and} \quad A_n = B \quad \text{for} \quad n = 2, 3, \dots$$

we infer that

$$A \cup B = \bigcup_{n=1}^{\infty} A_n \in \mathcal{M}, \qquad A \cap B = \bigcap_{n=1}^{\infty} A_n \in \mathcal{M}.$$

It follows from 2.4 and 2.2 that every σ-algebra of subsets of X also has properties (1) and (2).

All the above properties of a σ-algebra of sets $\mathcal{M}$ can be formulated, in brief, as follows: all the finite or countable set-theoretical operations performed on sets in $\mathcal{M}$ yield sets in $\mathcal{M}$.

As the simplest example of a σ-algebra of subsets of a set X we mention the class of all subsets of X.

For any given class $\mathcal{K}$ of subsets of a space X there are σ-algebras $\mathcal{M}$ of subsets of X, such that $\mathcal{K} \subset \mathcal{M}$ (e.g. the σ-algebra of all subsets of X has this property). It is easy to verify that the common part $\mathcal{B}$ of all those σ-algebras $\mathcal{M}$ is also a σ-algebra of subsets of X, and $\mathcal{K} \subset \mathcal{B}$. By definition, $\mathcal{B}$ is the smallest σ-algebra of subsets of X such that $\mathcal{K}$ is its subclass. In other words, for any σ-algebra $\mathcal{M}$ of subsets of X,

$$\text{if } \mathcal{K} \subset \mathcal{M}, \text{ then } \mathcal{B} \subset \mathcal{M}. \tag{3}$$

Suppose now that X is a metric space and $\mathcal{K}$ is the class of all open subsets of X. Let $\mathcal{B}$ be the smallest σ-algebra of subsets of X such that $\mathcal{K} \subset \mathcal{B}$. Sets in $\mathcal{B}$ are called *Borel sets* (more precisely, *Borel subsets of* X). The σ-algebra $\mathcal{B}$ is called the *algebra of Borel subsets of* X.

By definition, every open subset of X is a Borel set. Every closed set $A \subset X$ is also a Borel set because $A = X - B$ where B is open (see (a)). A subset A of a metric space X is said to be a G_δ-*set* if $A = \bigcap_{n=1}^{\infty} A_n$ where the A_n are open sets, $n = 1, 2, \dots$ A subset A of a metric space X is said to be an F_σ-*set* if $A = \bigcup_{n=1}^{\infty} A_n$ where the A_n are closed sets, $n = 1, 2, \dots$

It follows from (c′) and (b′) that every G_δ-set and every F_σ-set are Borel sets. It can be proved that, in general, there are Borel sets that are neither open sets, nor closed sets, nor G_δ-sets, nor F_σ-sets (this is so, e.g., in the case where $X = R^k$).

It follows directly from the de Morgan laws I § 2 (26) that if a subset A of a metric space X is a G_δ-set (an F_σ-set), then $X-A$ is an F_σ-set (G_δ-set).

Let Z be a subset of a metric space X. The set Z can be treated as a metric space with the same definition of the distance between points (see III § 4). Thus we can introduce the notion of Borel sets in the space Z. The following theorem establishes a connection between Borel sets in Z and Borel sets in X.

2.5. *Let $Z \subset X$. A subset A of Z is a Borel set in the space Z if and only if $A = Z \cap B$ where B is a Borel set in X.*

If Z is a Borel set in X, then A is a Borel set in Z if and only if A is a Borel set in X.

It is easy to verify that the class of all sets $B \subset X$ such that $Z \cap B$ is a Borel set in Z is a σ-algebra containing all open subsets in X (see III 4.1). Thus this class contains all Borel sets in X, i.e.,

$$\text{if } B \text{ is a Borel set in } X, \text{ then } A = Z \cap B \text{ is a Borel set in } Z.$$

On the other hand, the class of all sets $A = Z \cap B$ where B is a Borel set in X, is a σ-algebra of subsets of Z and contains the class of all sets open in Z (see III 4.1). Thus this class contains all Borel sets in Z, i.e.

$$\text{if } A \text{ is a Borel set in } Z, \text{ then } A = Z \cap B \text{ where } B \text{ is a Borel set in } X.$$

This proves the first part of 2.5. The second follows directly from the first.

EXERCISES

1. Let X be an infinite set. The class $\mathcal{M}$ composed of all finite sets $A \subset X$ and their complements $X-A$ is an algebra of sets but not a σ-algebra of sets.

2. Let X be an uncountable set. The class $\mathcal{M}$ composed of all countable sets $A \subset X$ and their complements $X-A$ is a σ-algebra of sets.

3. The class of all finite unions of bounded or unbounded intervals open from the left is an algebra of subsets of R but it is not a σ-algebra of sets. The smallest σ-algebra containing this algebra of sets coincides with the algebra of all Borel subsets of R.

The same is true if "open from the left" is replaced by "open from the right".

4. The class of all sets which are the union of a finite number of bounded or unbounded open intervals and a finite set of real numbers is an algebra of subsets of R but not a σ-algebra of sets. The smallest σ-algebra containing this algebra of sets coincides with the algebra of all Borel subsets of R.

5. The algebra of all Borel subsets of a metric space X is the smallest σ-algebra containing all closed subsets of X.

6. The algebra of all Borel subsets of R^k is the smallest σ-algebra containing all open bounded intervals (or all closed bounded intervals).

7. The union of a finite number of G_δ-sets is a G_δ-set. The intersection of a countably many G_δ-sets is a G_δ-set.

The intersection of a finite number F_σ-sets is an F_σ-set. The union of countably many F_σ-sets is an F_σ-set.

If A is a G_δ-set and B is an F_σ-set, then $A-B$ is an G_δ-set. If A is an F_σ-set and B is a G_δ-set, then $A-B$ is an F_σ-set.

Hint: Use I § 2 (25).

8. Every closed set is a G_δ-set. Every open set is an F_σ-set.

Hint. Use III § 3, exercise 6.

9. The set of all rational numbers is an F_σ-set in the space R of all real numbers but it is neither open nor closed. The set of all irrational numbers is a G_δ-set in R but it is neither open nor closed. (It can be proved that the set of all irrational numbers is not an F_σ-set in R and, consequently, the set of all rational numbers is not a G_δ-set).

10. Every G_δ-set (every bounded G_δ-set) is the intersection of a decreasing sequence of open sets (of open bounded sets). Every F_σ-set is the union of an increasing sequence of bounded closed sets.

Hint: Use I 2.4.

11. Let Z be a subset of a metric space X. A set $A \subset Z$ is a G_δ-set (an F_σ-set) in the space Z if and only if it is of the form $A = Z \cap B$ where B is a G_δ-set (an F_σ-set) in the space X.

12. Every k-dimensional hypersurface in R^l $(0 \leqslant k \leqslant l)$ is an F_σ-set in R^l.

13. For any class $\mathcal{K}$ of subsets of a set X there exists the smallest algebra $\mathcal{M}$ of subsets of X, such that $\mathcal{K} \subset \mathcal{M}$. If $\mathcal{K}$ is empty, then $\mathcal{M}$ contains only the empty set and the whole space X. If $\mathcal{K}$ is not empty, then $\mathcal{M}$ is the class of all sets which are finite unions of finite intersections of sets in $\mathcal{K}$ and of the complements of sets in $\mathcal{K}$. In that case, $\mathcal{K}$ is also the class of all sets which are finite intersections of finite unions of sets in $\mathcal{K}$ and of the complements of sets in $\mathcal{K}$.

§ 3. Measure. By a *measure* we mean any set function μ defined on a σ-algebra $\mathcal{M}$ of subsets of a space X, such that

(a) $0 \leqslant \mu(A) \leqslant \infty$;

(b) $\mu(0) = 0$;

(c) for every sequence of disjoint sets $A_1, A_2, \ldots$ in $\mathcal{M}$,

$$\mu(A_1 \cup A_2 \cup \ldots) = \mu(A_1) + \mu(A_2) + \ldots$$

The number $\mu(A)$, i.e. the value of the function μ at an $A \in \mathcal{M}$, is called the *measure of the set A*. Sets in the σ-algebra $\mathcal{M}$ on which the measure μ is defined, are said to be *μ-measurable*, or simply *measurable*.

By (a), a measure μ is an extended non-negative real function. By (b), the measure of the empty set is always equal to 0. Property (c) is called the *countable additivity* of measure.

Simple examples of measures will be given at the end of this section and in exercises 1 and 2. Examples of non-trivial measures will be given in § 6 and in IX § 1.

3.1. *Every measure μ on $\mathcal{M}$ has the following properties:*

$$\mu(A_1 \cup \ldots \cup A_m) = \mu(A_1) + \ldots + \mu(A_m) \tag{1}$$

for every finite sequence of disjoint sets $A_1, \ldots, A_m \in \mathcal{M}$; in particular

$$\mu(A \cup B) = \mu(A) + \mu(B) \tag{1'}$$

for any disjoint sets $A, B \in \mathcal{M}$.

(2) *If* $A \subset B$ $(A, B \in \mathcal{M})$, *then* $\mu(A) \leqslant \mu(B)$.

(3) *If* $A \subset B$ $(A, B \in \mathcal{M})$ *and* $\mu(A) < \infty$, *then* $\mu(B-A) = \mu(B) - \mu(A)$.

(4) $$\mu(A_1 \cup A_2 \cup \ldots) \leqslant \mu(A_1) + \mu(A_2) + \ldots$$

for every infinite sequence of sets $A_1, A_2, \ldots \in \mathcal{M}$.

(5) $$\mu(A_1 \cup \ldots \cup A_m) \leqslant \mu(A_1) + \ldots + \mu(A_m)$$

for every finite sequence $A_1, \ldots, A_m \in \mathcal{M}$; *in particular,*

(5′) $$\mu(A \cup B) \leqslant \mu(A) + \mu(B) \quad \textit{for all } A, B \in \mathcal{M}.$$

Property (1) is called the *finite additivity* of measure. Property (2) is called the *monotonicity* of measure; it asserts that every measure is a *non-decreasing* set function. Properties (4) and (5) are called the *countable* and *finite subadditivity* of measure.

To prove (1), it suffices to assume in (c) that $A_{m+1} = A_{m+2} = \ldots = 0$.

Applying (1′) to the disjoint sets A and $B-A$ we get

$$\mu(B) = \mu(A) + \mu(B-A),$$

which implies (2) since $\mu(B-A) \geqslant 0$. The last identity implies also (3).

To prove (4) let us assume

$$B_1 = A_1 \quad \text{and} \quad B_n = A_n - (A_1 \cup \ldots \cup A_{n-1}) \quad \text{for} \quad n = 2, 3, \ldots$$

The sets $B_1, B_2, \ldots$ are measurable, disjoint and

$$\bigcup_{n=1}^{\infty} A_n = \bigcup_{n=1}^{\infty} B_n, \quad B_n \subset A_n \quad \text{for} \quad n = 1, 2, \ldots$$

(see I 2.5). It follows from (c) and (2) that

$$\mu\Big(\bigcup_{n=1}^{\infty} A_n\Big) = \mu\Big(\bigcup_{n=1}^{\infty} B_n\Big) = \sum_{n=1}^{\infty} \mu(B_n) \leqslant \sum_{n=1}^{\infty} \mu(A_n)$$

since $\mu(B_n) \leqslant \mu(A_n)$.

(5) follows directly from (b) and (4) where $A_{m+1} = A_{m+2} = \ldots = 0$.

If $\mu(A) = 0$, we say that the set A is *of μ-measure zero* (or simply *of measure zero*).

3.2. *The union of a countable or finite sequence of sets of measure zero is of measure zero. Every measurable subset of a set of measure zero is of measure zero.*

This follows directly from (4), (5) and (2) (see also (a)).

Observe that, in general, not all subsets of a set of measure zero are measurable. If every subset of every set of measure zero is measurable, the measure is said to be *complete*.

The measure of a set remains unchanged if we add or subtract a set of measure zero. More precisely,

3.3. *If* $A \in \mathcal{M}$ *and* $\mu(B) = 0$, *then*

$$\mu(A \cup B) = \mu(A) \quad \textit{and} \quad \mu(A - B) = \mu(A)\,.$$

In fact, $\mu(A) \leqslant \mu(A \cup B) \leqslant \mu(A) + \mu(B) = \mu(A)$ by (2) and (5′), and therefore $\mu(A \cup B) = \mu(A)$. Similarly, $\mu(A - B) \leqslant \mu(A) = \mu(A - B) + \mu(A \cap B) \leqslant \mu(A - B) + \mu(B) = \mu(A - B)$ by (2), (1′), (2), and consequently $\mu(A - B) = \mu(A)$.

Property (c) can be generalized as follows:

3.4. *If* $A_1, A_2, \ldots$ *is a finite or infinite sequence of measurable sets and* $\mu(A_i \cap A_j) = 0$ *for* $i \neq j$, *then*

$$\mu(A_1 \cup A_2 \cup \ldots) = \mu(A_1) + \mu(A_2) + \ldots$$

The union B of all the sets $A_i \cap A_j$, where $i \neq j$, is a set of measure zero since it is the union of countably many sets of measure zero. Since

$$A_1 \cup A_2 \cup \ldots = B \cup (A_1 - B) \cup (A_2 - B) \cup \ldots$$

and since the sets on the right-hand side are disjoint, we have

$$\mu(A_1 \cup A_2 \cup \ldots) = \mu(B) + \mu(A_1 - B) + \mu(A_2 - B) + \ldots = \mu(A_1) + \mu(A_2) + \ldots$$

by 3.3.

If $\mu(A) < \infty$, we say that the set A is of *finite* μ*-measure* (or simply *of finite measure*). It follows from (2) that every measurable subset of a set of finite measure is a set of finite measure. It follows also from (2) that the measure μ is *finite* (i.e. $\mu(A) < \infty$ for every $A \in \mathcal{M}$) if and only if $\mu(X) < \infty$, i.e. if the whole space X has a finite measure.

We say that the measure μ is σ*-finite* if the space X is the union of a sequence of sets of finite measure. A measurable set A is said to be of σ*-finite measure* if A is the union of a sequence of sets of finite measure.

The following two theorems on monotonic sequences of measurable sets will be used several times in this book.

3.5. *The measure of the union of an increasing sequence of measurable sets is equal to the limit of the measures of these sets.*

In symbols,

$$\mu\Big(\bigcup_{n=1}^{\infty} A_n\Big) = \lim_{n \to \infty} \mu(A_n)$$

provided that $A_1 \subset A_2 \subset \ldots$

3.6. *The measure of the intersection of a decreasing sequence of measurable sets of finite measures is equal to the limit of the measures of these sets.*

In symbols,

$$\mu\Big(\bigcap_{n=1}^{\infty} B_n\Big) = \lim_{n\to\infty} \mu(B_n)$$

provided that $B_1 \supset B_2 \supset \ldots$ and $\mu(B_1) < \infty$.

To prove 3.5 let us observe that the sets

$$C_1 = A_1\,, \qquad C_n = A_n - A_{n-1} \quad \text{for} \quad n > 1$$

are measurable, disjoint and

$$A_1 \cup A_2 \cup \ldots = C_1 \cup C_2 \cup \ldots, \qquad A_n = C_1 \cup \ldots \cup C_n$$

(see I 2.6). Hence

$$\mu(A_1 \cup A_2 \cup \ldots) = \mu(C_1) + \mu(C_2) + \ldots = \lim_{n\to\infty}\big(\mu(C_1) + \ldots + \mu(C_n)\big) = \lim_{n\to\infty} \mu(A_n)$$

by (c) and (1).

To prove 3.6 let us observe that if the sets $B_1, B_2, \ldots$ satisfy the hypotheses of 3.6, then the sets

$$A_n = B_1 - B_n \qquad (n = 1, 2, \ldots)$$

satisfy the hypotheses of 3.5 and consequently $\mu(\bigcup_{i=1}^{\infty} A_n) = \lim_{n\to\infty} \mu(A_n)$. Hence by (3) and the de Morgan formula

$$\mu(B_1) - \mu\Big(\bigcap_{n=1}^{\infty} B_n\Big) = \mu\Big(B_1 - \bigcap_{n=1}^{\infty} B_n\Big) = \mu\Big(\bigcup_{n=1}^{\infty} A_n\Big) = \lim_{n\to\infty} \mu(A_n)$$

$$= \lim_{n\to\infty}\big(\mu(B_1) - \mu(B_n)\big) = \mu(B_1) - \lim_{n\to\infty} \mu(B_n)\,.$$

This implies 3.6.

Let X be any set and let μ_0 be the set function defined for all sets $A \subset X$ as follows

$$\mu_0(A) = \text{the number of elements in } A\,.$$

It is easy to see that μ_0 is a measure, called the *zero-dimensional measure on* X. All subsets of X are measurable. The measure μ_0 is finite if and only if the space X is finite. It is σ-finite if and only if X is countable.

EXERCISES

1. The function μ defined on the σ-algebra of all subsets of a set X as follows

$$\mu(A) = \begin{cases} 0 & \text{if } A \text{ is empty}\,, \\ \infty & \text{if } A \text{ is not empty}\,, \end{cases}$$

is a measure. The measure μ is σ-finite if and only if $X = 0$.

2. The function μ defined on the σ-algebra $\mathcal{M}$ defined in § 2, exercise 2, as follows

$$\mu(A) = \begin{cases} 0 & \text{if the set } A \text{ is countable,} \\ 1 & \text{if the set } X-A \text{ is countable,} \end{cases}$$

is a finite measure.

3. For any measure μ and any sequence (A_n) of measurable sets

$$\mu(\liminf_{n\to\infty} A_n) \leqslant \liminf_{n\to\infty} \mu(A_n)$$

(see I § 2, exercise 2). If $\mu(A_1 \cup A_2 \cup \ldots) < \infty$, then also

$$\mu(\limsup_{n\to\infty} A_n) \geqslant \limsup_{n\to\infty} \mu(A_n)\,.$$

Consequently, if (A_n) is a convergent sequence of measurable sets and $\mu(A_1 \cup A_2 \cup \ldots) < \infty$, then

$$\mu(\lim_{n\to\infty} A_n) = \lim_{n\to\infty} \mu(A_n)\,.$$

4. Let μ be a non-complete measure on a σ-algebra $\mathcal{M}$ of subsets of X. Let $\mathcal{M}'$ be the class of all sets $A \cup B$ where $A \in \mathcal{M}$ and B is a subset of a set of measure zero. The formula $\mu'(A \cup B) = \mu(A)$ uniquely defines a set function μ' on $\mathcal{M}'$. Prove that $\mathcal{M}'$ is a σ-algebra of subsets of X and that μ' is a complete measure.

The measure μ' is said to be the *completion* of μ.

§ 4. The volume of an interval. We say that closed intervals $I_1, I_2 \subset R^k$ *do not overlap* if they have no common interior points, i.e.

$$\operatorname{int} I_1 \cap \operatorname{int} I_2 = \operatorname{int}(I_1 \cap I_2) = 0\,.$$

In other words, I_1 and I_2 do not overlap if $I_1 \cap I_2$ is a degenerate interval. In particular, disjoint intervals do not overlap. Figures 4 and 5 show that (for $k = 2$) non-overlapping intervals can have common points.

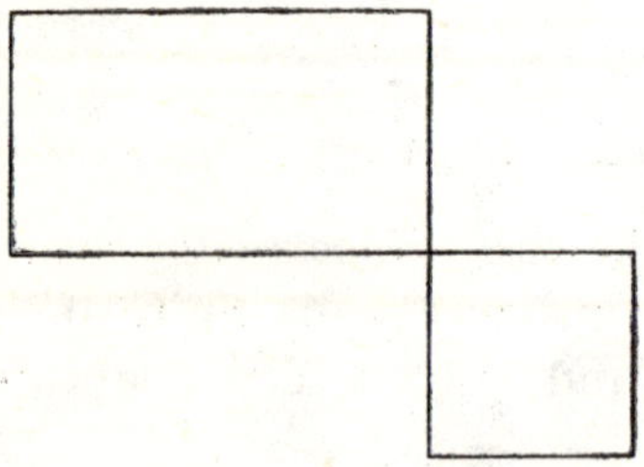

Fig. 4

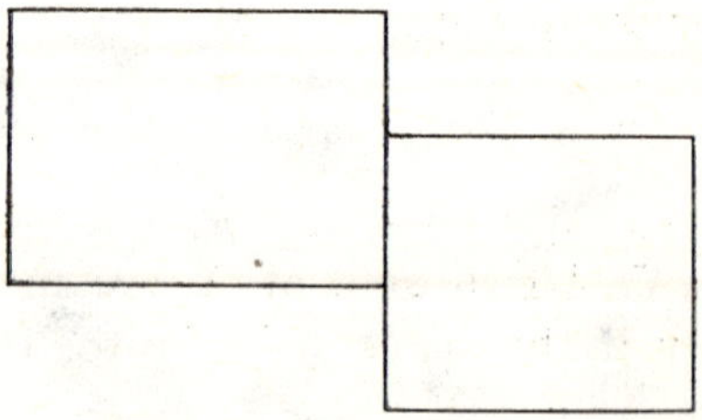

Fig. 5

In the case of $k = 1$, intervals I_1, I_2 do not overlap if and only if either they are disjoint or they have only one common point which is the end-point of one of the intervals and the initial point of the other.

By a *partition* of a closed interval $I \subset R^k$ we shall mean any finite set **P** of closed non-overlapping intervals whose union is equal to I.

Figures 6 and 7 give examples of partitions in the case of $k = 2$.

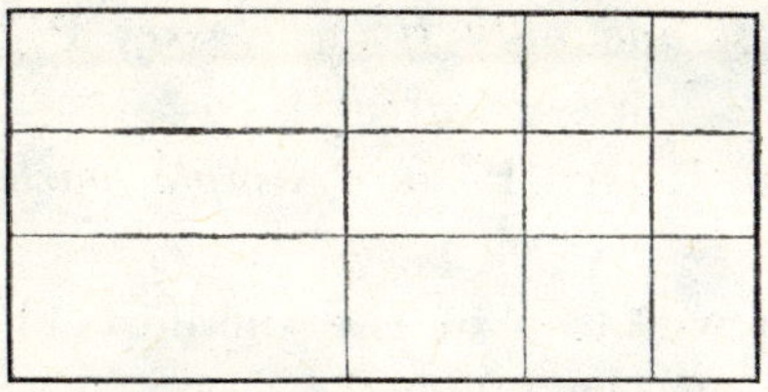

Fig. 6

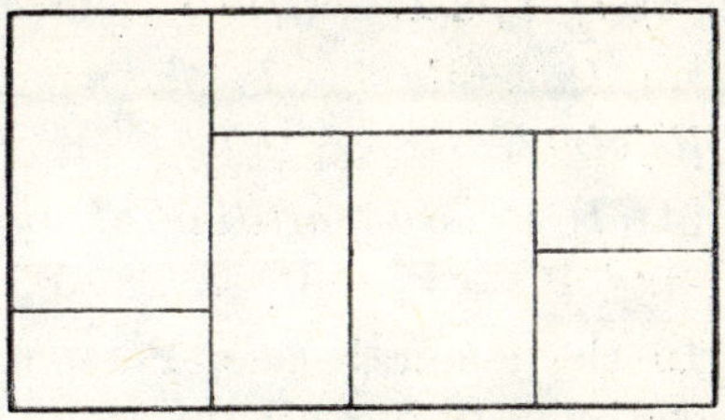

Fig. 7

Let

(1) $$I = \{(x_1, \ldots, x_k): a_i \leqslant x_i \leqslant \beta_i \text{ for } i = 1, \ldots, k\}$$

be a k-dimensional closed interval. Let $\gamma_{i,j}$ ($i = 1, \ldots, k$, $j = 0, 1, \ldots, r_i$) be real numbers such that

$$a_i = \gamma_{i,0} < \gamma_{i,1} < \ldots < \gamma_{i,r_i} = \beta_i .$$

The class **P** of all closed intervals

(2) $$\{(x_1, \ldots, x_k): \gamma_{i,j_i-1} \leqslant x_i \leqslant \gamma_{i,j_i} \text{ for } i = 1, \ldots, k\},$$

where $j_1, \ldots, j_k$ is a sequence of positive integers, $j_i \leqslant r_i$ for $i = 1, \ldots, k$, is a partition of I. Partitions of this form are called *simple*. In other words, simple partitions are obtained by intersecting I by a finite number of $(k-1)$-dimensional hyperplanes

$$\{(x_1, \ldots, x_k): x_i = \gamma_{i,j}\}$$

orthogonal to the coordinate axes. The partition in figure 6 is simple. The partition in figure 7 is not simple.

Let **P** be a finite class of closed k-dimensional intervals:

(3) $$I_j = \{(x_1, \ldots, x_k): a_{i,j} \leqslant x_i \leqslant \beta_{i,j} \text{ for } i = 1, \ldots, k\},$$

$j = 0, 1, \ldots, m$. Let, for $i = 1, \ldots, k$,

$$\gamma_{i,0} < \gamma_{i,1} < \ldots < \gamma_{i,r_i}$$

be the increasing sequence formed of all the numbers

$$a_{i,0}, a_{i,1}, \ldots, a_{i,m}, \beta_{i,0}, \beta_{i,1}, \ldots, \beta_{i,m} .$$

The symbol $\mathbf{P}^*$ will now denote the class of all the intervals (2). The following statements express fundamental properties of $\mathbf{P}^*$.

4.1. (a) *The class*

$$\mathbf{P}_j^* = \{J \in \mathbf{P}^* \colon J \subset I_j\}$$

is a simple partition of I_j.

(b) *If* $I_0 \subset I_1 \cup \ldots \cup I_m$, *then* $\mathbf{P}_0^* \subset \mathbf{P}_1^* \cup \ldots \cup \mathbf{P}_m^*$.

(c) *Intervals* I_{j_1}, I_{j_2} *do not overlap if and only if the classes* $\mathbf{P}_{j_1}^*, \mathbf{P}_{j_2}^*$ *are disjoint.*

(d) *If* $\mathbf{P}$ *is a partition of an interval* I, *then* $\mathbf{P}^*$ *is a simple partition of* I *and* $\mathbf{P}^* = \mathbf{P}_0^* \cup \mathbf{P}_1^* \cup \ldots \cup \mathbf{P}_m^*$.

In the general case, $\mathbf{P}^*$ is a simple partition of the smallest closed interval containing all intervals (3), i.e. of the interval

$$\{(x_1, \ldots, x_k) \colon \gamma_{i,0} \leqslant x_i \leqslant \gamma_{i,r_i} \text{ for } i = 1, \ldots, k\}.$$

Figures 8 and 9 illustrate the construction of $\mathbf{P}^*$ in the case of $k = 2$. The faces of intervals in $\mathbf{P}$ are printed by heavy lines.

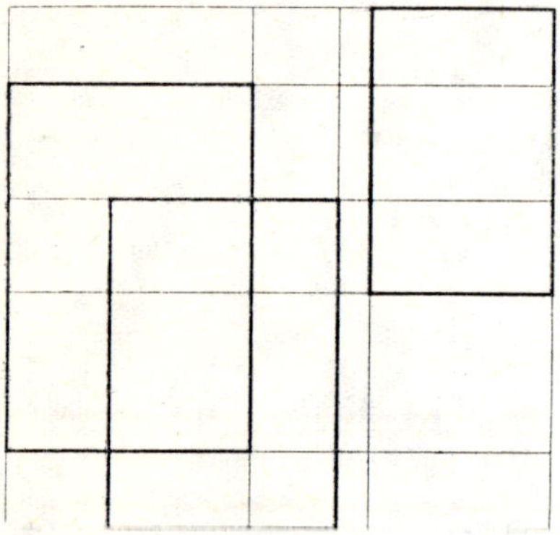

Fig. 8

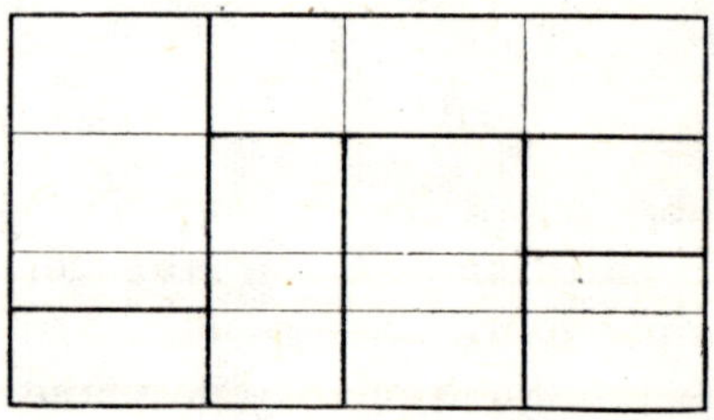

Fig. 9

By the *volume* of an k-dimensional interval (1) we mean the positive number

$$\operatorname{vol} I = (\beta_1 - \alpha_1) \ldots (\beta_k - \alpha_k),$$

i.e. the product of the lengths of the edges of I.

In the case where $k = 1$, I is a segment and $\operatorname{vol} I$ is its length. If $k = 2$, I is a rectangle and $\operatorname{vol} I$ is its area. If $k = 3$, I is a right parallelepiped and $\operatorname{vol} I$ is its ordinary volume. For $k > 3$, $\operatorname{vol} I$ is a natural generalization of the length of a segment, the area of a rectangle, and the volume of a right parallelepiped.

4.2. *For any partition* $\mathbf{P}$ *of* I,

$$\operatorname{vol} I = \sum_{J \in \mathbf{P}} \operatorname{vol} J.$$

First we shall prove 4.2 under the hypothesis that **P** is simple, i.e. that I is the interval (1) and **P** is the class of all the intervals (2). Then

$$\operatorname{vol} I = \prod_{i=1}^{k} (\beta_i - \alpha_i) = \prod_{i=1}^{k} \sum_{j=1}^{r_i} (\gamma_{i,j} - \gamma_{i,j-1})$$

$$= \sum_{(j_i)} \prod_{i=1}^{k} (\gamma_{i,j_i} - \gamma_{i,j_i-1}) = \sum_{J \in \mathbf{P}} \operatorname{vol} J .$$

Now let $\mathbf{P} = \{I_0, I_1, \dots, I_m\}$ be any partition of I. It follows from (a), (c) and (d) and from the part of 4.2 just proved that

$$\operatorname{vol} I = \sum_{J \in \mathbf{P}^*} \operatorname{vol} J = \sum_{j=0}^{m} \sum_{J \in \mathbf{P}_j^*} \operatorname{vol} J = \sum_{j=0}^{m} \operatorname{vol} I_j = \sum_{J \in \mathbf{P}} \operatorname{vol} J .$$

4.3. *For every finite sequence $I_0, I_1, \dots, I_m$ of closed intervals,*

if $I_0 \subset I_1 \cup \dots \cup I_m$, *then* $\operatorname{vol} I_0 \leqslant \operatorname{vol} I_1 + \dots + \operatorname{vol} I_m$.

Let **P** be the class composed of the intervals $I_0, I_1, \dots, I_m$ and let $\mathbf{P}^*$, $\mathbf{P}_j^*$ have the same meaning as previously. By 4.1 (a)-(b) and 4.2,

$$\operatorname{vol} I_0 = \sum_{J \in \mathbf{P}_0^*} \operatorname{vol} J \leqslant \sum_{j=1}^{m} \sum_{J \in \mathbf{P}_j^*} \operatorname{vol} J = \sum_{j=1}^{m} \operatorname{vol} I_j .$$

The following property of $\operatorname{vol} I$ follows directly from the definition.

4.4. *If* $I_1 \subset I_2$, *then* $\operatorname{vol} I_1 \leqslant \operatorname{vol} I_2$.

In this Chapter the empty set will also be considered as a closed interval (see II § 3, p. 50). We define

$$\operatorname{vol} 0 = 0 .$$

Observe that Theorems 4.3 and 4.4 remain true if some of the intervals under consideration are empty.

§ 5. The outer measure. By the *k-dimensional Lebesgue outer measure* (or simply *outer measure*) of a set $A \subset R^k$ we mean the least upper bound of all the numbers

(1) $$\sum_{n=1}^{\infty} \operatorname{vol} I_n$$

where (I_n) is any sequence of closed intervals covering A, i.e. such that

(2) $$A \subset \bigcup_{n=1}^{\infty} I_n .$$

The outer measure of A will be denoted by $|A|$. By definition,

(3) $$|A| \leqslant \sum_{n=1}^{\infty} \operatorname{vol} I_n$$

for every sequence (I_n) of closed intervals satisfying (2). Conversely, for every $\varepsilon > 0$ there is a sequence (I_n) of closed intervals satisfying (2) such that

(4) $$\sum_{n=1}^{\infty} \operatorname{vol} I_n \leqslant |A| + \varepsilon .$$

The outer measure of any set $A \subset R^k$ is a non-negative, finite or infinite real number:

(5) $$0 \leqslant |A| \leqslant \infty .$$

5.1. *The outer measure of the empty set is equal to* 0:

(6) $$|0| = 0 .$$

To obtain (6) it suffices to take in (3) $I_n = 0$ for $n = 1, 2, \ldots$

5.2. *If* $A \subset \bigcup_{n=1}^{\infty} A_n$, *then* $|A| \leqslant \sum_{n=1}^{\infty} |A_n|$.

Let $\varepsilon > 0$ be arbitrary. By (4) there exists a sequence $(I_{m,1}, I_{m,2}, \ldots)$ of closed intervals such that

$$A_m \subset \bigcup_{n=1}^{\infty} I_{m,n} \quad \text{and} \quad \sum_{n=1}^{\infty} \operatorname{vol} I_{m,n} \leqslant |A_m| + \varepsilon/2^m .$$

Hence

(7) $$A \subset \bigcup_{m=1}^{\infty} A_m \subset \bigcup_{m=1}^{\infty} \bigcup_{n=1}^{\infty} I_{m,n} ,$$

(8) $$\sum_{m=1}^{\infty} \sum_{n=1}^{\infty} \operatorname{vol} I_{m,n} \leqslant \sum_{m=1}^{\infty} |A_m| + \varepsilon .$$

The double sequence $(I_{m,n})$ can be arranged into an ordinary sequence of closed intervals. Since this sequence covers A by (7), we have by (3)

$$|A| \leqslant \sum_{m=1}^{\infty} \sum_{n=1}^{\infty} \operatorname{vol} I_{m,n} .$$

Hence, by (8),

$$|A| \leqslant \sum_{m=1}^{\infty} |A_m| + \varepsilon$$

for every positive ε. For $\varepsilon \to 0$ we get $|A| \leqslant \sum_{m=1}^{\infty} |A_m|$.

The following lemmas 5.3-5.8 are direct consequences of properties 5.1 and 5.2.

5.3. *If* $A \subset A_1 \cup \ldots \cup A_m$, *then* $|A| \leqslant |A_1| + \ldots + |A_m|$.

To obtain 5.3, take $A_n = 0$ for $n > m$ in 5.2.

5.4. *If* $A \subset B$, *then* $|A| \leqslant |B|$.

This follows from 5.3 where $m = 1$ and $A_1 = B$.

Theorem 5.4 states that the outer measure is a *monotonic non-decreasing* set function.

It follows from 5.4 that if $|B| < \infty$ and $A \subset B$, then $|A| < \infty$.

The properties expressed in Theorems 5.5 and 5.6 below are called the (finite) *subadditivity* and the *countable subadditivity* of outer measure.

5.5. *For any finite sequence of sets,*

$$|A_1 \cup \ldots \cup A_m| \leqslant |A_1| + \ldots + |A_m|.$$

5.6. *For any infinite sequence of sets,*

$$|A_1 \cup A_2 \cup \ldots| \leqslant |A_1| + |A_2| + \ldots$$

5.5 follows from 5.3 where $A = A_1 \cup \ldots \cup A_m$.

5.6 follows from 5.2 where $A = A_1 \cup A_2 \cup \ldots$

We say that a set $A \subset R^k$ is of *measure zero* (more precisely, of *k-dimensional Lebesgue measure zero*) if $|A| = 0$; in the next section we shall explain why we say here "measure" instead of "outer measure". For instance, the empty set is of measure zero (see 5.1). It follows from the monotonicity and the finite or countable subadditivity of outer measure that

5.7. *A subset of a set of measure zero is of measure zero. The union of a finite or infinite sequence of sets of measure zero is of measure zero.*

We now prove that sets which differ only by a set of measure zero have the same outer measure.

5.8. *If* $|B| = 0$, *then* $|A \cup B| = |A| = |A - B|$.

The proof is the same as that of the analogous statement § 3 (1) for measures.

It follows from the monotonicity and subadditivity that $|A| \leqslant |A \cup B| \leqslant |A| + |B| = |A|$ which proves that $|A \cup B| = |A|$. Similarly, it follows from 5.4 and 5.3 that $|A - B| \leqslant |A| \leqslant |A - B| + |B| = |A - B|$, which proves that $|A| = |A - B|$.

5.9. *For every closed interval* $I \subset R^k$,

$$|I| = \operatorname{vol} I .$$

The sequence of closed intervals

$$I_1 = I, \quad I_n = 0 \quad \text{for} \quad n = 2, 3, \ldots$$

is a covering of I. Therefore, by (3),

$$|I| \leqslant \operatorname{vol} I .$$

On the other hand, let (I_n) be any sequence of closed intervals that cover I, and let $\varepsilon > 0$ be arbitrary. We can find open intervals J_n such that

$$I_n \subset J_n \quad \text{and} \quad \operatorname{vol}\bar{J}_n \leqslant \operatorname{vol} I_n + \varepsilon/2^n$$

where $\bar{J}_n$ is the closure of J_n.

Consequently,

$$\sum_{n=1}^{\infty} \operatorname{vol}\bar{J}_n \leqslant \sum_{n=1}^{\infty} \operatorname{vol} I_n + \varepsilon .$$

Since the sequence (J_n) of open sets is a covering of the compact set I, there exists an integer m such that $I \subset J_1 \cup \ldots \cup J_m$ and consequently $I \subset \bar{J}_1 \cup \ldots \cup \bar{J}_m$. By 4.3,

$$\operatorname{vol} I \leqslant \sum_{n=1}^{m} \operatorname{vol}\bar{J}_n .$$

It follows from the last two inequalities that

$$\operatorname{vol} I \leqslant \sum_{n=1}^{\infty} \operatorname{vol} I_n + \varepsilon .$$

Since (I_n) was any sequence of closed intervals that cover I, we infer that

$$\operatorname{vol} I \leqslant |I| + \varepsilon .$$

Since $\varepsilon > 0$ is arbitrary, we infer that

$$\operatorname{vol} I \leqslant |I|$$

which, together with the converse inequality obtained earlier, proves that $|I| = \operatorname{vol} I$.

5.10. *For every degenerate interval I, $|I| = 0$.*

For every $\varepsilon > 0$ there exists a non-degenerate closed interval I' such that $I \subset I'$ and $\operatorname{vol} I' < \varepsilon$. We have $0 \leqslant |I| \leqslant |I'| = \operatorname{vol}|I'|$ and consequently $0 \leqslant |I| < \varepsilon$ for every positive ε. This proves 5.10.

5.11. *For every open (open from the left, open from the right) interval I, the outer measure $|I|$ is equal to the volume of the closure of I.*

Let J be the closure of I. The set $J - I$ is the union (or a subset of the union) of a finite number of degenerate intervals, viz. faces of J. By 5.7 and 5.10, $|J - I| = 0$. Assuming $A = I$ and $B = J - I$ in 5.8 we infer that $|I| = |J| = \operatorname{vol} J$ by 5.9.

It follows from 5.10 that every one-point subset of R^k is of measure zero. Hence it follows, by 5.7, that

5.12. *Every finite or countable subset of R^k has measure zero.*

EXERCISES

1. For every set $A \subset R^k$, $|A| = |A \cap (-\infty; a)| + |A \cap \langle a; \infty)|$.

2. If a set B is obtained from a set $A \subset R^k$ by a translation, then $|B| = |A|$.

3. Divide the interval $\langle 0; 1)$ into disjoint sets in such a way that two numbers x, y are in the same set if and only if $x - y$ is a rational number. Take a real number z from each of those sets. Let A be the set of all these z's. For every rational number w in $\langle 0; 1)$, let

$$B_w = \{x + w\colon (x \in A) \& (x + w < 1)\}, \qquad C_w = \{x + w - 1\colon (x \in A) \& (x + w > 1)\},$$

$$A_w = B_w \cup C_w .$$

Prove that $|A_w| = |A|$, the sets A_w are disjoint, and $\bigcup\limits_{w \in W} A_w = \langle 0; 1)$ where W is the set of all rational numbers in the interval $\langle 0; 1)$.

§ 6. Lebesgue-measurable sets. The Lebesgue measure. From the intuitive point of view the notion of the Lebesgue outer measure seems to be a good generalization of the notion of volume. However, a more exact analysis shows that the outer measure does not have properties which our intuition attributes to a generalization of the notion of volume. damely it can be proved that the Lebesgue outer measure is not an ad-Nitive set function, i.e. that the identity

$$|A \cup B| = |A| + |B| \quad \text{for disjoint } A, B$$

does not hold, in general. Consequently the Lebesgue outer measure is not a countably additive set function, i.e. the identity

$$|A_1 \cup A_2 \cup \ldots| = |A_1| + |A_2| + \ldots \quad \text{for disjoint sets } A_1, A_2, \ldots$$

does not hold, in general. Thus the Lebesgue outer measure is not a measure in the sense defined in § 3.

The cause of the difficulty is that the outer measure is defined for too large a class of sets. Sets for which the Lebesgue outer measure is not additive are very singular and, as a rule, they do not appear in mathematical investigations. If we restrict the notion of outer measure to a smaller class of sets, defined in a proper way, the phenomenon of non-additivity does not appear. To define this class of sets, we introduce the following definition.

We say that a set $A \subset R^k$ is *Lebesgue-measurable* (or simply *measurable*) if

$$(1) \qquad |Z \cap A| + |Z - A| = |Z| \quad \text{for every set } Z \subset R^k.$$

The intuitive meaning of this definition becomes clearer if we observe that condition (1) is equivalent to the following condition:

(2) $|Z_1|+|Z_2| = |Z_1 \cup Z_2|$ for any set $Z_1 \subset A$ and $Z_2 \subset R^k - A$.

In other words, A is measurable if identity (2) holds for any disjoint sets separated by A.

To obtain (2) from (1) it suffices to take $Z = Z_1 \cup Z_2$. To deduce (1) from (2) it suffices to take $Z_1 = Z \cap A$ and $Z_2 = Z - A$.

In this and the next section we shall examine properties of Lebesgue-measurable sets. We shall show that the class of all Lebesgue-measurable sets is very large, that it is a σ-algebra of subsets of R^k and that the Lebesgue outer measure restricted to that class is a measure in the sense from § 3 (see Theorems 6.5, 6.6, and 7.4).

The symbol $\mathcal{M}_k$ will stand for the class of all Lebesgue-measurable subsets of R^k. In this and the next sections, "measurable" always means "Lebesgue-measurable". All sets under consideration are subsets of R^k unless the contrary is explicitly stated.

It follows from the subadditivity of outer measure that

$$|Z \cap A| + |Z - A| \geqslant |Z|$$

for all subsets A and Z of R^k because Z is the union of $Z \cap A$ and $Z - A$. Thus, in order to show that a set $A \subset R^k$ is measurable, it suffices to show that

(1′) $|Z \cap A| + |Z - A| \leqslant |Z|$ for every set $Z \subset R^k$.

The whole space R^k is the simplest example of a measurable set:

$$|Z \cap R^k| + |Z - R^k| = |Z| + |0| = |Z| \quad \text{for every } Z \subset R^k.$$

In a similar way we can prove that the empty set is measurable. More generally,

6.1. *Every set of measure zero is measurable.*

If $|A| = 0$, then $|Z \cap A| = 0$ by the first part of 5.7, and $|Z - A| = |Z|$ by 5.8. This proves (1) for any set Z.

Other examples of measurable sets will be given in the next section.

6.2. *The class $\mathcal{M}_k$ of Lebesgue-measurable sets is an algebra of subsets of R^k.*

If A is measurable, then for every set $Z \subset R^k$

$$|Z \cap (R^k - A)| + |Z - (R^k - A)| = |Z - A| + |Z \cap A| = |Z|$$

by (1) since $Z \cap (R^k - A) = Z - A$ and $Z - (R^k - A) = Z \cap A$. The identity just proved shows that the set $R^k - A$ also has property (1), i.e. it is measurable.

Suppose now that A and B are measurable, i.e.

$$|Z \cap A|+|Z-A| = |Z| \quad \text{for every set } Z$$

and

$$|Z' \cap B|+|Z'-B| = |Z'| \quad \text{for every set } Z'.$$

Assuming $Z' = Z \cap A$ or $Z-A$ we get

$$|(Z \cap A) \cap B|+|(Z \cap A)-B| = |Z \cap A|,$$

$$|(Z-A) \cap B|+|(Z-A)-B| = |Z-A|.$$

Hence

$$|Z \cap A \cap B|+|(Z \cap A)-B|+|(Z-A) \cap B|+|(Z-A)-B| = |Z|.$$

Since the set $Z \cap (A \cup B)$ is the union of the sets $Z \cap A \cap B$, $(Z \cap A)-B$ and $(Z-A) \cap B$, the sum of the first three terms is not greater than $|Z \cap (A \cup B)|$ by the subadditivity of the outer measure. Moreover $(Z-A)-B = Z-(A \cup B)$. Hence

$$|Z \cap (A \cup B)|+|Z-(A \cup B)| \leqslant |Z|.$$

The union $A \cup B$ has property (1'), i.e. it is a measurable set.

We have proved that the class $\mathcal{M}_k$ has properties (a) and (b) from § 2. Thus it is an algebra of subsets of R^k.

6.3. *If $A_1, \ldots, A_m$ are disjoint measurable sets, then for every set Z*

$$\sum_{n=1}^{m} |Z \cap A_n| = |Z \cap \bigcup_{n=1}^{m} A_n|.$$

The proof is by induction on m. The case of $m = 1$ is obvious.

Suppose the lemma is true for $m-1 \geqslant 1$. We shall prove that it is true for m.

Replacing A in (1) by the measurable set A_m, and Z by $Z \cap \bigcup_{n=1}^{m} A_n$, we get

$$|Z \cap A_m|+|Z \cap \bigcup_{n=1}^{m-1} A_n| = |Z \cap \bigcup_{n=1}^{m} A_n|.$$

By the induction hypothesis,

$$\sum_{n=1}^{m-1} |Z \cap A_n| = |Z \cap \bigcup_{n=1}^{m-1} A_n|.$$

The last two equations yield the identity 6.3.

6.4. *The union of an infinite sequence (A_n) of disjoint measurable sets is measurable and*

$$|Z \cap \bigcup_{n=1}^{\infty} A_n| = \sum_{n=1}^{\infty} |Z \cap A_n| \tag{3}$$

for every set Z.

The inequality

$$|Z \cap \bigcup_{n=1}^{\infty} A_n| \leqslant \sum_{n=1}^{\infty} |Z \cap A_n|$$

follows from the countable subadditivity of the outer measure since $Z \cap \bigcup_{n=1}^{\infty} A_n = \bigcup_{n=1}^{\infty} Z \cap A_n$. On the other hand, for every positive integer m

$$|Z \cap \bigcup_{n=1}^{\infty} A_n| \geqslant |Z \cap \bigcup_{n=1}^{m} A_n| = \sum_{n=1}^{m} |Z \cap A_n|$$

by 5.4 and 6.3. As $m \to \infty$ we get

$$|Z \cap \bigcup_{n=1}^{\infty} A_n| \geqslant \sum_{n=1}^{\infty} |Z \cap A_n|,$$

which completes the proof of (3).

Since the set $\bigcup_{n=1}^{m} A_n$ is measurable by 6.2, we have

$$|Z \cap \bigcup_{n=1}^{m} A_n| + |Z - \bigcup_{n=1}^{m} A_n| = |Z|.$$

Hence, by 6.3 and 5.4,

$$\sum_{n=1}^{m} |Z \cap A_n| + |Z - \bigcup_{n=1}^{\infty} A_n| \leqslant |Z|$$

for every positive integer m. As $m \to \infty$ we get

$$\sum_{n=1}^{\infty} |Z \cap A_n| + |Z - \bigcup_{n=1}^{\infty} A_n| \leqslant |Z|,$$

i.e., by (3),

$$|Z \cap \bigcup_{n=1}^{\infty} A_n| + |Z - \bigcup_{n=1}^{\infty} A_n| \leqslant |Z|.$$

The set $\bigcup_{n=1}^{\infty} A_n$ has property (1′), i.e. it is measurable.

6.5. *The class $\mathcal{M}_k$ of all Lebesgue-measurable subsets of R^k is a σ-algebra of subsets of R^k.*

If sets A_n $(n = 1, 2, \ldots)$ are measurable, then the sets

$$B_1 = A_1 \quad \text{and} \quad B_n = A_n - (A_1 \cup \ldots \cup A_n) \quad \text{for} \quad n = 2, 3, \ldots$$

are measurable (by 6.2) and disjoint (see I 2.5). Thus their union is measurable by 6.4. However $\bigcup_{n=1}^{\infty} B_n = \bigcup_{n=1}^{\infty} A_n$ (see I 2.5) which proves that the union of any sequence of measurable sets A_n is measurable. Thus $\mathcal{M}_k$ has property (b′) from § 2. It also has property (a) from § 3, by 6.2. This proves that $\mathcal{M}_k$ is a σ-algebra of sets.

The Lebesgue outer measure $|A|$ of a Lebesgue-measurable subset A of R^k is called the *Lebesgue measure of* A (more precisely, the *k-dimensional Lebesgue measure of* A). By the *k-dimensional Lebesgue measure* (or the *Lebesgue measure*, for short) we shall mean the extended real function on $\mathcal{M}_k$ which assigns to every $A \in \mathcal{M}_k$ its Lebesgue measure $|A|$.

We see that the difference between the Lebesgue outer measure and the Lebesgue measure is as follows. The Lebesgue outer is an extended real function defined on the class of all subsets of R^k. The Lebesgue measure is the same function but restricted to $\mathcal{M}_k$.

Assuming $Z = R^k$ in 6.4 we get

(4) $$|A_1 \cup A_2 \cup \ldots| = |A_1| + |A_2| + \ldots$$

for every sequence of disjoint sets $A_n \in \mathcal{M}_k$. It follows from (4), 6.5 and 5.1 that all the conditions in the definition of measure are satisfied, i.e.

6.6. *The k-dimensional Lebesgue measure is a measure on the σ-algebra $\mathcal{M}_k$ of all Lebesgue-measurable subsets of R^k.*

If $|A| = 0$, then A is measurable by 6.1. Consequently, instead of "the Lebesgue outer measure of A" we can say in this case "the Lebesgue measure of A", i.e. the adjective "outer" can be omitted. This was the reason for adopting the convention (see § 5, p. 243) that sets of outer measure zero be called sets of measure zero.

It follows from 6.1 and the monotonicity of the outer measure that every subset of a set of measure zero is measurable, i.e. that

6.7. *The k-dimensional Lebesgue measure is complete.*

By 6.6, the Lebesgue measure has all the properties of measures from § 3. We shall formulate them again using the notation $|A|$ for the measure.

By 3.4 the identity (4) holds for every sequence of measurable sets such that $|A_i \cap A_j| = 0$ for $i \neq j$.

By § 3 (1) the Lebesgue measure is finitely additive, i.e.

(5) $$|A_1 \cup \ldots \cup A_m| = |A_1| + \ldots + |A_m|$$

for any disjoint measurable sets $A_1, \ldots, A_m$.

By § 3 (3), for any measurable sets A, B,

(6) $$\text{if } A \subset B \text{ and } |A| < \infty, \text{ then } |B - A| = |B| - |A|.$$

By 3.5 and 3.6, for any sequence (A_n) of measurable sets,

(7) $$\text{if } A_n \subset A_{n+1} \text{ for } n = 1, 2, \ldots, \text{ then } \left|\bigcup_{n=1}^{\infty} A_n\right| = \lim_{n\to\infty} |A_n|;$$

(8) $$\text{if } A_n \supset A_{n+1} \text{ and } |A_n| < \infty \text{ for } n = 1, 2, \ldots, \text{ then } \left|\bigcap_{n=1}^{\infty} A_n\right| = \lim_{n\to\infty} |A_n|.$$

Other properties of the Lebesgue measure, such as finite and countable additivity, monotonicity etc., were stated earlier (see 5.5, 5.6, 5.4). We recall once more that, by 5.7 or 3.2, the union of a finite or infinite sequence of sets of Lebesgue measure zero is a set of Lebesgue measure zero.

In the next section we shall show that the class $\mathcal{M}_k$ of Lebesgue measurable sets is rather large. It contains all sets which usually appear in mathematical analysis.

EXERCISES

1. The set A defined in § 5, exercise 3, is not measurable.

Hint. If A is measurable, then all the sets A_w are measurable and $1 = |\langle 0; \infty)| = \infty \cdot |A|$.

2. Show that the one-dimensional Lebesgue outer measure is not an additive set function, i.e. that there are disjoint sets of real numbers A, B such that $|A \cup B| \neq |A| + |B|$.

Hint. Use the sets A_w from § 5, exercise 3.

§ 7. A characterization of Lebesgue-measurable sets. We recall that if A, B are non-empty subsets of R^k, then (see III § 1)

$$\operatorname{dist}(A, B) = \inf\{|p-q|: (p \in A) \,\&\, (q \in B)\}.$$

7.1. *For any subsets A, B of R^k, if* $\operatorname{dist}(A, B) > 0$, *then*

$$|A \cup B| = |A| + |B|.$$

Let $\eta = \operatorname{dist}(A, B) > 0$. For any given $\varepsilon > 0$ there exists a sequence (I_n) of closed intervals such that

$$A \cup B \subset \bigcup_{n=1}^{\infty} I_n \quad \text{and} \quad \sum_{n=1}^{\infty} \operatorname{vol} I_n \leqslant |A \cup B| + \varepsilon \tag{1}$$

(see § 5 (4)). We may additionally suppose that

$$\operatorname{dia} I_n < \eta \quad \text{for} \quad n = 1, 2, \ldots$$

For if the diameter of an interval I_n is greater than η, we can partition I_n into a finite number of closed intervals with diameters $< \eta$, and replace I_n by all the intervals of this partition. The sum of the volumes remains the same by 4.2.

The hypothesis that $\operatorname{dia} I_n < \eta$ for all n implies that no interval I_n has non-empty intersections with both A and B, i.e. that the sets

$$U = \{n: I_n \cap A \neq 0\}, \quad V = \{n: I_n \cap B \neq 0\}$$

are disjoint. This implies that

$$\sum_{n \in U} \operatorname{vol} I_n + \sum_{n \in V} \operatorname{vol} I_n \leqslant \sum_{n=1}^{\infty} \operatorname{vol} I_n$$

since every term on the right-hand side is non-negative and appears at most once on the left-hand side. Since the sequence (I_n) is a covering of $A \cup B$, we have

$$A \subset \bigcup_{n \in U} I_n \quad \text{and} \quad B \subset \bigcup_{n \in V} I_n .$$

Hence

$$|A| \leqslant \sum_{n \in U} \operatorname{vol} I_n , \quad |B| \leqslant \sum_{n \in V} \operatorname{vol} I_n$$

(see § 5 (3)). It follows from the above inequalities that

$$|A|+|B| \leqslant |A \cup B|+\varepsilon \quad \text{for every } \varepsilon > 0 .$$

As $\varepsilon \to 0$ we get $|A|+|B| \leqslant |A \cup B|$. The converse inequality is also true since the outer measure is subadditive. This proves 7.1.

7.2. *Every interval (closed, open or half-open) is a measurable set.*

Let I be a closed interval and let $\varepsilon > 0$ be arbitrary. Let I' be a closed interval such that

(2) $$I' \subset \operatorname{int} I \quad \text{and} \quad \operatorname{vol} I - \operatorname{vol} I' < \varepsilon .$$

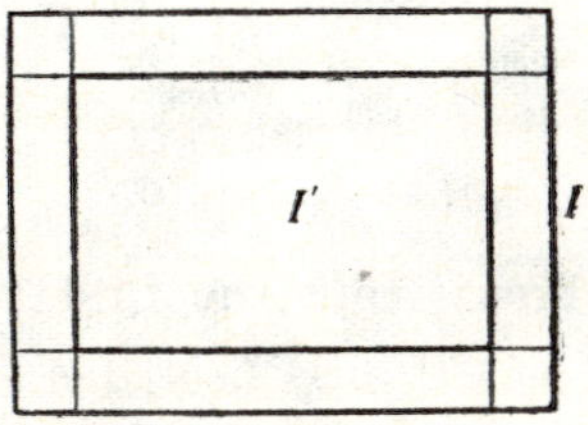

Fig. 10

The set $I - \operatorname{int} I'$ can be represented as the union of a finite number o non-overlapping intervals $I_1, \ldots, I_m$ (figure 10 illustrates the two-di mensional case). All the intervals $I', I_1, \ldots, I_m$ form a partition of I Hence, by 4.2,

$$\operatorname{vol} I = \operatorname{vol} I' + \operatorname{vol} I_1 + \ldots + \operatorname{vol} I_m .$$

Consequently, by the subadditivity of the outer measure,

$$|I - I'| \leqslant |I_1| + \ldots + |I_m| = \operatorname{vol} I_1 + \ldots + \operatorname{vol} I_m = \operatorname{vol} I - \operatorname{vol} I' < \varepsilon .$$

Let Z be a subset of R^k. We have

(3) $$|Z \cap I'| + |Z - I| = |(Z \cap I') \cup (Z - I)| \leqslant |Z| .$$

The inequality in (3) follows from the monotonicity of the outer measure. The equality in (3) is obvious if one of the sets $Z \cap I'$, $Z - I$ is empty. It follows from 7.1 if both sets are not empty, since the sets have a positive distance because they are subsets of the sets I' and $R^k - I$ whose distance is positive (see the first of conditions (2)).

On the other hand, it follows from the subadditivity and monotonicity of the outer measure that

$$|Z \cap I| \leqslant |Z \cap I'| + |Z \cap (I - I')| \leqslant |Z \cap I'| + |I - I'| \leqslant |Z \cap I'| + \varepsilon .$$

This and (3) imply that

$$|Z \cap I| + |Z - I| \leqslant |Z| + \varepsilon .$$

Since $\varepsilon > 0$ is arbitrary, we get

$$|Z \cap I| + |Z - I| \leqslant |Z|$$

for any set $Z \subset R^k$. This proves that I is measurable (see § 6 (1')).

Every open (or half-open) interval is of the form $I - B$ where I is a closed interval and B is the union (or a subset of the union) of the faces of I. By 5.3 and 5.10 the set B has measure zero. Since I and B are measurable, the set $I - B$ is also measurable.

7.3. *Every Lebesgue-measurable set A is the union of an increasing sequence (A_n) of measurable sets of finite measure. Thus the Lebesgue measure is σ-finite.*

It suffices to take $A_n = A \cap I_n$ where I_n is the interval

$$I_n = \{(x_1, \ldots, x_k) \colon |x_i| \leqslant n \text{ for } i = 1, \ldots, k\} .$$

The second part of 7.3 follows from the first (the case where $A = R^k$).

7.4. *All Borel subsets of R^k are Lebesgue-measurable. In particular, all open sets, all closed sets, all F_σ-sets and all G_δ-sets are measurable.*

By III 5.3, every open set is the union of a sequence of open rational intervals and therefore is measurable by 6.5 and 7.2. The class of all Borel subsets of R^k is the smallest σ-algebra containing all open sets. The class $\mathcal{M}_k$ of all measurable sets is a σ-algebra containing all open sets. Hence it follows that every Borel subset belongs to $\mathcal{M}_k$.

7.5. *The following conditions are equivalent for every set $A \subset R^k$:*

(a) *A is Lebesgue-measurable;*

(b) *for every $\varepsilon > 0$ there exists an open set G such that $A \subset G$ and $|G - A| < \varepsilon$;*

(c) *$A = H - B$ where H is a G_δ-set and $|B| = 0$;*

(d) *for every $\varepsilon > 0$ there exists a closed set F such that $F \subset A$ and $|A - F| < \varepsilon$;*

e) *$A = K \cup B$ where K is an F_σ-set and $|B| = 0$.*

First we shall prove that conditions (a), (b), (c) are equivalent.

(a) $\Rightarrow$ (b). Suppose that A is measurable and $\varepsilon > 0$.

Consider first the case where $|A| < \infty$. Let (I_n) be a sequence of closed intervals such that (see § 5 (4))

$$A \subset \bigcup_{n=1}^{\infty} I_n \quad \text{and} \quad \sum_{n=1}^{\infty} \operatorname{vol} I_n \leqslant |A| + \varepsilon/2 .$$

Let J_n be an open interval such that

$$I_n \subset J_n \quad \text{and} \quad |J_n| < \operatorname{vol} I_n + \varepsilon/2^{n+1} .$$

The set $G = J_1 \cup J_2 \cup \ldots$ is open and $A \subset G$. Moreover

$$|G| < |A| + \varepsilon$$

since

$$|G| \leqslant \sum_{n=1}^{\infty} |J_n| < \sum_{n=1}^{\infty} \operatorname{vol} I_n + \varepsilon/2 \leqslant |A| + \varepsilon .$$

Since the sets A and G are measurable and $A \subset G$, we have by § 6 (6)

$$|G - A| = |G| - |A| < \varepsilon .$$

The set G has all the properties mentioned in (b).

Now let A be any measurable set. The set A is the union of a sequence (A_n) of measurable sets of finite measure (see 7.3). By the part of the theorem just proved, each of the sets A_n has the property (b). Thus there exists an open set G_n such that

$$A_n \subset G_n \quad \text{and} \quad |G_n - A_n| < \varepsilon/2^n .$$

The set $G = G_1 \cup G_2 \cup \ldots$ is open and $A \subset G$. Since

$$G - A \subset \bigcup_{n=1}^{\infty} (G_n - A_n) ,$$

we have

$$|G - A| \leqslant \sum_{n=1}^{\infty} |G_n - A_n| < \sum_{n=1}^{\infty} \varepsilon/2^n = \varepsilon .$$

The set G has all the properties mentioned in (b).

(b) $\Rightarrow$ (c). Suppose A has property (b). Let G_n be an open set such that

$$A \subset G_n \quad \text{and} \quad |G_n - A| < \frac{1}{n} \quad (n = 1, 2, \ldots) .$$

The G_δ-set $H = G_1 \cap G_2 \cap \ldots$ and the set $B = H - A$ have the properties mentioned in (c): $A = H - B$ and $|B| = 0$ since

$$0 \leqslant |B| \leqslant |G_n - A| < 1/n \quad \text{for} \quad n = 1, 2, \ldots$$

(c) ⇒ (a). In fact, A is measurable since it is the difference of two measurable sets.

Now we shall prove the equivalence of (a), (d) and (e).

(a) ⇒ (d). If A is measurable, then $R^k - A$ is measurable, consequently it has the property (b), i.e. there exists an open set G such that

$$R^k - A \subset G \quad \text{and} \quad |G - (R^k - A)| < \varepsilon .$$

The closed set $F = R^k - G$ has all the properties mentioned in (d) since $F \subset A$ and $A - F = G - (R^k - A)$.

(d) ⇒ (e). If A has property (b), then there exists a closed set F_n such that

$$F_n \subset A \quad \text{and} \quad |A - F_n| < 1/n$$

($n = 1, 2, \ldots$). The F_σ-set $K = F_1 \cup F_2 \cup \ldots$ and the set $B = A - K$ have the properties mentioned in (e): $A = K \cup B$ and $|B| = 0$ since

$$0 \leqslant |B| \leqslant |A - F_n| < 1/n \quad \text{for} \quad n = 1, 2, \ldots$$

(e) ⇒ (a). Indeed, A is measurable since it is the union of two measurable sets.

EXERCISES

1. Let (s_n) be a sequence of positive numbers such that $\sum_{n=1}^{\infty} 2^{n-1} s_n \leqslant 1$. Remove from the interval $\langle 0; 1\rangle$ the open interval with center $\frac{1}{2}$ and length s_1. The remaining set is the union of two closed disjoint intervals. Denote the left one by I_0 and the right one by I_1. Suppose all the intervals $I_{a_1,\ldots,a_n}$ were defined for any sequence $(a_1, \ldots, a_n)$ composed of the numbers 0 and 1 only. Remove from $I_{a_1,\ldots,a_n}$ the open interval whose center is the center of $I_{a_1,\ldots,a_n}$ and whose length is equal to s_{n+1}. The remaining set is the union of two disjoint closed intervals. Denote the left one by $I_{a_1,\ldots,a_n,0}$ and the right one by $I_{a_1,\ldots,a_n,1}$.

Let D_n denote the union of all the intervals $I_{a_1,\ldots,a_n}$ and let $C = D_1 \cap D_2 \cap \ldots$ Prove that the closed set C does not contain any interval, $|D_n| = 1 - (s_1 + 2s_2 + \ldots + 2^n s_n)$, $|C| = 1 - \sum_{n=1}^{\infty} 2^{n-1} s_n$. If $s_n = 3^{-n}$, then C is identical with the Cantor set defined in I § 7, exercise 2. In the general case the set C defined above is homeomorphic with the Cantor set. Namely the function f which assigns the number $\sum_{n=1}^{\infty} \frac{2a_n}{3^n}$ to the point $x \in I_{a_1} \cap I_{a_1,a_2} \cap I_{a_1,a_2,a_3} \ldots$ is a homeomorphism from C onto the Cantor set.

2. Construct a measurable set $A \subset I$ such that the sets $I \cap A$ and $I - A$ have positive measure for every non-empty interval I.

Hint. It follows from exercise 1 that every interval I contains a closed set C such that C does not contain any interval and the measure of C is equal to any given non-negative number less than $|I|$.

3. If an open set $G \subset R^k$ is the union of a sequence of non-overlapping intervals I_n, $n = 1, 2, \ldots$, then $|G| = |I_1| + |I_2| + \ldots$

4. For every (measurable or non-measurable) set $A \subset R^k$ and for every $\varepsilon > 0$ there exists an open set B such that $A \subset B$ and $|B| \leqslant |A| + \varepsilon$.

5. For every (measurable or non-measurable) set $A \subset R^k$ there exists a G_δ-set B such that $A \subset B$ and $|A| = |B|$.

6. For every set $A \subset R^k$ there exists a G_δ-set B such that $A \subset B$ and $|A \cap I| = |B \cap I|$ for every interval I.

CHAPTER VII

GENERAL THEORY OF THE INTEGRAL

§ 1. Introduction. In this chapter we shall consider a fixed measure μ defined on a σ-algebra $\mathcal{M}$ of subsets of a space X.

The purpose of this chapter is to formulate a definition and to prove fundamental properties of the integral of a real function f defined on a set $A \subset X$ with respect to the measure μ. As we shall see in the following sections, it is essential for the theory of the integral that certain sets determined by these functions, e.g. sets of the form

$$\{p: f(p) > g(p)\}, \quad \{p: a < f(p) < b\}, \quad \text{etc.}$$

have to be measurable. For this reason we have to restrict our investigation to a class of functions called measurable functions which have the required properties. In § 2 we shall formulate the definition and fundamental properties of measurable functions. In § 3 we shall investigate a special kind of measurable functions, the simple functions. As their name indicates, they are really the simplest among all measurable functions. Simple functions will be useful in the formulation of the definition of the integral. §§ 5 and 6 are devoted to a presentation of the theory of the integral with respect to an arbitrary measure μ.

An examination of properties of the integral with respect to the Lebesgue measure in R^k will be the subject of Chapter VIII. In particular, we shall prove there that, in the case of $k = 1$ and for a continuous function under the sign of the integral, the notion of the integral introduced in this chapter coincides with the notion of the integral defined by means of the primitive function, i.e. with the integral considered as an operation converse to differentiation. We shall also give practical methods of calculating integrals of functions defined on rather simple subsets of a Euclidean space. In Chapter IX we shall develop a theory of integration with respect to the r-dimensional measure in R^k, $r \leqslant k$.

As we stated at the beginning, in what follows μ is a fixed measure on a σ-algebra $\mathcal{M}$ of subsets of a space X. By a convention from VI § 3, by a *measurable set* we shall always mean any set in $\mathcal{M}$.

All functions in this chapter are extended real functions; in other words, we admit ∞ and $-\infty$ as values. We say that a sequence (f_n) of extended real functions defined on a set A *converges* (or *is convergent*) if $\lim_{n\to\infty} f_n(p)$ exists for every $p \in A$ (the limit can be finite or infinite).

§ 2. Measurable functions. First we shall prove the following lemma.

2.1. *Let f be an extended real function defined on a measurable set A. The following conditions are equivalent:*

(a) *the set $\{p: f(p) < a\}$ is measurable for every real number a;*

(b) *the set $\{p: f(p) \leqslant a\}$ is measurable for every real number a;*

(c) *the set $\{p: f(p) > a\}$ is measurable for every real number a;*

(d) *the set $\{p: f(p) \geqslant a\}$ is measurable for every real number a.*

The proof is based on the fact that finite or countable set-theoretical operations performed on measurable sets yield measurable sets.

If f has property (a), it has property (b) since

$$\{p: f(p) \leqslant a\} = \bigcap_{m=1}^{\infty} \left\{p: f(p) < a + \frac{1}{m}\right\} \in \mathcal{M}.$$

If f has property (b), it has property (c) since

$$\{p: f(p) > a\} = A - \{p: f(p) \leqslant a\} \in \mathcal{M}.$$

If f has property (c), it has property (d) since

$$\{p: f(p) \geqslant a\} = \bigcap_{m=1}^{\infty} \left\{p: f(p) > a - \frac{1}{m}\right\} \in \mathcal{M}.$$

If f has property (d), it has property (a) since

$$\{p: f(p) < a\} = A - \{p: f(a) \geqslant a\} \in \mathcal{M}.$$

It follows from 2.1 that if an extended real function has one of the properties (a), (b), (c), (d), then it has all the remaining properties. We say then that the function f is *measurable* (more precisely, *μ-measurable* or *measurable with respect to the measure μ*). By the definition, the domain of any measurable function is always a measurable set.

The constant function

$$f(p) = c \quad \text{for} \quad p \in A$$

is the simplest example of a measurable function. Indeed,

$$\{p: f(p) < a\} = \begin{cases} 0 & \text{if} \quad a \leqslant c, \\ A & \text{if} \quad a > c, \end{cases}$$

i.e. f has property (a).

In what follows in this section we shall always assume that all given functions are defined on the same measurable set A.

2.2. *If f and g are measurable functions, then the sets*

$$\{p: f(p) < g(p)\}, \quad \{p: f(p) \leqslant g(p)\}, \quad \{p: f(p) = g(p)\}$$

are measurable.

Let Q be the set of all rational numbers. We have

$$\{p: f(p) < g(p)\} = \bigcup_{b \in Q} \big(\{p: f(p) < b\} \cap \{p: g(p) > b\}\big) \in \mathcal{M}$$

since Q is countable. It follows from the part of 2.2 just proved that

$$\{p: f(p) \leqslant g(p)\} = A - \{p: g(p) < f(p)\} \in \mathcal{M},$$

and consequently

$$\{p: f(p) = g(p)\} = \{p: f(p) \leqslant g(p)\} \cap \{p: g(p) \leqslant f(p)\} \in \mathcal{M}.$$

2.3. *If f is a measurable function, then for every finite or infinite number a the sets*

$$\{p: f(p) = a\}, \quad \{p: f(p) \neq a\}$$

are measurable.

The measurability of the first set follows from 2.2 where $g = a$. The second set is the difference between A and the first set. Therefore t is also measurable.

2.4. *If f is a measurable function on A, and B is a measurable subset of A, then the function $g = f|B$ is measurable.*

In fact,

$$\{p: g(p) < a\} = B \cap \{p: f(p) < a\} \in \mathcal{M}.$$

2.5. *If f is a function defined on $A = A_1 \cup A_2 \cup \ldots$ and all the functions $g_1 = f|A_1$, $g_2 = f|A_2$, ... are measurable, then f is measurable.*

Indeed,

$$\{p: f(p) < a\} = \{p: g_1(p) < a\} \cup \{p: g_2(p) < a\} \cup \ldots \in \mathcal{M}.$$

2.6. *If f is a measurable function on A and B is a measurable subset of A, then the function g defined on A by*

$$g(p) = \begin{cases} f(p) & \textit{for} \quad p \in B, \\ 0 & \textit{for} \quad p \in A - B \end{cases}$$

is measurable.

In fact, the function $g|B = f|B$ is measurable by 2.4, and the constant function $g|A-B = 0$ is measurable. Thus 2.6 follows from 2.5.

We recall the convention from I § 4:

$$0 \cdot (\pm\infty) = 0.$$

2.7. *If f is a measurable function and $c \in R$, then cf is measurable. If f and g are measurable functions, then $f+g$ and $f-g$ are also measurable.*

The first part of 2.7 is obvious for $c = 0$ since cf is then constant, $cf = 0$. If $c > 0$, then

$$\{p: cf(p) < a\} = \{p: f(p) < a/c\} \in \mathcal{M} .$$

If $c < 0$, then

$$\{p: cf(p) < a\} = \{p: f(p) > a/c\} \in \mathcal{M} .$$

This proves that cf is measurable.

The proof of the second part of 2.7 will be preceded by the following remark: if g is measurable, then $g+c$ is measurable. In fact,

$$\{p: g(p)+c < a\} = \{p: g(p) < a-c\} \in \mathcal{M} .$$

If f and g are measurable, then $f-g$ is measurable since

$$\{p: f(p)-g(p) < a\} = \{p: f(p) < g(p)+a\} \in \mathcal{M}$$

by the first part of 2.2.

If f and g are measurable, then the function $f+g = f-(-1)g$ is measurable by the parts of 2.7 just proved.

2.8. *Any linear combination $a_1f_1+\ldots+a_mf_m$ of measurable functions $f_1, \ldots, f_m$ is measurable.*

This follows directly from 2.7.

2.9. *The product of two measurable functions is measurable.*

First we observe that if f is measurable then its square f^2 is also measurable. This follows from the identity

$$\{p: f(p)^2 < a\} = \begin{cases} 0 & \text{if} \quad a \leqslant 0 , \\ \{p: f(p) < \sqrt{a}\} \cap \{p: f(p) > -\sqrt{a}\} \in \mathcal{M} & \text{if} \quad a > 0 . \end{cases}$$

Suppose now that f and g are finite measurable functions. The functions $(f+g)^2$ and $(f-g)^2$ are measurable. Consequently the function $fg = \frac{1}{4}\big((f+g)^2-(f-g)^2\big)$ is measurable.

Suppose f or g assumes infinite values. The set

$$B = \{p: f(p) = \infty\} \cup \{p: f(p) = -\infty\} \cup \\ \cup \{p: g(p) = \infty\} \cup \{p: g(p) = -\infty\}$$

is measurable.

The function $fg|A-B$ is measurable since it is the product of finite measurable functions $f|A-B$ and $g|A-B$. The set B is the union of a finite number of measurable sets $B_1, B_2, \ldots$ on which one of the functions f, g

has a constant infinite value and the second is positive or negative or equal to 0:

$$B_1 = \{p\colon f(p) = \infty\} \cap \{p\colon g(p) > 0\}\,,$$
$$B_2 = \{p\colon f(p) = \infty\} \cap \{p\colon g(p) < 0\}\,,$$
$$B_3 = \{p\colon f(p) = \infty\} \cap \{p\colon g(p) = 0\}\,,$$
$$\dots\dots\dots\dots\dots\dots\dots\dots$$

Each of the functions $fg|B_i$ is constant (equal to ∞ or $-\infty$ or 0) and therefore measurable. Since $A = (A-B) \cup B_1 \cup B_2 \cup \dots$, the function fg is measurable by 2.5.

The following notation will often be used in Chapters VII and VIII: if f is an extended real function, then

(1) $$f^+ = \max(f, 0) \quad \text{and} \quad f^- = \max(-f, 0)\,.$$

In other symbols,

(2) $$f^+(p) = \max\big(f(p), 0\big)\,, \quad f^-(p) = \max\big(-f(p), 0\big)\,.$$

The functions f^+ and f^- are called the *non-negative part* of f and the *non-positive part* of f, respectively. Observe that f^+ and f^- are non-negative, are defined on the same set as f, and

(3) $$f = f^+ - f^-\,, \quad |f| = f^+ + f^-\,.$$

2.10. *If f and g are measurable functions, then* $\max(f, g)$ *and* $\min(f, g)$ *are measurable functions.*

In particular, if f is a measurable function, then f^+ and f^- are measurable and, consequently, $|f|$ is measurable.

Indeed,

$$\{p\colon \max\big(f(p), g(p)\big) < a\} = \{p\colon f(p) < a\} \cap \{p\colon g(p) < a\} \in \mathcal{M}\,,$$
$$\{p\colon \min\big(f(p), g(p)\big) < a\} = \{p\colon f(p) < a\} \cup \{p\colon g(p) < a\} \in \mathcal{M}\,.$$

2.11. *If all the functions f_n $(n = 1, 2, \dots)$ are measurable, then the functions* $\sup\limits_n f_n$ *and* $\inf\limits_n f_n$ *are measurable.*

Indeed,

$$\{p\colon \sup_n f_n(p) > a\} = \bigcup_{n=1}^{\infty} \{p\colon f_n(p) > a\} \in \mathcal{M}\,,$$

$$\{p\colon \inf_n f_n(p) < a\} = \bigcup_{n=1}^{\infty} \{p\colon f_n(p) < a\} \in \mathcal{M}\,.$$

2.12. *If all the functions f_n $(n = 1, 2, \dots)$ are measurable, then the functions* $\limsup\limits_{n\to\infty} f_n$ *and* $\liminf\limits_{n\to\infty} f_n$ *are measurable.*

We have (see I § 4)

$$\limsup_{n\to\infty} f_n = \inf_n g_n \quad \text{where} \quad g_n = \sup_m f_{n+m} .$$

By 2.11, all the functions g_n are measurable and, consequently, $\limsup\limits_{n\to\infty} f_n$ is measurable. Since

$$\liminf_{n\to\infty} f_n = \sup_n h_n \quad \text{where} \quad h_n = \inf_m f_{n+m} ,$$

the function $\liminf\limits_{n\to\infty} f_n$ is measurable by the same argument.

2.13. *If all the functions f_n $(n = 1, 2, \ldots)$ are measurable, then the set B of all points p for which $\lim\limits_{n\to\infty} f_n(p)$ exists is measurable, and the function $\lim\limits_{n\to\infty} f_n$ is measurable.*

The measurability of the set B follows from 2.12, 2.2 and the identity

$$B = \{p\colon \limsup_{n\to\infty} f_n(p) = \liminf_{n\to\infty} f_n(p)\} .$$

The function $\lim\limits_{n\to\infty} f_n$ is defined on the measurable set B and is the restriction of the measurable function $\limsup\limits_{n\to\infty} f_n$ (or $\liminf\limits_{n\to\infty} f_n$) to the set B. Thus it is measurable by 2.4.

EXERCISES

1. Let $\mathcal{F}$ be a non-empty set of continuous real functions defined on a metric space X and let

$$g(p) = \sup\{f(p)\colon f \in \mathcal{F}\} , \qquad h(p) = \inf\{f(p)\colon f \in \mathcal{F}\} .$$

Prove that, for every number a, the sets

$$\{p\colon g(p) > a\} , \qquad \{p\colon h(p) < a\}$$

are open.

2. Let f be a real function defined on a metric space X, such that the set $\{p\colon f(p) > a\}$ is open for every number a. Prove that the set $\{p\colon f(p) < a\}$ is an F_σ-set for every number a.

3. Prove the following *Egoroff theorem*: If $\mu(A) < \infty$ and (f_n) is a sequence of finite measurable functions convergent on A to a finite measurable function f, then for every $\varepsilon > 0$ there exists a measurable set $B \subset A$ such that $\mu(A-B) < \varepsilon$ and $f_n|B \to f|B$ uniformly.

Hint. Let

$$B_{m,n} = \{p\colon \bigwedge_{j\geqslant n} |f_j(p) - f(p)| < 1/m\} .$$

Since $A = B_{m,1} \cup B_{m,2} \cup \ldots$, there exists an integer n_m such that $\mu(A - B_{m,n_m}) < \varepsilon/2^m$. Take $B = \bigcap\limits_{m=1}^{\infty} B_{m,n_m}$.

§ 3. Simple functions. Let B be a subset of a set A. The symbol $\chi_{B,A}$ will always denote the real function defined on A by

$$\chi_{B,A}(p) = \begin{cases} 1 & \text{if} \quad p \in B, \\ 0 & \text{if} \quad p \in A-B. \end{cases}$$

$\chi_{B,A}$ is called the *characteristic function of the subset B of A*. If the sets A and B are measurable, then the characteristic function $\chi_{B,A}$ is measurable since

$$\{p\colon \chi_{B,A}(p) > a\} = \begin{cases} 0 & \text{if} \quad a \geqslant 1, \\ B & \text{if} \quad 0 \leqslant a < 1, \\ A & \text{if} \quad a < 0. \end{cases}$$

If A is the whole space, $\chi_{B,A}$ is often denoted by χ_B.

By a *simple function* we shall mean any non-negative finite measurable function f assuming only a finite number of values. Since f is measurable, f is defined on a measurable set A.

For instance, if $A_1, \ldots, A_m$ are measurable subsets of a measurable set A and $a_1, \ldots, a_m$ are non-negative finite real numbers, then the formula

$$f = \sum_{i=1}^{m} a_i \chi_{A_i, A} \tag{1}$$

defines a simple function on A. In fact, the non-negative finite measurable (see 2.8) function f assumes only a finite number of values since only the numbers $a_{i_1} + \ldots + a_{i_j}$ $(1 \leqslant i_1 < i_2 < \ldots < i_j \leqslant m)$ and the number 0 can be values of f.

In particular, if the measurable sets $A_1, \ldots, A_m$ are disjoint and non-empty, and $A = A_1 \cup \ldots \cup A_m$, then the numbers $a_1, \ldots, a_m$ are the only values of f.

The function f can then be defined as follows:

$$f(p) = a_i \quad \text{for} \quad p \in A_i, \quad i = 1, \ldots, m.$$

Conversely, every simple function f defined on a non-empty measurable set A can be represented in the form (1) such that

(c_1) the sets $A_1, \ldots, A_m$ are measurable, disjoint and non-empty;

(c_2) $A = A_1 \cup \ldots \cup A_m$;

(c_3) $0 \leqslant a_1 < a_2 < \ldots < a_m$.

In fact, it is enough to take as $a_1, \ldots, a_m$ all values of f arranged as in (c_3) and to define

$$A_i = \{p\colon f(p) = a_i\} \quad \text{for} \quad i = 1, \ldots, m. \tag{2}$$

The sets A_i are measurable by 2.3.

Conversely, if the representation (1) of f satisfies conditions (c_1), (c_2), (c_3), then $a_1, \ldots, a_m$ must be the sequence of all values of f and (2) holds. This proves that every simple function can be represented in the form (1) with properties (c_1), (c_2), (c_3), in only one way. This unique representation (1) satisfying (c_1), (c_2), (c_3) will be called the *canonical representation* of the simple function f.

3.1. *Every non-negative measurable function f is the limit of a non-decreasing sequence of simple functions.*

Let

$$(3)\qquad f_n(p) = \begin{cases} \dfrac{m}{2^n} & \text{if} \quad \dfrac{m}{2^n} \leqslant f(p) < \dfrac{m+1}{2^n}, \quad m = 0, 1, \ldots, n\cdot 2^n - 1, \\ n & \text{if} \quad f(p) \geqslant n. \end{cases}$$

In other words,

$$f_n = \sum_{m=0}^{n\cdot 2^n - 1} \frac{m}{2^n}\cdot \chi_{A_{m,n}, A} + n\chi_{A_n, A}$$

where

$$A_{m,n} = \left\{p\colon \frac{m}{2^n} \leqslant f(p) < \frac{m+1}{2^n}\right\} = \left\{p\colon f(p) \geqslant \frac{m}{2^n}\right\} \cap \left\{p\colon f(p) < \frac{m+1}{2^n}\right\}$$

and

$$A_n = \{p\colon f(p) \geqslant n\}.$$

Since the sets $A_{m,n}$ and A_n are measurable, the function f_n is simple.

It follows from the definition that $f_n(p)$ is the greatest among the numbers

$$(4)\qquad 0, \frac{1}{2^n}, \frac{2}{2^n}, \ldots, \frac{n\cdot 2^n}{2^n}$$

which are $\leqslant f(p)$. Similarly, $f_{n+1}(p)$ is the greatest among the numbers

$$(5)\qquad 0, \frac{1}{2^{n+1}}, \frac{2}{2^{n+1}}, \ldots, \frac{(n+1)\cdot 2^{n+1}}{2^{n+1}}$$

which are $\leqslant f(p)$. Since all the numbers (4) appear in (5), we have $f_{n+1}(p) \geqslant f_n(p)$, i.e. the sequence (f_n) is non-decreasing.

If $f(p) < \infty$, then the value of $f_n(p)$ for $n > f(p)$ is calculated by the first part of the definition (3). Hence it follows that

$$0 \leqslant f(p) - f_n(p) < \frac{1}{2^n} \quad \text{for} \quad n > f(p)$$

which proves that $f_n(p) \to f(p)$. If $f(p) = \infty$, then $f(p) = n$ for $n = 1, 2, \ldots$ by the last part of the definition (3). Thus also $f_n(p) \to f(p)$. This proves that $f_n \to f$.

The following lemma on sequences of extended real functions defined on the same set A will be useful in § 6.

3.2. *If $f_n \nearrow f$ as $n \to \infty$ and, for every fixed n,*

$$f_{n,m} \nearrow f_n \quad \text{as} \quad m \to \infty,$$

then the sequence

$$g_n = \max(f_{1,n}, \dots, f_{n,n})$$

has the properties

$$g_n \leqslant f_n \quad \text{and} \quad g_n \nearrow f.$$

If the functions $f_{n,m}$ are simple, then the functions g_n are simple.

Since $f_{1,n} \leqslant f_{1,n+1}, \dots, f_{n,n} \leqslant f_{n,n+1}$, we have $g_n \leqslant g_{n+1}$. Since $f_{1,n} \leqslant f_1, \dots, f_{n,n} \leqslant f_n$ and $f_1 \leqslant f_2 \leqslant \dots \leqslant f_n$, we have $g_n \leqslant f_n$. Consequently $g_n \leqslant f$ since $f_n \leqslant f$. On the other hand, if $n \geqslant \max(n_1, m_1)$, then $g_n \geqslant f_{n_1, m_1}$ since

$$g_n = \max(f_{1,n}, \dots, f_{n_1,n}, \dots, f_{n,n}) \geqslant f_{n_1,n} \geqslant f_{n_1,m_1}.$$

Let p be any point in A and let c be any real number such that $c < f(p)$. There is a positive integer n_1 such that $f_{n_1}(p) > c$. There is also an integer m_1 such that $f_{n_1,m_1}(p) > c$. This implies that $c < g_n(p) \leqslant f(p)$ for every $n \geqslant \max(n_1, m_1)$. Thus $f(p) = \lim_{n\to\infty} g_n(p)$. The first part of 3.2 is proved. The second part follows from 2.10.

EXERCISES

1. If the function f in 3.1 is bounded, then the sequence of functions (3) converges to f uniformly.

2. Every measurable function is the limit of a sequence of finite measurable functions, each of which assumes only a finite number of values.

Every bounded measurable function is the uniform limit of a sequence of measurable functions, each of which assumes only a finite number of values.

3. Verify that, for every sequence (A_n) of subsets of a space X,

$$A = \limsup_{n\to\infty} A_n \text{ if and only if } \chi_A = \limsup_{n\to\infty} \chi_{A_n};$$

$$A = \liminf_{n\to\infty} A_n \text{ if and only if } \chi_A = \liminf_{n\to\infty} \chi_{A_n};$$

$$A = \lim_{n\to\infty} A_n \text{ if and only if } \chi_A = \lim_{n\to\infty} \chi_{A_n}.$$

§ 4. The integral of a simple function. We recall that, in addition to the natural definitions of algebraic operations on infinite numbers, we assumed in I § 4 the following convention:

$$0 \cdot \infty = 0 = \infty \cdot 0. \tag{1}$$

Let f be a simple function defined on a measurable set A. Let

$$(2) \qquad f = \sum_{i=1}^{m} a_i \chi_{A_i, A}$$

be the unique canonical representation of f. In other words, the sets $A_1, \ldots, A_m$ and the numbers $a_1, \ldots, a_m$ satisfy conditions (c_1), (c_2), (c_3) from § 3.

By the *integral* of the simple function (2) on the set A with respect to the measure μ we mean the non-negative, finite or infinite number

$$(3) \qquad \int_A f = \sum_{i=1}^{m} a_i \mu(A_i)\,.$$

Since the canonical representation (2) is uniquely determined by f, the number (3) is also uniquely determined by f.

For instance, the integral of the function identically equal to 0 is equal to 0:

$$(4) \qquad \int_A 0 = 0\,.$$

More generally, for every constant function $f = a$ on A

$$(5) \qquad \int_A a = a\mu(A)$$

since $f = a\chi_{A,A}$ is its canonical representation ($m = 1$).

For any measurable set $B \subset A$

$$(6) \qquad \int_A \chi_{B,A} = \mu(B)\,.$$

In fact, the identity

$$\chi_{B,A} = 0 \cdot \chi_{A-B,A} + 1 \cdot \chi_{B,A}$$

is the canonical representation of $\chi_{B,A}$ under the hypothesis that both the sets $B - A$ and B are non-empty. This implies (6). If $B = 0$, then (6) follows from (4). If $A - B = 0$, i.e. $B = A$, then (6) follows from (5).

Observe that for every simple function (2),

$$(7) \qquad \text{if } \mu(A) = 0, \text{ then } \int_A f = 0$$

by (3) and (1) since then $\mu(A_i) = 0$ for $i = 1, \ldots, m$.

We used the canonical representation of f in the definition of the integral (3) only in order to assure the uniqueness of the integral. Equation (3) holds under much weaker hypotheses about the representation of f as a linear combination of characteristic functions. The most general formulation of hypotheses under which (3) is true will be given on p. 268 as a corollary of 4.2. For now, we shall prove only the following remark:

If

$$f = \sum_{i=1}^{n} b_i \chi_{B_i, A}$$

is another representation of the simple function (2), such that

(8) $B_1, \ldots, B_n$ are disjoint measurable subsets of A,

(8′) $$A = B_1 \cup \ldots \cup B_n ,$$

then

(9) $$\int_A f = \sum_{i=1}^{n} b_i \mu(B_i) .$$

Conditions (8)-(8′) are, of course, weaker than § 3 (c_1)-(c_3).

To prove (9) let us observe that the function f can assume only the numbers $b_1, \ldots, b_n$ as its values. Some of them may not be values of f, then the corresponding sets B_j with the same indices j have to be empty. Without any loss of generality we may assume that

$$b_1 = b_2 = \ldots = b_{j_1} = a_1 ,$$

$$b_{j_1+1} = b_{j_1+2} = \ldots = b_{j_2} = a_2 ,$$

$$\ldots\ldots\ldots\ldots\ldots\ldots$$

$$b_{j_{m-1}+1} = b_{j_{m-1}+2} = \ldots = b_{j_m} = a_m$$

(where $a_1, \ldots, a_m$ are coefficients in the canonical representation (2) of f), and that the numbers $b_{j_m+1}, \ldots, b_n$ are not values of f, and consequently

$$B_{j_m+1} = \ldots = B_n = 0 .$$

Since (2) is the canonical representation of f, we have

$$A_1 = B_1 \cup \ldots \cup B_{j_1} ,$$

$$A_2 = B_{j_1+1} \cup \ldots \cup B_{j_2} ,$$

$$\ldots\ldots\ldots\ldots\ldots\ldots$$

$$A_m = B_{j_{m-1}+1} \cup \ldots \cup B_{j_m} .$$

Since the sets $B_1, \ldots, B_n$ are disjoint, we have

$$\sum_{i=1}^{n} b_i \mu(B_i) = \sum_{i=1}^{m} a_i \big(\mu(B_{j_{i-1}+1}) + \ldots + \mu(B_{j_i})\big) = \sum_{i=1}^{m} a_i \mu(A_i) = \int_A f .$$

Of course, we assume here that $j_0 = 0$.

We shall use identity (9) in the proof of the following theorem.

4.1. *For any simple functions f, g on A,*

(10) $$\int_A (f+g) = \int_A f + \int_A g .$$

Suppose that (2) is the canonical representation of f, and that

$$g = \sum_{j=1}^{n} b_j \chi_{B_j, A}$$

is the canonical representation of g. Thus $B_1, \dots, B_n$ are measurable disjoint sets whose union is A. Consequently the sets $A_i \cap B_j$ ($i = 1, \dots, m$, $j = 1, \dots, n$) are measurable and disjoint, and A is their union. Since f has the value a_i and g has the value b_j on the set $A_i \cap B_j$, we have

$$f+g = \sum_{i=1}^{m} \sum_{j=1}^{n} (a_i + b_j) \chi_{A_i \cap B_j, A}\,.$$

Replacing in (9) the function f by $f+g$, the numbers b_i by $a_i + b_j$ and the sets B_i by the sets $A_i \cap B_j$, we get

$$\int_A (f+g) = \sum_{i=1}^{m} \sum_{j=1}^{n} (a_i + b_j) \mu(A_i \cap B_j)\,.$$

Since $A_i = \bigcup_{j=1}^{n} A_i \cap B_j$ and $B_j = \bigcup_{i=1}^{m} A_i \cap B_j$ and all the terms are disjoint, we infer that

$$\begin{aligned}\sum_{i=1}^{m} \sum_{j=1}^{n} (a_i + b_j) \mu(A_i \cap B_j) &= \sum_{i=1}^{m} a_i \sum_{j=1}^{n} \mu(A_i \cap B_j) + \sum_{j=1}^{n} b_j \sum_{i=1}^{m} \mu(A_i \cap B_j) \\ &= \sum_{i=1}^{m} a_i \mu(A_i) + \sum_{j=1}^{n} b_j \mu(B_j) = \int_A f + \int_B g\,,\end{aligned}$$

which proves (10).

4.2. *For every simple function f on A and any number $c \geqslant 0$*

$$\int_A cf = c \int_A f\,.$$

In the case where $c = 0$, Theorem 4.2 follows from (4).

If $c > 0$ and (2) is the canonical representation of f, then

$$cf = \sum_{i=1}^{m} ca_i \chi_{A_i, A}$$

is the canonical representation of cf and consequently

$$\int_A cf = \sum_{i=1}^{m} ca_i \mu(A_i) = c \sum_{i=1}^{m} a_i \mu(A_i) = c \int_A f\,.$$

Identity (10) can be generalized, by induction, to any finite number of terms. Using this remark and 4.2 we shall now show that the identity (3) holds under the hypothesis that (2) holds and that $A_1, \ldots, A_m$ are any measurable subsets of A:

$$\int_A f = \sum_{i=1}^{m} \int_A a_i \chi_{A_i, A} = \sum_{i=1}^{m} a_i \int_A \chi_{A_i, A} = \sum_{i=1}^{m} a_i \mu(A_i)$$

(see (6)).

4.3. *For any simple functions f, g on A, if $f \leqslant g$, then $\int_A f \leqslant \int_A g$.*

In fact, $g = f + h$ where h is a simple function. Consequently, by 4.1,

$$\int_A g = \int_A f + \int_A h \geqslant \int_A f$$

since $\int_A h \geqslant 0$.

Let f be a simple function on a measurable set A and let B be a measurable subset of A. Generalizing the definition on p. 265, by the integral $\int_B f$ of f on A with respect to μ we shall mean the integral $\int_B f|B$ of $f|B$ on B with respect to μ. In symbols

$$\int_B f = \int_B f|B . \tag{11}$$

If $B = 0$, we assume the convention

$$\int_0 f = 0 \tag{12}$$

for every simple function f.

For instance, if the simple function f is defined by (2) where $A_1, \ldots, A_m$ are any measurable subsets of A, then

$$\int_B f = \sum_{i=1}^{m} a_i \mu(A_i \cap B) \tag{13}$$

for every measurable set $B \subset A$.

Indeed, $\chi_{A_i, A}|B = \chi_{A_i \cap B, B}$ and consequently $f|B = \sum_{i=1}^{m} a_i \chi_{A_i \cap B, B}$ which implies (13).

4.4. *For any simple function f defined on the union $A = B \cup C$ of disjoint measurable sets B, C,*

$$\int_{B \cup C} f = \int_B f + \int_C f . \tag{14}$$

Suppose that f is given by (2). Then we have the identity (13) and the analogous identity for C:

$$\int_C f = \sum_{i=1}^{m} a_i \mu(A_i \cap C) .$$

Consequently

$$\int_B f + \int_C f = \sum_{i=1}^{m} a_i\big(\mu(A_i \cap B) + \mu(A_i \cap C)\big) = \sum_{i=1}^{m} a_i \mu(A_i) = \int_A f .$$

Identity (14) can be generalized as follows: if f is a simple function on the union of disjoint measurable sets $A_1, \ldots, A_m$, then

$$\int_{A_1 \cup \ldots \cup A_m} f = \int_{A_1} f + \ldots + \int_{A_m} f . \tag{14'}$$

4.5. *If f, f_n $(n = 1, 2, \ldots)$ are simple functions on A and $f_n \nearrow f$, then $\int_A f_n \nearrow \int_A f$.*

The hypothesis $f_n \leqslant f_{n+1}$ implies that $\int_A f_n \leqslant \int_A f_{n+1}$, $n = 1, 2, \ldots$ Thus the sequence of the integrals has a limit. We have only to show that it is equal to $\int_A f$.

Consider first the case where the function f is constant: $f = c$. The case of $c = 0$ is obvious since then $f_n = 0$ for $n = 1, 2, \ldots$ Assume that $c > 0$. Let a be any number such that $0 < a < c$. Since $f_n \nearrow c$, the measurable sets

$$A_n = \{p \in A \colon f_n(p) > a\}$$

form an increasing sequence and $A = A_1 \cup A_2 \cup \ldots$ Consequently $\mu(A_n) \to \mu(A)$ by VI 3.5. Since $a\chi_{A_n, A} \leqslant f_n \leqslant c$, we have by 4.3

$$a\mu(A_n) = \int_A a\chi_{A_n, A} \leqslant \int_A f_n \leqslant \int_A c = c\mu(A) .$$

As $n \to \infty$ we get

$$a\mu(A) \leqslant \lim_{n\to\infty} \int_A f_n \leqslant c\mu(A) .$$

Since this inequality holds for every number $a < c$, we infer that

$$\lim_{n\to\infty} \int_A f_n = c\mu(A) = \int_A c .$$

Suppose now that f is any simple function on A. Let (2) be the canonical representation of f, i.e. $f|A_i = a_i$, the sets $A_1, \ldots, A_m$ are disjoint and measurable and A is their union. Since $f_n|A_i \nearrow a_i$, it follows from the part just proved that

$$\int_{A_i} f_n|A_i \nearrow a_i \mu(A_i) \quad \text{for} \quad i = 1, \ldots, m .$$

Adding with respect to i, we get $\int_A f_n \nearrow \int_A f$.

4.6. *If two non-decreasing sequences (f_n), (g_n) of simple functions converge to the same limit, then* $\lim_{n\to\infty} \int_A f_n = \lim_{n\to\infty} \int_A g_n$.

The hypotheses $f_n \leqslant f_{n+1}$, $g_n \leqslant g_{n+1}$ imply (see 4.3) that $\int_A f_n \leqslant \int_A f_{n+1}$ and $\int_A g_n \leqslant \int_A g_{n+1}$ for $n = 1, 2, \ldots$ Thus the sequences of integrals have limits. We have only to show that the limits are equal.

Let m be a positive integer and let $h_n = \min(f_n, g_m)$, $n = 1, 2, \ldots$ Since $h_n \nearrow g_m$ for $n \to \infty$ and the functions h_n, g_m are simple, we have $\int_A h_n \nearrow \int_A g_m$ by 4.5. Since $h_n \leqslant f_n$, we have $\int_A h_n \leqslant \int_A f_n$ for $n = 1, 2, \ldots$ Consequently $\int_A g_m \leqslant \lim_{n\to\infty} \int_A f_n$ for every positive integer m. Hence

$$\lim_{n\to\infty} \int_A g_n \leqslant \lim_{n\to\infty} \int_A f_n .$$

The converse inequality can be obtained by the same argument (replace f_n by g_n, and conversely). Thus both limits are equal.

§ 5. The general definition and fundamental properties of the integral. Let f be a non-negative measurable function defined on a measurable set A. By the *integral* of f on A with respect to the measure μ we mean the (finite or infinite) number

$$(1) \qquad \int_A f = \lim_{n\to\infty} \int_A f_n$$

where (f_n) is a non-decreasing sequence of simple functions converging to f on A. It follows from 3.1 that such a sequence always exists. It follows from 4.6 that the limit on the right hand side of (1) does not depend on the choice of the sequence (f_n). It follows from 4.5 that, in the case where f is a simple function, the integral of f defined by (1) coincides with the integral of f defined in § 4. Thus t e definition (1) is a generalization of the definition (3) from § 4.

By definition (1),

$$(2) \qquad \text{if } f \geqslant 0, \text{ then } 0 \leqslant \int_A f \leqslant \infty .$$

Now let f be any measurable function defined on a measurable set A. By the *integral* of f on A with respect to the measure μ we mean the (finite or infinite) number

$$(3) \qquad \int_A f = \int_A f^+ - \int_A f^-$$

where f^+ and f^- are the non-negative and the non-positive parts of f, respectively, i.e. they are the measurable non-negative functions

$$(4) \qquad f^+ = \max(f, 0) , \quad f^- = \max(-f, 0) .$$

The integral $\int\limits_A f$ exists if and only if the subtraction on the right-hand side of (3) is feasible, i.e. if at least one of the two integrals on the right-hand side is finite.

Note that if $f \geqslant 0$, then $f^- = 0$ and consequently $\int\limits_A f^- = 0$ by § 4 (4). This proves that the integral of $f \geqslant 0$ defined by (3) coincides with the integral of f defined by (1), i.e. that the definition (3) is a generalization of (1).

The integral of a non-negative measurable function always exists. The integral of a measurable function assuming both positive and negative values does not always exist.

Since

$$(-f)^+ = f^- \quad \text{and} \quad (-f)^- = f^+,$$

it follows from definition (2) that

$$\int\limits_A -f = -\int\limits_A f. \tag{5}$$

More precisely: if one of the integrals appearing in (5) exists, then the other integral also exists and the equation (5) holds.

Similarly as in § 4 (11), if f is a measurable function on a set A, and B is a measurable subset of A, then by the *integral* $\int\limits_B f$ of f on B we shall mean the integral of $f|B$ on B. In symbols,

$$\int\limits_B f = \int\limits_B f|B. \tag{6}$$

Thus the existence and the value of the integral $\int\limits_B f$ is independent of what values the function f assumes outside the set B.

Often, instead of the symbol $\int\limits_A f$ we use the more precise symbol $\int\limits_A f d\mu$. We always use it when we wish to indicate the measure, with respect to which we integrate the function f. Sometimes we also use the notation

$$\int\limits_A f(p)\,dp \quad \text{or} \quad \int\limits_A f(p)\,d\mu(p) \quad \text{or} \quad \int\limits_A f(p)\,\mu(dp).$$

We use the notation especially in the case where the integrated function is defined by a formula which, besides the variable p, also contains other variables and we wish to indicate that the expression under the sign of the integral is considered to be a function of p. In this chapter we shall use the abbreviated notation $\int\limits_A f$ for the integral. We recall that the integration is always with respect a fixed measure μ mentioned in § 1.

We say that a measurable function f is *integrable* on A if the integral $\int_A f$ exists and is finite. The case of integrable functions is most important in the theory of the integral. However the notion of the (finite or infinite) integral of a measurable non-negative function is also very useful and very convenient because there is no problem of the existence of the integral in this case.

The notion of the integral introduced in this section is called the *Lebesgue integral*. There are other notions of integral but they will not be discussed in this book.

5.1. *For every measurable function f, if $\mu(A) = 0$, then $\int_A f = 0$.*

In the case where f is a simple function, Theorem 5.1 was proved in § 4 (7). If it is a non-negative measurable function and (f_n) is a non-decreasing sequence of simple functions converging to f, then $\int_A f_n = 0$ for $n = 1, 2, \ldots$ and consequently $\int_A f = 0$ by (1). If f is any measurable function, then $\int_A f^+ = 0$ and $\int_A f^- = 0$ by the part of the theorem just proved. Thus $\int_A f = 0$ by (3).

5.2. *If B and C are disjoint measurable sets, then for every measurable function f*

$$\int_{B \cup C} f = \int_B f + \int_C f. \tag{7}$$

Identity (7) should be understood as follows: If the integral on the left-hand side exists, then the integrals on the right-hand side also exist, the addition is meaningful and equation (7) holds. Conversely, if the integrals on the right-hand side exist and the addition is meaningful, then the integral on the left-hand side exists and equation (7) holds.

Property (7) is called the *additivity of the integral* with respect to the set of integration.

In the case where f is simple, (7) was proved—see 4.4. If f is a non-negative measurable function and (f_n) is a non-decreasing sequence of simple functions, $f_n \nearrow f$, then by (1) and 4.4

$$\int_{B \cup C} f = \lim_{n\to\infty} \int_{B \cup C} f_n = \lim_{n\to\infty} \Big(\int_B f_n + \int_C f_n \Big) = \lim_{n\to\infty} \int_B f_n + \lim_{n\to\infty} \int_C f_n = \int_B f + \int_C f,$$

i.e. (7) is true in this case. If f is any measurable function, then by the part of 5.2 just proved we have

$$\int_{B \cup C} f^+ = \int_B f^+ + \int_C f^+, \qquad \int_{B \cup C} f^- = \int_B f^- + \int_C f^-.$$

Subtracting the second identity from the first, we get (7) by (3).

The above proofs of 5.1 and 5.2 are typical for the theory of the integral with the definition of integral assumed in this section. The characteristic property of these proofs is that they are divided into three steps. To prove a property of the integral, first we verify that it holds for simple functions. Using definition (1) we later prove that it holds for non-negative measurable functions. Finally, using (3), we infer that it holds for any measurable function.

5.3. *If B is a measurable subset of A and the integral $\int_A f$ exists, then the integral $\int_B f$ also exists. If f is integrable on A, it is also integrable on B.*

Theorem 5.3 follows directly from 5.2 where $C = A - B$.

5.4. *If $A \subset B$ and $f \geqslant 0$, then $\int_B f \leqslant \int_A f$.*

Theorem 5.4 follows directly from (7) where $C = A - B$ (see also (2)).

5.5. *If B is of measure zero, then*

$$\int_A f = \int_{A \cup B} f \quad \textit{and} \quad \int_A f = \int_{A-B} f.$$

Theorem 5.5 follows from 5.2 and 5.1.

5.6. *If $f \leqslant g$, then $\int_A f \leqslant \int_A g$* (under the hypothesis that the integrals exist).

The case where f and g are simple has been proved—see 4.3. Consider now the case where f and g are any non-negative measurable functions, $f \leqslant g$. Let (f_n) and (g_n) be non-decreasing sequences of simple functions converging to f and g, respectively. The functions $h_n = \min(f_n, g_n)$ are also simple. Since $h_n \leqslant g_n$, we have $\int_A h_n \leqslant \int_A g_n$. Since $h_n \nearrow f$ and $g_n \nearrow g$, we get

$$\int_A f = \lim_{n\to\infty} \int_A h_n \leqslant \lim_{n\to\infty} \int_A g_n = \int_A g.$$

Suppose now that f and g are any measurable functions. If $f \leqslant g$, then $f^+ \leqslant g^+$ and $f^- \geqslant g^-$. By the part of 5.6 just proved

$$\int_A f^+ \leqslant \int_A g^+ \quad \text{and} \quad \int_A f^- \geqslant \int_A g^-.$$

Subtracting the second inequality from the first we get $\int_A f \leqslant \int_A g$.

5.7. *If a function f is integrable on a set A, then the set*

$$\{p \in A \colon |f(p)| = \infty\}$$

has measure zero.

Let $B = \{p \in A \colon f(p) = \infty\}$. Since $f^+ \geqslant n\chi_{B,A}$, we have

$$n\mu(B) = \int\limits_A n\chi_{B,A} \leqslant \int\limits_A f^+ < \infty \quad \text{for} \quad n = 1, 2, \ldots$$

which proves that $\mu(B) = 0$. Analogously we prove that the set $C = \{p \in A \colon f(p) = -\infty\}$ has measure zero. The set mentioned in 5.7 is the union of B and C, therefore it has measure zero.

In what follows we shall say that *almost all* points of a set A have a property P (or that P holds *almost everywhere* on A) if the set of all $p \in A$ which do not have property P has measure zero. For instance, we shall say that a measurable function f defined on A is *almost everywhere finite* if the set of all points $p \in A$ such that $f(p) = \pm\infty$ has measure zero (thus Theorem 5.7 can be now formulated as follows: every integrable function is almost everywhere finite). We shall say that a measurable function f is *defined almost everywhere* on a set A if the set of all points $p \in A$ at which the value of f is not defined has measure zero. We shall say that two measurable functions f, g defined on a set A (or, more generally, defined almost everywhere on A) are *almost everywhere equal* if the set of all points p for which the equation $f(p) = g(p)$ does not hold is of measure zero; in this case we shall also say that the functions f, g are *equivalent*.

When writing $\int\limits_A f$, we have always assumed that f is a measurable function defined on the whole set A (or, more generally, that A is a subset of the domain of f). Now we shall generalize the definition of the integral $\int\limits_A f$ in such a way that it will also be valid in the case where the function f is defined almost everywhere on A. Such a generalization is very convenient in practice since often integrated functions are defined as the result of operations on other functions, and the operations may only be feasible almost everywhere. For instance, if f and g are integrable on A and assume infinite values, then the sum $f+g$ is not defined on the whole set A, in general, because it is not defined at points $p \in A$ such that one of the numbers $f(p)$, $g(p)$ is equal to ∞ and the other to $-\infty$. However it follows from 5.7 that the set of all points $p \in A$ at which the value of $f+g$ is not defined has measure zero since it is a subset of the union of two sets of measure zero

$$\{p \colon |f(p)| = \infty\} \quad \text{and} \quad \{p \colon |g(p)| = \infty\}\,.$$

Thus the sum $f+g$ is defined almost everywhere on A. The adoption of the proposed generalization of the integral will enable a simple formulation of Theorem 5.8 on the integral of the sum of two integrable functions (see also 6.7).

Suppose that f is a measurable function defined almost everywhere on a measurable set A. Let B be the set (of measure zero) of all $p \in A$ at which f is not defined. By the *integral* of the function f on the set A we shall mean the integral of this function on $A-B$. In symbols

$$\int_A f = \int_{A-B} f. \tag{8}$$

The function f defined almost everywhere on A can be always extended to a measurable function f_1 defined on the whole set A (assume e.g. $f_1(p) = f(p)$ for $p \in A-B$ and $f_1(p) = 0$ for $p \in B$). It follows from 5.5 that the integral $\int_A f$ defined by (8) is equal to the integral $\int_A f_1$. This statement can be considered as another equivalent definition of the integral: if f is defined almost everywhere on A, then the integral $\int_A f$ is the number $\int_A f_1$ where f_1 is any measurable extension of f over A.

The reader can verify without difficulty that all the theorems proved in this section are also true in the case where the integrated functions are defined almost everywhere on the sets of integration.

The following simple fact should also be mentioned: the integrals of equivalent functions are equal. In fact, if the set $B = \{p \in A: f(p) \neq g(p)\}$ has measure zero, then by 5.5

$$\int_A f = \int_{A-B} f = \int_{A-B} g = \int_A g.$$

In other words: the integral remains the same if the integrated function is modified in a set of measure zero. Consequently, when we deal with functions which are almost everywhere finite (in particular, with integrable functions) we can assume, without any loss of generality, that they are finite. In fact, if f is almost everywhere finite, then the measurable function (see 2.6)

$$g(p) = \begin{cases} f(p) & \text{if} \quad |f(p)| < \infty, \\ 0 & \text{if} \quad |f(p)| = \infty \end{cases}$$

is finite and equivalent to f and therefore has the same integral as f on each measurable set contained in the domain of these functions.

5.8. *If f and g are non-negative measurable functions, or are integrable on A, then*

$$\int_A (f+g) = \int_A f + \int_A g. \tag{9}$$

Consequently, the sum of two integrable functions is integrable.

If $f \geqslant 0$ and $g \geqslant 0$ and f_n, g_n $(n = 1, 2, \ldots)$ are simple functions such that $f_n \nearrow f$ and $g_n \nearrow g$, then $f_n + g_n \nearrow f+g$ and

$$\int_A (f+g) = \lim_{n\to\infty} \int_A (f_n+g_n) = \lim_{n\to\infty} \int_A f_n + \lim_{n\to\infty} \int_A g_n = \int_A f + \int_A g$$

by (1) and 4.1.

Consider now the case where f and g are integrable. By the remark just before 5.8 it suffices to consider the case where f and g are finite (for if they are not finite, we can replace them by equivalent finite functions). Let

$$A_1 = \{p \in A: (f(p) \geqslant 0)\&(g(p) \geqslant 0)\},$$
$$A_2 = \{p \in A: (f(p) < 0)\&(g(p) < 0)\},$$
$$A_3 = \{p \in A: (f(p) \geqslant 0)\&(g(p) < 0)\&(f(p)+g(p) \geqslant 0)\},$$
$$A_4 = \{p \in A: (f(p) \geqslant 0)\&(g(p) < 0)\&(f(p)+g(p) < 0)\},$$
$$A_5 = \{p \in A: (f(p) < 0)\&(g(p) \geqslant 0)\&(f(p)+g(p) \geqslant 0)\},$$
$$A_6 = \{p \in A: (f(p) < 0)\&(g(p) \geqslant 0)\&(f(p)+g(p) < 0)\}.$$

It follows from the part of 5.8 just proved that

$$\int_{A_1} (f+g) = \int_{A_1} f + \int_{A_1} g,$$
$$\int_{A_2} -(f+g) = \int_{A_2} -f + \int_{A_2} -g,$$
$$\int_{A_3} f = \int_{A_3} (f+g) + \int_{A_3} -g,$$
$$\int_{A_4} -g = \int_{A_4} -(f+g) + \int_{A_4} f,$$
$$\int_{A_5} g = \int_{A_5} (f+g) + \int_{A_5} -f,$$
$$\int_{A_6} -f = \int_{A_6} -(f+g) + \int_{A_6} g,$$

since the function under the integral on the left-hand side is the sum of the non-negative functions under the integrals on the right-hand side of the equation. Using identity (5) we infer that

$$\int_{A_i} (f+g) = \int_{A_i} f + \int_{A_i} g \quad \text{for} \quad i = 1, \ldots, 6.$$

Adding the equations we get (9) by 5.2.

Observe that if f and g are integrable on A, then $f-g$ is also integrable on A and

$$\int_A (f-g) = \int_A f - \int_A g. \tag{10}$$

This follows from (9) (where g should be replaced by $-g$) and (5).

Identity (5) can be now generalized as follows.

5.9. *For every real number* c

$$\int_A cf = c\int_A f. \tag{11}$$

If f is integrable on A, then cf is integrable on A.

Consider first the case where $c \geqslant 0$. If f is simple, (11) is true—see 4.2. If $f \geqslant 0$, let (f_n) be a non-decreasing sequence of simple functions convergent to f. Then $cf_n \nearrow cf$ and the cf_n are simple. Consequently

$$\int_A cf = \lim_{n\to\infty} \int_A cf_n = c \lim_{n\to\infty} \int_A f_n = c \int_A f,$$

i.e. (11) holds. If f is any measurable function, then

$$(cf)^+ = cf^+, \quad (cf)^- = cf^-$$

and, by the part just proved,

$$\int_A cf = \int_A (cf)^+ - \int_A (cf)^- = c\int_A f^+ - c\int_A f^- = c\int_A f.$$

If $c < 0$, then $(-c) > 0$ and, by the part just proved,

$$\int_A (-c)f = (-c)\int_A f.$$

Using (5), we get (11).

The second part of 5.9 follows from the first.

It follows from 5.8 and 5.9 that every linear combination of integrable functions is integrable and

$$\int_A (c_1 f_1 + \ldots + c_m f_m) = c_1 \int_A f_1 + \ldots + c_m \int_A f_m. \tag{12}$$

5.10. *If the integral $\int_A f$ exists, then $\left|\int_A f\right| \leqslant \int_A |f|$.*

A measurable function f on A is integrable on A if and only if $|f|$ is integrable on A.

Since $|f| = f^+ + f^-$, we have by 5.8

$$\int_A |f| = \int_A f^+ + \int_A f^-. \tag{13}$$

Hence

$$\int_A |f| \geqslant \left|\int_A f^+ - \int_A f^-\right| = \left|\int_A f\right|.$$

It follows from (13) that $|f|$ is integrable if both the integrals on the right-hand side of (13) are finite, i.e. if f is integrable (see (3)).

5.11. *If f is a measurable function on A, g is integrable on A and $|f| \leqslant g$, then f is integrable on A.*

Since $0 \leqslant \int_A |f| \leqslant \int_A g < \infty$, $|f|$ is integrable on A. By 5.10, f is integrable on A.

5.12. *Every bounded measurable function is integrable on any set of finite measure.*

Suppose that $|f| \leqslant M < \infty$. The constant function $g = M$ is integrable on A if $\mu(A) < \infty$, since $\int_A M = M\mu(A)$ (see § 4 (5)). By 5.11, f is integrable on A.

5.13. *If f is bounded and measurable on A, and g is integrable on A, then fg is integrable on A.*

Assume that $|f| \leqslant M < \infty$. Thus $|fg| \leqslant M \cdot |g|$. By 5.10 and 5.9, the function $M \cdot |g|$ is integrable. Thus fg is integrable by 5.11 and 2.9.

5.14. *If $f \geqslant 0$, then $\int_A f = 0$ if and only if $f = 0$ almost everywhere on A.*

If $f = 0$ almost everywhere, i.e. f is equivalent to the constant function equal to 0, then $\int_A f = \int_A 0 = 0$ (the hypothesis $f \geqslant 0$ is here superfluous).

If $f \geqslant 0$ and $\int_A f = 0$, then denoting

$$A_n = \left\{p \in A \colon f(p) > \frac{1}{n}\right\}$$

we have $f \geqslant \frac{1}{n} \cdot \chi_{A_n, A}$ and consequently

$$0 = \int_A f \geqslant \int_A \frac{1}{n} \cdot \chi_{A_n, A} = \frac{1}{n} \cdot \mu(A_n) \geqslant 0 \, .$$

This proves that $\mu(A_n) = 0$, $n = 1, 2, \ldots$ The set $\{p \in A \colon f(p) \neq 0\}$ is the union of all the sets $A_1, A_2, \ldots$, therefore it is of measure zero.

5.15. *If f is integrable on A, then the set $B = \{p \in A \colon f(p) \neq 0\}$ is of σ-finite measure.*

The set B is the union of the sets $B_n = \{p \in A \colon \ |f(p)| \geqslant 1/n\}$ $n = 1, 2, \ldots$ Since

$$\frac{1}{n} \mu(B_n) = \int_{B_n} \frac{1}{n} \leqslant \int_{B_n} |f| \leqslant \int_A |f| < \infty \, ,$$

we have $\mu(B_n) < \infty$. This proves that B is of σ-finite measure.

5.16. *If f is integrable on A, then for every $\varepsilon > 0$ there exists a $\delta > 0$ such that, for every measurable set $B \subset A$,*

$$\textit{if } \mu(B) < \delta, \textit{ then } \int_B |f| < \varepsilon \, .$$

Let (f_n) be a non-decreasing sequence of simple functions, $f_n \nearrow |f|$. There exists a positive integer n_0 such that

$$\int_A (|f| - f_{n_0}) = \int_A |f| - \int_A f_{n_0} < \varepsilon/2 \, .$$

Obviously $|f| - f_{n_0} \geqslant 0$. The function f_{n_0} is bounded: $f_{n_0} \leqslant M < \infty$. Let $\delta = \varepsilon/2M$. If $B \subset A$ and $\mu(B) < \delta$, then

$$\int_B |f| = \int_B (|f| - f_{n_0}) + \int_B f_{n_0} \leqslant \int_A (|f| - f_{n_0}) + \int_B M < \frac{\varepsilon}{2} + \frac{\varepsilon}{2M} M = \varepsilon .$$

EXERCISES

1. Let f be a measurable bounded function defined on a set A of finite measure. For every sequence $\mathfrak{p} = (A_1, \ldots, A_n)$ of disjoint measurable sets whose union is A, let

$$S(\mathfrak{p}) = \sum_{i=1}^{n} \sup\{f(p) \colon p \in A_i\} \cdot \mu(A_i), \quad s(\mathfrak{p}) = \sum_{i=1}^{n} \inf\{f(p) \colon p \in A_i\} \cdot \mu(A_i).$$

Prove that

$$\int_A f = \inf S(\mathfrak{p}) = \sup s(\mathfrak{p})$$

where "inf" and "sup" are extended over all sequences $\mathfrak{p}$ satisfying the above conditions.

2. Let f be integrable on a set A of finite measure. For every $\varepsilon > 0$ let

$$S(\varepsilon) = \sum_{n=-\infty}^{\infty} n\varepsilon\mu(A_n) \quad \text{where} \quad A_n = \{p \in A \colon n\varepsilon \leqslant f(p) < (n+1)\varepsilon\}.$$

Prove that $\lim_{\varepsilon \to 0} S(\varepsilon) = \int_A f$.

3. Let f be integrable on A and let $A_n = \{p \in A \colon |f(p)| \geqslant n\}$. Prove that $n\mu(A_n) \to 0$.

4. Let μ_0 be the zero-dimensional measure defined on the algebra of all subsets of a space X (see VI § 3, p. 237). A real function f defined on X is integrable on X if and only if $\sum_{p \in X} |f(p)| < \infty$, i.e. if f has the value 0 outside a countable set B and the sum of absolute values of $f(p)$ for $p \in B$ is a convergent series. Then $\int_X f d\mu_0 = \sum_{p \in X} f(p)$.

5. Prove that identity (9) holds under the hypothesis that the integrals on the right-hand side exist and the addition is feasible.

§ 6. Convergence theorems for the integral. The following *monotonic convergence theorem* is a generalization of Theorem 4.5.

6.1. *For any non-decreasing sequence of non-negative measurable functions,*

$$\int_A \lim_{n\to\infty} f_n = \lim_{n\to\infty} \int_A f_n .$$

Let $f = \lim_{n\to\infty} f_n$, and let $(f_{n,1}, f_{n,2}, \ldots)$ be a non-decreasing sequence of simple functions convergent to f_n $(n = 1, 2, \ldots)$. The sequence

$$g_n = \max(f_{1,n}, \ldots, f_{n,n})$$

of simple functions is non-decreasing and converges to f (see 3.2). Therefore $\int_A g_n \to \int_A f$. Since $g_n \leqslant f_n \leqslant f$, we have $\int_A g_n \leqslant \int_A f_n \leqslant \int_A f$ for $n = 1, 2, \ldots$ This implies that $\int_A f_n \to \int_A f$.

6.2. *For every non-increasing sequence of non-negative integrable functions,*

$$\int_A \lim_{n\to\infty} f_n = \lim_{n\to\infty} \int_A f_n .$$

Since $(f_1 - f_n)$ is a non-decreasing sequence of non-negative measurable functions, we have by 6.1

$$\int_A (f_1 - \lim_{n\to\infty} f_n) = \int_A f_1 - \lim_{n\to\infty} \int_A f_n .$$

Subtracting $\int_A f_1$ from both the sides of the equation and changing the sign we get the identity required.

The next theorem is called the *Fatou lemma.*

6.3. *For every sequence (f_n) of non-negative measurable functions,*

$$\liminf_{n\to\infty} \int_A f_n \geqslant \int_A \liminf_{n\to\infty} f_n .$$

Let $g_n = \inf(f_n, f_{n+1}, \ldots)$. By definition,

$$\lim_{n\to\infty} g_n = \liminf_{n\to\infty} f_n \quad \text{and} \quad \int_A g_n \leqslant \int_A f_n$$

since $g_n \leqslant f_n$. Since (g_n) is a non-decreasing sequence of non-negative measurable functions, we have by 6.1

$$\int_A \liminf_{n\to\infty} f_n = \int_A \lim_{n\to\infty} g_n = \lim_{n\to\infty} \int_A g_n \leqslant \liminf_{n\to\infty} \int_A f_n .$$

Now we shall prove the following *Lebesgue dominated convergence theorem.*

6.4. *If (f_n) is a convergent sequence of measurable functions on A, and $|f_n| \leqslant g$ where g is integrable on A, then*

(1)
$$\int_A \lim_{n\to\infty} f_n = \lim_{n\to\infty} \int_A f_n .$$

In particular the identity holds if the measure of A is finite and $|f_n| \leqslant M$ $(n = 1, 2, \ldots)$ for a constant $M < \infty$.

Let $f = \lim_{n\to\infty} f_n$. Since $|f_n| \leqslant g$, we have $f_n + g \geqslant 0$ and $g - f_n \geqslant 0$. Applying the Fatou lemma to tbe sequence $(f_n + g)$ we get

$$\liminf_{n\to\infty} \int_A (f_n + g) \geqslant \int_A (f + g) .$$

Hence, by subtracting $\int_A g$,

(2)
$$\liminf_{n\to\infty} \int_A f_n \geqslant \int_A f .$$

Applying the Fatou lemma to the sequence $(g-f_n)$ we get

$$\liminf_{n\to\infty} \int_A (g-f_n) \geqslant \int_A (g-f) .$$

Subtracting the number $\int_A g$ and multiplying the inequality by -1, we obtain

$$\limsup_{n\to\infty} \int_A f_n \leqslant \int_A f . \tag{3}$$

(1) follows directly from (2) and (3). The second part of 6.4 follows from the first where $g = M$ is a constant function.

By the convention on p. 274 we say that a sequence (f_n) *converges almost everywhere* if the set B of points p for which the limit $\lim_{n\to\infty} f_n(p)$ does not exist is of measure zero.

Observe that Theorem 6.4 remains true if (f_n) converges almost everywhere on A. Indeed, all the hypotheses of 6.4 are satisfied on the set $A-B$ and consequently (1) holds for the integrals on $A-B$. On the other hand, $\int_{A-B} f_n = \int_A f_n$, and the same for the limit function.

Now we shall prove the following *uniform convergence theorem.*

6.5. *The limit of a uniformly convergent sequence (f_n) of finite integrable functions on a set A of finite measure is integrable and*

$$\int_A \lim_{n\to\infty} f_n = \lim_{n\to\infty} \int_A f_n .$$

Let n_0 be an integer such that $|f_n-f_{n_0}| \leqslant 1$ for $n > n_0$. The function $g = |f_{n_0}|+1$ is integrable on A since it is the sum of two integrable functions (the constant function equal to 1 is integrable since the measure of A is finite). Moreover $|f_n| \leqslant g$ for $n > n_0$. Theorem 6.5 follows directly from 6.4.

Now we shall apply the result obtained to the theory of series of functions.

6.6. *For any sequence (g_n) of non-negative measurable functions,*

$$\int_A \sum_{n=1}^{\infty} g_n = \sum_{n=1}^{\infty} \int_A g_n .$$

Theorem 6.6 follows directly from 6.1 where

$$f_n = g_1 + \dots + g_n . \tag{4}$$

Now we shall prove the following *integral criterion for the convergence of series of functions.*

6.7. *For every sequence* (g_n) *of measurable functions on* A, *if*

$$\sum_{n=1}^{\infty} \int_A |g_n| < \infty ,$$

then the series $\sum_{n=1}^{\infty} g_n(p)$ *converges absolutely for almost every* $p \in A$, *the function* $\sum_{n=1}^{\infty} g_n$ *is integrable and*

$$\int_A \sum_{n=1}^{\infty} g_n = \sum_{n=1}^{\infty} \int_A g_n . \tag{5}$$

Let $g = \sum_{n=1}^{\infty} |g_n|$. By 6.6, $\int_A g = \sum_{n=1}^{\infty} \int_A |g_n| < \infty$, i.e. g is integrable and therefore almost everywhere finite. At every $p \in A$ for which $g(p) < \infty$, the series $\sum_{n=1}^{\infty} g_n(p)$ absolutely converges. Thus the function $\sum_{n=1}^{\infty} g_n$ is defined almost everywhere on A. It is integrable since $|\sum_{n=1}^{\infty} g_n| \leqslant g$ (see 5.11). To obtain (5) it suffices to apply the first part of Theorem 6.4 to the sequence (4). The inequality $|f_n| \leqslant g$ is obvious.

The following theorem states the *countable additivity of the integral* with respect to the set of integration.

6.8. *If* f *is a measurable function defined on the union* A *of a sequence* (A_n) *of disjoint measurable sets, then*

$$\int_A f = \sum_{n=1}^{\infty} \int_{A_n} f \tag{6}$$

provided that the integral on the left-hand side exists.

Consider first the case where $f \geqslant 0$. The functions

$$f_n(p) = \begin{cases} f(p) & \text{for} \quad p \in A_1 \cup \ldots \cup A_n , \\ 0 & \text{for} \quad p \in A - (A_1 \cup \ldots \cup A_n) \end{cases}$$

form a non-decreasing sequence of measurable functions convergent to f. By 6.1,

$$\int_{A_1} f + \ldots + \int_{A_n} f = \int_{A_1 \cup \ldots \cup A_n} f = \int_A f_n \to \int_A f$$

which proves (6).

If f is any measurable function, write (6) for f^+ and f^- and subtract the second identity from the first.

Theorem 6.8 is a generalization of 5.2. Observe that, in both theorems, the hypothesis that the sets are disjoint can be replaced by the weaker hypothesis that they are *almost disjoint*, i.e. that

$$\mu(A_i \cap A_j) = 0 \quad \text{for} \quad i \neq j . \tag{7}$$

More precisely, if A is the union of a finite or infinite sequence of measurable sets (A_n), and (7) holds, then

(8) $$\int_A f = \int_{A_1} f + \int_{A_2} f + \dots$$

To prove it, let us observe that the union B of all the sets $A_i \cap A_j$ $(i \neq j)$ has measure zero and therefore

$$\int_B f = 0 \quad \text{and} \quad \int_{A_i} f = \int_{A_i - B} f .$$

Since A is the union of the disjoint measurable sets $B, A_1 - B, A_2 - B, \dots,$ we have

$$\int_A f = \int_B f + \int_{A_1 - B} f + \int_{A_2 - B} f + \dots$$

which yields the identity required.

EXERCISES

1. If (f_n) is a sequence of measurable functions on A, and $|f_n| \leqslant g$ for $n = 1, 2, \dots$ where g is integrable on A, then

$$\int_A \limsup_{n\to\infty} f_n \geqslant \limsup_{n\to\infty} \int_A f_n \geqslant \liminf_{n\to\infty} \int_A f_n \geqslant \int_A \liminf_{n\to\infty} f_n .$$

2. Integrable functions $f_1, f_2, \dots$ on A are said to be *uniformly integrable* on A if for every $\varepsilon > 0$ there exists a $\delta > 0$ such that, for every measurable set $B \subset A$,

$$\text{if } \mu(B) < \delta, \text{ then } \int_B |f_n| < \varepsilon \text{ for } n = 1, 2, \dots$$

Prove that if (f_n) is a convergent sequence of uniformly integrable functions on a set A of finite measure, then

$$\int_A \lim_{n\to\infty} f_n = \lim_{n\to\infty} \int_A f_n .$$

Hint. Use the Egoroff theorem, § 2, exercise 3.

3. If $\int_A |f_n| \leqslant M < \infty$ for $n = 1, 2, \dots$, then the series $\sum_{n=1}^{\infty} a_n f_n$ converges absolutely almost everywhere for every absolutely convergent series $\sum_{n=1}^{\infty} a_n$.

CHAPTER VIII

THE INTEGRAL IN EUCLIDEAN SPACES

§ 1. Introduction. Integrals of continuous functions. In this chapter we shall examine exclusively the k-dimensional Lebesgue measure in the k-dimensional Euclidean space R^k. By measurable sets we shall always mean Lebesgue-measurable sets (see VI § 6), by measurable functions we shall always mean functions measurable with respect to the k-dimensional Lebesgue measure. The integration of functions will always be with respect to the k-dimensional Lebesgue measure. We recall that the Lebesgue measure of a set $A \subset R^k$ is denoted by $|A|$. The characteristic function χ_{A,R^k} of a set $A \subset R^k$ is denoted by χ_A for brevity.

Sometimes we shall examine, in addition to R^k, Euclidean spaces of other dimensions and the corresponding Lebesgue measures—this will be evident from the text.

As in Chapter VII, the integral of a function f on a set $A \subset R^k$ will be denoted by the symbol $\int\limits_A f$ or $\int\limits_A f(p)\,dp$. We shall also use the notation

$$(1) \qquad \int\limits_A \ldots \int f(x_1, \ldots, x_k)\,dx_1 \ldots dx_k\,.$$

The sign $\int$ at the beginning of (1) is repeated k times. Of course $x_1, \ldots, x_k$ denote here the coordinates of any point p in A. The symbols

$$(2) \qquad \int\limits_{-\infty}^{\infty} f \quad \text{or} \quad \int\limits_{-\infty}^{\infty} f(p)\,dp \quad \text{or} \quad \int\limits_{-\infty}^{\infty} \ldots \int f(x_1, \ldots, x_k)\,dx_1 \ldots dx_k$$

stand for the integral of f on the whole space R^k.

Of course, the integral with respect to the Lebesgue measure in R^k has all the properties of the integral from VII § 5 and § 6. We shall add two very important theorems to these properties, viz. the Fubini theorem and the theorem on integration by substitution. Both theorems have great theoretical significance and play an important role in the numerical calculation of integrals.

In practical applications of the theory of the integral in R^k, the case where the integrated function is continuous is most important. The following two theorems guarantee the measurability and integrability of continuous functions.

1.1. *Every continuous real function defined on a measurable set* $A \subset R^k$ *is measurable.*

1.2. *Every bounded continuous real function is integrable on every measurable set of finite measure. In particular, every continuous real function is integrable on every compact set.*

If f is continuous on A, the set $\{p \in A: f(p) < a\}$ is open in the metric space A since it is the inverse image of the open set $\{x: x < a\} \subset R$ (see III 8.2 (b)). Thus the set is of the form $A \cap G$, where G is an open subset of R^k. It is measurable since it is the intersection of two measurable sets. This proves 1.1. The first part of 1.2 follows directly from 1.1 and the general theorem VII 5.12, while the second part follows from the first since every continuous function on a compact set is bounded (see III 8.9).

Note that there exist measurable functions that are continuous at no point. The simplest example of such functions is given, for $k = 1$, by the characteristic function of the set of all rational numbers, defined on R.

In practical applications of the theory of integration in R^k, the sets on which we integrate have a rather simple structure. As a rule, they are closed or open sets with a rather simple boundary. For instance, in the case of the plane they are sets bounded by a finite number of curves; in the case of R^3 they are sets bounded by surfaces, etc. To distinguish a rather large class of open and closed sets appearing in practical applications of the notion of the integral, we introduce the following definition. We shall say that an open set $G \subset R^k$ is *regular* if its boundary $\mathrm{cl}\, G - G$ has measure zero. In practice we always integrate either on regular open sets or on their closures. Observe that it follows directly from the above definition and from VII 5.5 that

1.3. *If G is a regular open set, then for every real measurable function f*

$$\int_G f = \int_{\mathrm{cl}\, G} f .$$

Every open interval is an example of a regular open set. It is more difficult to prove that every open ball $\mathrm{B}(p, \eta)$ is regular (see p. 317). Other examples of regular open sets appearing in practical applications are given in Theorem 4.4.

Let A be a measurable set. By a *partition* of A we shall mean any finite sequence

$$\mathbf{P} = (A_1, \ldots, A_m)$$

of non-empty measurable sets such that

$$A = A_1 \cup \ldots \cup A_m \quad \text{and} \quad |A_i \cap A_j| = 0 \quad \text{for} \quad i \neq j .$$

The *diameter* of a partition $\mathbf{P}$ is the non-negative number

$$\operatorname{dia}\mathbf{P} = \max(\operatorname{dia} A_1, \ldots, \operatorname{dia} A_m) .$$

Let f be a real function on A and let $\mathbf{P}$ be a partition of A. By a *Riemann sum of f corresponding to the partition* $\mathbf{P}$ we shall mean any number of the form

$$S = \sum_{i=1}^{m} f(p_i)\cdot|A_i| \quad \text{where} \quad p_i \in A_i \quad \text{for} \quad i = 1, \ldots, m .$$

An infinite sequence $(\mathbf{P}_n)$ of partitions of A

$$\mathbf{P}_n = (A_{n,1}, \ldots, A_{n,m_n})$$

is said to be *regular* if $\operatorname{dia}\mathbf{P}_n \to 0$ for $n\to\infty$.

1.4. *Let f be a real function defined on a bounded measurable set A and let $(\mathbf{P}_n)$ be a regular sequence of partitions of A and, for every positive integer n, let S_n be a Riemann sum of f corresponding to $\mathbf{P}_n$. If f is uniformly continuous on A, then*

$$\lim_{n\to\infty} S_n = \int_A f .$$

In particular, the above identity holds if f is continuous and A is compact.

It suffices to prove the first part of the theorem. The second follows from the first since every continuous function on a compact set is uniformly continuous (III 8.6).

By definition,

$$S_n = \sum_{i=1}^{m_n} f(p_{n,i})\cdot|A_{n,i}| \quad \text{where} \quad p_{n,i} \in A_{n,i} .$$

For any given $\varepsilon > 0$ there exists a $\delta > 0$ such that for any $p, p' \in A$

$$|p-p'| < \delta \quad \text{implies} \quad |f(p)-f(p')| < \varepsilon .$$

Since $(\mathbf{P}_n)$ is regular, there is an integer n_0 such that

$$\operatorname{dia}\mathbf{P}_n < \delta \quad \text{for} \quad n > n_0 .$$

Consequently, for $n > n_0$,

$$\text{if } p \in A_{n,i}, \text{ then } |p-p_{n,i}| < \delta \text{ and } |f(p)-f(p_{n,i})| < \varepsilon .$$

This implies that

$$\Big|\int_{A_{n,i}} f - f(p_{n,i})\cdot|A_{n,i}|\Big| = \Big|\int_{A_{n,i}} f(p)\,dp - \int_{A_{n,i}} f(p_{n,i})\,dp\Big|$$
$$\leqslant \int_{A_{n,i}} |f(p)-f(p_{n,i})|\,dp \leqslant \int_{A_{n,i}} \varepsilon = \varepsilon\cdot|A_{n,i}| .$$

Hence, by VII § 6 (8),

$$\Big|\int_A f - S_n\Big| = \Big|\sum_{i=1}^{m_n}\Big(\int_{A_{n,i}} f - f(p_{n,i})\cdot|A_{n,i}|\Big)\Big| \leqslant \sum_{i=1}^{m_n}\Big|\int_{A_{n,i}} f - f(p_{n,i})\cdot|A_{n,i}|\Big| \leqslant \varepsilon\cdot|A|$$

for $n > n_0$. Since $\varepsilon > 0$ is arbitrary, this proves that $S_n \to \int_A f$.

Observe that Theorem 1.4 yields a method for an approximate numerical calculation of $\int_A f$. For this purpose it suffices to form a very fine partition $\mathbf{P}$ of A and to calculate a Riemann sum of f corresponding to $\mathbf{P}$. This is an approximate value of $\int_A f$.

Now we shall discuss the case of $k = 1$, i.e. the case of real functions of a real variable. In Theorem 1.5 below the integration is with respect to the one-dimensional Lebesgue measure. If I is an interval with end-points a, b $(a < b)$, and f is a function on this interval, we write

$$\int_a^b f \quad \text{or} \quad \int_a^b f(x)\,dx$$

instead of $\int_I f$. Since one-point sets have measure zero, it does not matter whether or not the points a, b are included in I (see VII 5.8). The above notation is also used in the case where the interval I is unbounded, i.e. where one or both of the numbers a, b are infinite.

1.5. *If f is a continuous function on the interval $\langle a; b\rangle$, and F is a primitive function of f, then*

$$\int_a^b f = F(b) - F(a)\,. \tag{3}$$

We recall that a function F is said to be a *primitive function* of f if $F' = f$ on the whole interval $\langle a; b\rangle$ (at the end-point the derivative should be understood as the one-sided derivative).

Formula (3) is often taken as a definition of the integral $\int_a^b f$ of a continuous function f on $\langle a; b\rangle$ in elementary courses in the theory of the integral. Theorem 1.5 states that, for continuous functions on $\langle a; b\rangle$, the Lebesgue integral defined in this book coincides with the elementary notion of the integral defined as the increase of a primitive function. The definition assumed by us is much more general than the elementary one because it can also be used for a very large class of non-continuous functions.

To prove 1.5, let $I_{n,i} = \Big\langle a + (i-1)\dfrac{b-a}{n};\, a + i\dfrac{b-a}{n}\Big\rangle$, for short. The sequence of partitions of $\langle a; b\rangle$

$$\mathbf{P}_n = (I_{n,1}, \ldots, I_{n,n})$$

is regular. By the mean-value theorem there exists a number $x_{n,i} \in I_{n,i}$ such that

$$F\left(a+i\frac{b-a}{n}\right)-F\left(a+(i-1)\frac{b-a}{n}\right)=f(x_{n,i})\cdot\frac{b-a}{n}.$$

Since

$$S_n=\sum_{i=1}^{n} f(x_{n,i})\cdot|I_{n,i}|=\sum_{i=1}^{n} f(x_{n,i})\cdot\frac{b-a}{n}$$

is a Riemann sum of f corresponding to $\mathbf{P}_n$, we infer from 1.4 that

$$S_n \to \int_a^b f.$$

On the other hand,

$$S_n=\sum_{i=1}^{n}\left(F\left(a+i\frac{b-a}{n}\right)-F\left(a+(i-1)\frac{b-a}{n}\right)\right)=F(b)-F(a)$$

$$\text{for} \quad n=1,2,\ldots$$

which proves (3).

The notation $\int_a^b f$, defined till now only in the case where $a<b$, can be generalized as follows:

$$\int_a^b f=-\int_b^a f \quad \text{if} \quad a>b,$$

$$\int_a^a f=0.$$

Observe that always

$$\int_a^c f=\int_a^b f+\int_b^c f.$$

EXERCISES

1. Prove the following *Lusin theorem*: a real function defined on a measurable set $A\subset R^k$ of finite measure is measurable if and only if for every $\varepsilon>0$ there exists a closed set $B\subset A$ such that $|A-B|<\varepsilon$ and $f|B$ is continuous.

Hint. Use VI 7.5 (d) to verify that every characteristic function, and consequently every simple function, has the property mentioned in the theorem. This implies, by the Egoroff theorem, that every measurable function has this property. Conversely, if B_n denotes the set B for $\varepsilon=1/n$ and $C=A-\bigcup_{n=1}^{\infty}B_n$, then $|C|=0$ and all the functions $f|C$, $f|B_n$ are measurable, which proves that f is measurable.

2. Let $A \subset R$ be measurable, and let $B \subset R$ be any set. Let f be a real function on $A \times B \subset R^2$ such that for every fixed $y \in B$ the function $f(\cdot, y)$ is integrable on A, and for every fixed $x \in A$ the function $f(x, \cdot)$ is continuous (for notation, see pp. 34–35). If there is a function g integrable on A such that $|f(x,y)| \leqslant g(x)$ for every point $(x, y) \in A \times B$, then the function

$$\varphi(y) = \int_A f(x, y)\,dx \qquad (y \in B)$$

is continuous on B. In particular, the function φ is continuous if f is bounded and A has finite measure.

Hint. Use VII 6.4.

3. Let A and B be open subsets of R, and let f be a continuous real function defined on the open set $A \times B \subset R^2$ such that the partial derivative $f_{|2}$ is continuous and bounded on $A \times B$. For every compact set $C \subset A$, the derivative of the function

$$\varphi(y) = \int_C f(x, y)\,dx \qquad (y \in B)$$

is equal to

$$\varphi'(y) = \int_C f_{|2}(x, y)\,dx\,.$$

In the traditional notation,

$$\frac{d}{dy}\int_C f(x, y)\,dx = \int_C \frac{\partial}{\partial y} f(x, y)\,dx\,.$$

Hint. Write the integral formula for $\dfrac{\varphi(y+\Delta y)-\varphi(y)}{\Delta y}$, use the mean-value theorem under the sign of the integral and apply VII 6.4.

4. Generalize exercises 2 and 3 for multi-dimensional Euclidean spaces.

5. By a *step function* we mean any linear combination $g = a_1\chi_{I_1} + \ldots + a_m\chi_{I_m}$ of the characteristic functions of intervals $I_1, \ldots, I_m \subset R^k$.

Prove that for every function f integrable on R^k and for any $\varepsilon > 0$ there exists a step function g such that $\int_{-\infty}^{\infty} |f-g| < \varepsilon$.

Hint. First consider the case where f is the characteristic function of an open set, of a G_δ-set, etc.

6. If f is integrable on R^k, then for every $\varepsilon > 0$ there is a $\delta > 0$ such that

$$\text{if } |\mathfrak{a}| < \delta, \text{ then } \int_{-\infty}^{\infty} |f(p+\mathfrak{a}) - f(p)|\,dp < \varepsilon\,.$$

Hint. First consider the case where f is a step function. Apply exercise 5.

§ 2. The Fubini theorem. In this section the k-dimensional Euclidean space R^k will always be regarded as the product of Euclidean spaces R^{k_1} and R^{k_2} where $k = k_1 + k_2$. According to a general procedure described in I § 6, we fix two finite sequences

$$\mathbf{i} = (i_1, \ldots, i_{k_1})\,, \qquad \mathbf{j} = (j_1, \ldots, j_{k_2})$$

of positive integers such that $i_1, \ldots, i_{k_1}, j_1, \ldots, j_{k_2}$ is a permutation of $1, \ldots, k$. Any point

$$p = (x_1, \ldots, x_k)$$

will be identified with the pair

$$p = (p_1, p_2),$$

where

$$p_1 = (x_{i_1}, \ldots, x_{i_k}) \in R^{k_1} \quad \text{and} \quad p_2 = (x_{j_1}, \ldots, x_{j_{k_2}}) \in R^{k_2}.$$

If $A_1 \subset R^{k_1}$ and $A_2 \subset R^{k_2}$, then $A_1 \times A_2$ is the set of all $p = (p_1, p_2) \in R^k$ such that $p_1 \in A_1$ and $p_2 \in A_2$. In particular,

$$A_1 \times R^{k_2} = \{(p_1, p_2) \colon p_1 \in A_1\}, \quad R^{k_1} \times A_2 = \{(p_1, p_2) \colon p_2 \in A_2\}.$$

We shall consider various measurable subsets of R^k, R^{k_1} and R^{k_2}. It is obvious that if $A \subset R^k$, $A \subset R^{k_1}$, or $A \subset R^{k_2}$, then $|A|$ denotes respectively the k-dimensional, k_1-dimensional, or k_2-dimensional Lebesgue measure of A.

Observe that if $I_1 \subset R^{k_1}$ and $I_2 \subset R^{k_2}$ are intervals (closed, open, open from the left, open from the right) then $I_1 \times I_2$ is also an interval (closed, open, open from the left, open from the right) and

$$|I_1 \times I_2| = |I_1| \cdot |I_2|. \tag{1}$$

This follows from the fact that the measure of an interval is equal to the product of the lengths of its edges.

2.1. *If $A_1 \subset R^{k_1}$ is closed, then $A_1 \times R^{k_2}$ is closed. If $A_1 \subset R^{k_1}$ is an F_σ-set, then $A_1 \times R^{k_2}$ is an F_σ-set.*

If $A_2 \subset R^{k_2}$ is closed, then $R^{k_1} \times A_2$ is closed. If $A_2 \subset R^{k_2}$ is an F_σ-set, then $R^{k_1} \times A_2$ is an F_σ-set.

Let $q_n = (p_{1,n}, p_{2,n}) \to q = (p_1, p_2) \in R^k$. Since the convergence of a sequence of points in a Euclidean space is equivalent to the convergence of the sequences of their coordinates, we infer that $p_{1,n} \to p_1$ and $p_{2,n} \to p_2$. If A_1 is closed and $q_n \in A \times R^{k_2}$, then $p_{1,n} \in A_1$ for $n = 1, 2, \ldots$ and consequently $p_1 \in A_1$, i.e. $q \in A_1 \times R^{k_2}$. This proves that $A_1 \times R^{k_2}$ is closed.

If A is an F_σ-set, i.e. $A = \bigcup_{n=1}^{\infty} A_{1,n}$ where the $A_{1,n}$ are closed, then the sets $A_{1,n} \times R^{k_2}$ are closed by the part of 2.1 just proved. Consequently

$$A \times R^{k_2} = \bigcup_{n=1}^{\infty} A_{1,n} \times R^{k_2}$$

is an F_σ-set.

The proof of the second part of 2.1 is analogous.

2.2. *If $A_1 \subset R^{k_1}$ has measure zero, then $A_1 \times R^{k_2}$ has measure zero. If $A_2 \subset R^{k_2}$ has measure zero, then $R^{k_1} \times A_2$ has measure zero.*

Let $I_{2,m}$ be the closed interval in R^{k_2} composed of all points, all of whose coordinates have absolute value $\leqslant m$ $(m = 1, 2, \ldots)$. By definition, $|I_{2,m}| = (2m)^{k_2}$.

First we shall prove that if $|A_1| = 0$, then $|A_1 \times I_{2,m}| = 0$. For any $\varepsilon > 0$, let $(I_{1,n})$ be a sequence of closed k_1-dimensional intervals such that

$$A \subset \bigcup_{n=1}^{\infty} I_{1,n} \quad \text{and} \quad \sum_{n=1}^{\infty} \operatorname{vol} I_{1,n} < \varepsilon/(2m)^{k_2}.$$

The closed interval $I_n = I_{1,n} \times I_{2,m}$ has the volume (see (1))

$$\operatorname{vol} I_n = \operatorname{vol} I_{1,n} \cdot \operatorname{vol} I_{2,m} = \operatorname{vol} I_{1,n} \cdot (2m)^{k_2}.$$

Thus $A \times I_{2,m} \subset \bigcup_{n=1}^{\infty} I_n$ and $\sum_{n=1}^{\infty} \operatorname{vol} I_n < \varepsilon$. Since $\varepsilon > 0$ is arbitrary, this proves that $|A \times I_{2,m}| = 0$.

Since $A_1 \times R^{k_2} = \bigcup_{m=1}^{\infty} A_1 \times I_{2,m}$ and all the summands have measure zero, the set $A_1 \times R^{k_2}$ has measure zero.

The proof of the second part of 2.2 is analogous.

2.3. *If A_1 is a measurable subset of R^{k_1} and A_2 is a measurable subset of R^{k_2}, then $A_1 \times A_2$ is a measurable subset of R^k.*

By VI 7.5, $A_1 = B_1 \cup C_1$ where B_1 is an F_σ-set and $|C_1| = 0$. Thus the set

$$A_1 \times R^{k_2} = (B_1 \times R^{k_2}) \cup (C_1 \times R^{k_2})$$

is measurable since it is the union of an F_σ-set and a set of measure zero, by 2.1 and 2.2.

Similarly we prove that the set $R^{k_1} \times A_2$ is measurable. The set $A_1 \times A_2$ is also measurable since

$$A_1 \times A_2 = (A_1 \times R^{k_2}) \cap (R^{k_1} \times A_2).$$

Let f be any function defined on a subset of R^k. According to I § 6, the value of f at a point $(p_1, p_2) \in R^k$ will be denoted by the symbol

$$f(p_1, p_2). \tag{2}$$

We recall that if p_1 is fixed, then (2) can be interpreted as a function of p_2 only (see I § 6); this function is denoted by $f(p_1, \cdot)$.

If f is a function defined on the whole space R^k, then the symbol f^* will denote the function

$$f^*(p_1) = \int_{R^{k_2}} f(p_1, p_2)\, dp_2 = \int_{R^{k_2}} f(p_1, \cdot) \tag{3}$$

defined on the set of all points $p_1 \in R^{k_1}$ for which the integration is feasible.

We shall prove the following *Fubini theorem.*

2.4. *If f is a measurable function on R^k and either $f \geqslant 0$ or f is integrable, then for almost all points $p_1 \in R^{k_1}$ the function $f(p_1, \cdot)$ is measurable; the function f^* is measurable and defined almost everywhere on R^{k_1}, and*

$$\int\limits_{R^k} f(p)\,dp = \int\limits_{R^{k_1}} f^*(p_1)\,dp_1\,, \tag{4}$$

that is,

$$\int\limits_{R^k} f(p)\,dp = \int\limits_{R^{k_1}} \Big(\int\limits_{R^{k_2}} f(p_1, p_2)\,dp_2 \Big)\,dp_1\,. \tag{4'}$$

On the left-hand side of (4′) we interpret f as a function of a point p in R^k and we integrate it with respect to the k-dimensional Lebesgue measure. On the right-hand side we interpret f as a function of a point $p_2 \in R^{k_2}$ and we integrate it with respect to the k_2-dimensional Lebesgue measure. The integration is feasible for almost all points $p_1 \in R^{k_1}$ and the result of the integration is a function f^* defined almost everywhere on R^{k_1}. We integrate this function with respect to the k_1-dimensional Lebesgue measure.

The Fubini theorem states that the integration in R^k can be reduced to a double integration in spaces of lower dimensions.

To prove the Fubini theorem let us observe that the following three remarks are true.

(A) A linear combination with non-negative coefficients of non-negative measurable functions satisfying the Fubini theorem satisfies the Fubini theorem.

(B) If $0 \leqslant f_n \nearrow f$ and if the functions f_n satisfy the Fubini theorem, then the function f satisfies the Fubini theorem.

(C) If $0 \leqslant f_n \searrow f$ and if all the functions f_n are integrable on R^k and satisfy the Fubini theorem, then the function f satisfies the Fubini theorem.

If $0 \leqslant f_n \nearrow f$, then by VII 6.1 $0 \leqslant f_n^* \nearrow f^*$ and

$$\int\limits_{R^{k_1}} f_n^* \to \int\limits_{R^{k_1}} f^*, \qquad \int\limits_{R^k} f_n \to \int\limits_{R^k} f\,.$$

Since, by hypothesis, $\int\limits_{R^k} f_n = \int\limits_{R^{k_1}} f_n^*$, as $n \to \infty$ we get (4). This proves (B). The proof of (C) is similar (use VII 6.2 instead of VII 6.1), and (A) is obvious.

The characteristic function χ_I of every interval $I \subset R^k$ open from the left satisfies the Fubini theorem. In fact, $I = I_1 \times I_2$ where I_1 and I_2 are intervals open from the left in R^{k_1} and R^{k_2}, respectively. By a simple calculation,

$$\chi_I^*(p_1) = \begin{cases} |I_2| & \text{if} \quad p_1 \in I_1\,, \\ 0 & \text{if} \quad p_1 \notin I_1\,. \end{cases}$$

Consequently,

$$\int_{R^k} \chi_I(p)\, dp = |I| = |I_1| \cdot |I_2| = \int_{R^{k_1}} \chi_I^*(p_1)\, dp_1 .$$

The characteristic function χ_A of the union

$$A = I_1 \cup \ldots \cup I_m \tag{5}$$

of a finite number of k-dimensional disjoint intervals open from the left satisfies the Fubini theorem by (A) since $\chi_A = \chi_{I_1} + \ldots + \chi_{I_m}$.

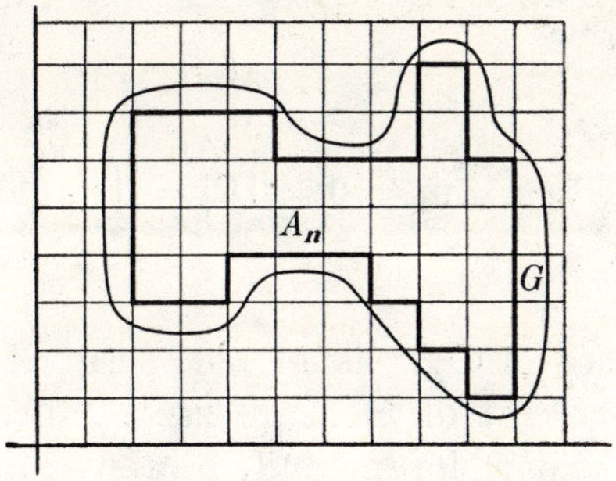

Fig. 11

Every bounded open set $G \subset R^k$ is the union of an increasing sequence (A_n) of sets of the form (5). Indeed, the $(k-1)$-dimensional hyperplanes perpendicular to the coordinate axes and intersecting the coordinate axes at points of the form $i/2^n$ $(i = 0, \pm 1, \pm 2, \ldots)$ divide the whole space R^k into disjoint intervals (cubes) open from the left; let A_n be the union of all the intervals which are subsets of G (see fig. 11 for the two-dimensional case). The sequence (A_n) has the property required. Since $\chi_{A_n} \nearrow \chi_G$, the function χ_G satisfies the Fubini theorem by means of (B).

Every bounded G_δ-set A is the intersection of a decreasing sequence (G_n) of bounded open sets. Since $\chi_{G_n} \searrow \chi_A$ and χ_{G_n} are integrable, it follows from (C) that χ_A satisfies the Fubini theorem.

Every G_δ-set A is the union of an increasing sequence of bounded G_δ-sets (A_n) (define e.g. $A_n = A \cap B(p, n)$, $n = 1, 2, \ldots$). Since $\chi_{A_n} \nearrow \chi_A$, the characteristic function of every G_δ-set A satisfies the Fubini theorem.

We shall now show that the characteristic function χ_B of any set $B \subset R^k$ of measure zero satisfies the Fubini theorem. The set B is a subset of a G_δ-set C of measure zero. Since

$$\int_{R^{k_1}} \chi_C^*(p_1)\, dp_1 = \int_{R^k} \chi_C(p)\, dp = |C| = 0 ,$$

by VII 5.14 the function χ_C^* is equal to 0 almost everywhere, i.e. there is a set $C_1 \subset R^{k_1}$ of measure zero such that for every point $p_1 \in R^{k_1} - C_1$

$$\int_{R^{k_2}} \chi_C(p_1, p_2)\, dp_2 = \chi_C^*(p_1) = 0 .$$

Using VII 5.14 once more, we infer that, for every fixed point $p_1 \in R^{k_1} - C_1$, the function $\chi_C(p_1, p_2)$ of p_2 is equal to 0 almost everywhere. Since $0 \leqslant \chi_B(p_1, p_2) \leqslant \chi_C(p_1, p_2)$, for every $p_1 \in R^{k_1} - C_1$ the function $\chi_B(p_1, p_2)$ of p_2 is equal to 0 almost everywhere and consequently

$$\chi_B^*(p_1) = \int_{R^{k_2}} \chi_B(p_1, p_2)\, dp_2 = 0 \quad \text{for} \quad p_1 \in R^{k_1} - C_1 .$$

This proves that

$$\int_{R^{k_1}} \chi_B^*(p_1)\, dp_1 = 0 = |B| = \int_{R^k} \chi_B(p)\, dp ,$$

i.e. (4) holds.

Any measurable set $A \subset R^k$ is of the form $A = C - B$, where C is a G_δ-set and B is its subset of measure zero. Since $\chi_A = \chi_C - \chi_B$ and χ_C and χ_B satisfy the Fubini theorem and, moreover, χ_B and χ_B^* are almost everywhere equal to zero, by writing (4) for χ_B and χ_C and subtracting these identities we get the analogous identity for χ_A.

Now, using (A), we infer that every simple function satisfies the Fubini theorem. Using (B) again we infer that every non-negative measurable function f satisfies the Fubini theorem (see VII 3.1).

If f is integrable on R^k then writing formulas (3) and (4) for the non-negative measurable functions f^+ and f^- and subtracting, we get the analogous formulas for f. Note that f^{+*} and f^{-*} are integrable, and therefore almost everywhere finite; therefore their difference exists almost everywhere.

The Fubini theorem is often formulated in the following more general form.

2.5. *If f is a measurable function on the product $A_1 \times A_2$ of measurable sets $A_1 \subset R^{k_1}$ and $A_2 \subset R^{k_2}$, and either f is non-negative or f is integrable, then for almost all points $p_1 \in A_1$ the function $f(p_1, \cdot)$ is measurable on A_2, the integral $\int_{A_2} f(p_1, p_2)\, dp_2$ exists and is a measurable function of p_1, and*

$$\int_{A_1 \times A_2} f(p)\, dp = \int_{A_1} \Big(\int_{A_2} f(p_1, p_2)\, dp_2 \Big)\, dp_1 . \tag{6}$$

To prove 2.5, let us extend the function f over the whole space R^k assuming

$$f(p_1, p_2) = 0 \quad \text{if} \quad (p_1, p_2) \notin A_1 \times A_2 .$$

Since

$$\int_{R^{k_2}} f(p_1, p_2)\,dp_2 = \begin{cases} 0 & \text{if} \quad p_1 \notin A_1, \\ \int_{A_2} f(p_1, p_2)\,dp_2 & \text{if} \quad p_1 \in A_1, \end{cases}$$

(6) follows from (4′).

The following supplement to 2.3 is a simple corollary of 2.5.

2.6. *If A_1 and A_2 are measurable subsets of R^{k_1} and R^{k_2} respectively, then the measure of the measurable set $A_1 \times A_2$ is equal to the product of the measures of A_1 and A_2:*

$$|A_1 \times A_2| = |A_1| \cdot |A_2| .$$

To obtain it, let us take $f = 1$ in 2.5. Then

$$|A_1 \times A_2| = \int_{A_1 \times A_2} 1\,dp = \int_{A_1} \Big(\int_{A_2} 1\,dp_2 \Big) dp_1 = \int_{A_1} |A_2|\,dp_1 = |A_2| \cdot |A_1| .$$

The integral on the right-hand side of (6) is usually denoted by the symbol

$$\int_{A_1} dp_1 \int_{A_2} f(p_1, p_2)\,dp_2 .$$

Using this notation and the notation from § 1 (1) we can write (6) as follows:

$$(7) \quad \int \!\!\dots\!\! \int_{A_1 \times A_2} f(x_1, \dots, x_k)\,dx_1 \dots dx_k$$
$$= \int \!\!\dots\!\! \int_{A_1} dx_{i_1} \dots dx_{i_{k_1}} \int \!\!\dots\!\! \int_{A} f(x_1, \dots, x_k)\,dx_{j_1} \dots dx_{j_{k_2}} .$$

In particular for $k_1 = 1$ and $k_2 = k-1$ we get the formula

$$\int_{-\infty}^{\infty} \!\!\dots\!\! \int f(x_1, \dots, x_k)\,dx_1 \dots dx_k$$
$$= \int_{-\infty}^{\infty} dx_i \int_{-\infty}^{\infty} \!\!\dots\!\! \int f(x_1, \dots, x_k)\,dx_1 \dots dx_{i-1}\,dx_{i+1} \dots dx_k$$

for every function f integrable on the whole space R^k. The integrated function

$$g(x_i) = \int_{-\infty}^{\infty} \!\!\dots\!\! \int f(x_1, \dots, x_k)\,dx_1 \dots dx_{i-1}\,dx_{i+1} \dots dx_k$$

is defined almost everywhere on the set R of all real numbers and is integrable on R.

Let j be a positive integer $\leqslant k$, $j \neq i$. Applying the Fubini theorem to the integral defining g, we get

$$g(x_i) = \int_{-\infty}^{\infty} dx_j \int_{-\infty}^{\infty} \cdots \int f(x_1, \ldots, x_k)\, dx_1 \ldots dx_{i-1} dx_{i+1} \ldots dx_{j-1} dx_{j+1} \ldots dx_k$$

(where $i < j$ is assumed to simplify the formulation). Consequently

$$\int_{-\infty}^{\infty} \cdots \int f(x_1, \ldots, x_k)\, dx_1 \ldots dx_k$$

$$= \int_{-\infty}^{\infty} dx_i \int_{-\infty}^{\infty} dx_j \int_{-\infty}^{\infty} \cdots \int f(x_1, \ldots, x_k)\, dx_1 \ldots dx_{i-1} dx_{i+1} \ldots dx_{j-1} dx_{j+1} \ldots dx_k .$$

Continuing the above procedure we prove the following corollary.

2.7. *Let $l_1, \ldots, l_k$ be a permutation of $1, \ldots, k$. For every function f integrable on the whole space R^k,*

$$\int_{-\infty}^{\infty} \cdots \int f(x_1, \ldots, x_k)\, dx_1 \ldots dx_k \tag{8}$$

$$= \int_{-\infty}^{\infty} dx_{l_1} \int_{-\infty}^{\infty} dx_{l_2} \ldots \int_{-\infty}^{\infty} dx_{l_{k-1}} \int_{-\infty}^{\infty} f(x_1, \ldots, x_k)\, dx_{l_k} .$$

This shows that integration with respect to the k-dimensional Lebesgue measure can be reduced to integration with respect to the one-dimensional Lebesgue measure.

EXERCISES

1. Calculate the integrals

$$\int_C \sin(x+y)\, dx\, dy , \qquad C = \{(x, y) \colon (0 \leqslant x \leqslant \pi) \& (0 \leqslant y \leqslant \pi/2)\} ;$$

$$\int_C (1+x+y+z)^2\, dx\, dy\, dz , \qquad C = \{(x, y, z) \colon \max(|x|, |y|, |z|) \leqslant 1\} .$$

2. If a function f_1 is integrable on $A_1 \subset R^{k_1}$, and a function f_2 is integrable on a set $A_2 \subset R^{k_2}$, then the function $f(p_1, p_2) = f_1(p_1) \cdot f_2(p_2)$ is integrable on the set $A_1 \times A_2$ and

$$\int_{A_1 \times A_2} f_1(p_1) f_2(p_2)\, d(p_1, p_2) = \int_{A_1} f_1(p_1)\, dp_1 \cdot \int_{A_2} f_2(p_2)\, dp_2 .$$

3. Let f be the following function on the square $I \times I$, where $I = \langle 0; 1 \rangle$,

$$f(x, y) = \begin{cases} 2^n & \text{if } \dfrac{1}{2^n} < x < \dfrac{3}{2^{n+1}} \text{ and } \dfrac{1}{2^n} < y < \dfrac{3}{2^{n+1}}, \\ 2^n & \text{if } \dfrac{3}{2^{n+1}} < x < \dfrac{1}{2^{n-1}} \text{ and } \dfrac{3}{2^{n+1}} < y < \dfrac{1}{2^{n-1}}, \\ -2^n & \text{if } \dfrac{1}{2^n} < x < \dfrac{3}{2^{n+1}} \text{ and } \dfrac{3}{2^{n+1}} < y < \dfrac{1}{2^{n-1}}, \\ -2^n & \text{if } \dfrac{3}{2^{n+1}} < x < \dfrac{1}{2^{n-1}} \text{ and } \dfrac{1}{2^n} < y < \dfrac{3}{2^{n+1}}, \\ 0 & \text{otherwise.} \end{cases}$$

Prove that both the iterated integrals

$$\int_I dx \int_I f(x,y)\,dy\,, \quad \int_I dy \int_I f(x,y)\,dx$$

exist and are equal (viz. equal to 0) but the integral $\iint_{I\times I} f(x,y)\,dx\,dy$ does not exist, since the integrals of f^+ and f^- are infinite.

§ 3. A geometrical interpretation of the integral. If $p=(x_1,\dots,x_k)\in R^k$ and if x is a real number, then (p,x) will denote in this section the point $(x_1,\dots,x_k,x)\in R^{k+1}$. In other words, the space R^{k+1} will be interpreted in this section as the Cartesian product $R^k\times R$.

By the *graph in* R^{k+1} of an extended real function f defined on a set $A\subset R^k$ we mean the set

$$W=\{(p,f(p))\in R^{k+1}\colon\ p\in A\}\subset R^{k+1}\,.$$

3.1. *The graph in* R^{k+1} *of any measurable function* f *defined on a set* $A\subset R^k$ *has measure zero.*

Of course, by "measure" we mean here the $(k+1)$-dimensional Lebesgue measure.

First we shall prove 3.1 under the additional hypothesis that the measure of A is finite. For any $\varepsilon>0$,

$$W\subset\bigcup_{n=-\infty}^{\infty} B_n\times I_n$$

where

$$B_n=\{p\in A\colon\ n\varepsilon\leqslant f(p)<(n+1)\varepsilon\}\quad\text{and}\quad I_n=\{x\colon\ n\varepsilon\leqslant x<(n+1)\varepsilon\}.$$

By 2.6, $|B_n\times I_n|=|B_n|\cdot\varepsilon$. Consequently

$$|W|\leqslant\sum_{n=-\infty}^{\infty}|B_n|\cdot\varepsilon=|A|\cdot\varepsilon$$

since A is the union of disjoint measurable sets B_n. Since $\varepsilon>0$ is arbitrary, we infer that $|W|=0$.

If $|A|=\infty$, then $A=\bigcup_{n=1}^{\infty}A_n$ where A_n are measurable sets of finite measure (see VI 7.3). Since the graph of f is the union of the graphs of the functions $f|A_n$, i.e. of sets of measure zero, it has measure zero.

3.2. *For every measurable function* f *on a set* $A\subset R^k$ *the following subsets of* R^{k+1} *are measurable*:

(1) $\{(p,x)\in R^{k+1}\colon\ x<f(p)\}\,,\quad \{(p,x)\in R^{k+1}\colon\ x>f(p)\}\,,$

(2) $\{(p,x)\in R^{k+1}\colon\ x\leqslant f(p)\}\,,\quad \{(p,x)\in R^{k+1}\colon\ x\geqslant f(p)\}\,.$

If g is another measurable function on A, then the sets

(3) $\{(p, x) \in R^{k+1}\colon g(p) < x < f(p)\}$, $\{(p, x) \in R^{k+1}\colon g(p) \leqslant x \leqslant f(p)\}$

are also measurable.

For any rational number w, let

$$A_w = \{p \in A\colon f(p) > w\}, \quad I_w = \{x\colon x < w\}.$$

The first of the sets (1) is the countable union of the sets $A_w \times I_w$ which are measurable by 2.3. Therefore it is measurable. The proof of the measurability of the second of the sets (1) is analogous.

The sets (2) are the unions of sets (1) and the graph of the function f in R^{k+1}. Therefore they are measurable.

The sets (3) are the intersections of the measurable sets

$$\{(p, x) \in R^{k+1}\colon x > g(p)\}, \quad \{(p, x) \in R^{k+1}\colon x < f(p)\}$$

or

$$\{(p, x) \in R^{k+1}\colon x \geqslant g(p)\}, \quad \{(p, x) \in R^{k+1}\colon x \leqslant f(p)\},$$

respectively, therefore they are measurable.

3.3. *For every measurable function f on a set $A \subset R^k$,*

(4) $$\int_A f = |S^+| - |S^-|$$

where

(5) $$S^+ = \{(p, x) \in R^{k+1}\colon 0 < x < f(p)\},$$
$$S^- = \{(p, x) \in R^{k+1}\colon 0 > x > f(p)\}.$$

Roughly speaking, S^+ is the set of all the points in R^{k+1} which are under the graph of f but over the set A, and S^- is the set of all the points in R^{k+1} which are over the graph of f but under the set A (see figure 12 for $k = 1$; the set A is represented as the heavy segment). Theorem 3.3 explains the geometric sense of the integral of f: it is the difference of the $(k+1)$-dimensional Lebesgue measures of S^+ and S^-.

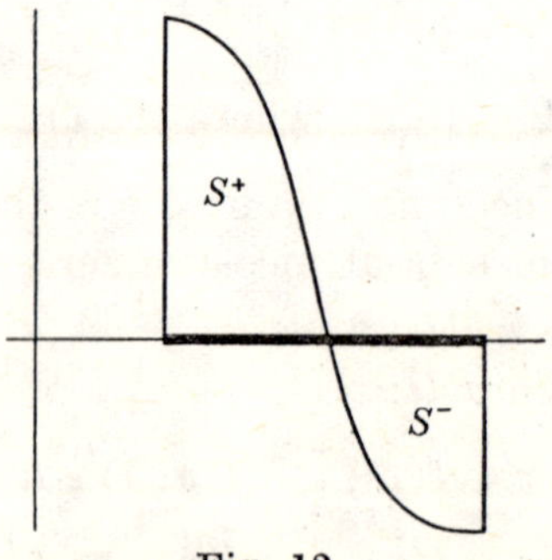

Fig. 12

Identity (4) should be understood as follows: the integral exists if and only if the subtraction on the right-hand side of (4) is feasible. Then equation (4) holds.

To prove 3.3, let us observe that the sets S^+, S^- are measurable by 3.2. Let

$$B = \{p\colon f(p) > 0\} \quad \text{and} \quad C = \{p\colon f(p) < 0\}\,.$$

By the Fubini theorem,

$$|S^+| = \int\limits_{B\times R} \chi_{S^+,B\times R}(p, x)\,d(p, x) = \int\limits_B dp \int\limits_{-\infty}^{\infty} \chi_{S^+,B\times R}(p, x)\,dx\,.$$

On the other hand, for $p \in B$,

$$\int\limits_{-\infty}^{\infty} \chi_{S^+,B\times R}(p, x)\,dx = f(p) = f^+(p)$$

since the integrated function of x is equal to 1 in the interval $(0; f(p))$ and it is equal to 0 outside this interval. Consequently

$$|S^+| = \int\limits_B f^+(p)\,dp = \int\limits_A f^+\,.$$

Replacing S^+ by S^-, and B by C everywhere, we prove in the same way that

$$|S^-| = \int\limits_A f^-\,.$$

(4) follows directly from the last two identities.

Note that identity (4) remains true if in the definition (5) of S^+ and S^- the signs $<$ and $>$ are replaced by $\leqslant$ and $\geqslant$. Then certain subsets of the graphs of the functions f and the function identically equal to 0 are added to the sets (5) but this does not change their measures by 3.1 and VI 3.3.

We shall prove the following corollary of 3.1.

3.4. *The l-dimensional Lebesgue measure of every k-dimensional hypersurface $H \subset R^l$, $k < l$, is equal to* 0.

By V 6.2, every point $q \in H$ has a neighbourhood U in H which is a graph of class C_1 (in the sense defined on p. 211). Since the metric space H is separable (see III 5.2 and III 5.3), by the Lindelöf theorem III 5.1 the set H is the union of a countable sequence of these neighbourhoods U. Since the union of a sequence of sets of measure zero is a set of measure zero, to prove 3.4 it suffices to show each neighbourhood U has the l-dimensional Lebesgue measure zero. To simplify the notation, let us assume that

$$U = \{(x_1, \ldots, x_k, \psi_1(x_1, \ldots, x_k), \ldots, \psi_s(x_1, \ldots, x_k))\colon (x_1, \ldots, x_k) \in G\}$$

where G is an open subset of R^k, and $\psi_1, \ldots, \psi_s$ are functions of class C_1 in G, $s = l-k$. The set U is a subset of the set

$$W = \{(x_1, \ldots, x_{l-1}, \psi_s(x_1, \ldots, x_k)) : (x_1, \ldots, x_k) \in G\}$$

which is the graph of the continuous function

$$f(p) = \psi_s(x_1, \ldots, x_k) \quad \text{for} \quad p = (x_1, \ldots, x_{l-1})$$

defined on the open set

$$\{(x_1, \ldots, x_{l-1}) : (x_1, \ldots, x_k) \in G\} \subset R^{l-1}.$$

By 3.1, $|W| = 0$. Consequently $|U| = 0$.

EXERCISES

1. Deduce Theorems VII 6.1 and 6.2 (for the Lebesgue measure) from 3.3 and VI 3.5 or VI 3.6, respectively. Deduce the Fatou lemma VII 6.3 from 3.3 and VI § 3, exercise 3.

2. Prove that for every measurable non-negative function f on a set $A \subset R^k$,

$$\int_A f = \int_0^\infty g$$

where $g(x) = |\{p \in A : f(p) \geqslant x\}|$.

3. For any non-negative measurable functions f, h on a set $A \subset R^k$,

$$\int_A fh = \int_0^\infty g$$

where $g(x) = \int_{A_x} h$ and $A_x = \{p \in A : f(p) \geqslant x\}$.

§ 4. The reduction of multi-dimensional integrals to iterated integrals on one-dimensional intervals. Before we formulate a general theorem on the reduction of multi-dimensional integrals to one-dimensional integrals, we shall examine its particular cases in R^2 and R^3.

A set $A \subset R^2$ is said to be *normal with respect to the second coordinate axis* if it is of the form

$$A = \{(x_1, x_2) : (a < x_1 < b) \& (g(x_1) < x_2 < h(x_1))\} \tag{1}$$

where $-\infty \leqslant a < b \leqslant \infty$ and g, h are measurable functions on $(a; b)$, $g \leqslant h$. If A is of the form (1), then for every function f integrable on A:

$$\iint_A f(x_1, x_2)\,dx_1 dx_2 = \int_a^b dx_1 \int_{g(x_1)}^{h(x_1)} f(x_1, x_2)\,dx_2 . \tag{2}$$

To prove (2) let us extend the function f to the whole plane R^2 assuming $f(x_1, x_2) = 0$ if $(x_1, x_2) \notin A$. By the Fubini theorem,

$$\iint_A f(x_1, x_2)\,dx_1 dx_2 = \iint_{-\infty}^{\infty} f(x_1, x_2)\,dx_1 dx_2 = \int_{-\infty}^{\infty} dx_1 \int_{-\infty}^{\infty} f(x_1, x_2)\,dx_2$$

$$= \int_a^b dx_1 \int_{-\infty}^{\infty} f(x_1, x_2)\,dx_2$$

since $f(x_1, x_2) = 0$ if $x_1 \leqslant a$ or $x_1 \geqslant b$. If $a < x_1 < b$, then $f(x_1, x_2) = 0$ for $x_2 \geqslant h(x_1)$ and for $x_2 \leqslant g(x_1)$, and consequently

$$\int_{-\infty}^{\infty} f(x_1, x_2)\,dx_2 = \int_{g(x_1)}^{h(x_1)} f(x_1, x_2)\,dx_2$$

which completes the proof of (2).

A set $A \subset R^2$ is said to be *normal with respect to the first coordinate axis* if it is of the form

$$A = \{(x_1, x_2)\colon\ (a < x_2 < b) \& (g(x_2) < x_1 < h(x_2))\}\,, \tag{3}$$

where $-\infty \leqslant a < b \leqslant \infty$ and g, h are measurable functions on $(a; b)$, $g \leqslant h$. If A is of the form (3), then for every function f integrable on A:

$$\iint_A f(x_1, x_2)\,dx_1\,dx_2 = \int_a^b dx_2 \int_{g(x_2)}^{h(x_2)} f(x_1, x_2)\,dx_1\,. \tag{4}$$

The proof of (4) is the same as that of (2).

We say that a set $A \subset R^2$ is *normal* if it is normal with respect to the first or second coordinate axis.

We see that if A is normal, the integration on A can be reduced to two integrations on one-dimensional intervals. In practice formulas (2) and (4) are used in the case where the functions f, g, h are continuous. As we prove later (see 6.2), the integrated functions in the integrals $\int_a^b$ in (2) and (4) are also continuous under certain hypotheses. Thus all the integrals on the right-hand side of (2) and (4) can be calculated by means of 1.5, i.e. by means of primitive functions. This method permits an easy calculation of integrals in many particular cases.

As an example we shall calculate the integral

$$\iint_A x_1 x_2\,dx_1\,dx_2\,, \qquad A = \{(x_1, x_2)\colon\ (0 < x_1 < 1) \& (x_1^2 < x_2 < x_1)\}\,.$$

By (2),

$$\iint_A x_1 x_2\,dx_1\,dx_2 = \int_0^1 dx_1 \int_{x_1^2}^{x_1} x_1 x_2\,dx_2\,.$$

First we calculate the integral $\int_{x_1^2}^{x_1} x_1 x_2\,dx_2$. We interpret the expression $x_1 x_2$ as a function of x_2 only. The function $\frac{1}{2}x_1 x_2^2$ is its primitive; therefore

$$\int_{x_1^2}^{x_1} x_1 x_2\,dx_2 = \tfrac{1}{2}x_1\cdot(x_1)^2 - \tfrac{1}{2}x_1\cdot(x_1^2)^2 = \tfrac{1}{2}(x_1^3 - x_1^5)\,.$$

Consequently

$$\iint_A x_1 x_2\,dx_1\,dx_2 = \int_0^1 \tfrac{1}{2}(x_1^3 - x_1^5)\,dx_1 = \tfrac{1}{2}(\tfrac{1}{4} - \tfrac{1}{6}) = \tfrac{1}{24}\,.$$

We shall now show how this method can be used to calculate integrals in R^3.

A set $A \subset R^3$ is said to be *normal with respect to the third coordinate axis* if it is of the form

$$(5) \qquad A = \{(x_1, x_2, x_3): ((x_1, x_2) \in B) \& (g_2(x_1, x_2) < x_3 < h_2(x_1, x_2))\}$$

where B is a normal subset of R^2 and g_2, h_2 are measurable functions on B, $g_2 \leqslant h_2$.

To fix the notation, suppose e.g. that B is normal with respect to the second coordinate axis, i.e.

$$B = \{(x_1, x_2): (a < x_1 < b) \& (g_1(x_1) < x_2 < h_1(x_1))\} \qquad (g_1 \leqslant h_1)$$

where a, b, g_1, h_1 satisfy the same hypotheses as in (1).

Analogously as in the case of R^2 we prove that for every function f integrable on A

$$\iiint_A f(x_1, x_2, x_3)\, dx_1 dx_2 dx_3 = \iint_B dx_1 dx_2 \int_{g_2(x_1,x_2)}^{h_2(x_1,x_2)} f(x_1, x_2, x_3)\, dx_3 .$$

Replacing in (2) the set A by B, and f by the function

$$F(x_1, x_2) = \int_{g_2(x_1,x_2)}^{h_2(x_1,x_2)} f(x_1, x_2, x_3)\, dx_3 ,$$

we obtain

$$\iint_B F(x_1, x_2)\, dx_1 dx_2 = \int_a^b dx_1 \int_{g_1(x_1)}^{h_1(x_1)} F(x_1, x_2)\, dx_2 .$$

Consequently we get the identity

$$(6) \quad \iiint_A f(x_1, x_2, x_3)\, dx_1 dx_2 dx_3 = \int_a^b dx_1 \int_{g_1(x_1)}^{h_1(x_1)} dx_2 \int_{g_2(x_1,x_2)}^{h_2(x_1,x_2)} f(x_1, x_2, x_3)\, dx_3 .$$

In the case where the set B is normal with respect to the first coordinate axis, identity (6) has an analogous form. In that case, x_1 should be replaced by x_2, and conversely.

The definition of a set *normal with respect to the first or the second coordinate axis* in R^3 is analogous. We leave to the reader the formulation of (6) in these cases.

A set $A \subset R^3$ is said to be *normal* if it is normal with respect to one of the coordinate axes. We see that if $A \subset R^3$ is normal, the integration on A can be reduced to three integrations on one-dimensional intervals.

The general definition of *normal subset of* R^k is by induction with respect to k. In the case of $k = 1$, the only normal sets are bounded or unbounded intervals $\{x\colon a < x < b\}$ $(-\infty \leqslant a < b \leqslant \infty)$. A set $A \subset R^k$ is said to be *normal with respect to the i-th coordinate axis* if it is of the form

$$(7) \quad A = \{(x_1, \ldots, x_k)\colon ((x_1, \ldots, x_{i-1}, x_{i+1}, \ldots, x_k) \in B)\& \\ \&(g(x_1, \ldots, x_{i-1}, x_{i+1}, \ldots, x_k) < x_i < h(x_1, \ldots, x_{i-1}, x_{i+1}, \ldots, x_k))\}$$

where B is a normal subset of R^{k-1} and g, h are measurable functions defined on B, $g \leqslant h$. A set $A \subset R^k$ is said to be *normal* if it is normal with respect to a coordinate axis.

In other words, a set $A \subset R^k$ is normal if there exists a permutation $i_1, \ldots, i_k$ of the integers $1, \ldots, k$ such that A is the set of all points $(x_1, \ldots, x_k)$ satisfying the following system of inequalities:

$$(\mathrm{n}_1) \qquad a < x_{i_1} < b\,,$$

$$(\mathrm{n}_2) \qquad g_1(x_{i_1}) < x_{i_2} < h_1(x_{i_1})\,,$$

$$\ldots\ldots\ldots\ldots\ldots$$

$$(\mathrm{n}_j) \qquad g_{j-1}(x_{i_1}, \ldots, x_{i_{j-1}}) < x_{i_j} < h_{j-1}(x_{i_1}, \ldots, x_{i_{j-1}})\,,$$

$$(\mathrm{n}_{j+1}) \qquad g_j(x_{i_1}, \ldots, x_{i_j}) < x_{i_{j+1}} < h_j(x_{i_1}, \ldots, x_{i_j})\,,$$

$$\ldots\ldots\ldots\ldots\ldots\ldots\ldots\ldots$$

$$(\mathrm{n}_{k-1}) \qquad g_{k-2}(x_{i_1}, \ldots, x_{i_{k-2}}) < x_{i_{k-1}} < h_{k-2}(x_{i_1}, \ldots, x_{i_{k-2}})\,,$$

$$(\mathrm{n}_k) \qquad g_{k-1}(x_{i_1}, \ldots, x_{i_{k-1}}) < x_{i_k} < h_{k-1}(x_{i_1}, \ldots, x_{i_{k-1}})\,.$$

where $-\infty \leqslant a < b \leqslant \infty$, and the functions g_j, h_j are measurable and defined on the set of all points $(x_{i_1}, \ldots, x_{i_j}) \in R^j$ which satisfy inequalities $(\mathrm{n}_1), \ldots, (\mathrm{n}_j)$; moreover $g_j \leqslant h_j$ $(j = 1, \ldots, k-1)$. More precisely, A is then normal with respect to the i_kth coordinate axis, i.e., it is of the form (7) where $i = i_k$ and B is the set of all points satisfying inequalities $(\mathrm{n}_1), \ldots, (\mathrm{n}_{k-1})$.

4.1. *For every integrable function f on a normal set $A \subset R^k$ defined by inequalities* (n_1)-(n_k),

$$(8) \quad \int_A \cdots \int f(x_1, \ldots, x_k)\, dx_1 \ldots dx_k \\ = \int_a^b dx_{i_1} \int_{g_1(x_{i_1})}^{h_1(x_{i_1})} dx_{i_2} \int_{g_2(x_{i_1}, x_{i_2})}^{h_2(x_{i_1}, x_{i_2})} dx_{i_3} \ldots \int_{g_{k-1}(x_{i_1}, \ldots, x_{i_{k-1}})}^{h_{k-1}(x_{i_1}, \ldots, x_{i_{k-1}})} f(x_1, \ldots, x_k)\, dx_{i_k}\,.$$

The proof of 4.1 is by induction on k, and the main idea of the proof is a repetition of that in the proof of (2), (4) and (6).

The case $k = 1$ is obvious. Suppose the theorem is true for $k-1$. We shall prove it for k. Extend f over the whole space R^k assuming

$f(p) = 0$ for $p \notin A$. As previously, let B be the set of all points $(x_1, \ldots, x_{i_k-1}, x_{i_k+1}, \ldots, x_k)$ satisfying all the inequalities (n_1)-(n_{k-1}). Thus

$$A = \{(x_1, \ldots, x_k): ((x_1, \ldots, x_{i_k-1}, x_{i_k+1}, \ldots, x_k) \in B) \& \\ \& (g_{k-1}(x_{i_1}, \ldots, x_{i_{k-1}}) < x_{i_k} < h_{k-1}(x_{i_1}, \ldots, x_{i_{k-1}}))\}.$$

By the Fubini theorem,

$$\int_A \cdots \int f(x_1, \ldots, x_k)\, dx_1 \ldots dx_k = \int_{-\infty}^{\infty} \cdots \int f(x_1, \ldots, x_k)\, dx_1 \ldots dx_k$$

$$= \int_{-\infty}^{\infty} \cdots \int dx_1 \ldots dx_{i_k-1} dx_{i_k+1} \ldots dx_k \int_{-\infty}^{\infty} f(x_1, \ldots, x_k)\, dx_{i_k}$$

$$= \int_B \cdots \int F(x_1, \ldots, x_{i_k-1}, x_{i_k+1}, \ldots, x_k)\, dx_1 \ldots dx_{i_k-1} dx_{i_k+1} \ldots dx_k,$$

where

$$F(x_1, \ldots, x_{i_k-1}, x_{i_k+1}, \ldots, x_k) = \int_{g_{k-1}(x_{i_1}, \ldots, x_{i_{k-1}})}^{h_{k-1}(x_{i_1}, \ldots, x_{i_{k-1}})} f(x_1, \ldots, x_k)\, dx_{i_k}.$$

Applying the induction hypothesis to the integral of F we get formula (8).

The case where the functions appearing in (8) satisfy the hypothesis of the following theorem is the most important since then all the functions integrated on the right-hand side of (8) are continuous and their integrals can be calculated by means of Theorem 1.5.

4.2. *If all the real functions $g_1, \ldots, g_{k-1}, h_1, \ldots, h_{k-1}$ are continuous and bounded, then for every continuous bounded real function f on the set A defined by inequalities (n_1)-(n_k) the functions*

$$f_j(x_{i_1}, \ldots, x_{i_j}) \\ = \int_{g_j(x_{i_1}, \ldots, x_{i_j})}^{h_j(x_{i_1}, \ldots, x_{i_j})} dx_{i_{j+1}} \int_{g_{j+1}(x_{i_1}, \ldots, x_{i_{j+1}})}^{h_{j+1}(x_{i_1}, \ldots, x_{i_{j+1}})} dx_{i_{j+2}} \cdots \int_{g_{k-1}(x_{i_1}, \ldots, x_{i_{k-1}})}^{h_{k-1}(x_{i_1}, \ldots, x_{i_{k-1}})} f(x_1, \ldots, x_k)\, dx_{i_k}$$

$(j = 1, \ldots, k-1)$ are continuous and bounded on the set B_j of all points $(x_{i_1}, \ldots, x_{i_j}) \in R^j$ satisfying inequalities (n_1)-(n_j).

Since

$$f_j(x_{i_1}, \ldots, x_{i_j}) = \int_{g_j(x_{i_1}, \ldots, x_{i_j})}^{h_j(x_{i_1}, \ldots, x_{i_j})} f_{j+1}(x_{i_1}, \ldots, x_{i_j}, x_{i_{j+1}})\, dx_{i_{j+1}}$$

(where $f_k(x_{i_1}, \ldots, x_{i_k}) = f(x_1, \ldots, x_k)$), in order to prove 4.2 it suffices to prove the following lemma which enables us to deduce, step by step, the continuity and the boundedness of $f_{k-1}, f_{k-2}, \ldots, f_1$.

4.3. *If g and h are bounded continuous real functions defined on a set $B\subset R^{k-1}$, $g\leqslant h$, and if f is a bounded continuous real function defined on the set*

$$A=\{(x_1,\ldots,x_k):\ ((x_1,\ldots,x_{k-1})\in B)\,\&\,(g(x_1,\ldots,x_{k-1})<x_k<h(x_1,\ldots,x_{k-1}))\},$$

then the real function

$$F(x_1,\ldots,x_{k-1})=\int\limits_{g(x_1,\ldots,x_{k-1})}^{h(x_1,\ldots,x_{k-1})} f(x_1,\ldots,x_k)\,dx_k$$

is continuous and bounded on B.

For the proof we shall interpret the space R^k as the product $R^{k-1}\times R$, i.e., if $p=(x_1,\ldots,x_{k-1})\in R^{k-1}$ and $x\in R$, then $(p,x)=(x_1,\ldots,x_{k-1},x)\in R^k$. Using this notation, we can write the definitions of A and F as follows

$$A=\{(p,x):\ (p\in B)\&(g(p)<x<h(p))\},$$

$$F(p)=\int\limits_{g(p)}^{h(p)} f(p,x)\,dx \quad \text{for} \quad p\in B.$$

Suppose that $g\geqslant a>-\infty$, $h\leqslant b<\infty$ and $|f|\leqslant M<\infty$. Let $I=\langle a;b\rangle$. Extend the function f over the set $B\times I$ by assuming $f(p,x)=0$ if $(p,x)\in B\times I-A$. Hence it follows that

$$F(p)=\int\limits_a^b f(p,x)\,dx.$$

Let $p,p_n\in B$ for $n=1,2,\ldots$, and $p_n\to p$. It follows from the continuity of f, g, h that

$$f(p_n,x)\to f(p,x)$$

for every point $x\in I$ except perhaps for $x=g(p)$ and $x=h(p)$, i.e. almost everywhere on I. Applying Theorem VII 6.4 to the sequence $(f(p_n,x))$ of functions of x, we infer that

$$\int\limits_a^b f(p_n,x)\,dx\to\int\limits_a^b f(p,x)\,dx,$$

i.e. that $F(p_n)\to F(p)$. This proves that F is continuous. F is bounded since

$$|F(p)|\leqslant\int\limits_a^b |f(p,x)|\,dx\leqslant\int\limits_a^b M\,dx=M(b-a)<\infty.$$

4.4. *If $A\subset R^k$ is a normal set defined by means of inequalities* (n_1)-(n_k) *and if all the real functions $g_1,\ldots,g_{k-1},h_1,\ldots,h_{k-1}$ are continuous, then the set A is regular open.*

If $p = (x_1, \ldots, x_k) \in A$, i.e. if the coordinates of p satisfy inequalities (n_1)-(n_k), then it follows from the continuity of the functions $g_1, \ldots, g_{k-1}$, $h_1, \ldots, h_{k-1}$ that the coordinates of every point in a ball $B(p, \eta)$ also satisfy (n_1)-(n_k), i.e. $B(p, \eta) \in A$. This proves that A is open.

The proof that A is regular is by induction on k. The case $k = 1$ is obvious (A is then an open, bounded or unbounded interval). Suppose the theorem is true for $k-1$. Then the open set B of all points $(x_1, \ldots, x_{i_k-1}, x_{i_k+1}, \ldots, x_k)$ whose coordinates satisfy inequalities (n_1)-(n_{k-1}) is regular, i.e. $|\mathrm{cl}B - B| = 0$. The set $\mathrm{cl}A - A$ is contained in the union of the following three sets: the graph of g_{k-1}, the graph of h_{k-1} and the Cartesian product of $\mathrm{cl}B - B$ with the i_kth coordinate axis. Each of these sets has measure zero (see 3.1 and 2.2 or 2.6). This proves that $|\mathrm{cl}A - A| = 0$, i.e. A is regular.

The reader can verify without difficulty that all the formulas given above and Theorems 4.2, 4.3 remain true if the sign $<$ is replaced by $\leqslant$ everywhere in the definition of a normal set A. Replacing everywhere $<$ by $\leqslant$ in the definition of A, we augment the set A by a set C of measure zero. The proof that C has measure zero is by induction on k and is similar to a part of the proof of 4.4.

EXERCISES

1. Calculate the integrals

$$\iint_A xy^2 dx\,dy\,, \qquad A = \{(x, y)\colon (a^2x^2 + b^2y^2 \leqslant 1)\,\&\,(x \geqslant 0)\}\,;$$

$$\iiint_A x(y^2+z^2)\,dx\,dy\,dz\,, \qquad A = \{(x, y, z)\colon (a^2x^2 + b^2y^2 + c^2z^2 \leqslant 1)\,\&\,(x \geqslant 0)\}\,.$$

2. Calculate the integral $\iint_A e^{x+y}\, x\,dx\,dy$ where A is the triangle with vertices $(-1, -1)$, $(1, 2)$, $(2, 1)$.

3. Find the three-dimensional Lebesgue measure of the set

$$\{(x, y, z)\colon (0 \leqslant y \leqslant x^2)\,\&\,(0 \leqslant x \leqslant z^3)\,\&\,(0 \leqslant z \leqslant 1)\,.$$

4. Verify that the function F in 4.3 is also continuous in the case where g and h are not bounded.

§ 5. Integration by substitution. The purpose of this section is to prove the following *theorem on integration by substitution*.

5.1. *Let Φ be a diffeomorphism from an open set $G \subset R^k$ into R^k. For every measurable function f defined on the open set $\Phi(G)$,*

$$\int_{\Phi(G)} f = \int_G (f \circ \Phi) \cdot |\Phi'|\,. \tag{1}$$

Identity (1) can also be written in the form

$$\int_{\Phi(G)} f(q)\,dq = \int_G f(\Phi(p)) \cdot |\Phi'(p)|\,dp\,. \tag{2}$$

Identities (1) or (2) should be understood as follows: if one of the integrals exists, then the second exists and they are equal.

We recall that $|\Phi'|$ is the absolute value of the scalar derivative (i.e. the jacobian) of the mapping Φ.

In the case where $k = 1$ and G is an open interval with end points a, b, and Φ is a real function φ (on the closure of this interval) with a continuous non-vanishing derivative, identity (1) can be written as follows:

$$(3) \qquad \int_{\varphi(a)}^{\varphi(b)} f = \int_a^b (f \circ \varphi) \cdot \varphi' ,$$

that is,

$$(4) \qquad \int_{\varphi(a)}^{\varphi(b)} f(y)\, dy = \int_a^b f(\varphi(x))\varphi'(x)\, dx .$$

Indeed, it suffices to consider only the case where $a < b$. If $\varphi' > 0$, then $\varphi' = |\Phi'|$ and $\Phi(G) = (\varphi(a); \varphi(b))$. Thus (4) is identical with (2). If $\varphi' < 0$, then $|\Phi'| = -\varphi'$ and $\Phi(G) = (\varphi(b); \varphi(a))$ and consequently $\int_{\varphi(a)}^{\varphi(b)} f = -\int_{\Phi(G)} f$. Thus (4) is identical with equation (2) multiplied by -1.

In the case $k \geqslant 1$, formula (1) (where $\Phi = (\varphi_1, \ldots, \varphi_k)$) can be written using the traditional notation as follows:

$$(5) \qquad \int \cdots \int_{\Phi(G)} f(y_1, \ldots, y_k)\, dy_1 \ldots dy_k$$
$$= \int \cdots \int_G f(\varphi_1(x_1, \ldots, x_k), \ldots, \varphi_k(x_1, \ldots, x_k)) \left| \frac{\partial(\varphi_1, \ldots, \varphi_k)}{\partial(x_1, \ldots, x_k)} \right| dx_1 \ldots dx_k .$$

It is easier to remember this formula in the form of the following formal identity (analogous to the formal identity $dy = \varphi'(x)dx$ from the theory of functions of one real variable):

$$(5') \qquad dy_1 \ldots dy_k = \left| \frac{\partial(\varphi_1, \ldots, \varphi_k)}{\partial(x_1, \ldots, x_k)} \right| dx_1 \ldots dx_k = \left| \frac{\partial(y_1, \ldots, y_k)}{\partial(x_1, \ldots, x_k)} \right| dx_1 \ldots dx_k .$$

It suffices to prove 5.1 in the case where $f \geqslant 0$. Indeed, if f is any measurable function, then writing (1) for f^+ and f^- and subtracting the second identity from the first we get (1) for f.

The proof of 5.1 given below is by induction with respect to k and is composed of a few lemmas.

A) Theorem 5.1 is true for $k = 1$.

The method of the proof is similar to that of the Fubini theorem. First we verify, by a simple calculation, that the theorem is true if f is the characteristic function of a bounded interval. Since every open set of real numbers is the union of a sequence of disjoint bounded intervals, we infer that 5.1 holds if f is the characteristic function of an open set.

Using the monotonic convergence theorems VII 6.1 and 6.2 we verify that 5.1 is true if f is the characteristic function of a bounded G_δ-set and, later, that it is true for the characteristic function of any G_δ-set. Since every set of measure zero is a subset of a G_δ-set of measure zero, we infer that (1) holds if f is the characteristic function of a set of measure zero. Since every measurable set is the difference of a G_δ-set and a set of measure zero, we infer that (1) also holds if f is the characteristic function of any measurable set and, consequently, if f is any simple function. Using the monotonic convergence theorem VII 6.1 again, we prove that 5.1 is true for every non-negative measurable function f.

To formulate next lemmas we shall introduce the following definitions. We shall say that a diffeomorphism Φ *satisfies the theorem on integration by substitution* (or Φ *satisfies* TIS, for short), if (1) holds for every non-negative measurable function f on the range $\Phi(G)$ of Φ. We shall say that a diffeomorphism *satisfies* TIS *on a set* $A \subset G$ if

$$\int_{\Phi(A)} f = \int_A (f \circ \Phi) \cdot |\Phi'| \tag{6}$$

for every non-negative measurable function f defined on $\Phi(A)$ (thus both A and $\Phi(A)$ have to be measurable sets). We shall say that a diffeomorphism Φ *satisfies* TIS *locally* if for every point p in the domain G of Φ there exists a neighbourhood $A \subset G$ of p such that Φ satisfies TIS on A.

B) If a diffeomorphism Φ satisfies TIS on a set A, $B \subset A$, and if both B and $\Phi(B)$ are measurable, then Φ satisfies TIS on B.

Let f be any non-negative measurable function on $\Phi(B)$. Extend f over $\Phi(A)$ by assuming that $f(q) = 0$ for $q \in \Phi(A) - \Phi(B) = \Phi(A - B)$. Then

$$\int_{\Phi(B)} f = \int_{\Phi(A)} f = \int_A (f \circ \Phi) \cdot |\Phi'| = \int_B (f \circ \Phi) \cdot |\Phi'|$$

since $f(\Phi(p)) = 0$ for $p \in A - B$.

C) If Φ satisfies TIS locally, then Φ satisfies TIS.

By hypothesis, every $p \in G$ has a neighbourhood on which Φ satisfies TIS. Using the Lindelöf theorem we can select a sequence (A_n) of these neighbourhoods such that $G = \bigcup_{n=1}^{\infty} A_n$. By B), Φ satisfies TIS on the disjoint Borel sets

$$B_1 = A_1, \quad B_n = A_n - (A_1 \cup \dots \cup A_{n-1}) \quad \text{for} \quad n > 1$$

whose images

$$\Phi(B_1) = \Phi(A_1), \quad \Phi(B_n) = \Phi(A_n) - (\Phi(A_1) \cup \dots \cup \Phi(A_{n-1})) \quad \text{for } n > 1$$

are also Borel sets (see V 5.1 (a)), and therefore are measurable. Thus for every non-negative measurable function f on $\Phi(G)$

$$\int_{\Phi(B_n)} f = \int_{B_n} (f \circ \Phi)|\Phi'| .$$

Adding all these equations we get (1) since $G = B_1 \cup B_2 \cup \ldots$

D) The composition $\Phi = \Phi_1 \circ \Phi_2$ of diffeomorphisms Φ_1, Φ_2 satisfying TIS satisfies TIS.

Assume that $\Phi_2(G)$ is a subset of the domain of Φ_1. Then

$$\begin{aligned}\int_{\Phi(G)} f = \int_{\Phi_1(\Phi_2(G))} f = \int_{\Phi_2(G)} f\big(\Phi_1(q)\big)|\Phi_1'(q)|\,dq \\ = \int_G f\big(\Phi_1(\Phi_2(p))\big)\big|\Phi_1'\big(\Phi_2(p)\big)\big| \cdot |\Phi_2'(p)|\,dp = \int_G (f \circ \Phi)|\Phi'|\end{aligned}$$

(see V 3.5 (24)).

We shall say that a diffeomorphism

$$\Phi = (\varphi_1, \ldots, \varphi_k)\colon\ G \to R^k \tag{7}$$

is *reducible* if there are positive integers $i, j \leqslant k$ such that

$$\varphi_j(x_1, \ldots, x_k) = x_i .$$

E) If TIS is true for $k-1$, then every reducible diffeomorphism (7) satisfies TIS.

To simplify the notation let us assume that $i = j = k$, i.e. that

$$\varphi_k(x_1, \ldots, x_k) = x_k . \tag{8}$$

Interpret R^k as the Cartesian product $R^{k-1} \times R$. Thus if $p = (x_1, \ldots x_{k-1}) \in R^{k-1}$ and $x \in R$, then $(p, x) = (x_1, \ldots, x_{k-1}, x)$. The formula

$$\Phi_x(p) = \big(\varphi_1(p, x), \ldots, \varphi_{k-1}(p, x)\big)$$

defines, for every fixed x, a one-to-one mapping $\Phi_x\colon G_x \to R^{k-1}$ of class C_1 where

$$G_x = \{p \in R^{k-1}\colon (p, x) \in G\}$$

is an open subset of R^{k-1}. The mapping Φ_x, which depends of course on the parameter x, is defined for those x for which $G_x \neq 0$. It follows from (8) that

$$|(\Phi_x)'(p)| = |\Phi'(p, x)| \neq 0 \tag{9}$$

which proves that Φ_x is a diffeomorphism. By definition,

$$\Phi(p, x) = \big(\Phi_x(p), x\big) . \tag{10}$$

By C) it suffices to show that Φ satisfies TIS locally, e.g., that it satisfies TIS on every open interval $I \times J \subset G$ where I is an open interval in R^{k-1} and J is a one-dimensional open interval. It follows from (10) that

$$\Phi(I \times J) = \{(q, x)\colon (q \in \Phi_x(I)) \& (x \in J)\} \subset R^{k-1} \times J .$$

Let f be a non-negative measurable function on $\Phi(I \times J)$. Extend f on the product $R^{k-1} \times J$, taking

$$f(q, x) = 0 \quad \text{for} \quad (q, x) \in R^{k-1} \times J - \Phi(I \times J) .$$

By the Fubini theorem,

$$\int_{\Phi(I\times J)} f(q, x)\, d(q, x) = \int_{R^{k-1}\times J} f(q, x)\, d(q, x) = \int_J dx \int_{R^{k-1}} f(q, x)\, dq$$
$$= \int_J dx \int_{\Phi_x(I)} f(q, x)\, dq$$

since, for fixed $x \in J$, we have $f(q, x) = 0$ if $q \notin \Phi_x(I)$. By the TIS in R^{k-1} for the diffeomorphism Φ_x,

$$\int_{\Phi_x(I)} f(q, x)\, dq = \int_I f(\Phi_x(p), x) \cdot |(\Phi_x)'(p)|\, dp = \int_I f(\Phi(p, x)) \cdot |\Phi'(p, x)|\, dp$$

by (9) and (10). Using the Fubini theorem again we get

$$\int_{\Phi(I\times J)} f(q, x)\, d(q, x) = \int_J dx \int_I f(\Phi(p, x)) \cdot |\Phi'(p, x)|\, dp$$
$$= \int_{I\times J} f(\Phi(p, x)) \cdot |\Phi'(p, x)|\, d(p, x) ,$$

that is,

$$\int_{\Phi(I\times J)} f = \int_{I\times J} (f \circ \Phi) \cdot |\Phi'|$$

which we had to prove.

F) For $k > 1$, every diffeomorphism (7) is locally the composition of two reducible diffeomorphisms.

Let q_0 be a fixed point in the domain G of Φ. Since the determinant $|\Phi'(q_0)|$ is not equal to 0, at least one partial derivative $\varphi_{i|j}(q_0)$ is different from 0. To simplify the notation, let us assume that $i = j = k$, i.e. $\varphi_{k|k}(q_0) \neq 0$.

Interpret R^k as the Cartesian product $R^{k-1} \times R$. Thus, if $p = (x_1, \dots, x_{k-1}) \in R^{k-1}$ and $x \in R$, then $(p, x) = (x_1, \dots, x_{k-1}, x) \in R^k$. By hypothesis

(11) $$\varphi_{k|k}(p_0, x_0) \neq 0 \quad \text{where} \quad (p_0, x_0) = q_0 .$$

Let $y_0 = \varphi_k(p_0, x_0)$.

It follows from (11) that in a neighbourhood of the point (p_0, y_0) the equation

$$\varphi_k(p, x) - y = 0$$

can be solved with respect to x, i.e. there exists a real function ψ of the class C_1, defined in a neighbourhood of (p_0, y_0) such that

$$\varphi_k(p, \psi(p, y)) = y$$

and $\psi(p_0, y_0) = x_0$. In other words, for every point p which is sufficiently near to p_0, the function $\psi(p, y)$ of the real variable y is the inverse of the function $\varphi_k(p, x)$ of the real variable x restricted to a neighbourhood of x_0. Consequently

$$\psi(p, \varphi_k(p, x)) = x \tag{12}$$

on a neighbourhood U of the point $q_0 = (p_0, x_0)$.

Let

$$\psi_i(p, y) = \varphi_i(p, \psi(p, y)) \quad \text{for} \quad i = 1, \ldots, k-1,$$

and

$$\Phi_1(p, y) = (\psi_1(p, y), \ldots, \psi_{k-1}(p, y), y),$$

$$\Phi_2(p, x) = (p, \varphi_k(p, x)).$$

It follows from (12) that

$$\Phi(p, x) = \Phi_1(\Phi_2(p, x)) \quad \text{for} \quad (p, x) \in U. \tag{13}$$

The mappings Φ_1, Φ_2 are of class C_1. It follows from (13) that the mappings $\Phi_1|\Phi_2(U)$ and Φ_2 are one-to-one and are regular, i.e., their scalar derivatives do not vanish since the product of their scalar derivatives at corresponding points is equal to the scalar derivative of Φ (see V 3.5 (23)). This implies that they are diffeomorphisms (see V 5.2). It follows directly from their definitions that they are reducible. Thus $\Phi|U$ is the composition of the two reducible diffeomorphisms $\Phi_1|\Phi_2(U)$ and Φ_2.

G) If TIS is true for $k-1$, then it is true for k $(k > 1)$.

By C), it suffices to show that every diffeomorphism $\Phi: G \to R^k$ $(G \subset R^k)$ satisfies locally TIS. By F), every point $q \in G$ has a neighbourhood U such that $\Phi|U$ is the composition of two reducible diffeomorphisms Φ_1, Φ_2. By E), Φ_1 and Φ_2 satisfy TIS. By D), their composition $\Phi|U$ satisfies TIS, which proves that Φ satisfies TIS locally.

Theorem 5.1 follows directly from A) and G).

We shall formulate a few corollaries of 5.1.

5.2. *If $\Phi: G \to R^k$ $(G \subset R^k)$ is a diffeomorphism, then for every measurable function f on $\Phi(G)$, the composition $f \circ \Phi$ is also measurable.*

The image and the inverse image of a measurable set by the diffeomorphism Φ is a measurable set. The image and the inverse image by Φ of any set of measure zero is of measure zero.

Theorem 5.1 states that if $f \geqslant 0$ is measurable, then $(f \circ \Phi) \cdot |\Phi'|$ is measurable. Multiplying this function by the continuous (and therefore measurable) function $|\Phi'|^{-1}$ we infer that $f \circ \Phi$ is measurable. If f is any measurable function on $\Phi(G)$, then $f^+ \circ \Phi$ and $f^- \circ \Phi$ are measurable by the part of 5.2 just proved. Thus $f \circ \Phi = f^+ \circ \Phi - f^- \circ \Phi$ is measurable.

Let $A \subset \Phi(G)$ and $B = \Phi^{-1}(A)$. By definition, $\chi_{B,G} = \chi_{A,\Phi(G)} \circ \Phi$. If A is measurable, then $\chi_{A,\Phi(G)}$ is measurable and $\chi_{B,G}$ is measurable. Consequently the set $B = \{p\colon \chi_{B,G}(p) = 1\}$ is measurable. If $|A| = 0$, then $\int\limits_G \chi_{B,G} \cdot |\Phi'| = 0$. This implies, by VII 5.14, that $\chi_{B,G} = 0$ almost everywhere, i.e. $|B| = 0$.

Replacing the diffeomorphism Φ by Φ^{-1} in the part of 5.2 just proved, we infer that if $B \subset G$ is measurable, then the set $A = \Phi(B) = (\Phi^{-1})^{-1}(B)$ is measurable, and if B is of measure zero, then A is of measure zero.

Observe also that, under the hypotheses of 5.1, the identity (6) holds for every measurable set $A \subset G$ and for every measurable function on $\Phi(A)$. This follows directly from 5.2 and 5.1 (extend f over the whole set $\Phi(G)$ assuming $f(p) = 0$ for $p \in \Phi(G) - \Phi(A)$).

5.3. *If $\Phi\colon G \to R^k$ $(G \subset R^k)$ is a diffeomorphism, then for every measurable set $A \subset G$*

$$|\Phi(A)| = \int\limits_A |\Phi'| \,.$$

Theorem 5.3 follows directly from (1), where $f = \chi_{\Phi(A),\Phi(G)}$.

Diffeomorphisms $\Phi\colon G \to R^k$ $(G \subset R^k)$ such that

$$|\Phi'| = 1 \tag{14}$$

preserve not only the measurability of sets but also the measure of measurable sets, that is, for every measurable set $A \subset G$,

$$|\Phi(A)| = |A| \,. \tag{15}$$

This follows directly from 5.3. It is easy to show that only diffeomorphisms with property (14) have property (15) for all measurable subsets of G.

As the simplest example of diffeomorphisms with property (14) we mention *isometries* in R^k, i.e. mappings

$$\Phi = (\varphi_1, \ldots, \varphi_k)\colon R^k \to R^k$$

where

$$\begin{aligned} \varphi_1(x_1, \ldots, x_k) &= a_1 + a_{1,1}x_1 + \ldots + a_{1,k}x_k \,, \\ &\cdots\cdots\cdots\cdots\cdots \\ \varphi_k(x_1, \ldots, x_k) &= a_k + a_{k,1}x_1 + \ldots + a_{k,k}x_k \end{aligned}$$

and the matrix

$$\mathfrak{A} = \begin{bmatrix} a_{1,1} & \dots & a_{1,k} \\ \dots & \dots & \dots \\ a_{k,1} & \dots & a_{k,k} \end{bmatrix}$$

is orthogonal, i.e. $\mathfrak{A} \cdot \mathfrak{A}^* = \mathfrak{I}$, where $\mathfrak{I}$ is the unit matrix. Since $/\mathfrak{A}/ = /\mathfrak{A}^*/$ and $/\mathfrak{A}/ \cdot /\mathfrak{A}^*/ = /\mathfrak{A} \cdot \mathfrak{A}^*/ = /\mathfrak{I}/ = 1$, we have $/\mathfrak{A}/ = \pm 1$. Since $\Phi' = \mathfrak{A}$, we have $|\Phi'| = |\mathfrak{A}| = 1$.

Subsets A, B of R^k are said to be *congruent* if there is an isometry Φ such that $\Phi(A) = B$. It follows from 5.2 and 5.3 that congruent sets have the same measure. More precisely, if one of them is measurable, so is the other and their measures are equal.

In many books the theorem on integration by substitution is formulated for closed sets instead for open sets. We shall now quote the version of 5.1 for closed sets.

5.4. *Let Φ be a continuous mapping from the closure* cl G *of an regular open set $G \subset R^k$ into R^k. If $\Phi|G$ is a diffeomorphism and if the open set $\Phi(G)$ is regular, then for every measurable function f on* cl$\Phi(G)$

$$\int_{\mathrm{cl}\,\Phi(G)} f = \int_{\mathrm{cl}\,G} (f \circ \Phi) \cdot |\Phi'| \,. \tag{16}$$

The left-hand side of (16) is equal to the left-hand side of (1), and the right-hand side of (16) is equal to the right-hand side of (1). Thus (16) follows directly from (1).

EXERCISES

1. Calculate the integral $\int\limits_{-\infty}^{\infty}\int e^{-x^2-y^2} dx\, dy$.

Hint. Use the substitution $x = u\cos v$, $y = u \sin v$.

2. Calculate the integral $\int\limits_{-\infty}^{\infty} e^{-x^2} dx$.

Hint. Use § 2, exercise 3, where $f_1(x) = f_2(x) = e^{-x^2}$.

3. Generalize the results of exercises 1 and 2 as follows:

$$\int\limits_{-\infty}^{\infty} \dots \int e^{-(x_1^2+\dots+x_k^2)} dx_1 \dots dx_k = \pi^{k/2} \,.$$

4. Calculate the integrals:

$$\int\limits_{-\infty}^{\infty}\int (1+x^2+y^2)^{-2} dx\, dy;$$

$$\iint\limits_A \cos(x^2+y^2)\, dx\, dy \,, \quad A = \{(x, y)\colon x^2+y^2 \leqslant \pi/4\} \,;$$

$$\iiint\limits_A z(4+x^2+y^2)^{-1} dx\, dy\, dz \,, \quad A = \{(x, y, z)\colon (x^2+y^2 \leqslant 4) \& (0 \leqslant z \leqslant 1)\} \,.$$

§ 6. Spherical coordinates. The measure of a ball. We recall that the real numbers u, v are called the *spherical coordinates* of a point $p = (x_1, x_2)$ in the plane R^2 if

$$x_1 = u\cos v\,,$$
$$x_2 = u\sin v\,.$$

Usually one requires additionally that $u \geqslant 0$. The geometrical sense of u and v is well known: u is the distance from p to the point $p_0 = (0, 0)$, and v is the angle between the first coordinate axis and the segment joining p_0 and p.

The real numbers u, v_1, v_2 are the *spherical coordinates* of a point $p = (x_1, x_2, x_3) \in R^3$ if

$$x_1 = u\cos v_1 \cos v_2\,,$$
$$x_2 = u\sin v_1 \cos v_2\,,$$
$$x_3 = u\sin v_2\,.$$

Usually one reguires additionally that $u \geqslant 0$. The geometrical sense of u, v_1 and v_2 is well known: u is the distance from p to $p_0 = (0, 0, 0)$, v_2 is the angle between the segment joining p and p_0 and the plane $x_3 = 0$, and v_1 is the angle between the first coordinate axis and the projection of this segment on this plane.

More generally, the real numbers $u, v_1, \dots, v_{k-1}$ are said to be the *spherical coordinates* of a point $p = (x_1, \dots, x_k) \in R^k$ if

$$\begin{aligned}
x_1 &= u\cos v_1 \cos v_2 \cos v_3 \dots \cos v_{k-1}\,,\\
x_2 &= u\sin v_1 \cos v_2 \cos v_3 \dots \cos v_{k-1}\,,\\
x_3 &= u\sin v_2 \cos v_3 \dots \cos v_{k-1}\,,\\
x_4 &= u\sin v_3 \cos v_4 \dots \cos v_{k-1}\,,\\
&\dots\dots\dots\dots\dots\dots\\
x_{k-1} &= u\sin v_{k-2} \cos v_{k-1}\,,\\
x_k &= u\sin v_{k-1}\,.
\end{aligned} \tag{1}$$

Spherical coordinates $u, v_1, \dots, v_{k-1}$ in R^k have a simple geometrical interpretation similar to that in R^2 and R^3, but this interpretation is a little more difficult to formulate. The geometrical sense of spherical coordinates in R^k will not play a role in what follows. We shall not interpret formulas (1) as a change of a coordinate system in R^k. We shall regard (1) as a mapping $\Phi_k\colon R^k \to R^k$ which assigns to any point $(u, v_1, \dots, v_{k-1}) \in R^k$ the point $p = \Phi_k(u, v_1, \dots, v_{k-1}) = (x_1, \dots, x_k)$ whose coordinates $x_1, \dots, x_k$ are defined by formulas (1).

The practical meaning of the mapping Φ_k is that it appears very often in applications of Theorem 5.1 on integration by substitution, in the calculation of certain integrals. More precisely, to calculate an integral $\int_{G_0} f$ $(G_0 \subset R^k)$ we often use identity § 5 (1) where $\Phi = \Phi_k|G$. The mapping Φ_k is not one-to-one on the whole space R^k and, therefore, it is not a diffeomorphism; but for suitably chosen open sets $G \subset R^k$ the restriction $\Phi_k|G$ can be a diffeomorphism onto G_0.

We shall calculate the scalar derivative (i.e. the jacobian) of the mapping Φ_k.

For this purpose, let us observe that Φ_k can be represented in the form

$$\Phi_k(u, v_1, \dots, v_{k-1}) = \big(u\varphi_1(v_1, \dots, v_{k-1}), \dots, u\varphi_k(v_1, \dots, v_{k-1})\big),$$

where $\varphi_1, \dots, \varphi_k$ are certain real functions (trigonometrical polynomials) which can be easily calculated from formulas (1). Using the above notation, we can write the analogous formula for $\Phi_{k+1}\colon R^{k+1} \to R^{k+1}$ as follows:

$$\begin{aligned} &\Phi_{k+1}(u, v_1, \dots, v_k) \\ &\quad = \big(u\varphi_1(v_1, \dots, v_{k-1})\cos v_k, \dots, u\varphi_k(v_1, \dots, v_{k-1})\cos v_k, u\sin v_k\big). \end{aligned}$$

Let $q = (v_1, \dots, v_{k-1})$ for short. By a general convention from IV § 2, $\varphi_{i|j}(q)$ denotes the partial derivative of φ_i with respect to the jth coordinate (i.e., with respect to v_j) at a point q.

By a simple calculation,

$$\begin{aligned} &|\Phi'_{k+1}(u, v_1, \dots, v_{k-1}, v_k)| \\ &= \begin{vmatrix} \varphi_1(q)\cos v_k & u\varphi_{1|1}(q)\cos v_k & \dots & u\varphi_{1|k-1}(q)\cos v_k & -u\varphi_1(q)\sin v_k \\ \dots & \dots & \dots & \dots & \dots \\ \varphi_k(q)\cos v_k & u\varphi_{k|1}(q)\cos v_k & \dots & u\varphi_{k|k-1}(q)\cos v_k & -u\varphi_k(q)\sin v_k \\ \sin v_k & 0 & \dots & 0 & u\cos v_k \end{vmatrix} \\ &= (-1)^{k-1}\sin^2 v_k \cos^{k-1} v_k \begin{vmatrix} u\varphi_{1|1}(q) & \dots & u\varphi_{1|k-1}(q) & u\varphi_1(q) \\ \dots & \dots & \dots & \dots \\ u\varphi_{k|1}(q) & \dots & u\varphi_{k|k-1}(q) & u\varphi_k(q) \end{vmatrix} + \\ &\qquad + u\cos^{k+1} v_k \begin{vmatrix} \varphi_1(q) & u\varphi_{1|1}(q) & \dots & u\varphi_{1|k-1}(q) \\ \dots & \dots & \dots & \dots \\ \varphi_k(q) & u\varphi_{k|1}(q) & \dots & u\varphi_{k|k-1}(q) \end{vmatrix} \\ &= u\cos^{k-1} v_k \begin{vmatrix} \varphi_1(q) & u\varphi_{1|1}(q) & \dots & u\varphi_{1|k-1}(q) \\ \dots & \dots & \dots & \dots \\ \varphi_k(q) & u\varphi_{k|1}(q) & \dots & u\varphi_{k|k-1}(q) \end{vmatrix}. \end{aligned}$$

The last determinant is the scalar derivative $|\Phi'_k(u, v_1, \dots, v_{k-1})|$. Thus we have the inductive formula

$$|\Phi'_{k+1}(u, v_1, \dots, v_k)| = u \cdot |\Phi'_k(u, v_1, \dots, v_{k-1})| \cdot \cos^{k-1} v_k.$$

By a simple calculation,

$$|\Phi_2'(u, v_1)| = u .$$

Thus

$$(2) \qquad |\Phi_k'(u, v_1, \ldots, v_{k-1})| = u^{k-1} \cos v_2 \cos^2 v_3 \ldots \cos^{k-2} v_{k-1} .$$

The easy inductive proof is left to the reader.

We shall use identity (2) to calculate the k-dimensional Lebesgue measure $V_k(\varrho)$ of the k-dimensional open ball B with center $p_0 = (0, \ldots, 0)$ and radius ϱ:

$$B = B(p_0, \varrho) = \{p\colon |p - p_0| < \varrho\} \subset R^k .$$

Let

$$I = \{(u, v_1, \ldots, v_{k-1})\colon (0 < u < \varrho) \& (0 < v_1 < 2\pi) \& (-\pi/2 < v_2 < \pi/2) \& \\ \& (-\pi/2 < v_3 < \pi/2) \& \ldots \& (-\pi/2 < v_{k-1} < \pi/2)\} .$$

It is easy to verify that the mapping $\Phi = \Phi_k | I$ is one-to-one. It follows from (2) that the scalar derivative of Φ does not vanish. Thus Φ is a diffeomorphism. Moreover, $\Phi(I) \subset B$ and the set $B - \Phi(I)$ is a subset of the set

$$\{(x_1, \ldots, x_k)\colon (x_1 = 0) \text{ or } (x_2 = 0) \text{ or } \ldots \text{ or } (x_k = 0)\} ,$$

i.e. the union of the $(k-1)$-dimensional hyperplanes given by the equations

$$x_1 = 0, \; x_2 = 0, \; \ldots, \; x_k = 0 .$$

Thus $|B - \Phi(I)| = 0$. Using 5.3 we get

$$V_k(\varrho) = |B| = |\Phi(I)| = \int_I |\Phi'|$$

$$= \int_0^{\varrho} du \int_0^{2\pi} dv_1 \int_{-\pi/2}^{\pi/2} dv_2 \int_{-\pi/2}^{\pi/2} dv_3 \ldots \int_{-\pi/2}^{\pi/2} u^{k-1} \cos v_2 \cos^2 v_3 \ldots \cos^{k-2} v_{k-1}\, dv_{k-1}$$

by the Fubini theorem. Hence

$$V_k(\varrho) = \frac{2^{k-1} \varrho^k \pi}{k} a_1 a_2 \ldots a_{k-2} \quad \text{where} \quad a_i = \int_0^{\pi/2} \cos^i x\, dx \quad (i = 1, 2, \ldots) .$$

Let us introduce the following notation

$$0!! = 1 , \quad 1!! = 1 , \quad i!! = i \cdot (i-2)!! \quad \text{for } i > 1.$$

In other words, $i!!$ is the product of all the positive integers i, $i-2$, $i-4$, $i-6$, ...

As is known (see e.g. [10], p. 226),

$$a_i = \begin{cases} \dfrac{(i-1)!!}{i!!} & \text{if } i \text{ is odd,} \\[2ex] \dfrac{(i-1)!!}{i!!} \cdot \dfrac{\pi}{2} & \text{if } i \text{ is even} . \end{cases}$$

Consequently

(3) $$\mathrm{V}_k(\varrho) = \frac{2^{\mathrm{E}\frac{k+1}{2}}\pi^{\mathrm{E}\frac{k}{2}}}{k!!}\varrho^k$$

where $\mathrm{E}x$ denotes the greatest integer $\leqslant x$. In other symbols,

(3′) $$\mathrm{V}_{2m}(\varrho) = \frac{\pi^m}{m!}\varrho^{2m},$$

(3″) $$\mathrm{V}_{2m+1}(\varrho) = \frac{2^{m+1}\cdot\pi^m}{(2m+1)!!}\varrho^{2m+1}$$

for $m = 1, 2, \ldots$

Note also the following inductive formula

$$\mathrm{V}_k(\varrho) = \frac{2\pi}{k}\varrho^2\mathrm{V}_{k-2}(\varrho)\,.$$

We have

$$\mathrm{V}_1(\varrho) = 2\varrho\,, \qquad \mathrm{V}_2(\varrho) = \pi\varrho^2,$$

and consequently

$$\mathrm{V}_3(\varrho) = \frac{2\pi}{3}\cdot 2\varrho = \frac{4}{3}\pi\varrho^3, \qquad \mathrm{V}_4(\varrho) = \frac{2\pi}{4}\cdot\pi\varrho^4 = \tfrac{1}{2}\pi^2\varrho^4,$$

$$\mathrm{V}_5(\varrho) = \frac{2\pi}{5}\cdot\frac{4\pi}{3}\varrho^5 = \frac{8}{15}\pi^2\varrho^5, \quad \mathrm{V}_6(\varrho) = \frac{2\pi}{6}\cdot\frac{1}{2}\pi^2\varrho^6 = \frac{1}{6}\pi^3\varrho^6,$$

. . . .

The boundary $\mathrm{cl}B - B$ of the open ball B is the union of the graphs of the continuous functions

$$x_k = \sqrt{1-(x_1^2+\ldots+x_{k-1}^2)} \quad \text{and} \quad x_k = -\sqrt{1-(x_1^2+\ldots+x_{k-1}^2)}\,,$$

and therefore it is of measure zero (see 3.1). This proves that the open ball B is regular and that the k-dimensional measure of the closed ball

$$\mathrm{cl}B = \{p\colon |p-p_0| \leqslant \varrho\} \subset R^k$$

is equal to $\mathrm{V}_k(\varrho)$.

Of course, the hypothesis that p_0 is the point $(0, \ldots, 0)$ is inessential (see the remark at the end of § 5 on measure of congruent sets).

Looking for further applications we shall prove one formula more. Let

$$q = (v_1, \ldots, v_{k-1}) \quad \text{and} \quad (u, q) = (u, v_1, \ldots, v_{k-1})\,,$$

for short. Let $\varrho > 0$ be a fixed number and let

$$\Psi_k(q) = \Phi_k(\varrho, q)\,.$$

We shall calculate the modulus of the derivative of the mapping Ψ_k: $R^{k-1} \to R^k$, viz. we shall prove that

$$|\Psi'_k(q)| = \varrho^{k-1}|\cos v_2 \cos^2 v_3 \ldots \cos^{k-2} v_{k-1}|\,, \tag{4}$$

that is,

$$|\Psi'_k(q)| = |\Phi'_k(\varrho, q)|\,. \tag{4'}$$

The proof is by induction with respect to k. We verify by a direct calculation that (4) is true for $k = 2$. Suppose that (4) holds for $k > 2$. We shall prove that the analogous formula for $k+1$ holds:

$$|\Psi'_{k+1}(v_1, \ldots, v_k)| = \varrho^k|\cos v_2 \cos^2 v_3 \ldots \cos^{k-2} v_{k-1} \cos^{k-1} v_k|\,. \tag{5}$$

Using the same notation as previously we verify that $|\Psi'_{k+1}(v_1, \ldots, v_k)|^2$ is the sum of the squares of the determinants

$$\begin{vmatrix} \varrho\varphi_{1|1}(q)\cos v_k & \ldots & \varrho\varphi_{1|k-1}(q)\cos v_k & -\varrho\varphi_1(q)\sin v_k \\ \cdots & \cdots & \cdots & \cdots \\ \varrho\varphi_{i-1|1}(q)\cos v_k & \ldots & \varrho\varphi_{i-1|k-1}(q)\cos v_k & -\varrho\varphi_{i-1}(q)\sin v_k \\ \varrho\varphi_{i+1|1}(q)\cos v_k & \ldots & \varrho\varphi_{i+1|k-1}(q)\cos v_k & -\varrho\varphi_{i+1}(q)\sin v_k \\ \cdots & \cdots & \cdots & \cdots \\ \varrho\varphi_{k|1}(q)\cos v_k & \ldots & \varrho\varphi_{k|k-1}(q)\cos v_k & -\varrho\varphi_k(q)\sin v_k \\ 0 & \ldots & 0 & \varrho\cos v_k \end{vmatrix} \tag{6}$$

$(i = 1, \ldots, k)$ and of the square of the determinant

$$\begin{vmatrix} \varrho\varphi_{1|1}(q)\cos v_k & \ldots & \varrho\varphi_{1|k-1}(q)\cos v_k & -\varrho\varphi_1(q)\sin v_k \\ \cdots & \cdots & \cdots & \cdots \\ \varrho\varphi_{k|1}(q)\cos v_k & \ldots & \varrho\varphi_{k|k-1}(q)\cos v_k & -\varrho\varphi_k(q)\sin v_k \end{vmatrix}. \tag{7}$$

Expanding the determinants (6) along the last row and factoring out the terms $\cos v_k$ we see that the sum of the squares of the determinants (6) is equal to $\varrho^2|\Psi'_k(q)|^2\cos^{2k} v_k$. Factoring out all the terms $\cos v_k$ and $\sin v_k$ from the determinant (7) and transposing the last column to the first we see that the determinant (7) is equal to $\pm\varrho/\Phi'_k(\varrho, q)/\cos^{k-1} v_k \sin v_k$. Hence

$$|\Psi'_{k+1}(v_1, \ldots, v_k)|^2 = \varrho^2|\Psi'_k(q)|^2\cos^{2k} v_k + \varrho^2|\Phi'_k(\varrho, q)|^2\cos^{2k-2} v_k \sin^2 v_k\,.$$

This proves (5) by (2) and the induction hypothesis (4).

EXERCISES

1. Calculate the integrals

$$\iiiint\limits_{-\infty}^{\infty} (1+x^2+y^2+z^2+t^2)^{-3}dx\,dy\,dz\,dt\,;$$

$$\iiiint\limits_{A} e^{-(x^2+y^2+z^2+t^2)}dx\,dy\,dz\,dt\,, \qquad A = \{(x, y, z, t)\colon x^2+y^2+z^2+t^2 \leqslant a^2\}\,.$$

2. Calculate the k-dimensional Lebesgue measure of the k-dimensional ellipsoid

$$\left\{(x_1, \ldots, x_k)\colon \frac{x_1^2}{a_1^2}+\ldots+\frac{x_k^2}{a_k^2} \leqslant 1\right\}.$$

§ 7. Convolution. All functions in this section are defined on the whole space R^k. A measurable function f is said to be *locally integrable* if it is integrable on every interval $I \subset R^k$. We say that a function f *vanishes outside an interval* if the set $\{p\colon f(p) \neq 0\}$ is a subset of an interval (i.e. it is bounded). The symbol $\int_{-\infty}^{\infty}$ denotes the integral on the whole space R^k. If

$$p = (x_1, \ldots, x_k), \quad q = (y_1, \ldots, y_k)$$

then by $p-q$ we shall mean in this section the point

$$p-q = (x_1-y_1, \ldots, x_k-y_k).$$

The *convolution* of functions f, g is the function denoted by the symbol $f * g$, defined by the formula

$$f * g(p) = \int_{-\infty}^{\infty} f(p-q)g(q)\,dq. \tag{1}$$

The function $f * g$ is defined only on the set of all points p for which the integral (1) exists (the set may be empty).

For every fixed p, the mapping $\Phi(q) = p-q$ is a diffeomorphism from R^k onto itself, and $/\Phi'/ = (-1)^k$. Consequently $|\Phi'| = 1$. Performing the substitution Φ in the integral on the right-hand side of (1) (i.e. substituting $\Phi(q)$ for q), we get the identity

$$f * g(p) = \int_{-\infty}^{\infty} f(p-q)g(q)\,dq = \int_{-\infty}^{\infty} f(q)g(p-q)\,dq = g * f(p), \tag{2}$$

by 5.1. Thus

$$f * g = g * f. \tag{2'}$$

The convolution is an operation similar to multiplication. Identity (2′) states that the convolution is commutative. Under additional hypotheses, which will not be quoted here, the convolution is associative, i.e.

$$(f * g) * h = f * (g * h).$$

It is obvious that the convolution is distributive, i.e.

$$(f \pm g) * h = f * g \pm g * h,$$
$$f * (g \pm h) = f * g \pm f * h.$$

Moreover, for every $a \in R$,

$$a(f * g) = (af) * g = f * (ag).$$

The notion of convolution plays an important role in modern mathematics. In this section we shall prove only a few fundamental properties of the convolution which will be necessary for the proof of the approximation theorem 7.2.

For the convolution $f * g$ to be defined everywhere or almost everywhere on R^k, certain additional hypotheses on f and g are necessary. In this section we shall examine only the case where f is locally integrable and g is continuous and vanishes outside an interval. We shall prove that, under these hypotheses,

7.1. *The convolution $f * g$ is a continuous function defined on the whole space R^k. If f is of class C_1, then $f * g$ is of class C_1 and*

$$(f * g)_{|i} = f_{|i} * g . \tag{3}$$

If g is of class C_1, then $f * g$ is of class C_1 and

$$(f * g)_{|i} = f * g_{|i} . \tag{4}$$

Consequently, if f or g is of class C_m ($m = 1, 2, \ldots, \infty$), then $f * g$ is of class C_m.

By hypothesis there exists an interval I such that $g(q) = 0$ if $q \notin I$. Hence

$$f * g(p) = \int_I f(p-q)g(q)\,dq .$$

Since f is locally integrable, the function $f(p-q)$ of the variable q is integrable on I. The function g is bounded, $|g| \leqslant M$. This proves that the function integrated is integrable (see VII 5.13), i.e. $f * g$ is a finite real function defined for every p in R^k. If $p_n \to p$, then the sequence $f(q)g(p_n - q)$ of functions of q converges to the function $f(q)g(p-q)$ and $|f(q)g(p_n-q)| \leqslant M \cdot |f(q)|$. Moreover, there exists an interval I_0 such that all the functions $f(q)g(p_n-q), f(q)g(p-q)$ of q are equal 0 for $q \notin I_0$. Since the function $M \cdot |f(q)|$ is integrable on I_0, we have

$$f * g(p_n) = \int_{-\infty}^{\infty} f(q)g(p_n-q)\,dq = \int_{I_0} f(q)g(p_n-q)\,dq \to \int_{I_0} f(q)g(p-q)\,dq$$

$$= \int_{-\infty}^{\infty} f(q)g(p-q)\,dq = f * g(p)$$

by the Lebesgue dominated convergence theorem (VII 6.4). This proves that $f * g$ is continuous.

Let $\mathfrak{e}_i$ be the unit vector of the ith coordinate axis.

Suppose that f is of class C_1. Let p_0 be any point, and let Δt_n be any sequence of real numbers converging to 0. Let J be an interval containing all the points p_0 and $p_0 + \Delta t_n \mathfrak{e}_i$, $n = 1, 2, \ldots$ Since $f_{|i}$ is continuous and the set $\{p-q: (p \in J)\&(q \in I)\}$ is bounded, there exists a number K such that $|f_{|i}(p-q)| \leqslant K < \infty$ for $p \in J$ and $q \in I$. We have

$$\frac{f(p_0+\Delta t_n e_i - q) - f(p_0 - q)}{\Delta t_n} \cdot g(q) \to f_{|i}(p_0-q)g(q)$$

and all the functions of q on the left-hand side are bounded by the constant KM since, by the mean value theorem, the fractional factor is equal to $f_{|i}(p_n-q)$ for a $p_n \in J$. Hence by the Lebesgue dominated convergence theorem

$$\frac{f*g(p_0+\Delta t_n e_i) - f*g(p_0)}{\Delta t_n}$$

$$= \int_I \frac{f(p_0+\Delta t_n e_i - q) - f(p_0-q)}{\Delta t_n} \cdot g(q)\,dq \to \int_I f_{|i}(p_0-q)g(q)\,dq = f_{|i}*g(p_0).$$

This proves (3). It follows from (3) and the first part of 7.1 (where f should be replaced by $f_{|i}$) that all the first partial derivatives (3) of $f*g$ are continuous, i.e. $f*g$ is of class C_1.

The proof of (4) is similar. The final part of 7.1 follows from the parts just proved, by induction with respect to m.

For every open interval

$$I = \{(x_1, \ldots, x_k): a_i < x_i < b_i \text{ for } i = 1, \ldots, k\}$$

there exists a non-negative function g of class C_∞ such that $\int_{-\infty}^{\infty} g = 1$ and $g(p) = 0$ for $p \notin I$. To give an example of such a function g, let us observe that the function

$$\varphi(x) = \begin{cases} 0 & \text{for} \quad x \leqslant 0, \\ e^{-1/x} & \text{for} \quad x > 0 \end{cases}$$

is of class C_∞. The function $\varphi(x-a)\varphi(b-x)$ is of class C_∞, is positive in the interval $(a; b)$ and vanishes everywhere outside this interval. Thus the function

$$\psi(x_1, \ldots, x_k) = \prod_{i=1}^{k} \varphi(x_i - a_i)\varphi(b_i - x_i)$$

is of class C_∞, is positive in I and vanishes everywhere outside I. The function $g = \psi / \int_{-\infty}^{\infty} \psi$ has all the properties required.

Now let (g_n) be a sequence of non-negative functions of class C_∞ such that g_n vanishes everywhere outside a ball B_n with center $(0, \ldots, 0)$ and radius ϱ_n, and

$$\int_{-\infty}^{\infty} g_n = 1 \quad \text{and} \quad \varrho_n \to 0.$$

Under these hypotheses,

7.2. *For every continuous function f on R^k, the sequence $(f * g_n)$ of functions of class C_∞ converges to f uniformly on every interval. If f is of class C_1, then also $(f * g_n)_{|i} \to f_{|i}$ uniformly on every interval.*

The functions $f * g_n$ are of class C_∞ by the last part of 7.1.

Let I be any interval. By III 8.7, for every $\varepsilon > 0$ there exists a $\delta > 0$ such that

$$|p_1 - p| < \delta \text{ and } p \in I \quad \text{imply} \quad |f(p_1) - f(p)| < \varepsilon .$$

Let n_0 be an integer such that $\varrho_n < \delta$ for $n > n_0$. Hence, for $p \in I$ and $n > n_0$,

$$|f * g_n(p) - f(p)| = \Big|\int_{B_n} (f(p-q) - f(p)) g_n(q)\, dq\Big|$$

$$\leqslant \int_{B_n} |f(p-q) - f(p)| g_n(q)\, dq \leqslant \int_{B_n} \varepsilon g_n(q)\, dq = \varepsilon ,$$

which proves the first part of 7.2.

If f is of class C_1, then by (3) and the part of 7.2 just proved

$$(f * g_n)_{|i} = f_{|i} * g_n \to f_{|i}$$

uniformly on every interval I.

Observe that for any intervals

$$I = \{(x_1, \ldots, x_k): a_i \leqslant x_i \leqslant b_i \text{ for } i = 1, \ldots, k\},$$

$$J = \{(x_1, \ldots, x_k): \alpha_i < x_i < \beta_i \text{ for } i = 1, \ldots, k\},$$

such that $I \subset J$, i.e.

$$-\infty < \alpha_i < a_i < b_i < \beta_i < \infty ,$$

there exists a non-negative function φ of class C_∞ with the property

$$\varphi(p) = \begin{cases} 0 & \text{if} \quad p \notin J, \\ 1 & \text{if} \quad p \in I. \end{cases} \tag{5}$$

Namely, if ψ_i, ζ_i are non-negative functions of class C_∞ on R such that

$$\psi_i(x) = 0 \quad \text{if} \quad x \notin (\alpha_i; a_i), \quad \int_{-\infty}^{\infty} \psi_i = 1 ,$$

$$\zeta_i(x) = 0 \quad \text{if} \quad x \notin (b_i; \beta_i), \quad \int_{-\infty}^{\infty} \zeta_i = 1 ,$$

then the functions

$$\overline{\psi}_i(x) = \int_{-\infty}^{x} \psi_i , \quad \overline{\zeta}_i(x) = \int_{x}^{\infty} \zeta_i$$

are non-negative, of class C_∞, and

$$\bar{\psi}_i(x) = \begin{cases} 0 & \text{if} \quad x \leqslant a_i , \\ 1 & \text{if} \quad x \geqslant \alpha_i , \end{cases} \qquad \bar{\zeta}_i(x) = \begin{cases} 1 & \text{if} \quad x \leqslant b_i , \\ 0 & \text{if} \quad x \geqslant \beta_i . \end{cases}$$

The non-negative function

$$\varphi = \prod_{i=1}^{k} \bar{\psi}_i \bar{\zeta}_i$$

is of class C_∞ and has the property (5).

EXERCISES

1. If f is a measurable function defined on R^k, then $f(p-q)$, considered as a function of a point $(p, q) \in R^k \times R^k = R^{2k}$, is measurable on R^{2k}.

Hint. Examine, step by step, the case where f is the characteristic function of an open set, of a G_δ -set, etc.

If f and g are measurable on R^k, then $f(p-q)g(q)$ is measurable on R^{2k}.

2. If f and g are integrable on R^k, then the convolution $f * g$ is defined almost everywhere on R^k, is integrable on R^k and

$$\int_{-\infty}^{\infty} f*g = \int_{-\infty}^{\infty} f \cdot \int_{-\infty}^{\infty} g , \qquad \int_{-\infty}^{\infty} |f*g| \leqslant \int_{-\infty}^{\infty} |f| \cdot \int_{-\infty}^{\infty} |g| .$$

Hint. First treat the case where $f, g \geqslant 0$; use the Fubini theorem to interchange the order of integration.

3. If functions f, g, h are integrable on R^k, then $f * (g * h) = (f * g) * h$.

4. If f is integrable on R^k and g is bounded, then $f * g$ is a continuous real function on R^k.

Hint. Use § 1, exercise 5.

5. If f, g are functions defined on R, equal to 0 on $(-\infty; 0)$ and integrable on every bounded interval $(0; a)$, then $f * g$ vanishes on $(-\infty; 0)$ and is integrable on every bounded interval $(0; a)$. Moreover,

$$f * g(x) = \int_0^x f(x-t) g(t) = \int_0^x f(t) g(x-t)\, dt .$$

6. Under the hypotheses and notation of 7.2, if f is of class C_m, then $(f * g_n)^{(m} \to f^{(m)}$ uniformly on every interval.

§ 8. The indefinite integral. The improper integral. In this section we shall examine exclusively integrals of functions defined on sets of real numbers.

Let $I \subset R$ be an interval, bounded or unbounded, closed or open, or half-open. A measurable function f defined on I is said to be *locally integrable* if it is integrable on every bounded closed interval contained in I. Of course, every function integrable on I is locally integrable (see VII 5.3).

If f is locally integrable on I, then the formula

$$F(u, v) = \int_u^v f \tag{1}$$

defines a real function on the set $I \times I \subset R^2$.

8.1. *If f is locally integrable on I, then the function* (1) *is continuous on $I \times I$.*

Consequently, for every fixed $x_0 \in I$ and any number c, the function

$$g(x) = \int_{x_0}^x f + c \qquad (x \in I) \tag{2}$$

is continuous on I.

It suffices to prove the first part of 8.1. The second part follows directly from the first.

For this purpose it is enough to show that for every bounded closed interval $I_0 \subset I$ the function F is continuous on $I_0 \times I_0$. Since f is integrable on I_0, for every $\varepsilon > 0$ there exists a $\delta > 0$ such that for every measurable set $B \subset I_0$,

$$\text{if } |B| < \delta, \text{ then } \Big|\int_B f\Big| \leqslant \int_B |f| < \varepsilon/2$$

(see VII 5.16). Hence, for any $u, u_1, v, v_1 \in I_0$, if $|u - u_1| < \delta$ and $|v - v_1| < \delta$, then

$$|F(u, v) - F(u_1, v_1)| = \Big|\int_u^{u_1} f + \int_{v_1}^v f\Big| \leqslant \Big|\int_u^{u_1} f\Big| + \Big|\int_{v_1}^v f\Big| < \varepsilon/2 + \varepsilon/2 = \varepsilon,$$

which proves the continuity of F.

Every function (2) is called an *indefinite Lebesgue integral* of f. It is easy to prove that

$$g'(x) = f(x) \tag{3}$$

at every point x at which f is continuous. Thus $g' = f$ in the case where f is continuous. One can show by simple examples that (3) does not hold, in general, if f is not continuous at x. On the other hand, one can prove that (3) always holds almost everywhere on I.

Let f be a locally integrable function on an open (bounded or unbounded) interval $(a; b)$. If the integral

$$\lim_{\substack{v \to b- \\ u \to a+}} \int_u^v f \tag{4}$$

exists, it is called the *improper* Lebesgue integral of f on $(a; b)$. As we shall prove, the improper integral is a generalization of the ordinary integral which will here be called the *proper* integral. Namely,

8.2. *If the proper integral* $\int_a^b f$ *exists, then the improper integral of f on* $(a; b)$ *also exists, and both integrals are equal.*

It suffices to prove that if $u_n \searrow a$, $v_n \nearrow b$, $u_1 < v_1$, then $\int_{u_n}^{v_n} f \to \int_a^b f$. It follows from VII 6.8 that

$$\int_a^b f = \sum_{n=1}^{\infty} \int_{u_{n+1}}^{u_n} f + \int_{u_1}^{v_1} f + \sum_{n=1}^{\infty} \int_{v_n}^{v_{n+1}} f$$

$$= \lim_{n\to\infty} \Big(\int_{u_n}^{u_{n-1}} f + \ldots + \int_{u_2}^{u_1} f + \int_{u_1}^{v_1} f + \int_{v_1}^{v_2} f + \ldots + \int_{v_{n-1}}^{v_n} f \Big) = \lim_{n\to\infty} \int_{u_n}^{v_n} f .$$

The improper integral of f on $(a; b)$ will be denoted by the same symbol $\int_a^b f$ as the proper integral. By 8.2, this notation does not lead to misunderstanding.

The most important case is where the improper integral $\int_a^b f$ is finite. We say then that this integral is *convergent*.

It follows from 8.2 that in the case of non-negative measurable functions f, both notions, the proper integral and the improper integral, coincide since the integral of a non-negative measurable function always exists. In the case of measurable functions which do not have constant sign the two notions do not coincide.

8.3. *A convergent improper integral* $\int_a^b f$ *is proper if and only if* $\int_a^b |f| < \infty$.

If the improper integral is proper, i.e. if f is integrable on $(a; b)$, then $|f|$ is integrable on $(a; b)$ by VII 5.10, i.e. $\int_a^b |f| < \infty$. Conversely, if $\int_a^b |f| < \infty$, i.e. if $|f|$ is integrable on $(a; b)$, then f is integrable on $(a; b)$ by VII 5.10, and the proper and improper integrals on $(a; b)$ coincide by 8.2.

EXERCISES

1. The function $f(x) = (-1)^n n$ for $1/(n+1) < x < 1/n$ $(n = 1, 2, \ldots)$ has an improper integral on $(0; 1)$ but does not have a proper integral.

2. The improper Lebesgue integral $\int_0^\infty \frac{\sin x}{x}\, dx$ is convergent but it is not a proper integral.

3. Let f be a real function defined on the set

$$I = \{(x, y)\colon (a < x < b) \& (\alpha \leqslant y \leqslant \beta)\} \quad (-\infty < \alpha < \beta < \infty)$$

such that for every y in the interval $\langle\alpha;\beta\rangle$ the function $f(\cdot, y)$ has a convergent improper integral on $(a; b)$. We say that the improper integral $\int_a^b f(x, y)$ *converges uniformly* if for every $\varepsilon > 0$ there exist numbers u_0, v_0 $(a < u_0 < v_0 < b)$ such that the inequalities $a < u < u_0$, $v_0 < v < b$ imply

$$\left|\int_u^v f(x, y)\,dx - \int_a^b f(x, y)\,dx\right| < \varepsilon \quad \text{for every } y \text{ in } \langle\alpha;\beta\rangle .$$

Prove that if the integral $\varphi(y) = \int_a^b f(x, y)\,dx$ converges uniformly and if the function f is continuous, then the function φ is continuous.

Hint. Consider the functions $\varphi_n(y) = \int_{u_n}^{v_n} f(x, y)\,dx$ where $u_n \searrow a$, $v_n \nearrow b$; use § 1, exercise 2 and III 9.2.

4. Let f be a continuous function defined on the set I in exercise 3. If the integral $\int_a^b f(x, y)\,dx$ converges uniformly, then the improper integral $\int_a^b \psi(x)\,dx$ of the function $\psi(x) = \int_\alpha^\beta f(x, y)\,dy$ $(a < x < b)$ exists and

$$\int_a^b dx \int_\alpha^\beta f(x, y)\,dy = \int_\alpha^\beta dy \int_a^b f(x, y)\,dx .$$

Hint. Examine the sequence (φ_n) from exercise 2 and interchange the order of integration in the iterated integral $\int_\alpha^\beta \varphi_n(y)$, using the Fubini theorem. Apply VII 6.5.

5. Let f be a continuous function defined on the set I from exercise 3. If the improper integral $\varphi(y) = \int_a^b f(x, y)\,dx$ exists for every y in $\langle\alpha;\beta\rangle$, the partial derivative $f_{|2}$ is continuous and the improper integral $\int_a^b f_{|2}(x, y)\,dx$ converges uniformly, then the function φ is of class C_1 and

$$\varphi'(y) = \int_a^b f_{|2}(x, y)\,dx .$$

In the traditional notation,

$$\frac{d}{dy}\int_a^b f(x, y)\,dx = \int_a^b \frac{\partial}{\partial y} f(x, y)\,dx.$$

Hint. Introduce the functions φ_n from exercise 3, use § 1, exercise 3 and the known theorem on differentation of sequences of functions of one real variable.

6. If a function f is integrable on every interval $(a; v)$ where $a < v < b$, then the improper integral $\int_a^b f$ exists if and only if the limit $\lim\limits_{v\to b-} \int_a^v f$ exists. Then the improper integral is equal to this limit.

Similarly, if f is integrable on every interval $(u; b)$ where $a < u < b$, then the improper integral $\int_a^b f$ exists if and only if the limit $\lim\limits_{u\to a+} \int_u^b f$ exists. Then the improper integral is equal to this limit.

CHAPTER IX

INTEGRALS ON HYPERSURFACES AND POLYHEDRA

§ 1. The k-dimensional measure on k-dimensional hypersurfaces. By Theorem VIII 3.4, the l-dimensional Lebesgue measure of a k-dimensional regular hypersurface in R^l $(k < l)$ is equal to 0. This is consistent with intuition. The two-dimensional measure of a curve in the plane should be equal to 0, and similarly, the three-dimensional measure of a curve or a surface in R^3 should be equal to 0, etc. As we stated in VI § 1, our intuition also suggests that a special one-dimensional measure should be introduced for subsets of curves in such a way that it would be a generalization of the intuitive notion of length. A two-dimensional measure should be defined for subsets of surfaces in such a way that it would be a generalization of the intuitive notion of area, etc. This problem will be solved by Theorem 1.1 below which yields a general definition of a k-dimensional measure on any k-dimensional regular hypersurface in R^l, and on p. 331 we shall show that the formal definition from Theorem 1.1 is consistent with our intuition. The problem is restricted to regular hypersurfaces only, in order to avoid singularities which appear when the hypothesis of regularity is abandoned (see exercises 11 and 12).

1.1. *Let H be a k-dimensional regular hypersurface in R^l, $k \leqslant l$. There exists exactly one measure μ_H defined on a σ-algebra $\mathcal{M}_H$ of subsets of H, such that*

(a) *a set $A \subset H$ is in $\mathcal{M}_H$ if and only if, for every diffeomorphism Φ: $G \to H$ (where G is an open subset of R^k), the set $\Phi^{-1}(A)$ is Lebesgue-measurable;*

(b) *for every diffeomorphism Φ: $G \to H$ $(G \subset R^k)$, if $A \subset \Phi(G)$ and $A \in \mathcal{M}_H$, then*

$$\mu_H(A) = \int_{\Phi^{-1}(A)} |\Phi'| \,. \tag{1}$$

The measure μ_H is complete. All Borel subsets of the metric space H are in $\mathcal{M}_H$.

The integral on the right-hand side of (1) is with respect to the k-dimensional Lebesgue measure.

The unique measure μ_H will be called the *k-dimensional measure on the hypersurface H*. According to a general convention in VI § 3, sets in $\mathcal{M}_H$ will be called *μ_H-measurable*, or simply *measurable* if it is clear that the measurability in question is with respect to μ_H.

Note that if $k = l$, then H is an open subset of R^k (see p. 210). Assuming in (b) that Φ is the identity mapping from H onto itself, we infer that μ_H coincides with the k-dimensional Lebesgue measure restricted to Lebesgue-measurable subsets of H.

To prove 1.1, first observe that condition (a) uniquely defines the class $\mathcal{M}_H$. Namely, $\mathcal{M}_H$ is the class of all sets $A \subset H$ such that the set $\Phi^{-1}(A)$ is Lebesgue-measurable for every diffeomorphism $\Phi: G \to R^k$ $(G \subset R^k)$. It follows from I 3.1 (5), (6), (8) that the class $\mathcal{M}_H$ so defined is a σ-algebra of subsets of H.

If a set A is open in the space H, then for every diffeomorphism $\Phi: G \to H$ the set $\Phi^{-1}(A)$ is open in G (see III 8.2 (b)), i.e. it is open (III 4.2). This proves that every open subset of the space H is in $\mathcal{M}_H$. The class of all Borel subsets of the space H is the smallest σ-algebra containing all sets open in H. The class $\mathcal{M}_H$ is a σ-algebra containing all sets open in H. This proves that the first class is contained in the second one, i.e. that every Borel subset of the space H belongs to $\mathcal{M}_H$.

Now we shall describe a construction of a measure satisfying condition (b).

Since H is a regular k-dimensional hypersurface, every point $q \in H$ has a neighbourhood U in H which is a diffeomorphic image of an open set $G \subset R^k$. The space H is separable, by III 5.2 and III 5.3. Since it is the union of these open sets U, it is the union of a countable sequence of these sets, by the Lindelöf theorem III 5.1. In other words, there exists a sequence of diffeomorphisms

$$\Phi_n: G_n \to H \qquad (G_n \subset R^k)$$

such that

$$H = \bigcup_{n=1}^{\infty} \Phi_n(G_n)\,.$$

The disjoint sets

$$B_1 = \Phi_1(G_1)\,, \qquad B_n = \Phi_n(G_n) - \big(\Phi_1(G_1) \cup \ldots \cup \Phi_{n-1}(G_{n-1})\big) \qquad \text{for} \qquad n > 1$$

belong to $\mathcal{M}_H$ (because they are Borel sets) and

$$H = \bigcup_{n=1}^{\infty} B_n$$

(see I 2.5). For any set $A \in \mathcal{M}_H$, let

$$C_n = B_n \cap A \in \mathcal{M}_H \qquad (n = 1, 2, \ldots) \tag{2}$$

and let

$$\mu_H(A) = \sum_{n=1}^{\infty} \int_{\Phi_n^{-1}(C_n)} |\Phi_n'| \,. \tag{3}$$

The function μ_H just defined on $\mathcal{M}_H$ is a measure. In fact, it is non-negative and vanishes on the empty set. It is also countably additive. Indeed, if $A = \bigcup_{m=1}^{\infty} A_m$, where (A_m) is a sequence of disjoint sets in $\mathcal{M}_H$, then denoting $C_{m,n} = A_m \cap B_n$, we have

$$\Phi_n^{-1}(C_n) = \bigcup_{m=1}^{\infty} \Phi_n^{-1}(C_{m,n})$$

and the sets on the right-hand side are disjoint. Consequently

$$\mu_H(A) = \sum_{n=1}^{\infty} \int_{\Phi_n^{-1}(C_n)} |\Phi_n'| = \sum_{n=1}^{\infty}\sum_{m=1}^{\infty} \int_{\Phi_n^{-1}(C_{m,n})} |\Phi'|$$

$$= \sum_{m=1}^{\infty}\sum_{n=1}^{\infty} \int_{\Phi^{-1}(C_{m,n})} |\Phi_n'| = \sum_{m=1}^{\infty} \mu_H(A_m) \,.$$

The measure μ_H has property (b). Indeed, if $A \subset \Phi(G)$, where $A \in \mathcal{M}_H$ and Φ: $G \to R^k$ is a diffeomorphism ($G \subset R^k$), then denoting $\Lambda_n = \Phi_n^{-1} \circ \Phi$, we have

$$\Lambda_n\big(\Phi^{-1}(C_n)\big) = \Phi_n^{-1}(C_n) \,.$$

By the theorem on integration by substitution VIII 5.1

$$\int_{\Phi_n^{-1}(C_n)} |\Phi_n'| = \int_{\Phi^{-1}(C_n)} (|\Phi_n'| \circ \Lambda_n)\cdot|\Lambda_n'| = \int_{\Phi^{-1}(C_n)} |\Phi'|$$

(see V 3.5). Adding these equations ($n = 1, 2, \ldots$) we get (1).

Condition (b) uniquely determines the measure μ_H. In fact, if μ_H is a measure on $\mathcal{M}_H$ with property (b) and $A \in \mathcal{M}_H$, then defining C_n by (2) we have $C_n \subset \Phi_n(G_n)$ and, by (1),

$$\mu_H(C_n) = \int_{\Phi_n^{-1}(C_n)} |\Phi_n'| \,.$$

Adding all the equations ($n = 1, 2, \ldots$) we get (3) since A is the union of the disjoint sets C_n. We see that every measure satisfying (b) coincides with the measure defined by (3).

It follows from (3) and (b) (see also VII 5.14) that $\mu_H(A) = 0$ if and only if the k-dimensional Lebesgue measure of the set $\Phi^{-1}(A)$ is equal to 0 for every diffeomorphism Φ: $G \to H$ ($G \subset R^k$). Since the Lebesgue measure is complete, this implies that μ_H is complete.

The last remark on the structure of sets of μ_H-measure zero will be used in the proof of the following theorem.

1.2. *If H is a regular k-dimensional hypersurface, and if $H' \subset H$ is a regular hypersurface of dimension $<k$, then $\mu_H(H') = 0$.*

In fact, for every diffeomorphism Φ: $G \to H$ $(G \subset R^k)$ the set $\Phi^{-1}(H')$ is either empty or a regular hypersurface of dimension $<k$ (see V 6.3) and therefore its k-dimensional Lebesgue measure is equal to zero (see VIII 3.4).

1.3. *An extended real function f defined on a set $A \subset \mathcal{M}_H$ of a k-dimensional hypersurface H is μ_H-measurable if and only if, for every diffeomorphism Φ: $G \to H$ $(G \subset R^k)$, the composition $f \circ \Phi$ is measurable with respect to the k-dimensional Lebesgue measure. In particular, every real continuous function f on $A \in \mathcal{M}_H$ is measurable.*

If $A \in \mathcal{M}_H$ and $A \subset \Phi(G)$, where Φ: $G \to H$ is a diffeomorphism $(G \subset R^k)$, then for every μ_H-measurable function f on A

$$\int_A f\,d\mu_H = \int_{\Phi^{-1}(A)} (f \circ \Phi) \cdot |\Phi'| \,. \tag{4}$$

The integral on the right-hand side of (4) is with respect to the k-dimensional Lebesgue measure.

Identity (4) should be understood as follows: if one of the integrals in (4) exists, then the other also exists, and they are equal.

The first part of 1.3 follows from (a) and the following identity:

$$\{p\colon f \circ \Phi(p) < a\} = \Phi^{-1}(\{q\colon f(q) < a\}) \,.$$

It follows from (b) that (4) is true in the case where f is the characteristic function of a measurable set $B \subset A$. Consequently (4) is true if f is a simple function. This implies, by VII 3.1, that (4) is true for every measurable non-negative function. Representing f as the difference $f^+ - f^-$ of the non-negative measurable functions f^+ and f^- (see p. 260), we infer that (4) is true for every measurable f.

1.4. *Let H' and H be regular k-dimensional hypersurfaces in R^l and $H' \subset H$. Then*

$$\mathcal{M}_{H'} = \{A \in \mathcal{M}_H\colon A \subset H'\} \quad \textit{and} \quad \mu_{H'} = \mu_H | \mathcal{M}_{H'} \,, \tag{5}$$

that is, for every set $A \subset H'$,

$$\mu_{H'}(A) = \mu_H(A) \,.$$

We recall that H' is open in H by V 6.1. Consequently $H' \in \mathcal{M}_H$. It is easy to verify that the class $\mathcal{M}_{H'}$ and the set function $\mu_{H'}$ defined by (5) satisfy conditions (a) and (b) where H is replaced by H'. This proves 1.4 since (a) and (b) uniquely determine the σ-algebra of measurable subsets and the measure on H'.

Now we shall explain the geometrical meaning of the measure μ_H. Suppose Φ: $G \to H$ is a diffeomorphism ($G \subset R^k$). Let I be a closed interval, $I \subset G$. By 1.1 (b)

$$\mu_H^{*}(\Phi(I)) = \int_I |\Phi'| \,. \tag{6}$$

By VIII 1.4, the integral in (6) is the limit of Riemann sums corresponding to a regular sequence of partitions of I. For simplicity we shall examine only partitions of I onto closed subintervals. Let

$$\mathbf{P} = (I_1, \dots, I_m)$$

be such a partition and let

$$S = \sum_{i=1}^{m} |\Phi'(p_i)| \cdot |I_i| \qquad (p_i \in I_i)$$

be a corresponding Riemann sum. It is known from geometry that the number $|\Phi'(p_i)| \cdot |I_i|$ is the k-dimensional volume of the parallelepiped

$$\{\Phi(p_i) + \Phi'(p_i) \cdot (p - p_i) \colon \ p \in I_i\} \tag{7}$$

which is a part of the hyperplane tangent to H at the point p_i (the reader can check this, without difficulty, in the case where e.g. $l = 3$ and $k = 1$ or 2). We know from the Taylor formula (see V 2.1 (19)-(20)) that if p is close to p_i, then $\Phi(p)$ is almost equal to $\Phi(p_i) + \Phi'(p_i) \cdot (p - p_i)$. Thus if the diameter of I_i is very small, the set (7) is a good approximation to the set $\Phi(I_i)$, and the number $|\Phi'(p_i)| \cdot |I_i|$ is a good approximation to what our intuition suggests that we call the k-dimensional area (or volume) of the set $\Phi(I_i)$. Consequently the Riemann sum S is a good approximation to the k-dimensional area (volume) of the set $\Phi(I) = \Phi(I_1) \cup \dots \cup \Phi(I_m)$. Thus our intuition suggests that we define the k-dimensional area (or volume) of $\Phi(I)$ as the limit of a sequence of Riemann sums, i.e. as the integral (6). This argument shows that the formal definition of μ_H from 1.1 is consistent with the intuitive notion of k-dimensional measure on H.

As an illustration we shall calculate the k-dimensional measure $S_k(\varrho) = \mu_H(H)$ of the k-dimensional sphere

$$H = \{p \in R^{k+1} \colon \ |p - p_0| = \varrho\}$$

with radius $\varrho > 0$. It follows easily from 1.1 that the k-dimensional measure is invariant under translations. Therefore it suffices to calculate it in the case where $p_0 = (0, \dots, 0)$.

Let us take as the open set G in 1.1 (b) the open interval

$$\{(v_1, \dots, v_k) \colon \ (0 < v_1 < 2\pi) \& (-\pi/2 < v_2 < \pi/2) \& (-\pi/2 < v_3 < \pi/2) \& \\ \& \dots \& (-\pi/2 < v_k < \pi/2)\} ,$$

and let $\Phi = \Psi_{k+1}|G$ where Ψ_{k+1} is the mapping from VIII § 6, pp. 317-318, i.e.

$$\Psi_{k+1}(v_1, \ldots, v_k) = (x_1, \ldots, x_{k+1}),$$

where

$$\begin{aligned} x_1 &= \varrho \cos v_1 \cos v_2 \ldots \cos v_k, \\ x_2 &= \varrho \sin v_1 \cos v_2 \ldots \cos v_k, \\ x_3 &= \varrho \sin v_2 \cos v_3 \ldots \cos v_k, \\ x_4 &= \varrho \sin v_3 \cos v_4 \ldots \cos v_k, \\ &\ldots\ldots\ldots\ldots\ldots\ldots \\ x_k &= \varrho \sin v_{k-1} \cos v_k, \\ x_{k+1} &= \varrho \sin v_k. \end{aligned}$$

By definition, $\Phi(G) \subset H$ and $H - \Phi(G)$ is contained in the union of the intersections of H with the k-dimensional hyperplanes defined by the equations

$$x_1 = 0, \; \ldots, \; x_{k+1} = 0$$

respectively. Thus the set $H - \Phi(H)$ is a subset of the union of $k+1$ $(k-1)$-dimensional hypersurfaces and, therefore, $\mu_H(H - \Phi(G)) = 0$ by 1.2. Consequently

$$S_k(\varrho) = \mu_H(\Phi(G)) = \int_G |\Phi'| = \int_G |\Psi'_{k+1}|.$$

By VIII § 6 (5)

$$|\Psi'_{k+1}(v_1, \ldots, v_k)| = \varrho^k \cos v_2 \cos^2 v_3 \ldots \cos^{k-1} v_k$$

for $(v_1, \ldots, v_k) \in G$. By the Fubini theorem,

$$S_k(\varrho) = \varrho^k \int_0^{2\pi} dv_1 \int_{-\pi/2}^{\pi/2} dv_2 \int_{-\pi/2}^{\pi/2} dv_3 \ldots \int_{-\pi/2}^{\pi/2} \cos v_2 \cos^2 v_3 \ldots \cos^{k-1} v_k \, dv_k.$$

Comparing this formula with the analogous formula for the k-dimensional measure $V_k(\varrho)$ of the k-dimensional ball (see p. 316), we get the formula

(8) $$S_k(\varrho) = \frac{k+1}{\varrho} V_{k+1}(\varrho).$$

Consequently

(9) $$S_k(\varrho) = \frac{2^{1+E\frac{k}{2}} \pi^{E\frac{k+1}{2}}}{(k-1)!!} \varrho^k$$

and

$$S_k(\varrho) = \frac{2\pi\varrho^2}{k-1} S_{k-2}(\varrho).$$

In particular

$$S_1(\varrho) = 2\pi\varrho, \qquad S_2(\varrho) = 4\pi\varrho^2,$$
$$S_3(\varrho) = 2\pi^2\varrho^3, \qquad S_4(\varrho) = \tfrac{8}{3}\pi^2\varrho^4,$$
$$S_5(\varrho) = \pi^3\varrho^5, \qquad S_6(\varrho) = \tfrac{16}{15}\pi^3\varrho^6,$$
$$\ldots\ldots \qquad \ldots\ldots$$

We shall also discuss two particular cases occurring most frequently in practical applications.

A) Let $G = (a; b) \subset R$ be a bounded or unbounded open interval of real numbers and let

$$\Phi = (\varphi_1, \ldots, \varphi_l): G \to R^l$$

be a diffeomorphism. The set

$$H = \{\Phi(t): t \in G\}$$

is a regular curve in R^l. Each set of the form

(10) $$A = \{\Phi(t): t_1 \leqslant t \leqslant t_2\} \quad (a < t_1 < t_2 < b)$$

is called a *regular arc*. The one-dimensional measure $\mu_H(A)$ is called the *length of the arc* A. By (1), the length of A is equal to the integral

(11) $$\int_{t_1}^{t_2} |\Phi'| = \int_{t_1}^{t_2} \sqrt{(\varphi_1'(t))^2 + \ldots + (\varphi_l'(t))^2}\, dt\,.$$

Introducing the indefinite integral

(12) $$s(t) = \int_{t_0}^{t} |\Phi'| = \int_{t_0}^{t} \sqrt{(\varphi_1'(\tau))^2 + \ldots + (\varphi_l'(\tau))^2}\, d\tau + c \quad \text{for} \quad t \in G,$$

where t_0 is a fixed number in G, we see that the length (11) of A is equal to the increase $s(t_2) - s(t_1)$ of the function s. By definition (12) and VIII § 8 (3),

(13) $$s' = |\Phi'| = \sqrt{(\varphi_1')^2 + \ldots + (\varphi_l')^2}\,.$$

B) Let

$$\Phi = (\varphi_1, \varphi_2, \varphi_3): G \to R^3$$

be a diffeomorphism from an open set $G \subset R^2$ into R^3. The set

$$H = \{\Phi(p): p \in G\}$$

is a surface in R^3. The two-dimensional measure $\mu_H(H)$, i.e. the *area of the surface* H, is equal to the integral

$$(14)\qquad \int_G |\Phi'| = \int_G \sqrt{\begin{vmatrix} \varphi_{2|1} & \varphi_{2|2} \\ \varphi_{3|1} & \varphi_{3|2} \end{vmatrix}^2 + \begin{vmatrix} \varphi_{3|1} & \varphi_{3|2} \\ \varphi_{1|1} & \varphi_{1|2} \end{vmatrix}^2 + \begin{vmatrix} \varphi_{1|1} & \varphi_{1|2} \\ \varphi_{2|1} & \varphi_{2|2} \end{vmatrix}^2}$$

$$= \int_G |\Phi_{|1} \vee \Phi_{|2}| = \int_G \sqrt{\begin{vmatrix} \Phi_{|1}\Phi_{|1} & \Phi_{|1}\Phi_{|2} \\ \Phi_{|2}\Phi_{|1} & \Phi_{|2}\Phi_{|2} \end{vmatrix}}$$

with respect to the two-dimensional Lebesgue measure.

In the traditional notation, the area of H is equal to the integral

$$(14')\qquad \iint_G \sqrt{\left(\frac{\partial(\varphi_2, \varphi_3)}{\partial(u_1, u_2)}\right)^2 + \left(\frac{\partial(\varphi_3, \varphi_1)}{\partial(u_1, u_2)}\right)^2 + \left(\frac{\partial(\varphi_1, \varphi_2)}{\partial(u_1, u_2)}\right)^2}\, du_1 du_2$$

$$= \iint_G |\Phi_{u_1} \vee \Phi_{u_2}| = \iint_G \sqrt{\begin{vmatrix} \frac{\partial \Phi}{\partial u_1}\frac{\partial \Phi}{\partial u_1} & \frac{\partial \Phi}{\partial u_1}\frac{\partial \Phi}{\partial u_2} \\ \frac{\partial \Phi}{\partial u_2}\frac{\partial \Phi}{\partial u_1} & \frac{\partial \Phi}{\partial u_2}\frac{\partial \Phi}{\partial u_2} \end{vmatrix}}\, du_1 du_2,$$

where u_1, u_2 are the coordinates of any point $p \in G$.

In V § 6 we introduced the notion of the zero-dimensional hypersurface H in R^l. By μ_H we then mean the zero-dimensional measure defined on all sets $A \subset H$ by the formula

$$\mu_H(A) = \text{the number of elements in } A$$

(see VI § 3, p. 237). Only the case where A is a finite set will be considered in this chapter. If $A = \{p_1, \dots, p_n\}$ is finite ($p_i \neq p_j$ for $i \neq j$), then any real function f on A is measurable and

$$(15)\qquad \int_A f d\mu_H = f(p_1) + \dots + f(p_n).$$

EXERCISES

1. Calculate the two-dimensional measure (the area) of the torus T defined in V § 6, pp. 213-214.

2. Calculate the two-dimensional measure of the torus T_0 defined in V § 6, exercise **3.**

3. Calculate the two-dimensional measure of the open Möbius strip M_0 defined in V § 6, exercise 4.

4. Calculate the area of the surface

$$\{(u\cos v, u\sin c, cv) \colon (u_1 < u < u_2) \& (v_1 < v < v_2)\} \subset R^3$$

where $0 < u_1 < u_2$ and $v_1 < v_2$. Calculate the integral of the function

$$f(x_1, x_2, x_3) = x_3\sqrt{x_1^2 + x_2^2 + c^2}$$

on this surface.

5. Let Ψ be an isometric mapping of R^l onto itself. For every k-dimensional hypersurface $H \subset R^l$ the set $\Psi(H)$ is also a k-dimensional hypersurface. For every set $A \subset H$,

$$A \in \mathcal{M}_H \text{ if and only if } \Psi(A) \in \mathcal{M}_{\Psi(H)} .$$

Then

$$\mu_H(A) = \mu_{\Psi(H)}(\Psi(A)) .$$

Hint. Use V § 1, exercise 5.

6. Let $\Phi\colon G \to R^l$ where $G \subset R^k$, and $\Lambda = (\lambda_1, \ldots, \lambda_k)\colon (a\,; b) \to G$ (where $-\infty \leqslant a < b \leqslant \infty$) be diffeomorphisms, and let $\Psi = \Phi \circ \Lambda$. The length of the regular arc $A = \{\Psi(t)\colon t_1 \leqslant t \leqslant t_2\}$ is given by the integral

$$\int_{t_1}^{t_2} \sqrt{\sum_{i,j=1}^{k} g_{i,j}(\Lambda(t))\,\lambda_i'(t)\,\lambda_j'(t)\,dt} ,$$

where $g_{i,j} = \Phi_{|i}\Phi_{|j}$ is the scalar product of the partial derivatives of Φ.

7. Let $\Psi = (\psi_1, \ldots, \psi_l)$ be a mapping of class C_1 from a bounded or unbounded interval $(a\,; b) \subset R$ into R^l. The length of the regular arc

$$A = \{(x, \psi_1(x), \ldots, \psi_l(x))\colon x_1 \leqslant x \leqslant x_2\} \subset R^{l+1} \qquad (a < x_1 < x_2 < b)$$

is given by the integral

$$\int_{x_1}^{x_2} \sqrt{1+|\Psi'|^2} = \int_{x_1}^{x_2} \sqrt{1+\psi_1'(x)^2+\ldots+\psi_l'(x)^2}\, dx .$$

8. Let ψ be a real function of class C_1 defined on an open set $G \subset R^2$. The area of the surface

$$H = \{(x, y, \psi(x, y))\colon (x, y) \in G\} \subset R^3$$

is given by the integral

$$\int_G \sqrt{1+|\psi'|^2} = \int_G \sqrt{1+(\psi_{|1})^2+(\psi_{|2})^2} = \iint_G \sqrt{1+\psi_x(x, y)^2+\psi_y(x, y)^2}\, dx\, dy$$

with respect to the two-dimensional Lebesgue measure.

For every function f on H

$$\int_H f\, d\mu_H = \iint_G f(x, y, \psi(x, y)) \sqrt{1+\psi_x(x, y)^2+\psi_y(x, y)^2}\, dx\, dy .$$

9. Let ψ be a real function of class C_1 defined on an open set $G \subset R^k$. For any set $A \subset G$, let

$$A^* = \{(x_1, \ldots, x_k, \psi(x_1, \ldots, x_k))\colon (x_1, \ldots, x_k) \in A\} \subset R^{k+1} .$$

It follows from V § 6, p. 211 that the set $H = G^*$, i.e. the graph of the function ψ, is a k-dimensional regular hypersurface in R^{k+1}.

Verify that $A^* \in \mathcal{M}_H$ if and only if A is Lebesgue-measurable. Then

$$\mu_H(A^*) = \int_A \sqrt{1+(\psi_{|1})^2+\ldots+(\psi_{|k})^2} = \int_A \sqrt{1+\psi'\cdot\psi'} ,$$

where the integration is with respect to the k-dimensional Lebesgue measure. For every extended real function f on A^*,

$$\int_{A^*} f\,d\mu_H = \int_A \dots \int f(x_1, \dots, x_k, \psi(x_1, \dots, x_k)) \sqrt{1+\sum_{i=1}^{k} (\psi_{|i}(x_1, \dots, x_k))^2}\, dx_1 \dots dx_k\,.$$

More precisely, the integral on the left-hand side exists if and only if the integral on the right-hand side exists. Then both integrals are equal.

10. Let $\Psi = (\psi_1, \dots, \psi_l)$ be a mapping of class C_1 from an open set $G \subset R^k$ into R^l. If $p = (x_1, \dots, x_k)$ and $q = (x_{k+1}, \dots, x_{k+l})$, let (p, q) be the point $(x_1, \dots, x_{k+l}) \in R^{k+l}$. For any set $A \subset G$, let

$$A^* = \{(p, \Psi(p)) \in R^{k+l}: p \in A\}\,.$$

By V § 6 (p. 211), the set $H = G^*$, i.e. the graph of the mapping Ψ, is a k-dimensional regular hypersurface in R^{k+l}.

Prove that $A^* \in \mathcal{M}_H$ if and only if A is Lebesgue-measurable. Then

$$\mu_H(A^*) = \int_A J$$

(the integral with respect to the k-dimensional Lebesgue measure) where

$$J = \sqrt{1+\sum_{n=1}^{m} \sum_{i_1<\dots<i_n} \sum_{j_1<\dots<j_n} \begin{vmatrix} \psi_{i_1|j_1} & \dots & \psi_{i_1|j_n} \\ \cdot & \cdots & \cdot \\ \psi_{i_n|j_1} & \dots & \psi_{i_n|j_n} \end{vmatrix}^2}\,,$$

$m = \min(k, l)$, $i_1, \dots, i_n$ is any increasing sequence of positive integers $\leqslant l$ and $j_1, \dots, j_n$ is any increasing sequence of positive integers $\leqslant k$. In other symbols,

$$J = \sqrt{|\mathfrak{I}+\mathfrak{G}|}\,,$$

where $\mathfrak{I}$ is the k-dimensional unit matrix, and $\mathfrak{G}$ is the Gram matrix

$$\mathfrak{G} = \begin{bmatrix} \Psi_{|1}\Psi_{|1} & \dots & \Psi_{|1}\Psi_{|k} \\ \cdot & \cdots & \cdot \\ \Psi_{|k}\Psi_{|1} & \dots & \Psi_{|k}\Psi_{|k} \end{bmatrix}.$$

For every function f on A^*,

$$\int_{A^*} f\,d\mu_H = \int_A \dots \int f(x_1, \dots, x_k, \psi_1(x_1, \dots, x_k), \dots, \psi_l(x_1, \dots, x_k)) J(x_1, \dots, x_k)\, dx_1 \dots dx_k\,.$$

More precisely, the integral on the left-hand side exists if and only if the integral on the right-hand side exists. Then they are equal.

Hint. Introduce the diffeomorphism $\Phi(p) = (p, \Psi(p))$ from G onto H. Use II 5.2 (21) and II § 5 (18).

11. Let T be a closed triangle in R^2, let S be the two-dimensional Lebesgue measure of T, and let $(s_0, s_1, s_2, \dots)$ be a sequence of positive numbers such that

$$s = S - \sum_{n=0}^{\infty} 2^n s_n > 0\,.$$

Let q_1, q_2, q_3 be the vertices of T (see figure 13).

Remove from T a triangle with area s_0 (without two edges) whose base lies on the segment joining q_1 and q_2, and whose vertex is q_3, in such a way that the remaining closed set A is the union of two triangles T_0 and T_1 with the same area (see figure 14 and 15).

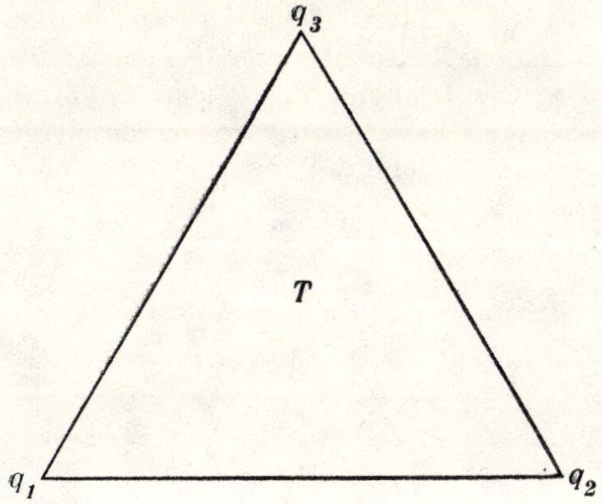

Fig. 13

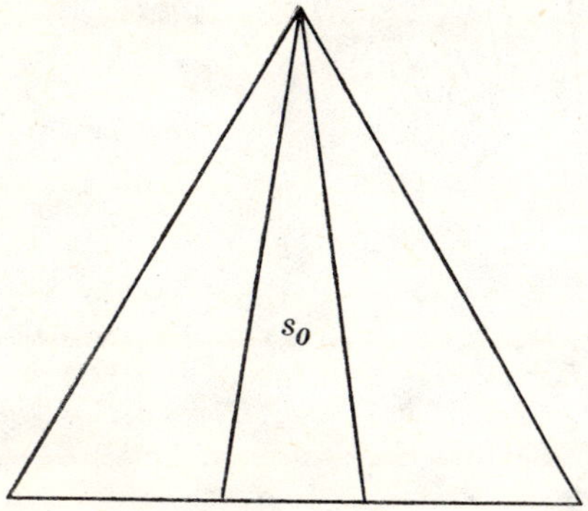

Fig. 14

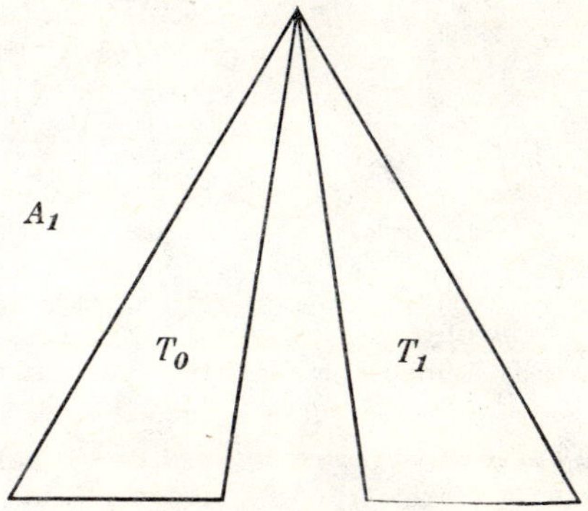

Fig. 15

Remove from each of the triangles T_0, T_1 a triangle with area s_1, such that the remaining closed set A_2 is the union of four triangles with the same area, which are situated as in figure 16.

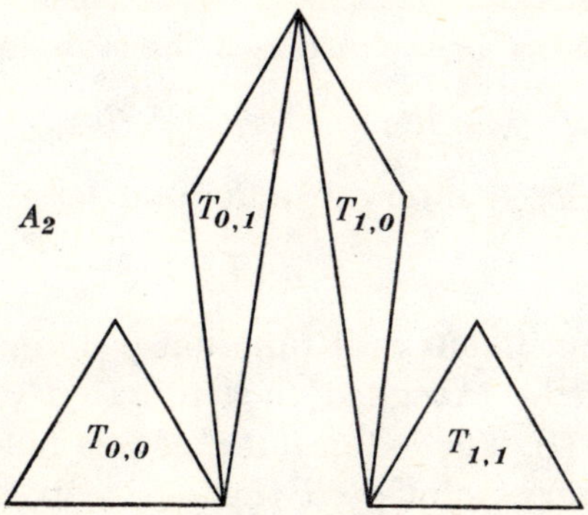

Fig. 16

Continuing the procedure described above, we define a decreasing sequence (A_n) of closed connected sets such that

(a) A_n is the union of 2^n triangles $T_{i_1,\dots,i_n}$ (where $i_1, \dots, i_n$ are equal to 0 or 1) with the same area;

(b) the triangles $T_{i_1,\dots,i_{n-1},0}$ and $T_{i_1,\dots,i_{n-1},1}$ are obtained from the triangle $T_{i_1,\dots,i_{n-1}}$ by removing a triangle (without two edges) with area s_{n-1};

(c) for every fixed n, the triangles $T_{i_1,\dots,i_n}$ can be arranged in a sequence such that two neighbouring triangles have exactly one common point (a vertex), and two non-neighbouring triangles are disjoint.

Figure 17 shows the set A_3.

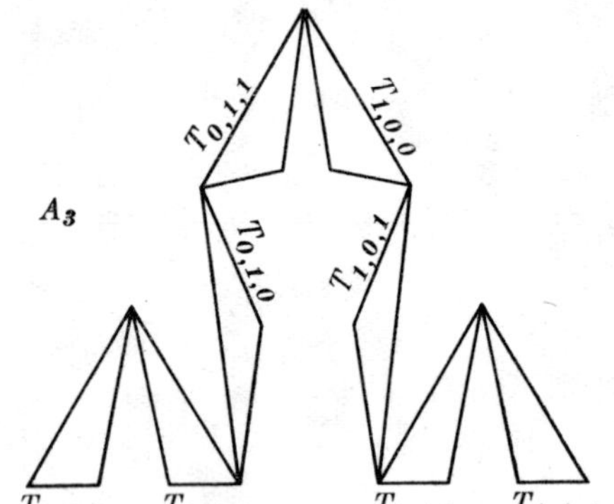

Fig. 17

The intersection $A = A_1 \cap A_2 \cap \dots$ is not empty since $q_1, q_2 \in A$. Prove that the set A is homeomorphic with the unit interval $I = \langle 0; 1\rangle$. More precisely, there exists a homeomorphism Φ of I onto A such that $\Phi(0) = q_1$ and $\Phi(1) = q_2$. The set $H - \{q_1, q_2\}$ is a curve in the sense defined in V § 6, p. 209. Verify that the two-dimensional Lebesgue measure of the arc A, and of the curve H, is positive, viz. it is equal to s.

Of course the curve H is not regular.

12. Let H be the curve defined in exercise 11. The set $H \times H$ is a surface (in R^4) whose four-dimensional Lebesgue measure is positive. For every $(l-2)$-dimensional open interval I $(l > 2)$, the set $H \times I$ is an $(l-1)$-dimensional hypersurface (in R^l) whose l-dimensional Lebesgue measure is positive.

§ 2. Functions of class C_m on a closed interval.

Let

$$I = \{(x_1, \dots, x_k) \in R^k\colon\ a_i \leqslant x_i \leqslant \beta_i \text{ for } i = 1, \dots, k\} \tag{1}$$

be a non-degenerate closed interval in R^k and let

$$I^\circ = \operatorname{int} I\,. \tag{2}$$

We say that a continuous real function φ defined on I is *of class* C_m $(1 \leqslant m < \infty)$ if it can be extended to a function ψ of class C_m, defined on an open set G $(I \subset G)$. The extension ψ of φ is not uniquely determined by φ. However, for every point $p \in I$ the derivatives $\psi^{(r)}(p)$ $(r = 1, \dots, m)$ are uniquely determined by φ. Indeed, if $p \in I^\circ$, then $\psi^{(r)}(p) = (\varphi|I^\circ)^{(r)}(p)$

since $\psi|I^\circ = \varphi|I^\circ$. If $p \in I - I^\circ$, then there exists a sequence $p_n \in I^\circ$ converging to p. Consequently

$$\psi^{(r)}(p) = \lim_{n\to\infty} \psi^{(r)}(p_n) = \lim_{n\to\infty} (\varphi|I^\circ)^{(r)}(p_n) .$$

We see that the matrix field $\psi^{(r)}|I$ is uniquely determined by φ, viz. it is the only continuous extension of $(\varphi|I^\circ)^{(r)}$ over the whole closed interval I. The matrix field $\psi^{(r)}|I$ will be called the *r-th derivative* (or the *derivative of order* r) of φ and will be denoted by the symbol $\varphi^{(r)}$. The $(i_1, \ldots, i_r)$-th coordinate of $\varphi^{(r)}$ will be called the *partial derivative of* φ *of order* r with respect to the $(i_1, \ldots, i_r)$-coordinates and will be denoted by the symbol $\varphi_{|i_1,\ldots,i_r}$, as in the case of functions defined on an open set. By definition,

(3) $$\varphi^{(r)} = [\varphi_{|i_1,\ldots,i_r}]_{i_1,\ldots,i_r\leqslant k} ,$$

(4) $$\varphi^{(r)}|I^\circ = (\varphi|I^\circ)^{(r)} ,$$

(5) $$\varphi_{|i_1,\ldots,i_r}|I^\circ = (\varphi|I^\circ)_{|i_1,\ldots,i_r} ;$$

namely $\varphi^{(r)}$ and $\varphi_{|i_1,\ldots,i_r}$ are the only continuous extensions of $(\varphi|I^\circ)^{(r)}$ and $(\varphi|I^\circ)_{|i_1,\ldots,i_r}$ to the closed interval I, respectively. Observe that, for $p \in I^\circ$, the partial derivative $\varphi_{|i_1,\ldots,i_r}(p)$ is the result of a consecutive differentiation of φ with respect to certain coordinates, as was described in IV § 2, pp. 132-133. The same is true for the partial derivative $\varphi_{|i_1,\ldots,i_r}(p)$ where $p \in I - I^\circ$ but some of the derivatives with respect to the coordinates should now be interpreted as one-sided derivatives.

For every $p \in I$ the matrix $\varphi^{(r)}(p) = [\varphi_{i_1,\ldots,i_r}(p)]$ is symmetric. In fact, it follows from IV 5.4 that this is so for $p \in I^\circ$. It follows from the continuity of the partial derivatives that this also holds for $p \in I - I^\circ$.

Every real function φ of class C_m on I $(1 \leqslant m < \infty)$ can be extended to a function of class C_m, defined on the whole space R^k. Indeed, if $\psi\colon G \to R$ is an extension of φ of class C_m to an open set G, then there exists an open interval J such that $I \subset J$ and $\mathrm{cl}J \subset G$, and there exists a real function φ_0 of class C_∞ defined on R^k such that φ_0 is equal to 1 on I and is equal to 0 outside J (see VIII § 7 (5)). The function

$$\psi_0(p) = \begin{cases} \varphi(p)\varphi_0(p) & \text{if} \quad p \in G , \\ 0 \quad \text{if} & p \notin G \end{cases}$$

is an extension of φ to R^k. It is of class C_m because it is of class C_m on the open sets G and $R^k - \mathrm{cl}J$ whose union is R^k. Namely it is equal to $\varphi\varphi_0$ on the first of the sets, and is equal to 0 on the second set.

Generalizing the definition introduced at the beginning of this section, we say that a continuous mapping

(6) $$\Phi = (\varphi_1, \ldots, \varphi_l)$$

from the closed interval I (see (1)) into R^l is *of class* C_m (or that a continuous matrix field

$$\Phi = [\varphi_{j_1,\dots,j_s}]_{j_1\leqslant k_1,\dots,j_s\leqslant k_s} \tag{7}$$

defined on I is *of class* C_m) if Φ can be extended to a mapping into R^l of class C_m (or to a matrix field of class C_m) which is defined on an open set G ($I \subset G$), $1 \leqslant m < \infty$. Then Φ can be extended to a mapping of class C_m (or to a matrix field of class C_m) which is defined on the whole space R^k. The proof of the last statement is the same as for real functions.

It follows directly from the definition that a mapping (6) on I is of class C_m if and only if all its coordinates $\varphi_1, \dots, \varphi_l$ are of class C_m. Analogously to the case of mappings defined on open sets (see V § 2), the matrix field

$$\Phi^{(r)} = [\varphi_{j|i_1,\dots,i_r}]_{j\leqslant l, i_1,\dots,i_r\leqslant k} \qquad (r = 1, \dots, m)$$

will be called the *r-th derivative* (or the *derivative of order r*) of Φ and the vector field

$$\Phi_{|i_1,\dots,i_r} = [\varphi_{1|i_1,\dots,i_r}, \dots, \varphi_{l|i_1,\dots,i_r}],$$

where $i_1, \dots, i_r$ are fixed positive integers $\leqslant k$, will be called the *r-th partial derivative* (or the *partial derivative of order r*) *of* (6) *with respect to the* $(i_1, \dots, i_r)$*-coordinates*. By definition, the derivative $\Phi^{(r)}$ is the only continuous extension of $(\Phi|I^\circ)^{(r)}$ to the closed interval I, and the partial derivative $\Phi_{|i_1,\dots,i_r}$ is the only continuous extension of the partial derivative $(\Phi|I^\circ)_{|i_1,\dots,i_r}$ to I.

Similarly, it follows directly from the definition that a matrix field (7) is of class C_m if and only if all its coordinates $\varphi_{j_1,\dots,j_s}$ are of class C_m. Analogously to the case of matrix fields defined on open sets, the matrix field

$$\Phi^{(r)} = [\varphi_{j_1,\dots,j_s|i_1,\dots,i_r}]_{j_1\leqslant k_1,\dots,j_s\leqslant k_s, i_1,\dots,i_r\leqslant k}$$

will be called the *r-th derivative* (or the *derivative of order r*) of Φ, and the matrix field

$$\Phi_{|i_1,\dots,i_r} = [\varphi_{j_1,\dots,j_s|i_1,\dots,i_r}]_{j_1\leqslant k_1,\dots,j_s\leqslant k_s},$$

where $i_1, \dots, i_r$ are fixed positive integers $\leqslant k$, will be called the *r-th partial derivative* (or the *partial derivative of order r*) *of* (7) *with respect to the* $(i_1, \dots, i_r)$*-coordinates*. By definition, the derivative $\Phi^{(r)}$ is the only continuous extension of $(\Phi|I^\circ)^{(r)}$ to the closed interval I, and the partial derivative $\Phi_{|i_1,\dots,i_r}$ is the only continuous extension of the partial derivative $(\Phi|I^\circ)_{|i_1,\dots,i_r}$ to I.

Following the definition from V § 5 we say that a mapping (6) on a closed interval (1) is *regular* if it is of class C_1 and $|\Phi'(p)| \neq 0$ for every

$p \in I$. We say that a mapping (6) on a closed interval (1) is a *diffeomorphism* if it is regular and is a homeomorphism.

2.1. *Every regular mapping of class C_m defined on a closed bounded interval I can be extended to a regular mapping of class C_m, defined on an open interval containing I.*

2.2. *Every diffeomorphism of class C_m defined on a closed bounded interval I can be extended to a diffeomorphism of class C_m, defined on an open interval containing I.*

Suppose that Φ is a regular mapping defined on a closed bounded interval (1). The mapping Φ has an extension Ψ of class C_m defined on the whole space R^k. Let

$$I_n = \left\{(x_1, \ldots, x_k)\colon\ a_i - \frac{1}{n} < x_i < \beta_i + \frac{1}{n}\right\}.$$

The set $G = \{p\colon |\Psi'(p)| \neq 0\}$ is open since $|\Psi'|$ is a continuous function. The interval I is a subset of G since $|\Psi'(p)| = |\Phi'(p)| \neq 0$ for $p \in I$. It follows from III 6.4 that

$$I_n \subset G \quad \text{for} \quad n \geqslant n_0 .$$

The mapping $\Psi|I_n$, where $n \geqslant n_0$, satisfies the conditions of Theorem 2.1.

Suppose now that Φ is a diffeomorphism. We shall prove that there exists an integer $n \geqslant n_0$ such that $\Psi|I_n$ is one-to-one and, therefore, is a diffeomorphism (see V 5.2). Consequently $\Psi|I_n$ is the extension of Φ, required in 2.2.

Suppose that such an integer n does not exist, i.e. for every positive integer n there are points p_n, q_n with the properties

$$p_n \neq q_n , \quad p_n, q_n \in I_n , \quad \Psi(p_n) = \Psi(q_n) .$$

Since the sequences (p_n) and (q_n) are bounded, there exists an increasing sequence (s_n) of positive integers such that

$$p_{s_n} \to p \in I \quad \text{and} \quad q_{s_n} \to q \in I .$$

Since $\Phi(p) = \Psi(p) = \lim\limits_{n\to\infty} \Psi(p_{s_n}) = \lim\limits_{n\to\infty} \Psi(q_{s_n}) = \Psi(q) = \Phi(q)$, we have $q = p$ since Φ is one-to-one. It follows from V 5.1 that the point p has a neighbourhood $U \subset I_{n_0}$ such that the regular mapping $\Psi|U$ is one-to-one. This leads to a contradiction since for n sufficiently large the points p_{s_n}, q_{s_n} are in U and $p_{s_n} \neq q_{s_n}$, $\Psi(p_{s_n}) = \Psi(q_{s_n})$.

Now let $J = \langle a; b\rangle$ be a one-dimensional closed interval and let $J^\circ = (a; b)$ be its interior.

A continuous real function φ defined on J is of class C_m $(1 \leqslant m < \infty)$ if and only if the function $\varphi|J^\circ$ is of class C_m and the one-sided limits

$$a_r = \lim_{x\to a+} \varphi^{(r)}(x) , \quad \beta_r = \lim_{x\to b-} \varphi^{(r)}(x) , \quad r = 1, \ldots, m$$

exist. The necessity of this condition is obvious. To prove the sufficiency it is enough to verify that the function ψ defined as follows

$$\psi(x) = \begin{cases} \varphi(a) + \dfrac{a_1(x-a)}{1!} + \ldots + \dfrac{a_m(x-a)^m}{m!} & \text{if} \quad x < a, \\ \varphi(x) \quad \text{if} \quad a \leqslant x \leqslant b, & \\ \varphi(b) + \dfrac{\beta_1(x-b)}{1!} + \ldots + \dfrac{\beta_m(x-b)^m}{m!} & \text{if} \quad x > b, \end{cases}$$

is an extension of φ of class C_m over the space R. Observe that the limits a_r and β_r are then the derivatives of the order r of φ at the points a and b respectively, in the sense of the definition adopted in this section.

Applying the above statement to the coordinates of the mapping (6) defined on a one-dimensional interval, we get the following theorem.

2.3. *A continuous mapping Φ from an one-dimensional closed interval J into R^l is of class C_m $(1 \leqslant m < \infty)$ if and only if the mapping $\Phi|J^\circ$ is of class C_m and all the derivatives $(\Phi|J^\circ)', \ldots, (\Phi|J^\circ)^{(m)}$ can be extended to the whole interval J in a continuous manner.*

We shall now prove the following modification of a particular case of Theorem V 5.4 on three mappings.

2.4. *Let Φ be a diffeomorphism from a bounded closed interval $I \subset R^k$ into R^l $(l \geqslant k)$, let Ψ be a mapping from a bounded closed one-dimensional interval $J \subset R$ into $\Phi(I)$, and let $\Lambda = \Phi^{-1} \circ \Psi \colon J \to I$.*

(a) *If Ψ is of class C_1, then Λ is of class C_1 and for every point $p \in J$*

$$\Lambda'(p) = \big(\Phi'(\Lambda(p))^* \cdot \Phi'(\Lambda(p))\big)^{-1} \cdot \Phi'(\Lambda(p))^* \cdot \Psi'(p)\,. \tag{8}$$

If Φ and Ψ are of class C_m, then Λ is of class C_m.

(b) *If Ψ is regular, then Λ is regular.*

(c) *If Ψ is a diffeomorphism, then Λ is a diffeomorphism.*

By 2.2 we may assume that the diffeomorphism Φ is defined not only on I but also on an open set G containing I.

Suppose that Ψ is of class C_1. Applying Theorem V 5.4 to the diffeomorphism Φ on G and to the mappings Ψ and Λ reduced to the open set $G' = \operatorname{int} J$ we assert that (8) holds for every point $p \in G'$. The expression on the right-hand side of (8) is a continuous function on the whole closed interval J. Thus $(\Lambda|G')'$ has a continuous extension over the whole closed interval J. This proves that Λ is of class C_1 and that (8) holds for every $p \in J$.

The proof of the remaining parts of 2.4 is analogous to that of the corresponding parts of V 5.4.

Theorem 2.4 can be generalized to the case of a mapping Ψ defined on a $\varkappa$-dimensional closed interval J (then the sign $\cdot$ in (8) should be replaced by $\bullet$; the inequality $\varkappa \leqslant k$ must be postulated in (b) and (c)), but the proof is then more difficult.

2.5. *If Φ is a mapping of class C_m from a closed interval $I \subset R^k$ into R^l, and Λ is a mapping of class C_m from a closed interval $J \subset R^{\varkappa}$ into I, then the composition $\Psi = \Phi \circ \Lambda$ is of class C_m.*

If $\varkappa \leqslant k \leqslant l$, and Φ and Λ are regular mappings (diffeomorphisms), then Ψ is a regular mapping (diffeomorphism).

To prove the first part of 2.5, extend Φ and Λ to mappings of class C_m defined on the whole spaces R^k and $R^{\varkappa}$ respectively. By V 3.4 the composition of these mappings is of class C_m. The mapping Ψ is the restriction of the composition to J; therefore it is of class C_m.

The proof of the second part of 2.5 is analogous to that in V 5.5.

EXERCISE

1. Prove that the words "on an open interval containing I" can be replaced in 2.1 and 2.2 by the words "on the whole Euclidean space".

§ 3. Faces of an interval. The following notation and definitions will be systematically used in the next sections.

Let

$$(1) \qquad I = \{(x_1, \ldots, x_k)\colon \alpha_i \leqslant x_i \leqslant \beta_i \text{ for } i = 1, \ldots, k\}$$

be a bounded closed non-degenerate interval in R^k. As in § 2, the symbol I° will denote the interior of I:

$$(2) \qquad I^\circ = \{(x_1, \ldots, x_k)\colon \alpha_i < x_i < \beta_i \text{ for } i = 1, \ldots, k\}\,.$$

The sets

$$(3) \qquad I_{j,-1} = \{(x_1, \ldots, x_k)\colon x_j = \alpha_j \text{ and } \alpha_i \leqslant x_i \leqslant \beta_i \text{ for } i \neq j\}\,,$$

$$(4) \qquad I_{j,1} = \{(x_1, \ldots, x_k)\colon x_j = \beta_j \text{ and } \alpha_i \leqslant x_i \leqslant \beta_i \text{ for } i \neq j\}$$

($j = 1, \ldots, k$) are the $(k-1)$-*dimensional closed faces* of I (see II § 2 (7)). The sets

$$(5) \qquad I^\circ_{j,-1} = \{(x_1, \ldots, x_k)\colon x_j = \alpha_j \text{ and } \alpha_i < x_i < \beta_i \text{ for } i \neq j\}\,,$$

$$(6) \qquad I^\circ_{j,1} = \{(x_1, \ldots, x_k)\colon x_j = \beta_j \text{ and } \alpha_i < x_i < \beta_i \text{ for } i \neq j\}$$

are the $(k-1)$-*dimensional open faces* of I.

More generally, for every sequence $j_1, \ldots, j_r$ of different positive integers $\leqslant k$ and for any sequence $\eta_1, \ldots, \eta_r$ composed of the numbers -1 and 1, the sets

$$(7) \qquad I_{j_1,\eta_1;\ldots;j_r,\eta_r} = \{(x_1, \ldots, x_k)\colon x_{j_s} = \gamma_{j_s,\eta_s} \text{ for } s = 1, \ldots, r$$
$$\text{and } \alpha_i \leqslant x_i \leqslant \beta_i \text{ for } i \neq j_1, \ldots, j_r\}\,,$$

where

(8) $$\gamma_{j_s,\eta_s} = \begin{cases} \alpha_{j_s} & \text{if} \quad \eta_s = -1, \\ \beta_{j_s} & \text{if} \quad \eta_s = +1, \end{cases}$$

are the $(k-r)$-*dimensional closed faces* of I, and the sets

(9) $$I^\circ_{j_1,\eta_1;\ldots;j_r,\eta_r} = \{(x_1, \ldots, x_k): x_{j_s} = \gamma_{j_s} \text{ for } s = 1, \ldots, r \text{ and } \alpha_i < x_i < \beta_i \text{ for } i \neq j_1, \ldots, j_r\},$$

where the numbers γ_{j_s,η_s} are defined by (8), are the $(k-r)$-*dimensional open faces* of I. In particular (the case where $r = k$), the one-point sets of the form

$$\{(\gamma_1, \ldots, \gamma_k)\} \quad \text{where} \quad \gamma_i = \alpha_i \text{ or } \beta_i \text{ for } i = 1, \ldots, k,$$

that is, the one-point sets formed from *vertices* of I, are simultaneously the *zero-dimensional closed faces* and the *zero-dimensional open faces* of I.

It is convenient to assume that the whole interval I is its k-*dimensional closed face*, and that I° is its k-*dimensional open face*.

Observe that the closed face (7) does not depend on the ordering of the pairs of lower indices. In particular,

(10) $$I_{j_1,\eta_1;j_2,\eta_2} = I_{j_2,\eta_2;j_1,\eta_1} \quad \text{for} \quad j_1 \neq j_2.$$

Similarly the open face (9) does not depend on the ordering of pairs of lower indices. In particular

(11) $$I^\circ_{j_1,\eta_1;j_2,\eta_2} = I^\circ_{j_2,\eta_2;j_1,\eta_1} \quad \text{for} \quad j_1 \neq j_2.$$

3.1. *The intersection of two closed faces of I is either empty or is a face of I.*

Namely the intersection

(12) $$I_{j_1,\eta_1;\ldots;j_r,\eta_r} \cap I_{j'_1,\eta'_1;\ldots;j'_s,\eta'_s}$$

is empty if there exist positive integers i and i' such that

$$j_i = j'_{i'} \quad \text{and} \quad \eta_i = -\eta'_{i'}.$$

In the opposite case the set (12) is the face

(13) $$I_{j''_1,\eta''_1;\ldots;j''_n,\eta''_n}$$

such that the set of pairs j'', η'' appearing in (13) is the union of the sets of pairs of lower indexes appearing in (12).

Observe that the closed interval I is the union of all its open faces of all dimensions $\leqslant k$ and that the open faces are disjoint. Every closed face is the closure of exactly one open face, viz. the open face which has the same lower indices in the above notation.

The set $bI = I - I^\circ$ is the *boundary* of I. By definition, bI is the union of all $(k-1)$-dimensional closed faces of I. The set bI is also the union of all open faces of dimension $<k$ (in this union all summands are disjoint).

Observe also that all open and closed faces of I are connected (see III 11.1).

The union S_s of all s-dimensional closed faces of I $(0 \leqslant s \leqslant k)$ is called the *s-dimensional skeleton* of I. The s-dimensional skeleton S_s is also the union of all open faces of dimension $\leqslant s$. For $s > 0$, the skeleton S_s is connected (we do not prove this because we shall not use this fact in what follows).

If $0 \leqslant s < k$, the union S_s° of all s-dimensional open faces of I is not connected by III 11.5 since s-dimensional open faces are non-empty disjoint open sets which are open in the space S_s°. Since they are connected, the s-dimensional open faces are the components of S_s°.

We shall also assume the following notation. If $j_1, \ldots, j_r$ is any sequence of different positive integers $\leqslant k$ $(1 \leqslant r < k)$, let

$$i_1 < i_2 < \ldots < i_{k-r}$$

be the sequence formed of all the remaining $k-r$ positive integers $\leqslant k$. The symbol $I_{j_1,\ldots,j_r}$ will denote the following non-degenerate closed interval in R^{k-r}:

$$(14) \qquad I_{j_1,\ldots,j_r} = \{(x_{i_1}, \ldots, x_{i_{k-r}}):\ a_{i_s} \leqslant x_{i_s} \leqslant \beta_{i_s} \text{ for } s = 1, \ldots, k-r\}.$$

According to the convention (2) the symbol $I^\circ_{j_1,\ldots,j_r}$ will denote the interior of $I_{j_1,\ldots,j_r}$, i.e. the open interval

$$(14') \qquad I^\circ_{j_1,\ldots,j_r} = \{(x_1, \ldots, x_{k-1}):\ a_{i_s} < x_s < \beta_{i_s} \text{ for } s = 1, \ldots, k-r\}.$$

In particular

$$(15) \qquad I_j = \{(x_1, \ldots, x_{j-1}, x_{j+1}, \ldots, x_k):\ a_i \leqslant x_i \leqslant \beta_i \text{ for } i = 1, \ldots, j-1, j+1, \ldots, k\},$$

$$(15') \qquad I^\circ_j = \{(x_1, \ldots, x_{j-1}, x_{j+1}, \ldots, x_k):\ a_i < x_i < \beta_i \text{ for } i = 1, \ldots, j-1, j+1, \ldots, k\}.$$

Let the sequences $j_1, \ldots, j_r$ and $i_1, \ldots, i_{k-r}$ have the same meaning as above, and let $\eta_1, \ldots, \eta_r$ be a sequence formed of the numbers -1 and 1. Let γ_{j_s,η_s} $(s = 1, \ldots, r)$ be the numbers defined by (8). Then the formula

$$(16) \qquad \Lambda_{j_1,\eta_1;\ldots;j_r,\eta_r}(y_1, \ldots, y_{k-r}) = (x_1, \ldots, x_k),$$

where

$$(17) \qquad x_{j_s} = \gamma_{j_s,\eta_s} \text{ for } s = 1, \ldots, r \quad \text{and} \quad x_{i_s} = y_s \text{ for } s = 1, \ldots, k-r,$$

defines a mapping from $I_{j_1,\ldots,j_r}$ on the face $I_{j_1,\eta_1;\ldots;j_r,\eta_r}$. The mapping (16) is called the *natural mapping* of (14) onto (7).

In the case where $r = k$, by $I_{j_1,\ldots,j_k}$ and by $I^\circ_{j_1,\ldots,j_k}$ we shall mean the set $\{0\}$ composed of one element: the number zero. The set $\{0\}$ will be considered as the only *zero-dimensional interval* which is simultaneously open and closed. By $\Lambda_{j_1,\eta_1;\ldots;j_k,\eta_k}$ we shall mean the only mapping of $I_{j_1,\ldots,j_k}$ onto the one-point set $I_{j_1,\eta_1;\ldots;j_k,\eta_k}$ according to the general definition (16)-(17).

For any mapping $\Phi\colon I \to R^l$ we shall assume the following notation

(18) $$\Phi_{j_1,\eta_1;\ldots;j_r,\eta_r} = \Phi \circ \Lambda_{j_1,\eta_1;\ldots;j_r,\eta_r} \qquad (1 \leqslant r \leqslant k)\,.$$

By definition, (18) is a mapping of the $(k-r)$-dimensional interval $I_{j_1,\ldots,j_r}$ into R^l.

Observe that the interval $I_{j_1,\ldots,j_r}$ does not depend on the ordering of the indices $j_1, \ldots, j_r$. Similarly the mappings $\Lambda_{j_1,\eta_1;\ldots;j_r,\eta_r}$ and $\Phi_{j_1,\eta_1;\ldots;j_r,\eta_r}$ do not depend on the ordering of pairs j, η of indices. In particular we have

(19) $$I_{j_1,j_2} = I_{j_2,j_1}, \qquad \Lambda_{j_1,\eta_1;j_2,\eta_2} = \Lambda_{j_2,\eta_2;j_1,\eta_1}, \qquad \Phi_{j_1,\eta_1;j_2,\eta_2} = \Phi_{j_2,\eta_2;j_1,\eta_1}$$

for $j_1 \neq j_2$.

By definitions (16) and (18),

(20) $$\Phi_{j_1,\eta_1;\ldots;j_r,\eta_r}(y_1, \ldots, y_{k-r}) = \Phi(x_1, \ldots, x_k)\,,$$

where $x_1, \ldots, x_k$ are defined by (17). Thus the mapping (18) is obtained from Φ by fixing the variables $x_{j_1}, \ldots, x_{j_s}$, i.e. it is a function of the remaining variables $x_{i_1}, \ldots, x_{i_{k-r}}$ only.

It follows from (20) that if Φ is of class C_m, then the mapping (18) is also of class C_m (the partial derivatives of (18) coincide with certain partial derivatives of Φ). It is less evident that if Φ is regular, then (18) is also regular. To prove this, it suffices to show that if

$$p = (y_1, \ldots, y_{k-r}) \in I_{j_1,\ldots,j_r} \quad \text{and} \quad \mathfrak{a} = [a_1, \ldots, a_{k-r}] \neq 0\,,$$

then

(21) $$\Phi'_{j_1,\eta_1;\ldots;j_r,\eta_r}(p)\cdot \mathfrak{a} \neq 0\,.$$

This follows from the fact that the vector on the left-hand side of (21) is equal to

$$\Phi'(q)\cdot \mathfrak{b} \neq 0\,,$$

where $q = (x_1, \ldots, x_k)$ is the point whose coordinates are defined by (17), and $\mathfrak{b} = [b_1, \ldots, b_k]$ is a non-zero vector whose coordinates are defined as follows:

$$b_{j_s} = 0 \quad \text{for} \quad s = 1, \ldots, r\,, \qquad b_{i_s} = a_s \quad \text{for} \quad s = 1, \ldots, k-r\,.$$

It follows directly from the definition that if Φ is one-to-one, then the mapping (18) is one-to-one. If Φ is a homeomorphism, then (18) is a homeomorphism. Consequently,

3.2. *If Φ is a diffeomorphism of class* C_m, *then the mapping* (18) *is a diffeomorphism of class* C_m.

The following formulas are particular cases of (18):

(22) $$\Phi_{j,\eta}(y_1, \ldots, y_{k-1}) = (y_1, \ldots, y_{j-1}, \gamma_{j,\eta}, y_j, \ldots, y_{k-1}),$$

(23) $$\Phi_{j_1,\eta_1;j_2,\eta_2}(y_1, \ldots, y_{k-2})$$
$$= \Phi(y_1, \ldots, y_{j_1-1}, \gamma_{j_1,\eta_1}, y_{j_1}, \ldots, y_{j_2-2}, \gamma_{j_2,\eta_2}, y_{j_2-1}, \ldots, y_{k-2}) \quad (j_1 < j_2)$$

§ 4. Cells. A set $C \subset R^l$ is said to be a *k-dimensional cell* $(1 \leqslant k \leqslant l)$ if it is a diffeomorphic image of a bounded non-degenerate closed interval $I \subset R^k$, i.e. if there exists a diffeomorphism Φ on I such that $\Phi(I) = C$. If there exists a diffeomorphism Φ of class C_m from I onto C, we say that C is a k-dimensional cell *of class* C_m.

By the *zero-dimensional cell* in R^l we shall mean any one-point subset of R^l. We shall assume that every zero-dimensional cell is of class C_m for every m. To unify the definition and notation, we shall treat a zero-dimensional cell C as a diffeomorphic image $C = \Phi(I)$ of the zero-dimensional interval $I = \{0\}$ composed of the number zero only (see the convention on p. 346). By a *diffeomorphism* on the zero-dimensional interval I we mean any mapping from I into R^l.

It follows from the definition that every k-dimensional cell C $(0 \leqslant k \leqslant l)$ is a compact connected set (see III 8.3 and III 11.1-11.2).

Every non-degenerate closed interval I in R^k is an example of a k-dimensional cell in R^k (take as Φ the identity mapping from I onto itself!). All $(k-r)$-dimensional closed faces $I_{j_1,\eta_1;\ldots;j_r,\eta_r}$ of I (see § 3 (7)) are examples of $(k-r)$-dimensional cels in R^k (take as Φ the natural mapping $\Lambda_{j_1,\eta_1;\ldots;j_r,\eta_r}$ defined in § 3 (16)).

Figures 18 and 19 present examples of one-dimensional and two-dimensional cells in the plane R^2 and figures 20, 21 and 22 present examples of one-, two- and three-dimensional cells in the space R^3.

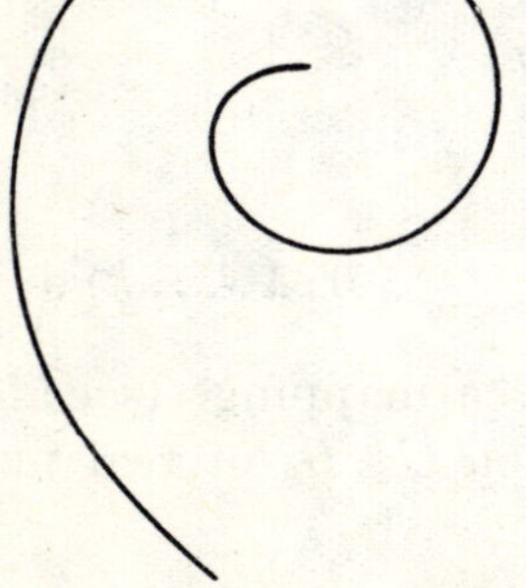

Fig. 18

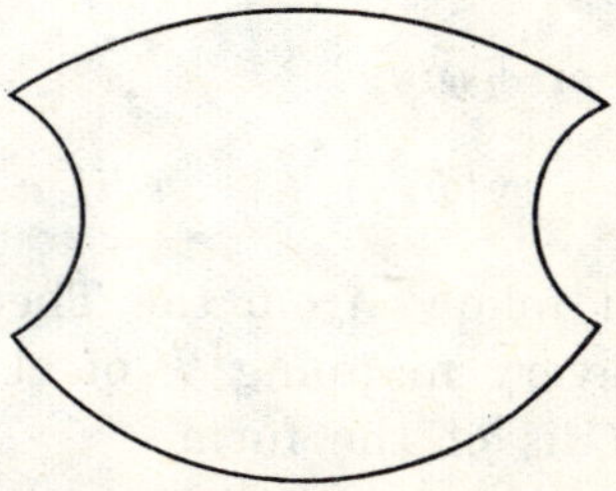

Fig. 19

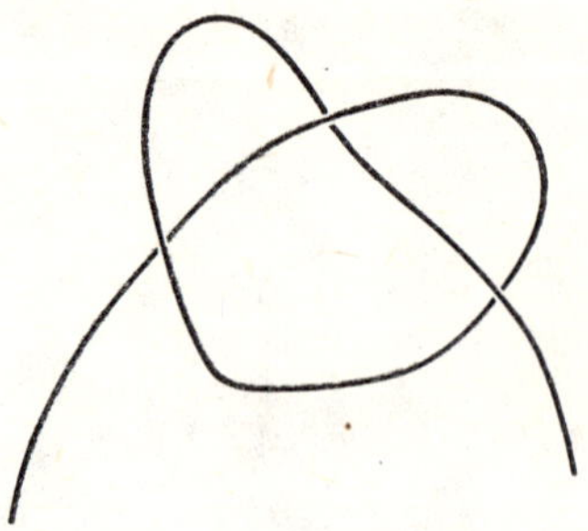

Fig. 20

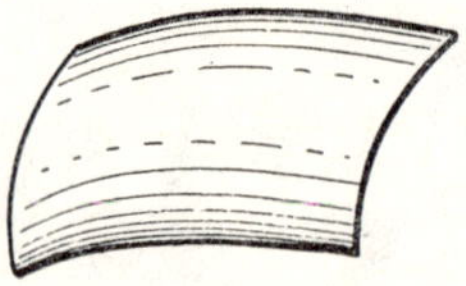

Fig. 21

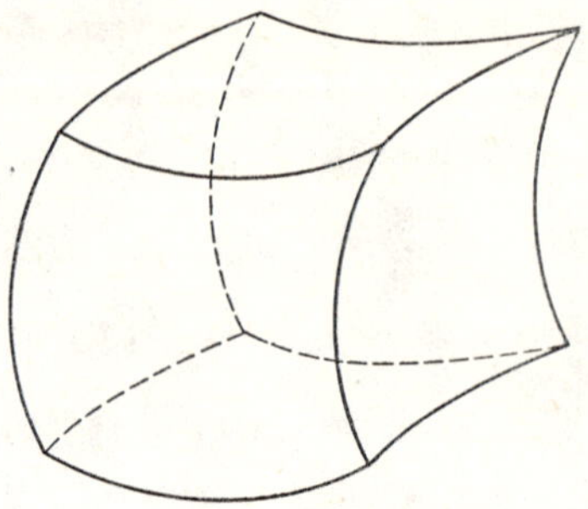

Fig. 22

It follows from 2.2 that every k-dimensional cell in R^l is a subset of a regular k-dimensional hypersurface in R^l.

Every k-dimensional cell C in R^l, $k < l$, has l-dimensional Lebesgue measure zero (see VIII 3.4), therefore it does not contain a ball, i.e. its topological interior $\operatorname{int} C$ in the space R^l is empty. However we can introduce another natural notion which can be called the geometric interior of C and which is always a non-empty set. For this purpose we first introduce the following notation.

For any point q in a subset Z of a Euclidean space the symbol

(1) $$\operatorname{tv}(q, Z)$$

will denote the maximum number of linearly independent vectors tangent at q to Z (see V § 6, p. 214).

For instance, if I is a closed interval in R^k (see § 3 (1)), then

(2) $$\operatorname{tv}(p, I) \leqslant k \quad \text{for every } p \in I .$$

More precisely,

(3) $$\operatorname{tv}(p, I) = s \text{ if and only if } p \in S_s^\circ \quad (s = 0, 1, \ldots, k) ,$$

where S_s° is the union of all s-dimensional open faces of I (see § 3, p. 345).

4.1. *If $C = \Phi(I)$ is a k-dimensional cell in R^l $(0 \leqslant k \leqslant l)$, then*

(4) $$\operatorname{tv}(q, C) \leqslant k \quad \textit{for every } q \in C .$$

More precisely,

(5) $$\operatorname{tv}(\Phi(p), C) = s \textit{ if and only if } p \in S_s^\circ \quad (s = 0, 1, \ldots, k) .$$

It follows from the Theorem V 5.4 on three mappings (see also 2.2) that every mapping Ψ of class C_1 from an open set G_0 of real numbers into C is of the form

$$\Psi = \Phi \circ \Lambda$$

where $\Lambda = \Phi^{-1} \circ \Psi$ is a mapping of class C_1 from G_0 into I. Since $\Psi'(t) = \Phi'(\Lambda(t)) \cdot \Lambda'(t)$, a vector $\mathfrak{b}$ is tangent to C at a point $q = \Phi(p)$ if and only if

(6) $$\mathfrak{b} = \Phi'(p) \cdot \mathfrak{a}$$

where $\mathfrak{a}$ is a vector tangent at p to I. Since $|\Phi'(p)| \neq 0$, the linear mapping (6) transforms linearly independent vectors $\mathfrak{a}$ onto linearly independent vectors $\mathfrak{b}$. Hence it follows that

(7) $$\operatorname{tv}(\Phi(p), C) = \operatorname{tv}(p, I) .$$

(7), (2) and (3) imply (4) and (5).

Let, as previously, $C = \Phi(I)$ be a k-dimensional cell in R^l $(0 \leqslant k \leqslant l)$. The set

$$C^\circ = \{q \in C\colon \operatorname{tv}(q, C) = k\}$$

will be called the *geometrical interior* of the cell C. It follows from 4.1 that

(8) $$C^\circ = \Phi(I^\circ) ,$$

i.e. that the geometrical interior of the cell C is the image of the interior I° of I, given by the diffeomorphism Φ.

In the case where the cell C is a closed interval I in R^k, the set C° just defined coincides with the set I° defined in § 3 (2), i.e. the geometrical interior is then identical to the topological interior. More generally, the geometrical interior of any k-dimensional cell in R^k coincides with the topological interior.

The geometrical interior C° of a zero-dimensional cell C (i.e. a one-point set C) coincides with C. The equation $C^\circ = C$ holds only for zero-dimensional cells.

By the *geometrical boundary* of a k-dimensional cell $C = \Phi(I)$ in R^l $(0 \leqslant k \leqslant l)$ we mean the set

(9) $$\mathrm{b}C = \{q \in C\colon \operatorname{tv}(q, C) < k\} = C - C^\circ .$$

It follows from 4.1 that

(10) $$\mathrm{b}C = \Phi(\mathrm{b}I) ,$$

i.e. the geometrical boundary of C is the image, by the diffeomorphism Φ, of the boundary of I (see § 3, p. 345).

In the case where C is a closed interval I in R^k, the set $\mathrm{b}C$ just defined coincides with the set $\mathrm{b}I$ defined in § 3, p. 345, i.e. the geometrical boundary of the cell coincides with its topological boundary. More generally, the geometrical boundary of any k-dimensional cell in R^k coincides with its topological boundary, defined in III § 3, p. 108.

The geometrical boundary of any zero-dimensional cell C (i.e. of any one-point set C) is empty. The equation $\mathrm{b}C = 0$ holds only for zero-dimensional cells.

As we stated on p. 348, every k-dimensional cell C in R^l $(0 \leqslant k \leqslant l)$ is a subset of a regular k-dimensional hypersurface H in R^l. One can prove that the geometrical interior of C is the topological interior of C in the metric space H and, similarly, the geometrical boundary of C is the topological boundary of C in the metric space H.

Let, as previously, $C = \Phi(I)$ be a k-dimensional cell in R^l $(0 \leqslant k \leqslant l)$. For any non-negative integer $s \leqslant k$ the components of the set

$$C_s^\circ = \{q \in C \colon \operatorname{tv}(q, C) = s\} \tag{11}$$

are called *s-dimensional open faces* of C. The closures of the s-dimensional open faces of C will be called the *s-dimensional closed faces* of C. The union C_s of all k-dimensional closed faces is called the *k-dimensional skeleton* of the cell C.

It follows from 4.1 (5) that

$$C_s^\circ = \Phi(S_s^\circ)$$

where S_s° is the union of all s-dimensional open faces of the interval I. The homeomorphism Φ transforms the components of S_s° onto the components of C_s° (see III 11.4). We know that the s-dimensional open faces of I are the components of S_s°. Hence it follows that a set $A \subset C$ is an s-dimensional open face of C if and only if A is the image (by the diffeomorphism Φ) of an s-dimensional open face of the interval I. In other words, the sets

$$\Phi(I^\circ_{j_1,\eta_1;\ldots;j_r,\eta_r}), \qquad r = k-s \tag{12}$$

(see § 3 (9)) are all the s-dimensional open faces of the cell C.

Since $\Phi(\mathrm{cl}B) = \mathrm{cl}\Phi(B)$ for every set $B \subset I$, we infer that a set $A \subset C$ is an s-dimensional closed face of C if and only if A is the image (by the diffeomorphism Φ) of an s-dimensional closed face of I. In other words, the sets

$$\Phi(I_{j_1,\eta_1;\ldots;j_r,\eta_r}), \qquad r = k-s \tag{13}$$

(see § 3 (7)) are all the s-dimensional closed faces of the cell C.

It follows from (12) and (13) (the case where Φ is the identity mapping on I) that if the cell C is a k-dimensional interval I in R^k, then the notions of s-dimensional open and closed faces of C and of the s-dimensional skeleton of C coincide with the analogous notions introduced in § 3 for the interval I.

Any s-dimensional closed face (13) of a cell $C = \Phi(I)$ is an s-dimensional cell. Indeed, the set (13) coincides with the set

$$(13') \qquad \Phi_{j_1,\eta_1;\ldots;j_r,\eta_r}(I_{j_1,\ldots,j_r}), \quad r = k-s,$$

where

$$(14) \qquad \Phi_{j_1,\eta_1;\ldots;j_r,\eta_r} = \Phi \circ \Lambda_{j_1,\eta_1;\ldots;j_r,\eta_r}$$

(see § 3 (18)). If Φ is of class C_m, then the diffeomorphism (13′) is also of class C_m. Thus every s-dimensional closed face of a k-dimensional cell of class C_m is an s-dimensional cell of class C_m ($s = 0, 1, \ldots, k$).

Observe also that the s-dimensional open face (12) is the geometrical interior of the s-dimensional cell (13), i.e.

$$(15) \qquad \Phi(I_{j_1,\eta_1;\ldots;j_r,\eta_r})^\circ = \Phi(I^\circ_{j_1,\eta_1;\ldots;j_r,\eta_r}).$$

It follows from (10) that the geometrical boundary bC of a k-dimensional cell $C = \Phi(I)$ is the union of all the $(k-1)$-dimensional closed faces

$$(16) \qquad \Phi(I_{j,\eta}) = \Phi_{j,\eta}(I_j), \quad j = 1, \ldots, k, \ \eta = \pm 1,$$

where

$$(17) \qquad \Phi_{j,\eta} = \Phi \circ \Lambda_{j,\eta}$$

according to (14).

It is convenient to assume that the whole k-dimensional cell C is the only k-dimensional closed face of C, and that its geometrical interior C° is the only k-dimensional open face of C (see the analogous definitions for an interval I in § 3, p. 344).

4.2. *The intersection of two closed faces of a cell C is either the empty set or it is a closed face of C.*

This follows directly from 3.1 (see also I § 3 (9)).

4.3. *If C_1 is a closed face of a cell C_2, and C_2 is a closed face of a cell C_3, then C_1 is a closed face of the cell C_3. If C_1 and C_2 are closed faces of a cell C_3 and $C_1 \subset C_2$, then C_1 is a closed face of C_2.*

In the case where C_3 is a closed k-dimensional interval in R^k, Theorem 4.3 results directly from the definition of closed faces of an interval (§ 3). Hence it follows, by (13), that it is true for any cell $C_3 = \Phi(I)$ where Φ is a diffeomorphism on I.

4.4. *Let C_1' be a face of a cell C_1, and let C_2' be a face of a cell C_2. If the set $C_0 = C_1 \cap C_2$ is simultaneously a face of C_1 and of C_2, then the set $C_0' = C_1' \cap C_2'$ is either empty or is simultaneously a face of C_1' and C_2'.*

Suppose that the set

$$C_0' = (C_1' \cap C_0) \cap (C_2' \cap C_0)$$

is not empty. By 4.2 the set $C_1' \cap C_0$ is a closed face of C_1 and consequently it is a closed face of C_0 by the second part of 4.3. By the same argument, the set $C_2' \cap C_0$ is a closed face of C_0. Replacing C by C_0 in 4.2 we infer that C_0' is a closed face of C_0. Thus C_0' is a closed face of $C_1' \cap C_0$ by the second part of 4.3 and, consequently, C_0' is a closed face of C_1' by the first part of 4.3. By the same argument C_0' is a closed face of C_2'.

EXERCISES

1. Let φ and ψ be real functions of class C_1 defined on an interval $\langle a; b\rangle \subset R$, $\varphi < \psi$. Prove that the set

$$\{(x, y)\colon (a \leqslant x \leqslant b) \& (\varphi(x) \leqslant y \leqslant \psi(x))\}$$

is a cell.

2. Let C be a k-dimensional cell in R^k and let φ and ψ be real functions of class C_1 defined on an open set G, $C \subset G \subset R^k$, $\varphi < \psi$. Prove that the set

$$\{(x_1, \ldots, x_k, x_{k+1})\colon ((x_1, \ldots, x_k) \in C) \& (\varphi(x_1, \ldots, x_k) \leqslant x_{k+1} \leqslant \psi(x_1, \ldots, x_k))\}$$

is a $(k+1)$-dimensional cell.

3. If C_1 is a k_1-dimensional cell in R^{l_1} and C_2 is a k_2-dimensional cell in R^{l_2}, then $C_1 \times C_2$ is a (k_1+k_2)-dimensional cell in $R^{l_1} \times R^{l_2} = R^{l_1+l_2}$.

4. Let Ψ be a mapping of class C_1 from a closed interval $I \subset R^k$ into R^l. The set

$$\{(p, \Psi(p)) \in R^{k+l}\colon p \in I\}$$

is a k-dimensional cell in R^{k+l}.

5. Give examples of diffeomorphisms which map the interval

$$\{(x, y)\colon (|x| \leqslant 1) \& (|y| \leqslant 1)\}$$

onto the two-dimensional cells C represented in fig. 23, fig. 24 (heavy contour) and fig. 25.

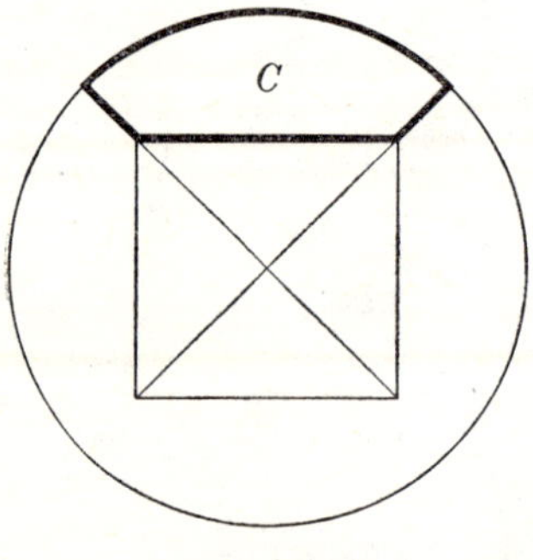

Fig. 23

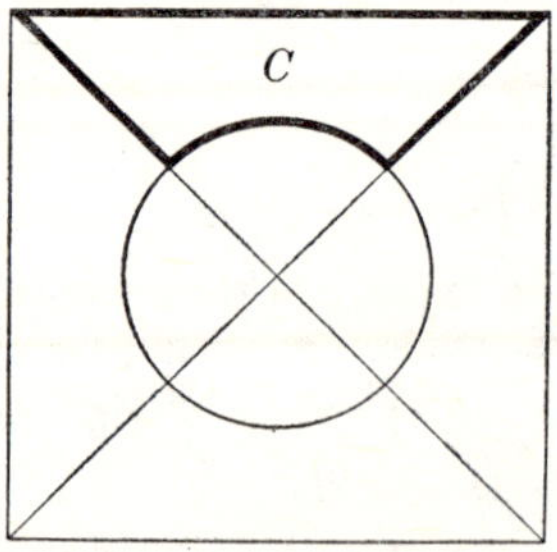

Fig. 24

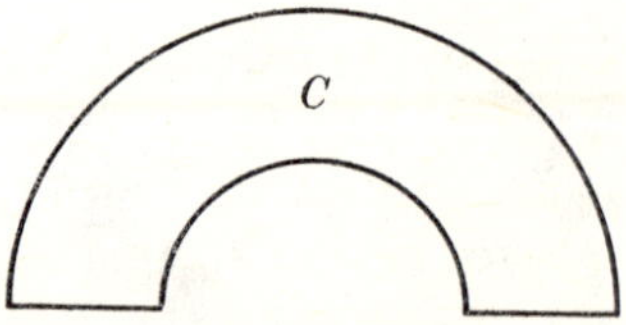

Fig. 25

6. Give an example of a diffeomorphism which maps a three-dimensional interval onto the three-dimensional analogue of the cell in fig. 23, which is obtained by replacing the circle by a sphere, and the square by a cube.

Generalize to the k-dimensional case.

Make the same generalization concerning the cell from fig. 24.

7. The set C shown in fig. 26, i.e. the set bounded by pieces of four externally tangent circles in the plane, is not a two-dimensional cell.

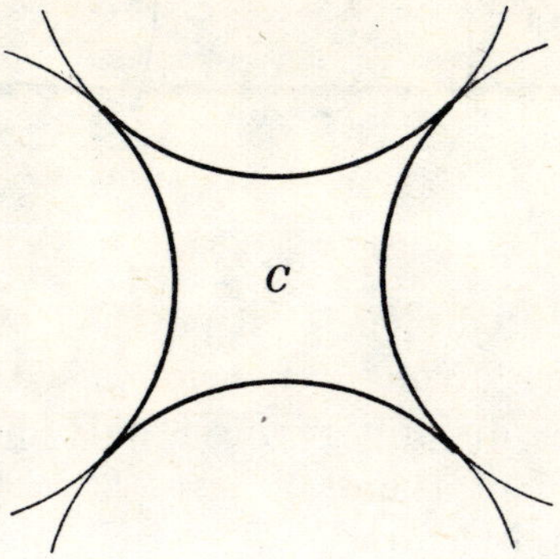

Fig. 26

§ 5. Polyhedra. A subset P of R^l is said to be a *k-dimensional regular polyhedron* (or simply a *k-dimensional polyhedron*) if there exists a finite set D of k-dimensional cells

(1) $$D = \{C_1, \ldots, C_n\}$$

such that

(2) $$P = C_1 \cup \ldots \cup C_n ,$$

and

(3) if $i \neq j$, then $C_i \cap C_j$ is empty or is a common r-dimensional closed face of C_i and C_j $(0 \leqslant r < k)$.

The integer k in the above definition satisfies the inequality $0 \leqslant k \leqslant l$.

Every finite set (1) of k-dimensional cells satisfying conditions (2) and (3) is called a *cellular decomposition* of the polyhedron P.

It follows from the definition that a set $P \subset R^l$ is a zero-dimensional polyhedron if and only if it is finite. The only cellular decomposition of a zero-dimensional polyhedron P is the set of all one-point subsets of P.

If $k > 0$, every k-dimensional polyhedron has infinitely many cellular decompositions.

Every k-dimensional cell C is a k-dimensional polyhedron. The one-element set $D = \{C\}$ is a cellular decomposition of C. If $k > 0$, the cell C can be represented, in infinitely many ways, as the union of a finite number of smaller cells satisfying conditions (2) and (3). Figure 27 illustrates this in the case of a two-dimensional cell on the plane R^2.

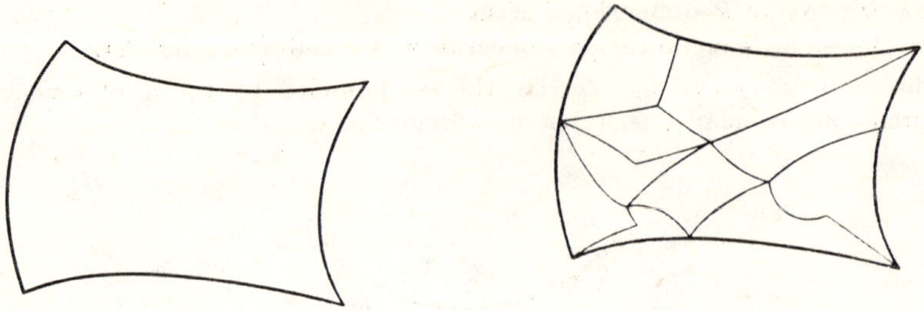

Fig. 27

A set $P \subset R^k$ is a one-dimensional polyhedron if and only if it is the union of a finite number of one-dimensional cells, i.e. regular arcs, such that two arcs are either disjoint or have only one common point which is an end-point of each of them. Figures 28 and 29 yield examples of some one-dimensional polyhedra.

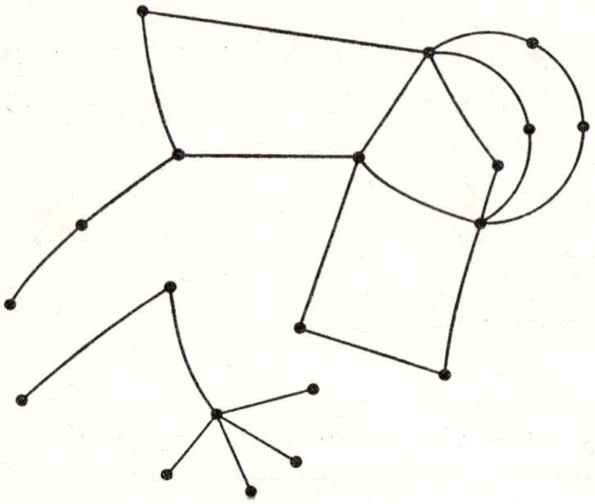

Fig. 28

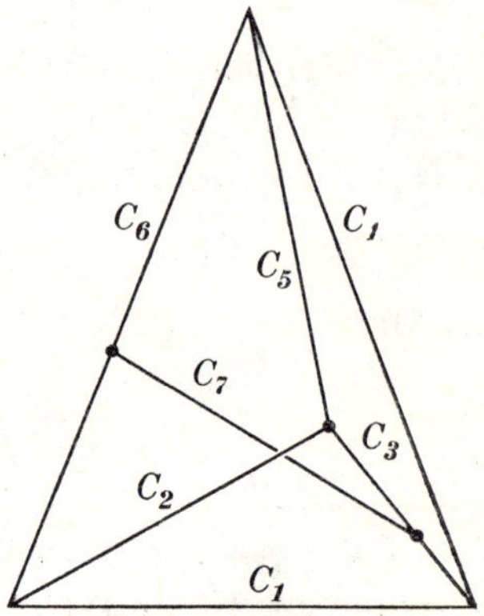

Fig. 29

In particular, the one-dimensional sphere, i.e. the circle $\{p\colon |p - p_0| = \varrho\}$ in the plane R^2 is a one-dimensional polyhedron—see figure 30.

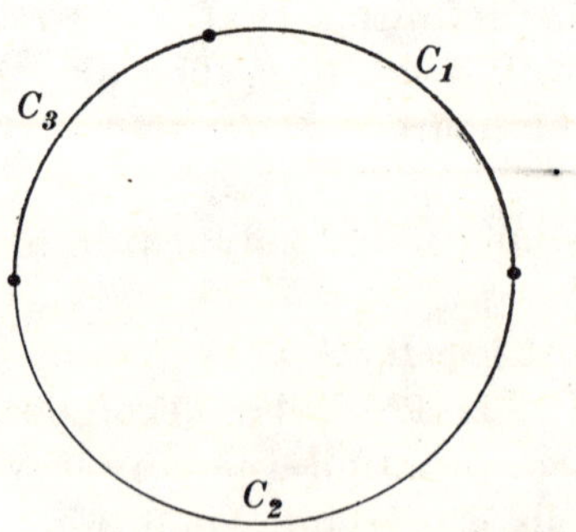

Fig. 30

The set shown in figure 31 is the union of two one-dimensional cells, viz. the graphs of the functions $f(x) = x^2 \sin(1/x)$ $(f(0) = 0)$ and $g(x) = 0$ for $0 \leqslant x \leqslant a$, but it is not a polyhedron.

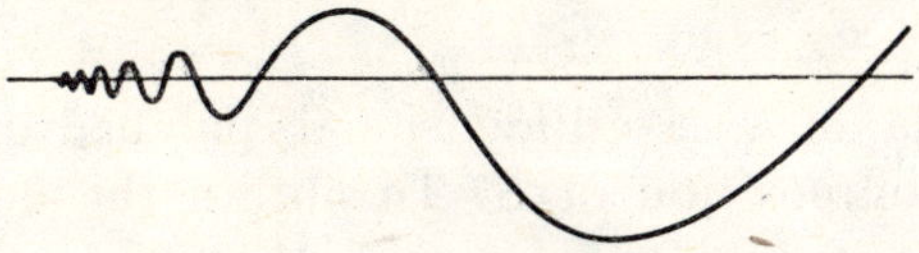
Fig. 31

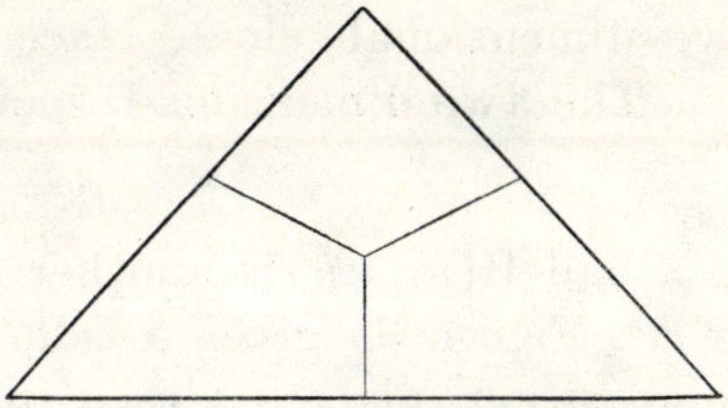
Fig. 32

Every triangle on the plane is a two-dimensional polyhedron. Figure 32 gives the simplest cellular decomposition of the triangle.

More generally, every subset of the plane R^2 which is the union of a finite number of triangles is a two-dimensional polyhedron. In particular every polygon in the plane (i.e. the contour plus all points inside it) is a two-dimensional polyhedron. Figure 33 shows how we can construct a cellular decomposition of a plane polygon.

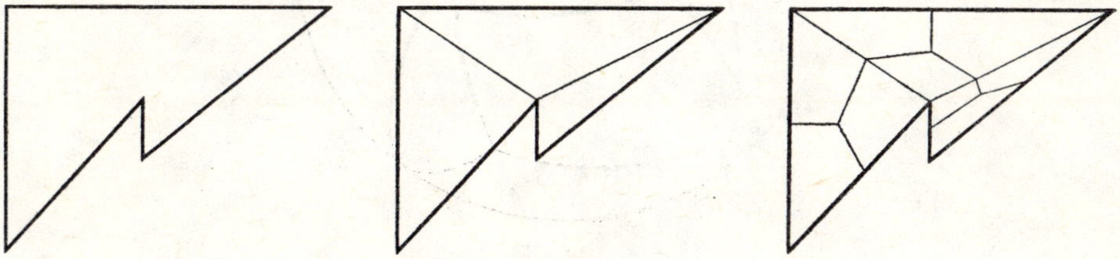
Fig. 33

Every disc $\{p \in R^2 \colon |p - p_0| \leqslant \varrho\}$ in the plane is a two-dimensional polyhedron but not a two-dimensional cell. Figure 34 shows a simple cellular decomposition of the disc.

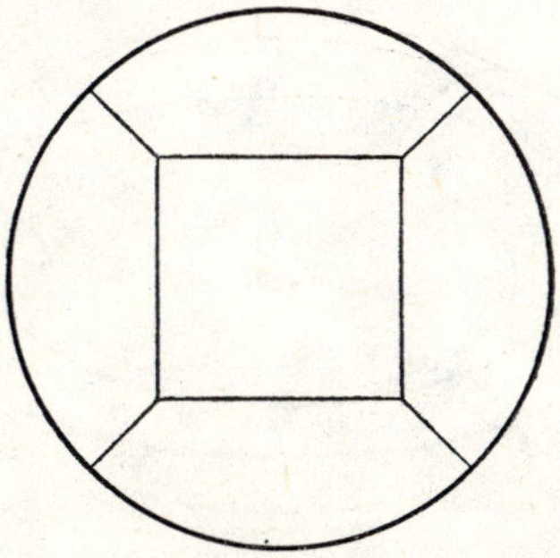
Fig. 34

Similarly any ellipse with its interior is a two-dimensional polyhedron in the plane R^2.

In the three-dimensional space R^3 the boundary of any closed interval I is an example of a two-dimensional polyhedron. The set of all two-dimensional closed faces is the simplest cellular decomposition.

The two-dimensional sphere in R^3, i.e. the boundary

$$S = \{p\colon\ |p - p_0| = \varrho\}$$

of a ball $\mathrm{B}(p_0, \varrho)$, is another example of a two-dimensional polyhedron in R^3. Figure 35 gives a cellular decomposition of S. To obtain the decomposition, place a closed interval I inside $\mathrm{B}(p_0, \varrho)$ such that p_0 is an interior point of I. The projections of all two-dimensional faces of I onto S from the point p_0 give a cellular decomposition of S.

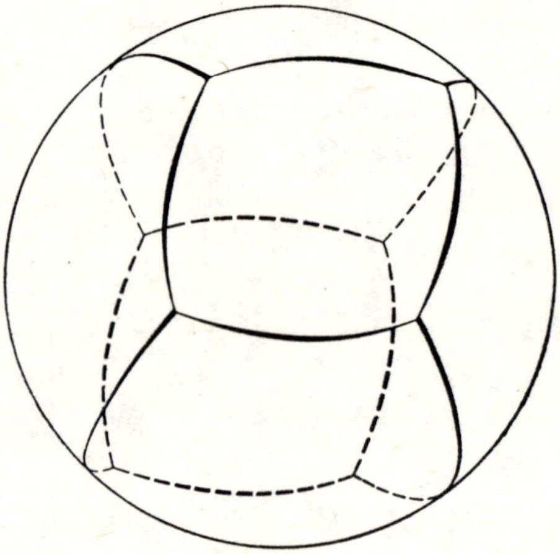

Fig. 35

Similarly we prove that every ellipsoid in R^3 is a two-dimensional polyhedron in R^3.

Every torus (see fig. 1 on p. 214) is another interesting example of a two-dimensional polyhedron in R^3. Fig. 36 gives a cellular decomposition of the torus.

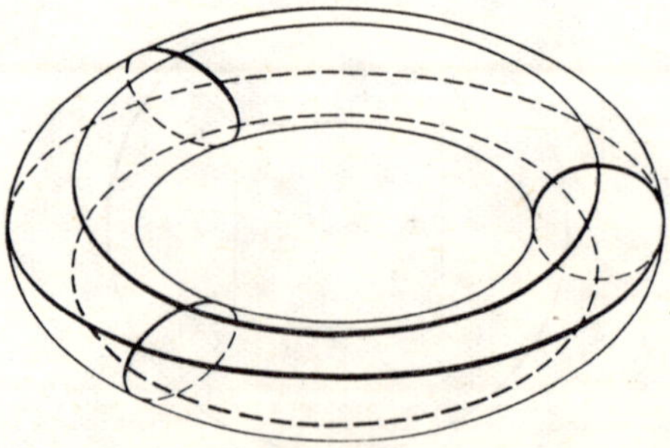

Fig. 36

The closed ball $\{p: |p-p_0| \leqslant \varrho\}$ in R^3 is an example of a three-dimensional polyhedron in R^3. Fig. 37 shows a method of construction of a cellular decomposition of the closed ball. To obtain the decomposition, place a closed interval I inside the ball, such that p_0 is in interior point of I. The one-dimensional faces of I, their projections from p_0 onto the sphere $S = \{p: |p-p_0| = \varrho\}$, and the segments joining the vertices of I with their projections from p_0 onto S are the one-dimensional faces of the cells in the cellular decomposition shown in fig 37.

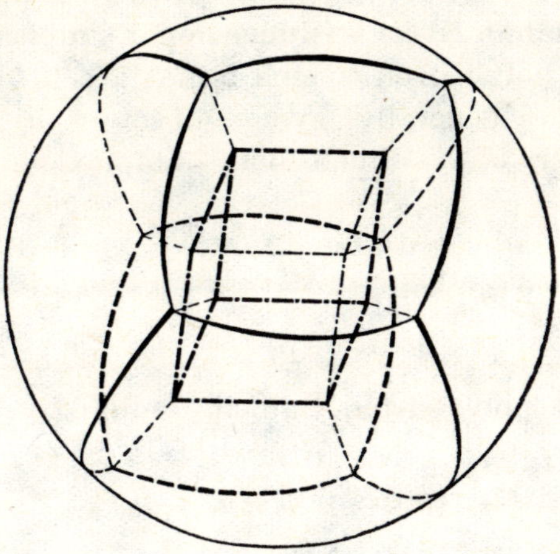

Fig. 37

Similarly one can prove that the full ellipsoid (i.e. the surface and all points inside it) is a three-dimensional polyhedron in R^3. Also the full torus is a three-dimensional polyhedron.

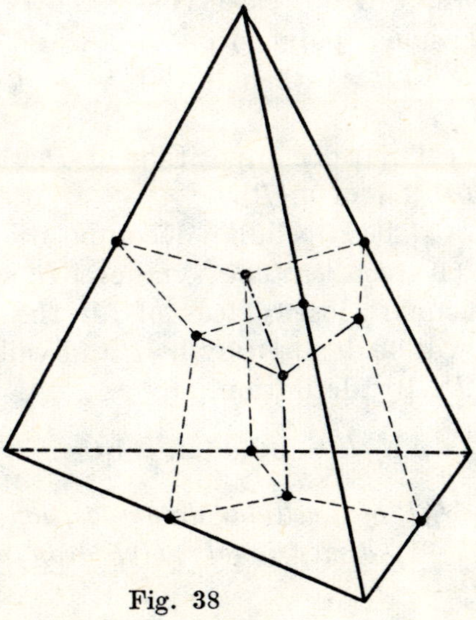

Fig. 38

Every pyramid is also an example of a three-dimensional polyhedron (which is not a cell) in R^3. Fig. 38 shows the edges of a simple cellular decomposition of a pyramid.

More generally, every set $P \subset R^3$ which is the union of a finite number of pyramids is a three-dimensional polyhedron.

A common generalization of the notion of segment, triangle and pyramid is given by the notion of simplex. A *k-dimensional simplex* in R^l ($k \leqslant l$) is the smallest convex set containing a given set of $k+1$ linearly independent points (i.e. points which do not lie on a $(k-1)$-dimensional hyperplane). Every k-dimensional simplex in R^l is an example of a k-dimensional polyhedron (which is not a cell for $k > 1$) in R^l. One can prove that, more generally, every subset of R^l which is the union of a finite number of k-dimensional simplexes is a k-dimensional polyhedron.

By the method indicated above for $k = 2$ and 3, one can prove that the k-dimensional closed ball

$$\{p \in R^k\colon\ |p-p_0| \leqslant \varrho\}$$

is a k-dimensional polyhedron, and its boundary, i.e. the $(k-1)$-dimensional sphere

$$\{p \in R^k\colon\ |p-p_0| = \varrho\}$$

is a $(k-1)$-dimensional polyhedron in R^k. Analogous remarks are true for the ellipsoid and the full ellipsoid in R^k.

One can prove that every compact regular k-dimensional hypersurface in R^l is a k-dimensional polyhedron. The proof of the existence of a cellular decomposition for any compact regular hypersurface is difficult and, therefore, will not be given here.

If P is a k-dimensional polyhedron in R^l, and $\Phi\colon G \to R^s$ is a diffeomorphism, $P \subset G \subset R^l$, $l \leqslant s$, then $\Phi(P)$ is a k-dimensional polyhedron in R^s.

Indeed, if (1) is a cellular decomposition of P, then $\{\Phi(C_1), \ldots, \Phi(C_n)\}$ is a cellular decomposition of $\Phi(P)$.

Let $D = \{C_1, \ldots, C_n\}$ be a cellular decomposition of a k-dimensional polyhedron P in R^l ($1 \leqslant k \leqslant l$). The symbol $\mathrm{b}D$ will denote the set of all the $(k-1)$-dimensional closed faces of all the cells $C_1, \ldots, C_n$. The union of all the faces will be denoted by $\mathrm{b}_D P$ and will be called the *skeleton of the decomposition D*. By definition,

$$\mathrm{b}_D P = \mathrm{b}C_1 \cup \ldots \cup \mathrm{b}C_n\,. \tag{4}$$

5.1. *The skeleton $\mathrm{b}_D P$ of a cellular decomposition D of a k-dimensional polyhedron P is a $(k-1)$-dimensional polyhedron and $\mathrm{b}D$ is a cellular decomposition of $\mathrm{b}_D P$.*

It is evident that bD satisfies the condition analogous to (2). To prove that bD is a cellular decomposition, it suffices to show that the common part of two different $(k-1)$-dimensional closed faces of cells C_{m_1} and C_{m_2} in D is their common face of dimension $<k-1$, provided it is not void. If $m_1 = m_2$, this follows from 4.2. If $m_1 \neq m_2$, this follows from 4.4, since D has property (3).

5.2. *All closed faces of cells in a cellular decomposition D of a polyhedron P are compact and, therefore, closed in the space P. All open faces of cells in D are differences of two closed sets, and therefore are Borel subsets of the space P. All open faces of cells in D are disjoint.*

The first part follows from III 8.3 and III 6.1. The second part follows from the first since every open face is the difference of its closure (i.e. a closed face) and its geometrical boundary (i.e. the union of a finite number of closed faces). The final part follows from (3).

Let P be a k-dimensional polyhedron in R^l $(1 \leqslant k \leqslant l)$. A point $q \in P$ is said to be *regular* if it has a neighbourhood U in the metric space P, such that U is a diffeomorphic image of an open set $G \subset R^k$. In other words, a point $q \in P$ is regular if there exists a diffeomorphism $\Phi\colon G \to P$ $(G \subset R^k)$ such that the set $U = \Phi(G)$ is a neighbourhood of q in P. If $q \in P$ is not regular, q is said to be *singular*.

The set of all regular points of P will be denoted by re$P$. By definition, re$P$ is the union of all open sets U in P which are diffeomorphic images of open subsets of R^k. This proves that re$P$ is an open subset of the space P. Consequently the set

$$\mathrm{si}\,P = P - \mathrm{re}\,P \tag{5}$$

of all singular points of P is closed in P. Observe that in the case where P is a cell C, then

$$\mathrm{re}\,C = C^\circ \quad \text{and} \quad \mathrm{si}\,C = \mathrm{b}C\,.$$

Let D be a cellular decomposition (1) of a polyhedron P. Then

$$C_1^\circ \cup \ldots \cup C_n^\circ \subset \mathrm{re}\,P\,. \tag{6}$$

Indeed, it follows from (2) and (3) that for $j = 1, \ldots, n$

$$C_j^\circ = P - (C_1 \cup \ldots \cup C_{j-1} \cup \mathrm{b}C_j \cup C_{j+1} \cup \ldots \cup C_n)\,,$$

which proves that the set C_j° is open in P because it is the complement of a closed set. Since C_j° is a diffeomorphic image of an open interval in R^k, each point of C_j° is regular, i.e. it belongs to re$P$.

Passing to complements in the space P, we can write (6) in the following form

$$\mathrm{si}\,P \subset \mathrm{b}_D P \tag{6'}$$

by (2) and (3).

In the case where P is a zero-dimensional polyhedron, i.e. a finite subset of R^l, every point of P is called *regular*. Consequently, in that case

$$\text{re}\,P = P \quad \text{and} \quad \text{si}\,P = 0\,. \tag{7}$$

The set $\text{re}\,P$ of all regular points of a k-dimensional polyhedron P is a regular k-dimensional hypersurface. In fact, every point $q \in \text{re}\,P$ has a neighbourhood U in P which is a diffeomorphic image of an open subset of R^k. Since $U \subset \text{re}\,P$, U is a neighbourhood of q in $\text{re}\,P$. One can prove that $\text{re}\,P$ is the greatest k-dimensional regular hypersurface which is a subset of P.

By 1.1, the k-dimensional complete measure $\mu_{\text{re}\,P}$ is defined on the σ-algebra $\mathcal{M}_{\text{re}\,P}$ of measurable subsets of $\text{re}\,P$. We extend the measure $\mu_{\text{re}\,P}$ to a measure μ_P on a σ-algebra $\mathcal{M}_P$ of subsets of P as follows:

For every set $A \subset P$,

$$A \in \mathcal{M}_P \text{ if and only if } A \cap \text{re}\,P \in \mathcal{M}_{\text{re}\,P}\,; \tag{8}$$

for every set $A \in \mathcal{M}_P$

$$\mu_P(A) = \mu_{\text{re}\,P}(A \cap \text{re}\,P)\,. \tag{9}$$

More precisely, condition (8) uniquely defines a class $\mathcal{M}_P$ of subsets of P. It follows directly from the definition that $\mathcal{M}_P$ is a σ-algebra of subsets of P. Equation (9) defines a set function μ_P on $\mathcal{M}_P$ which satisfies all the conditions (a), (b), (c) in VI § 3, i.e. it is a measure. It follows from the completeness of $\mu_{\text{re}\,P}$ that the measure μ_P is also complete. It follows from (8), the final remark in 1.1 and from VI 2.5 that every Borel subset of P belongs to $\mathcal{M}_P$. It follows from (8) and (9) that

$$\text{if } A \in \mathcal{M}_{\text{re}\,P}, \text{ then } A \in \mathcal{M}_P \text{ and } \mu_P(A) = \mu_{\text{re}\,P}(A)\,; \tag{10}$$

$$\text{if } A \subset \text{si}\,P, \text{ then } A \in \mathcal{M}_P \text{ and } \mu_P(A) = 0\,. \tag{11}$$

Observe that if the polyhedron P is a regular k-dimensional hypersurface H, then the σ-algebra $\mathcal{M}_P$ and the measure μ_P just defined coincide with the σ-algebra $\mathcal{M}_H$ and the measure μ_H defined in § 1.

Let D be a cellular decomposition (1) of a k-dimensional polyhedron P in R^l. We shall prove that μ_P has the following property, stronger than (11):

$$\text{if } A \subset \text{b}_D P, \text{ then } A \in \mathcal{M}_P \text{ and } \mu_P(A) = 0\,. \tag{11'}$$

Since the measure μ_P is complete, it suffices to show that $\mu_P(\text{b}_D P) = 0$. Since the set $\text{b}_D P$ is the union of a finite number of open faces of dimensions $< k$ of the cells of the decomposition D, and each of these open faces is a regular hypersurface of dimension $< k$, the equation $\mu_P(\text{b}_D P) = 0$ follows from the following property of the measure μ_P:

(12) $\mu_P(H) = 0$ for every regular hypersurface $H \subset P$ of a dimension less than the dimension of P.

Indeed, $H = (H \cap \operatorname{re} P) \cup (H \cap \operatorname{si} P)$. Since the set $H \cap \operatorname{re} P$ is open in the space H, it is either empty or a regular hypersurface of the same dimension as H. Hence it follows, by 1.2, that $\mu_P(H \cap \operatorname{re} P) = \mu_{\operatorname{re} P}(H \cap \operatorname{re} P) = 0$. By (11), $\mu_P(H \cap \operatorname{si} P) = 0$. This proves that $\mu_P(H) = 0$.

The following theorem expresses a fundamental property of the measure μ_P on a k-dimensional polyhedron P $(1 \leqslant k \leqslant l)$.

5.3. *Let $D = \{C_1, \ldots, C_n\}$ be a cellular decomposition of a k-dimensional polyhedron P, $C_i \neq C_j$ for $i \neq j$, $C_i = \Phi_i(I_i)$ where Φ_i is a diffeomorphism on a closed interval $I_i \subset R^k$, $i = 1, \ldots, n$. For every extended real function f on P, measurable with respect to μ_P (in particular, for every continuous function f on P),*

(13)
$$\int_P f\, d\mu_P = \sum_{i=1}^{n} \int_{I_i} (f \circ \Phi_i) \cdot |\Phi_i'| \, .$$

(13) should be interpreted as follows: the integral on the left-hand side exists if and only if all the integrals on the right-hand side exist and the addition is feasible. The integrals on the right-hand side are with respect to the k-dimensional Lebesgue measure in R^k.

To prove (13), let us observe that P is the union of the disjoint measurable sets $\mathrm{b}_D P, C_1^\circ, \ldots, C_n^\circ$, the first of which has measure zero by (11′). Thus

$$\int_P f\, d\mu_P = \sum_{i=1}^{\infty} \int_{C_i^\circ} f\, d\mu_P \, .$$

By 1.3,

$$\int_{C_i^\circ} f\, d\mu_P = \int_{C_i^\circ} f\, d\mu_{\operatorname{re} P} = \int_{I_i^\circ} (f \circ \Phi_i) \cdot |\Phi_i'| = \int_{I_i} (f \circ \Phi_i) \cdot |\Phi_i'| \, ,$$

which proves (1.3).

In particular, for the constant function $f = 1$,

(14)
$$\mu_P(P) = \int_P 1\, d\mu_P = \sum_{i=1}^{n} \int_{I_i} |\Phi_i'| < \infty \, .$$

Thus the measure μ_P is finite.

Every continuous function on P is μ_P measurable since, for every real number a, the set $\{p : f(p) < a\}$ is open in P and, therefore, measurable. Since P is compact, f is bounded. By VII 5.12, f is integrable on P.

If D_1 is a non-empty subset of a cellular decomposition of a k-dimensional polyhedron P, then the union P_1 of all cells in D_1 is a k-dimensional polyhedron and D_1 is a cellular decomposition of P_1.

Moreover,

(15) $$\text{if } A \in \mathcal{M}_{P_1}, \text{ then } A \in \mathcal{M}_P \text{ and } \mu_{P_1}(A) = \mu_P(A).$$

If P is a zero-dimensional polyhedron, i.e. P is a finite subset of a Euclidean space, then μ_P is the zero-dimensional measure defined on the algebra of all subsets of P (see VI § 3, p. 237 and § 1, p. 334).

If P is a one-dimensional polyhedron, the integral $\int_P f d\mu_P$ is called a *curvilinear integral*. If P is a two-dimensional polyhedron, the integral is called a *surface integral*.

The traditional notation for curvilinear and surface integral does not differ essentially from the notation used in this section but other symbols are used instead of $d\mu_P$. In the case of a curvilinear integral, the symbol ds in usually written instead of $d\mu_P$; the letter s usually denotes the arc length (see § 1 A)) and this notation for curvilinear integrals aims to emphasize that the integration is with respect to the arc length which replaces, in elementary courses, the notion of one-dimensional measure on curves and one-dimensional polyhedra. Thus a curvilinear integral is traditionally denoted by the symbol

(16) $$\int_P f ds \quad \text{or} \quad \int_P f(x_1, \ldots, x_l) ds .$$

In elementary courses of integration only two types of one-dimensional polyhedra are taken into consideration, viz. polyhedra homeomorphic with a one-dimensional interval (arcs) and polyhedra homeomorphic with a circle (closed curves).

In the case of surface integrals the symbols $d\sigma$ is often used to emphasize that the integration is with respect to the two-dimensional measure. Thus a surface integral on a two-dimensional polyhedron P is denoted by the symbol

(17) $$\iint_P f d\sigma \quad \text{or} \quad \iint_P f(x_1, \ldots, x_l) d\sigma .$$

The letter σ used here has nothing in common with the letter σ in expressions like σ-algebra, σ-finite, etc. used in this book. Many authors use an other letter instead of σ in the notation (17).

In elementary courses in integration only the case of $l = 2$ or 3 is discussed for curvilinear integrals, and the case of $l = 3$ for surface integrals.

EXERCISES

1. Calculate the integral of the function $f(x_1, x_2, x_3) = x_1^2 + x_2^2 + x_3^2$ on the torus T defined in V § 6, pp. 213-214.

2. Calculate the integral of the function $f(x_1, x_2, x_3, x_4) = (x_1^2 + x_2^2)(x_3^2 + x_4^2)$ on the torus T_0 defined in V § 6, exercise 3.

3. By a *closed Möbius strip* (or simply a *Möbius strip*) we mean every set homeomorphic with the set

$$\overline{M} = \{\Phi(u, v) \colon (|u| \leqslant \varrho_0) \& (|v| \leqslant \pi)\} \subset R^3,$$

where

$$\Phi(u,v)=\left(\left(\varrho+u\cos\frac{v}{2}\right)\cos v,\left(\varrho+u\cos\frac{v}{2}\right)\sin v,u\sin\frac{v}{2}\right)$$

and $\varrho_0<\varrho$ (see V § 6, exercise 4). We shall be interested only in the case where a Möbius strip is a regular polyhedron.

Verify that the Möbius strip $\overline{M}$ (which is the closure of the open Möbius strip from V § 6, exercise 4) is a two-dimensional polyhedron. Give examples of cellular decompositions of $\overline{M}$ composed of 3 or 4 cells only.

4. Verify that the set

$$\overline{M}_0=\{\Phi(u,v)\colon (|u|\leqslant\varrho_0)\&(|v|\leqslant\pi)\}\subset R^4,$$

where

$$\Phi(u,v)=\left(\varrho\cos v,\varrho\sin v,u\cos\frac{v}{2},u\sin\frac{v}{2}\right)$$

(see V § 6, exercise 4), is a Möbius strip. Give a cellular decomposition of the two-dimensional polyhedron $\overline{M}_0$. Calculate the integral of the function $f(x_1,x_2,x_3,x_4)=x_1+x_2+x_3+x_4$ on this polyhedron.

5. Calculate the surface integral of the function $f(x_1,x_2,x_3)=x_1+2x_2-3x_3$ on the two-dimensional polyhedron which is the union of all the two-dimensional faces of the pyramid whose vertices are $(1,0,1)$, $(0,1,1)$, $(1,1,0)$, $(-1,-1,-1)$.

6. Calculate the integral of the function $f(x_1,x_2,x_3,x_4)=x_1-x_2+x_3-x_4$ on the three-dimensional pyramid (simplex) P in R^4 whose vertices are $p_1=(1,0,0,0)$, $p_2=(0,1,0,0)$, $p_3=(0,0,1,0)$, $p_4=(0,0,0,1)$.

Hint. P is the set of all points of the form $p_1+\alpha(p_2-p_1)+\beta(p_3-p_1)+\gamma(p_4-p_1)$ where $\alpha,\beta,\gamma\geqslant 0$ and $\alpha+\beta+\gamma\leqslant 1$.

7. Calculate the integral of the function $f(x_1,x_2,x_3,x_4)=x_1x_2x_3x_4$ on the union of all the three-dimensional closed faces of the four-dimensional interval

$$\{(x_1,x_2,x_3,x_4)\colon 0\leqslant x_1,x_2,x_3,x_4\leqslant 1\}.$$

8. Let $P\subset R^3$ be the union of the following rectangles:

$$\begin{aligned}
P_1&=\{(x_1,x_2,x_3)\colon (-1\leqslant x_1\leqslant 1)\&(0\leqslant x_2\leqslant 1)\&(x_3=0)\},\\
P_2&=\{(x_1,x_2,x_3)\colon (x_1=1)\&(0\leqslant x_2\leqslant 1)\&(0\leqslant x_3\leqslant 1)\},\\
P_3&=\{(x_1,x_2,x_3)\colon (0\leqslant x_1\leqslant 1)\&(0\leqslant x_2\leqslant 1)\&(x_3=1)\},\\
P_4&=\{(x_1,x_2,x_3)\colon (0\leqslant x_1\leqslant 1)\&(x_2=1)\&(1\leqslant x_3\leqslant 2)\},\\
P_5&=\{(x_1,x_2,x_3)\colon (-1\leqslant x_1\leqslant 1)\&(0\leqslant x_2\leqslant 1)\&(x_3=2)\},\\
P_6&=\{(x_1,x_2,x_3)\colon (x_1=-1)\&(0\leqslant x_2\leqslant 1)\&(0\leqslant x_3\leqslant 2)\}.
\end{aligned}$$

Prove that P is a Möbius strip (see exercise 3) and $\{P_1,\ldots,P_6\}$ is its cellular decomposition. Calculate the integral of the function $f(x_1,x_2,x_3)=(x_1^2+x_3)\sin\pi x_2$ on P.

§ 6. The orientation of a cell. We recall that (see II § 10, p. 96) by a *l-dimensional unit k-vector field* we mean a function on a set X which assigns, to every q in X, an l-dimensional unit k-vector.

By an *orientation* of a k-dimensional cell C in R^l $(1\leqslant k\leqslant l)$ we shall mean a continuous l-dimensional unit k-vector field T defined on C such that, for every $q\in C^\circ$, $T(q)$ is a unit k-vector tangent at q to the hypersurface C°.

The cell C has exactly two orientations which differ only by sign.

Indeed, assume that $C = \Phi(I)$ where Φ is a diffeomorphism from a k-dimensional interval

$$(1) \qquad I = \{(x_1, \dots, x_k)\colon \alpha_i \leqslant x_i \leqslant \beta_i \text{ for } i = 1, \dots, k\}$$

onto C. It follows from V 6.5 that

$$(2) \qquad T(q) = \frac{\Phi_{|1}(p) \wedge \dots \wedge \Phi_{|k}(p)}{|\Phi'(p)|} \quad \text{for} \quad q = \Phi(p) \in C$$

is an orientation of C. This orientation is said to be *determined by the diffeomorphism* Φ. By definition,

$$(2') \qquad T \circ \Phi = \frac{\Phi_{|1} \wedge \dots \wedge \Phi_{|k}}{|\Phi'|} \quad \text{or} \quad T = \frac{\Phi_{|1} \wedge \dots \wedge \Phi_{|k}}{|\Phi'|} \circ \Phi^{-1}.$$

It follows from the definition of an orientation that if T is an orientation of C, then the matrix field $-T$ is also an orientation of C, different from T. Orientations of C that differ only by sign are said to be ***opposite***.

On the other hand, if T_1 and T are orientations of C, then either $T_1 = T$ or $T_1 = -T$. Indeed, for every point $q \in C^\circ$, unit k-vectors tangent to C° at q can differ only by sign (see II § 10 (17)). Thus, for every point $q \in C^\circ$

$$T_1(q) = \varphi(q)\,T(q) \quad \text{where} \quad \varphi(q) = 1 \text{ or } -1 .$$

It follows from the continuity of the non-vanishing matrix fields T_1 and T that the real function φ defined on C° is continuous (see III § 8 D)). The set C° is connected since it is a homeomorphic image of the open interval I°, and the function φ assumes only values 1 or -1. By III 11.3, the function φ is constant. If $\varphi = 1$, then $T_1(q) = T(q)$ for every q in C° and, by continuity, also for every $q \in C$. If $\varphi = -1$, then $T_1(q) = -T(q)$ for every $q \in C$.

Both orientations of C are determined by diffeomorphisms. In fact, suppose that Φ is a diffeomorphism from the interval (1) onto C and T is the orientation (2) determined by Φ. Then the mapping

$$\Psi(x_1, x_2, \dots, x_k) = \Phi(-x_1, x_2, \dots, x_k)$$

is a diffeomorphism from the interval

$$\{(x_1, \dots, x_k)\colon -\beta_1 \leqslant x_1 \leqslant -\alpha_1 \text{ and } \alpha_i \leqslant x_i \leqslant \beta_i \text{ for } i = 2, \dots, k\}$$

onto C. Since

$$\Psi_{|1}(x_1, x_2, \dots, x_k) = -\Phi_{|1}(-x_1, x_2, \dots, x_k),$$

and

$$\Psi_{|i}(x_1, x_2, \dots, x_k) = \Phi_{|i}(-x_1, x_2, \dots, x_k) \quad \text{for} \quad i = 2, \dots, k,$$

the diffeomorphism Ψ determines the orientation $-T$.

We have to introduce the following general definition. Let Z be a subset of a Euclidean space and let $q \in Z$. We say that a vector $\mathfrak{b}$ in this space is *half-tangent to Z at q* if one of the following conditions is satisfied:

(a) there exists a mapping Ψ of class C_1 from a closed interval $\langle a_1; b_1\rangle$ into Z such that $\Psi(b_1) = q$ and $\Psi'(b_1) = \mathfrak{b}$;

(b) there exists a mapping Ψ of class C_1 from a closed interval $\langle a_2; b_2\rangle$ into Z such that $\Psi(a_2) = q$ and $\Psi'(a_2) = \mathfrak{b}$.

A vector $\mathfrak{b}$ is tangent to Z at q if and only if conditions (a) and (b) are both satisfied. In fact, if (a) and (b) hold, we may assume that $b_1 = a_2$. Then both mappings Ψ from (a) and (b) defines a mapping Ψ on $\langle a_1; b_2\rangle$ of class C_1 such that q and $\mathfrak{b}$ are respectively the values of this mapping and of its derivative at the point b_1 of the interval $(a_1; b_2)$. Thus $\mathfrak{b}$ is tangent to Z at q. Conversely, if $\mathfrak{b}$ is tangent to Z at q, i.e. if there exists a mapping Ψ from an interval $(t_1; t_2)$ into Z such that $\Psi(t_0) = q$ and $\Psi'(t_0) = \mathfrak{b}$ for a t_0 in this interval, then assuming $t_1 < a_1 < b_1 = t_0 = a_2 < b_2 < t_2$ we see that the restrictions of Ψ to the intervals $\langle a_1; b_1\rangle$ and $\langle a_2; b_2\rangle$ satisfy conditions (a) and (b) respectively.

A vector $\mathfrak{b}$ is said to be *out-tangent* to Z at q if it has property (a) but does not have property (b), i.e. is not tangent. A vector $\mathfrak{b}$ is said to be *in-tangent* to Z at q if it has property (b) but does not have property (a), i.e. is not tangent. Every vector half-tangent to Z at q is either tangent or out-tangent or in-tangent to Z at q but is never in two of the above categories simultaneously.

If a vector $\mathfrak{b}$ has property (a), then the vector $-\mathfrak{b}$ has property (b). In fact, if (a) holds for a mappings Ψ in $\langle a_1; b_1\rangle$, then the mapping $\overline{\Psi}(t) = \overline{\Psi}(-t)$ defined on the interval $\langle -b_1; -a_1\rangle$ is of class C_1 and $\overline{\Psi}(-b_1) = q$, $\overline{\Psi}'(-b_1) = -\mathfrak{b}$.

Similarly we prove that if a vector $\mathfrak{b}$ has property (b), then the vector $-\mathfrak{b}$ has property (a).

Hence it follows that a vector $\mathfrak{b}$ is out-tangent (in-tangent) to Z at q if and only if the vector $-\mathfrak{b}$ is in-tangent (out-tangent) to Z at q. A vector $\mathfrak{b}$ is tangent to Z at q if and only if $-\mathfrak{b}$ is tangent to Z at q.

If Z is a closed interval (1) in R^k and if $p \in I^\circ$, then every k-dimensional vector is tangent to I at p, and no vector is out-tangent or in-tangent to I at p. If p is in the boundary $\mathrm{b}I$ of I, then not all vectors $\mathfrak{a}$ are tangent to I at p. Namely, if $p \in I^\circ_{j_1,\eta_1;\dots;j_r,\eta_r}$ (see § 3 (9)), then a vector

$$\mathfrak{a} = [a_1, \dots, a_k]$$

is tangent to I at p if and only if all the numbers $a_{j_1}, \dots, a_{j_r}$ are equal to 0.

Hence it follows that, for every point $p \in I$, any linear combination of vectors tangent to I at p is tangent to I at p.

Note that if p belongs to a $(k-1)$-dimensional open face $I^{\circ}_{j,\eta}$ (see § 3 (5)-(6)), then a vector $\mathfrak{a} = [a_1, \ldots, a_k]$ is

$$\text{tangent to } I \text{ at } p \text{ if and only if } a_j = 0\,;$$

$$\text{out-tangent to } I \text{ at } p \text{ if and only if } \eta a_j > 0\,;$$

$$\text{in-tangent to } I \text{ at } p \text{ if and only if } \eta a_j < 0\,.$$

Hence it follows that every vector $\mathfrak{a} \in V_k$ is half-tangent to I at $p \in I^{\circ}_{j,\eta}$.

6.1. *Let Φ be a diffeomorphism from the interval* (1) *onto a k-dimensional cell $C \subset R^k$.*

A vector $\mathfrak{b} \in V_l$ is half-tangent (tangent, out-tangent, in-tangent) to C at a point $q = \Phi(p)$ if and only if it is of the form

$$\mathfrak{b} = \Phi'(p)\cdot\mathfrak{a} = a_1\Phi_{|1}(p) + \ldots + a_k\Phi_{|k}(p) \tag{3}$$

where $\mathfrak{a} = [a_1, \ldots, a_k]$ is a vector half-tangent (tangent, out-tangent, in-tangent) to I at p.

Any linear combination of vectors tangent to C at q is tangent to C at q.

Out-tangent and in-tangent vectors exist only at points belonging to the geometrical boundary $\mathrm{b}C$ of C.

If $p \in I^{\circ}_{j,\eta}$, then the vector (3) *is tangent (out-tangent, in-tangent) to C at the point $q = \Phi(p)$ if and only if $a_j = 0$ ($\eta a_j > 0$, $\eta a_j < 0$). Every vector $\mathfrak{b}$ of the form* (3) *where $\mathfrak{a} = [a_1, \ldots, a_k] \in V_k$ is half-tangent to C at $q \in \Phi(I^{\circ}_{j,\eta})$. Consequently any linear combination of vectors half-tangent to C at $q \in \Phi(I^{\circ}_{j,\eta})$ is half-tangent to C at q.*

It follows from 2.4 that every mapping Ψ of class C_1 from an interval $\langle a; b\rangle$ into C is of the form $\Psi = \Phi \circ \Lambda$ where $\Lambda = \Phi^{-1} \circ \Psi$ is a mapping of class C_1 from $\langle a; b\rangle$ into I. The converse statement is also true (see 2.5).

If $q = \Phi(p)$ and $\Psi(b) = q$, i.e. $\Lambda(b) = p$, then $\Psi'(b) = \Phi'(p)\cdot\Lambda'(b)$. Thus a vector $\mathfrak{b}$ is of the form $\mathfrak{b} = \Psi'(b)$ for a mapping $\Psi\colon \langle a; b\rangle \to C$ ($\Psi(b) = q$) if and only if it is of the form $\mathfrak{b} = \Phi'(p)\mathfrak{a}$, where $\mathfrak{a} = \Lambda'(b)$ for a mapping $\Lambda\colon \langle a; b\rangle \to I$ ($\Lambda(b) = p$). In other words:

$\mathfrak{b}$ has property (a) with respect to C and $q = \Phi(p)$ if and only if $\mathfrak{b}$ is of the form (3), where $\mathfrak{a}$ has property (a) with respect to I and p.

Similarly,

$\mathfrak{b}$ has property (b) with respect to C and $q = \Phi(p)$ if and only if $\mathfrak{b}$ is of the form (3) where $\mathfrak{a}$ has property (b) with respect to I and p.

Consequently $\mathfrak{b}$ is half-tangent or tangent or out-tangent or in-tangent to C at $q = \Phi(p)$ if and only if it is of the form (3) where $\mathfrak{a}$ is respectively half-tangent or tangent or out-tangent or in-tangent to I at p.

The remaining part of 6.1 follows directly from what was just proved and from the remarks on half-tangent vectors to I, preceding 6.1.

In the case where $k = l$, every vector $\mathfrak{b} \in V_k$ is of the form (3). Hence it follows that every vector in R^k is half-tangent to a k-dimensional cell $C \subset R^k$ at every point $q \in C$.

Modifying the definition in V § 6 (p. 215), a k-dimensional hyperplane H through a point q of a k-dimensional cell $C \subset R^l$ is said to be *tangent to C at q* if every vector half-tangent to C at q is parallel to H. Every k-vector parallel to H is said to be *half-tangent* to C at q. Similarly as in V 6.5 we assert that if T is an orientation of C, then $T(q)$ is a unit k-vector half-tangent to C at q, and every k-vector half-tangent to C at q is of the form $cT(q)$ where $c \neq 0$ (see II § 8). In the case where $q \in C^\circ$, the notion of the k-vector half-tangent to C at q coincides with the notion of the k-vector tangent to C° at q. In the case where C is a one-dimensional cell, the above definition of a 1-vector half-tangent to C at q coincides with the definition of a vector half-tangent to the set C at q (see p. 365).

Now let C' be a $(k-1)$-dimensional closed face of a k-dimensional cell C in R^l $(2 \leqslant k \leqslant l)$ and let τ and T be orientations of C' and C respectively. Let $\mathfrak{b}$ be a vector out-tangent to C at a point $q \in C'^\circ$, i.e. at a point of the geometric interior of C'. The $(k-1)$-vector $\tau(q)$ is of the form

$$\tau(q) = \mathfrak{b}_1 \wedge \ldots \wedge \mathfrak{b}_{k-1}$$

where $\mathfrak{b}_1, \ldots, \mathfrak{b}_{k-1}$ are linearly independent vectors tangent to C' at q, and therefore tangent to C at q. The vector $\mathfrak{b}$ is not a linear combination of $\mathfrak{b}_1, \ldots, \mathfrak{b}_{k-1}$ since every linear combination of vectors tangent to C is tangent to C but the out-tangent vector $\mathfrak{b}$ is not tangent. Thus the vectors $\mathfrak{b}, \mathfrak{b}_1, \ldots, \mathfrak{b}_{k-1}$ are linearly independent and half-tangent to C at q. Consequently the k-vector

$$\mathfrak{b} \wedge \tau(q) = \mathfrak{b} \wedge \mathfrak{b}_1 \wedge \ldots \wedge \mathfrak{b}_{k-1}$$

is half-tangent to C at q, i.e. it is of the form $cT(q)$. This proves that

$$\mathfrak{b} \wedge \tau(q) = cT(q)$$

where c is a real number different from 0.

We say that orientations T and τ of a k-dimensional cell C and its $(k-1)$-dimensional closed face C' are *consistent* $(k \geqslant 2)$ if for every point $q \in C'^\circ$ and for every vector $\mathfrak{b}$ out-tangent to C at q

$$\mathfrak{b} \wedge \tau(q) = cT(q) \quad \text{for a number } c > 0 \tag{4}$$

which depends on $\mathfrak{b}$ and q.

It follows from the definition that if the orientations T and τ are consistent, then the orientations $-T$ and $-\tau$ are also consistent but the orientations $T, -\tau$ are not consistent, and the orientations $-T, \tau$ are not consistent.

For a given orientation T of C there is an orientation τ of C' which is consistent with T. To prove this, suppose that Φ is a diffeomorphism from an interval (1), and $C' = \Phi(I_{j,\eta}) = \Phi_{j,\eta}(I_j)$ where $I_{j,\eta}$, $\Phi_{j,\eta}$ and I_j are defined as in § 3.

6.2. *The orientation T of $C = \Phi(I)$ determined by the diffeomorphism Φ is consistent with the orientation $\tau = (-1)^{j-1}\eta T_{j,\eta}$ where $T_{j,\eta}$ is the orientation of $C' = \Phi_{j,\eta}(I_j)$ determined by the diffeomorphism $\Phi_{j,\eta}$.*

Let q be a point in C'°, i.e. $q = \Phi(p)$ where $p = (x_1, \ldots, x_k) \in I^{\circ}_{j,\eta}$, or equivalently $q = \Phi_{j,\eta}(p^*)$ where $p^* = (x_1, \ldots, x_{j-1}, x_{j+1}, \ldots, x_k) \in I^{\circ}_j$. It follows from the definition of $\Phi_{j,\eta}$ (see § 3 (22)) that

$$(\Phi_{j,\eta})_{|i}(p^*) = \begin{cases} \Phi_{|i}(p) & \text{for} \quad i = 1, \ldots, j-1\,, \\ \Phi_{|i+1}(p) & \text{for} \quad i = j, \ldots, k-1\,. \end{cases}$$

Hence it follows that

$$T_{j,\eta}(q) = c_1\Phi_{|1}(p)\wedge\ldots\wedge\Phi_{|j-1}(p)\wedge\Phi_{|j+1}(p)\wedge\ldots\wedge\Phi_{|k}(p)$$

where

$$c_1 = |\Phi'_{j,\eta}(p^*)|^{-1} > 0\,.$$

Let $\mathfrak{b}$ be a vector out-tangent to C at q. By 6.1,

$$\mathfrak{b} = a_1\Phi_{|1}(p) + \ldots + a_k\Phi_{|k}(p) \qquad \text{where} \qquad \eta a_j > 0\,.$$

Hence

$$\begin{aligned} \mathfrak{b}\wedge\tau(q) &= c_1(-1)^{j-1}\eta\big(a_1\Phi_{|1}(p) + \ldots + a_k\Phi_{|k}(p)\big)\wedge \\ &\qquad\qquad \wedge\Phi_{|1}(p)\wedge\ldots\wedge\Phi_{|j-1}(p)\wedge\Phi_{|j+1}(p)\wedge\ldots\wedge\Phi_{|k}(p) \\ &= c_1(-1)^{j-1}\eta a_j\Phi_{|j}(p)\wedge\Phi_{|1}(p)\wedge\ldots\wedge\Phi_{|j-1}(p)\wedge\Phi_{|j+1}(p)\wedge\ldots\wedge\Phi_{|k}(p) \\ &= c_1\eta a_j\Phi_{|1}(p)\wedge\ldots\wedge\Phi_{|j-1}(p)\wedge\Phi_{|j}(p)\wedge\Phi_{|j+1}(p)\wedge\ldots\wedge\Phi_{|k}(p) = cT(q) \end{aligned}$$

where

$$c = c_1\eta a_j|\Phi'(p)| > 0\,.$$

This proves that T and τ are consistent.

Now let C be a zero-dimensional cell in R^l, i.e. a one-point set. By an *orientation* of the zero-dimensional cell C we shall mean a function T on C, equal to 1 or to -1. Just as in the case of cells with positive dimension, a zero-dimensional cell has two orientations which differ by sign, i.e. are *opposite.* The orientation equal to $+1$ is called *positive,* the orientation equal to -1 is called *negative.*

Now we shall define when an orientation T of a one-dimensional cell C is consistent with an orientation τ of a zero-dimensional face $C' = \{q_0\}$ of C. Let $\mathfrak{b}$ be a vector out-tangent to C at q_0. We say that the orientations T and τ are *consistent* if

$$T(q) = c\tau(q)\mathfrak{b} \qquad \text{for a number } c > 0\,.$$

Observe that all vectors half-tangent to C at q_0 are parallel to the vector $T(q_0)$, i.e. they differ from $T(q_0)$ only by a scalar factor.

In other words, the orientations T and τ of a one-dimensional cell C and a zero-dimensional face $C' = \{q_0\}$ are consistent if and only if one of the following conditions is satisfied:

(5) $T(q_0)$ is out-tangent to C at q_0 and $\tau(q_0) = 1$;

(5') $T(q_0)$ is in-tangent to C at q_0 and $\tau(q_0) = -1$.

Two k-dimensional cells C_1 and C_2 in R^l $(1 \leqslant k \leqslant l)$ are *adherent* if their intersection $C' = C_1 \cap C_2$ is a common $(k-1)$-dimensional closed face of C_1 and C_2. Let T_1 and T_2 be orientations of adherent cells C_1 and C_2. We say that the orientations T_1, T_2 are *consistent* if the following condition is satisfied: if τ is an orientation of C' consistent with T_1, then $-\tau$ is consistent with T_2.

Now we shall discuss the case where $l = k$, i.e. the case of k-dimensional cells in R^k. We know from II §§ 9-10 that there are only two unit k-vectors in R^k, viz.

$$\mathrm{sgn} = [\mathrm{sgn}(i_1, \dots, i_k)]_{i_1, \dots, i_k \leqslant k}$$

and $-\mathrm{sgn}$. If T is an orientation of a k-dimensional cell $C \subset R^k$, then either

$$T(q) = \mathrm{sgn} \quad \text{for every } q \in C$$

or

$$T(q) = -\mathrm{sgn} \quad \text{for every } q \in C\,.$$

In the first case T is said to be the *positive orientation* of C. In the second case T is said to be the *negative orientation* of C.

6.3. *Let C' be a $(k-1)$-dimensional closed face of a k-dimensional cell C in R^k, and let τ be the orientation of C' which is consistent with the positive orientation of C. For every point $q \in C'^\circ$ the vector $\mathfrak{n} = \tau(q)^\vee$ dual to $\tau(q)$ is orthogonal to C'° at q and out-tangent to C at q. Moreover there exists a number $\delta > 0$ such that*

$$q + t\mathfrak{n} \begin{cases} \notin C & \text{if} \quad 0 < t < \delta\,, \\ \in C^\circ & \text{if} \quad 0 > t > -\delta\,. \end{cases} \tag{6}$$

The definition of a vector dual to a $(k-1)$-vector in R^k was given in II § 9 (30)-(31) (see also II § 10 (21)-(22)).

It follows from II § 10 (24) or II 10.2 that $\mathfrak{n}$ is orthogonal to C'° at q. Thus $\mathfrak{n}$ is not tangent at q to C'°. Since every vector in R^k is half-tangent to C at q (see a remark on p. 367), the vector $\mathfrak{n}$ is either out-tangent or in-tangent to C at q. By II § 9 (34), $\mathfrak{n} \wedge \tau(q) = \mathrm{sgn}$ and consequently $(-\mathfrak{n}) \wedge \tau(q) = -\mathrm{sgn}$. Since the orientation τ is consistent with the positive orientation sgn of C, the vector $-\mathfrak{n}$ is not out-tangent, i.e. it is in-tangent. This proves that $\mathfrak{n}$ is out-tangent.

To prove the final part of 6.3 let us assume that $C = \Phi(I)$ where Φ is a diffeomorphism defined on an open set $G \subset R^k$ containing the interval I (see 2.2). Assume that $C' = \Phi(I_{j,\eta})$. Thus $q = \Phi(p)$ where p is a point in $I^\circ_{j,\eta}$. Let

$$\Psi(t) = q + t\mathfrak{n} \quad \text{for} \quad |t| < \delta_0$$

where δ_0 is such a small positive number that $\Psi(t) \in \Phi(G)$ for $|t| < \delta_0$ (the set $\Phi(G)$ is open by V 5.1 (a); therefore such a number δ_0 exists). By Theorem V 5.4 on three mappings, $\Psi = \Phi \circ \Lambda$ where

$$\Lambda = \Phi^{-1} \circ \Psi = (\lambda_1, \ldots, \lambda_k)$$

is a mapping of class C_1 of the interval $(-\delta_0; \delta_0)$ into G. By definition,

$$\Lambda(0) = p \quad \text{and} \quad \mathfrak{n} = \Psi'(0) = \Phi'(p) \cdot \Lambda'(0) .$$

Since $\mathfrak{n}$ is out-tangent, we infer from 6.1 that $\eta\lambda'_j(0) > 0$. Hence it follows that there exists a number $\delta > 0$ $(\delta < \delta_0)$ such that

$$\Lambda(t) \begin{cases} \notin I & \text{if} \quad 0 < t < \delta , \\ \in I^\circ & \text{if} \quad 0 > t > -\delta . \end{cases}$$

This proves (6).

6.4. *The positive orientations of two adherent k-dimensional cells C_1, C_2 in R^k are consistent.*

Let $C' = C_1 \cap C_2$ be the common $(k-1)$-dimensional closed face of C_1 and C_2 and let τ_1 and τ_2 be orientations of C' which are consistent with the positive orientations of C_1 and C_2, respectively. It suffices to prove that $\tau_1(q) = -\tau_2(q)$ for $q \in C'^\circ$. Suppose that this is not true, i.e. that $\tau_1(q) = \tau_2(q)$ for a point $q \in C'^\circ$. Denoting $\mathfrak{n} = \tau_1(q)^\vee = \tau_2(q)^\vee$ we infer from 6.3 that there is a number $\delta > 0$ such that $q + t\mathfrak{n} \in C_1^\circ \cap C_2^\circ$ if $0 > t > -\delta$. Thus $C_1^\circ \cap C_2^\circ \neq 0$ contradicting the hypothesis that C_1, C_2 are adherent.

6.5. *Let D be a cellular decomposition of a k-dimensional polyhedron P in R^k. Every $(k-1)$-dimensional closed face of one of the cells in D is the common face of at most two cells in D.*

Suppose that a $(k-1)$-dimensional cell C' is a face of three different k-dimensional cells C_1, C_2, C_3 in D. Let τ_1, τ_2, τ_3 be the orientations of C' that are consistent with the positive orientations of C_1, C_2, C_3 respectively. By 6.4, the positive orientations of C_1, C_2, C_3 are consistent. Thus $\tau_1 = -\tau_2, \tau_2 = -\tau_3, \tau_3 = -\tau_1$. Consequently $\tau_1 = -\tau_1$, which is impossible.

EXERCISES

1. Let T_1 and T_2 be orientations of two adherent k-dimensional cells C_1 and C_2 in R^l. Suppose that $C_1 \cap C_2 \cap \operatorname{re}(C_1 \cup C_2) \neq 0$. The orientations T_1 and T_2 are consistent if and only if $T_1(q) = T_2(q)$ for every point (or for a point) $q \in C_1 \cap C_2 \cap \operatorname{re}(C_1 \cup C_2)$.

2. Let C_1 and C_2 be adherent two-dimensional cells of a cellular decomposition of a two-dimensional polyhedron $P \subset R^3$. Let $C' = C_1 \cap C_2$. Suppose that $C' \cap \mathrm{re}P \neq 0$. Let T_1 and T_2 be orientations of C_1 and C_2 and let $\mathfrak{n}_1 = T_1^\vee$ and $\mathfrak{n}_2 = T_2^\vee$. The orientations T_1 and T_2 are consistent if and only if $\mathfrak{n}_1(q) = \mathfrak{n}_2(q)$ for a point (or for every point) $q \in C' \cap \mathrm{re}P$.

3. Let M be the open Möbius strip defined in V § 6, exercise 4. Prove that there is no continuous vector field $\mathfrak{n}$ on M such that, for every $q \in M$, $\mathfrak{n}(q)$ is a unit vector orthogonal to M at q.

Prove the following property of the closed Möbius strip $\overline{M} = \mathrm{cl}\, M$ (see § 5, exercise 3): If D is a cellular decomposition of $\overline{M}$, then for any choice of orientations of all cells in D there exist two adherent cells whose orientations are not consistent.

4. Suppose that two adherent k-dimensional cells $C_1, C_2 \subset R^l$ are subsets of the range $\Phi(A)$ of a diffeomorphism $\Phi\colon A \to R^l$, where A is a closed interval or an open set in R^k. The orientations T_i $(i = 1, 2)$ of C_1 and C_2, defined by the formula

$$T_i(q) = |\Phi'(p)|^{-1}\Phi_{|1}(p) \wedge \ldots \wedge \Phi_{|k}(p) \quad \text{for} \quad q = \Phi(p) \in C_i$$

are consistent.

5. Let p be a point in the $(k-r)$-dimensional open face $I^\circ_{j_1,\eta_1;\ldots;j_r,\eta_r}$ of the interval (1). A vector $\mathfrak{a} = [a_1, \ldots, a_k]$ is out-tangent (in-tangent) to I at p if and only if $\eta_{j_i} a_{j_i} > 0$ (if $\eta_{j_i} a_{j_i} < 0$) for $i = 1, \ldots, r$.

Formulate a similar necessary and sufficient condition for a vector $\mathfrak{b} \in V_l$ to be out-tangent (in-tangent) to a k-dimensional cell $C = \Phi(I) \subset R^l$ at a point q belonging to an open $(k-r)$-dimensional face of C.

§ 7. Oriented cells. Chains. It is easy to guess that an oriented cell is a cell C plus a fixed orientation T of C. Thus, from a logical point of view, an oriented cell should be defined as a pair (C, T) composed of a cell C and its orientation T. Since C is uniquely determined by T (viz. C is the domain of the matrix field T), we could even identify the notion of an oriented cell with its orientation T, i.e. we could call T an oriented cell. By such a definition, an oriented cell would be a matrix field defined on a cell C. This definition, although completely correct, would be inconvenient because in this book we shall have to perform some algebraic operations on oriented cells, for instance, to add oriented cells which are orientations of distinct cells, etc. For this purpose it will be convenient to assume that an oriented cell is a matrix field defined on the whole Euclidean space, i.e. that it is an extension of an orientation T. The extension will be defined in the simplest way: at all the points where T is not defined the extension is equal to 0. Consequently we adopt the following definition of an oriented cell.

Let T be an orientation of a k-dimensional cell $C \subset R^l$ $(0 \leqslant k \leqslant l)$. By definition, T is a continuous k-vector field on C in the case where $k > 0$, and a real function on C for $k = 0$. The function

$$(1) \qquad K(q) = \begin{cases} T(q) & \text{for} \quad q \in C, \\ 0 & \text{for} \quad q \in R^l - C \end{cases}$$

will be called an *oriented cell with support C*. The symbol 0 in (1) denotes, of course, the zero k-vector in R^l in the case of $k > 0$, and the number zero in the case of $k = 0$. By definition, a k-dimensional cell in R^l is a l-dimensional k-vector field on R^l in the case of $k > 0$, and a real function on R^l for $k = 0$. Moreover, the set $C = \{q\colon K(q) \neq 0\}$ is a k-dimensional cell, viz. the support of K, and $K|C$ is an orientation of C. The orientation $T = K|C$ will be called the *orientation of the oriented cell K*.

For any k-dimensional cell there exist exactly two oriented cells whose support is C. They correspond to two opposite orientations of C. Namely, if T is an orientation of C, then (1) defines one of the two oriented cells with support C. The function $-K$ is another oriented cell with support C. It corresponds to the orientation $-T$, i.e. it is defined by (1) with T replaced by $-T$.

If Φ is a diffeomorphism from a k-dimensional closed interval I onto a k-dimensional cell C ($k > 0$) and T is the orientation determined by Φ, then the corresponding oriented cell (1) will be denoted by the symbol $\langle\Phi\rangle$.

We assumed in § 4 the convention that a zero-dimensional cell C will be considered as a diffeomorphic image $C = \Phi(I)$ of the zero-dimensional interval $I = \{0\}$ composed only of the number zero. In this case $\langle\Phi\rangle$ will always denote the positively oriented zero-dimensional cell with support C, i.e. the real function equal to 1 on the one-point set $C = \Phi(I)$, and equal to 0 outside this set.

The support of an oriented cell K will be denoted by $|K|$.

By a *k-dimensional chain* in R^l ($0 \leqslant k \leqslant l$) we shall mean any linear combination

(2) $$\Gamma = a_1 K_1 + \ldots + a_n K_n$$

where $a_1, \ldots, a_n$ are real numbers, $K_1, \ldots, K_n$ are k-dimensional oriented cells and, moreover, the set of their supports

(3) $$D = \{|K_1|, \ldots, |K_n|\}$$

is a cellular decomposition of a k-dimensional polyhedron.

The case where $a_1 = \ldots = a_n = 0$ is admitted; we say then that (2) is the *zero-chain*. The zero chain will be denoted by 0.

Any non-zero k-dimensional Γ can always be represented in the form (2) where all the coefficients $a_1, \ldots, a_n$ are different from zero, and the cells $|K_1|, \ldots, |K_n|$ are mutually distinct. Under these additional hypotheses the oriented cells $K_1, \ldots, K_n$ are determined by Γ uniquely up to their sign and, consequently, their supports $|K_1|, \ldots, |K_n|$ are uniquely determined by Γ. The union $|K_1| \cup \ldots \cup |K_n|$ is called the *support of the chain Γ* and is denoted by $|\Gamma|$. The set (3) is a cellular decomposition of the k-dimensional polyhedron $|\Gamma|$.

If (2) is a representation of a chain Γ as a linear combination of oriented cells with different supports, where some of the coefficients $a_1, \ldots, a_n$ may be equal to zero, then $|\Gamma|$ is the union of all those cells $|K_j|$ for which $a_j \neq 0$.

The empty set is the support of the zero chain.

A zero-dimensional chain is a real function defined on the whole space but equal to zero outside a finite set. The support of this chain is the finite set of points at which the function is different from 0.

If $k > 0$, a k-dimensional chain Γ in R^l is an l-dimensional k-vector field which is equal to zero outside a k-dimensional polyhedron, viz. the support $P = |\Gamma|$ of Γ. On the geometrical interior of each of the cells of a celluar decomposition D of P the k-vector field Γ is a constant multiple of an orientation of the cell. On the $(k-1)$-dimensional skeleton $\mathrm{b}_D P$ of the decomposition D, the k-vector field Γ is not continuous. However the values of Γ on $\mathrm{b}_D P$ do not play an essential role. In what follows chains Γ will be used to define certain integrals on $P = |\Gamma|$ with respect to the measure μ_P. The set $\mathrm{b}_D P$ has measure zero, therefore the values of Γ on $\mathrm{b}_D P$ do not play any role in the integration.

For this reason we shall not distinguish k-dimensional chains which differ only on a set whose k-dimensional measure is equal to zero. More precisely, we shall say, according to a general definition from VII § 5, p. 274, that two k-dimensional chains Γ_1 and Γ_2 are *equivalent* and we shall write

$$\Gamma_1 \equiv \Gamma_2$$

if they have the same support $P = |\Gamma_1| = |\Gamma_2|$ and the set $\{q\colon \Gamma_1(q) \neq \Gamma_2(q)\}$ has k-dimensional measure μ_P equal to zero.

Zero-dimensional chains are equivalent if and only if they are equal. This follows from the fact that every one-point subset of the support of a zero-dimensional chain has positive zero-dimensional measure, viz. measure 1.

For $k > 0$ it is easy to give examples of different but equivalent chains. For this purpose we shall introduce the following definition.

Let D be the cellular decomposition (3) of the support $|\Gamma|$ of a k-dimensional chain Γ $(k > 0)$ defined by (2), and let

$$\tilde{D} = \{\tilde{C}_1, \ldots, \tilde{C}_m\}$$

be another cellular decomposition of the polyhedron $|\Gamma|$, such that

(4) every k-dimensional cell $\tilde{C}_i$ is a subset of (exactly one) k-dimensional cell $|K_j|$ of the decomposition D.

If integers i and j satisfy (4), let

$$\tilde{T}_i = K_j|\tilde{C}_i\,, \qquad a_i = a_j$$

and let $\tilde{K}_i$ denote the oriented cell with support $\tilde{C}_i$ and orientation $\tilde{T}_i$. The chain

(5) $$\tilde{\Gamma} = a_1\tilde{K}_1 + \ldots + a_m\tilde{K}_m$$

is said to be a *refinement* of Γ. The chains Γ and $\tilde{\Gamma}$ have the same support. The set

$$\{q\colon \Gamma(q) \neq \tilde{\Gamma}(q)\}$$

is a subset of the $(k-1)$-dimensional skeleton of the decomposition D. Therefore its k-dimensional measure is equal to 0. Thus the chains Γ and $\tilde{\Gamma}$ are equivalent but, in general, they are not equal.

EXERCISE

1. If a chain $\Gamma = K_1 + \ldots + K_m$ is a refinement of a k-dimensional oriented cell K in R^l, and $|K_i|$ and $|K_j|$ are adherent, then the orientations of K_i and K_j are consistent.

§ 8. The boundary of an oriented cell and of a chain. We say that a $(k-1)$-dimensional oriented cell K_0 is an *oriented face* of a k-dimensional oriented cell K $(k > 0)$ if the support C_0 of K_0 is a closed $(k-1)$-dimensional face of the support C of K, and the orientation $T_0 = K_0|C_0$ of K_0 is consistent with the orientation $T = K|C$ of K. It follows from § 6 that every closed $(k-1)$-dimensional face C_0 of C has exactly one orientation T_0 consistent with the orientation T of K. Therefore the $(k-1)$-dimensional oriented cell

$$K_0(q) = \begin{cases} T_0(q) & \text{for} \quad q \in C_0, \\ 0 & \text{elsewhere} \end{cases}$$

is the only oriented $(k-1)$-dimensional face of K, whose support is C_0.

Since the support C of K has exactly $2k$ closed $(k-1)$-dimensional faces, and each of them is the support of exactly one oriented $(k-1)$-dimensional face of K, we infer that the k-dimensional cell K has exactly $2k$ oriented $(k-1)$-dimensional faces

(1) $$K_{0,1}, \ldots, K_{0,2k}.$$

The $(k-1)$-dimensional chain

(2) $$\mathrm{b}K = K_{0,1} + \ldots + K_{0,2k}$$

will be called the *boundary of the oriented cell* K.

It follows from 6.2 and § 6 (5) that if $K = \langle\Phi\rangle$, where Φ is a diffeomorphism from a closed interval $I \subset R^k$ onto the support C of the oriented k-dimensional cell K, then

(3) $$(-1)^{j-1}\eta\langle\Phi_{j,\eta}\rangle \quad (j = 1, \ldots, k, \ \eta = \pm 1)$$

are all oriented $(k-1)$-dimensional faces of K. Thus the chain

$$\mathrm{b}\langle\Phi\rangle = \sum_{j=1}^{k} \sum_{\eta=\pm1} (-1)^{j-1}\eta\langle\Phi_{j,\eta}\rangle \tag{4}$$

is the boundary of the oriented k-dimensional cell $\langle\Phi\rangle$.

If (1) are all oriented $(k-1)$-dimensional faces of an oriented k-dimensional cell K, then $-K_{0,1}, \ldots, -K_{0,2k}$ are all oriented $(k-1)$-dimensional faces of the oriented k-dimensional cell $-K$. This follows from the fact that if T_0 is the orientation of a closed $(k-1)$-dimensional face C_0 of $C = |K|$ which is consistent with the orientation $T = K|C$ of K, then $-T_0$ is the orientation of C_0 which is consistent with the orientation $-T$ of $-K$.

Consequently,

$$\mathrm{b}(-K) = -\mathrm{b}K \tag{5}$$

for every oriented cell K.

Now let

$$\Gamma = a_1 K_1 + \ldots + a_n K_n \tag{6}$$

be a k-dimensional chain in R^l $(k > 0)$. We assume that the cells $|K_1|, \ldots, |K_n|$ are different from each other and that they form a cellular decomposition D of a k-dimensional polyhedron P. By the ***boundary of the k-dimensional chain*** (6) we mean the $(k-1)$-dimensional chain

$$\mathrm{b}\Gamma = a_1 \mathrm{b}K_1 + \ldots + a_n \mathrm{b}K_n\,. \tag{7}$$

To verify the correctness of the above definition let us observe that the right-hand side of (7) is a linear combination of $(k-1)$-dimensional oriented faces of $K_1, \ldots, K_n$. Their supports form a cellular decomposition $\mathrm{b}D$ of the $(k-1)$-dimensional polyhedron $\mathrm{b}_D P$ (see 5.1). Thus the expression on the right-hand side of (7) is really a $(k-1)$-dimensional chain. The representation of Γ in the form (6) is not unique. However, if $a_j \neq 0$, then the oriented cell K_j is determined by Γ uniquely up to the sign. If K_j is replaced in (6) by $-K_j$, then a_j must be replaced by $-a_j$. Hence it follows (see (5)) that the jth summand on the right-hand side of (7) remains the same. In the case where $a_j = 0$, the value of (7) is the same if we add or delete the jth summand in (6) or if we replace K_j by $-K_j$. This proves that the right-hand side of (7) does not depend on the representation (6) of Γ.

Definition (7) is a generalization of definition (2).

It follows directly from (7) that for every real number a and every chain Γ,

$$\mathrm{b}(a\Gamma) = a\mathrm{b}\Gamma\,. \tag{8}$$

In particular, for $a = -1$,

(9) $$\mathrm{b}(-\Gamma) = -\mathrm{b}\Gamma\,.$$

If Γ_1, Γ_2 and $\Gamma_1+\Gamma_2$ are k-dimensional chains, then

(10) $$\mathrm{b}(\Gamma_1+\Gamma_2) = \mathrm{b}\Gamma_1+\mathrm{b}\Gamma_2\,.$$

Similarly, if Γ_1, Γ_2 and $\Gamma_1-\Gamma_2$ are k-dimensional chains, then

(11) $$\mathrm{b}(\Gamma_1-\Gamma_2) = \mathrm{b}\Gamma_1-\mathrm{b}\Gamma_2\,.$$

A k-dimensional chain Γ $(k>0)$ is said to be a *k-dimensional cycle* (or simply a *cycle*) if $\mathrm{b}\Gamma = 0$, i.e. if its boundary is the zero chain.

8.1. *The boundary* $\mathrm{b}\Gamma$ *of every k-dimensional chain* Γ $(k>1)$ *is a $(k-1)$-dimensional cycle, i.e.*

(12) $$\mathrm{bb}\Gamma = 0\,.$$

If Γ is of the form (6), then by (7), (8) and (10),

$$\mathrm{bb}\Gamma = a_1\mathrm{bb}K_1+\ldots+a_n\mathrm{bb}K_n\,.$$

Hence it follows that it suffices to show that $\mathrm{bb}K = 0$ for every k-dimensional oriented cell K $(k>1)$. For this purpose let us assume that $K = \langle\Phi\rangle$, where Φ is a diffeomorphism defined on a closed interval $I\subset R^k$. We shall use the notation from § 3 (22)-(23) and the formula (4) for the boundary of an oriented cell. Applying (4) to the oriented cell $\langle\Phi_{j,\eta}\rangle$ we get

$$\mathrm{b}\langle\Phi_{j,\eta}\rangle = \sum_{\zeta=\pm1}\zeta\cdot\big(\langle\Phi_{j,\eta;1,\zeta}\rangle-\langle\Phi_{j,\eta;2,\zeta}\rangle+\ldots+(-1)^{j-2}\langle\Phi_{j,\eta;j-1,\zeta}\rangle+$$
$$+(-1)^{j-1}\langle\Phi_{j,\eta;j+1,\zeta}\rangle+\ldots+(-1)^{k-2}\langle\Phi_{j,\eta;k,\zeta}\rangle\big)\,.$$

Hence by (4) and (7)

(13) $$\mathrm{bb}K = \sum_{\eta,\zeta=\pm1}\ \sum_{\substack{i,j=1\\ i\neq j}}^{k} a_{j,\eta;i,\zeta}\langle\Phi_{j,\eta;i,\zeta}\rangle$$

where

$$a_{j,\eta;i,\zeta} = \begin{cases}(-1)^{j-1}\eta(-1)^{i-1}\zeta & \text{for}\quad i<j\,,\\ (-1)^{j-1}\eta(-1)^{i-2}\zeta & \text{for}\quad i>j\,,\end{cases}$$

and consequently

$$a_{j,\eta;i,\zeta}+a_{i,\zeta;j,\eta} = 0 \quad \text{for}\quad i\neq j\,.$$

Since $\Phi_{j,\eta;i,\zeta} = \Phi_{i,\zeta;j,\eta}$ (see § 3 (19)) we have $\langle\Phi_{j,\eta;i,\zeta}\rangle = \langle\Phi_{i,\zeta;j,\eta}\rangle$ and

$$a_{j,\eta;i,\zeta}\langle\Phi_{j,\eta;i,\zeta}\rangle+a_{i,\zeta;j,\eta}\langle\Phi_{i,\zeta;j,\eta}\rangle = 0\,.$$

Grouping in (13) the terms which differ only by the ordering of pairs of indices, we infer that $\mathrm{bb}K = 0$.

To illustrate the introduced notions by means of some examples and figures we shall assume the following conventions.

Let K be a one-dimensional oriented cell with support C shown in figure 39 and with an orientation T. By definition, T is a continuous field of unit vectors tangent to C. A few tangent unit vectors $T(q)$ are shown in figure 40. The same is shown in figure 41 for the opposite orientation $-T$.

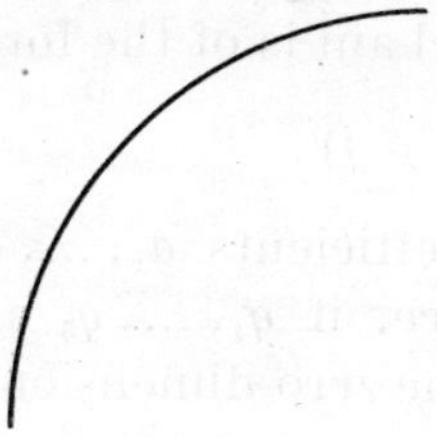

Fig. 39

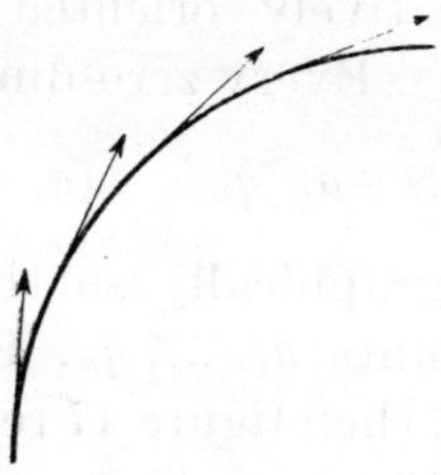

Fig. 40

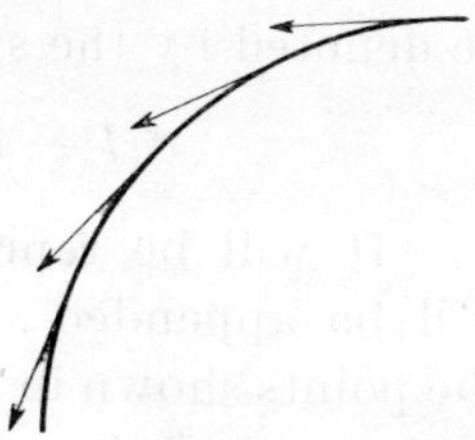

Fig. 41

To indicate graphically the chosen orientation T of the cell C, we shall draw an arrow showing the direction of the tangent unit vectors $T(q)$. For instance, figure 42 represents the orientation from figure 40 (i.e. the oriented cell K), and figure 43 represents the opposite orientation $-T$ (i.e. the oriented cell $-K$).

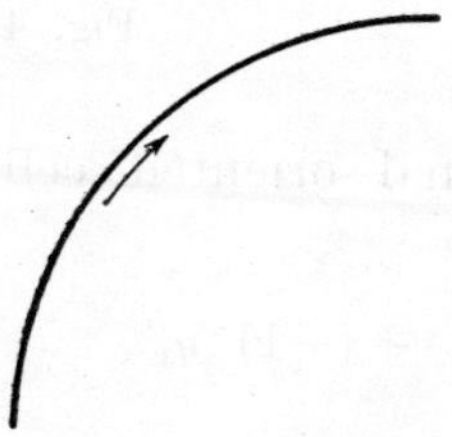

Fig. 42

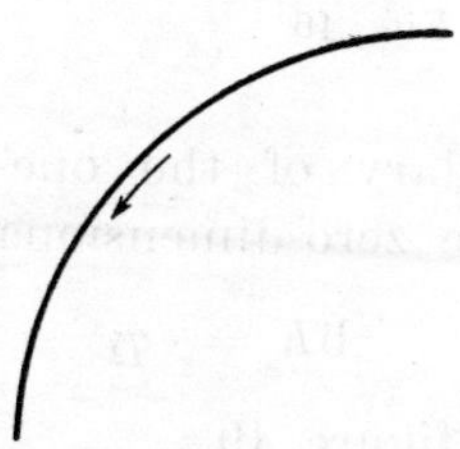

Fig. 43

A one-dimensional chain

$$\Gamma = a_1 K_1 + \ldots + a_n K_n \qquad (|K_i| \neq |K_j| \text{ for } i \neq j)$$

will be represented graphically in such a way that the coefficients $a_1, \ldots, a_n$ will be printed close to the oriented cells $K_1, \ldots, K_n$. For instance, if $K_1, \ldots, K_4$ are the oriented cells in figure 44, then figure 45 is a graphical representation of the chain $2K_1 - 3K_2 + 4K_3 - K_4$.

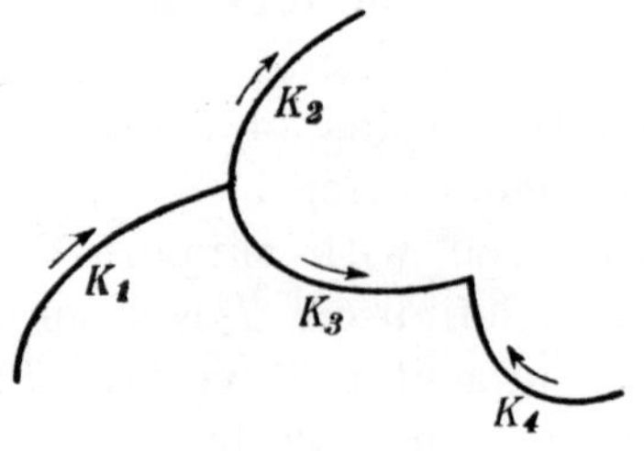

Fig. 44

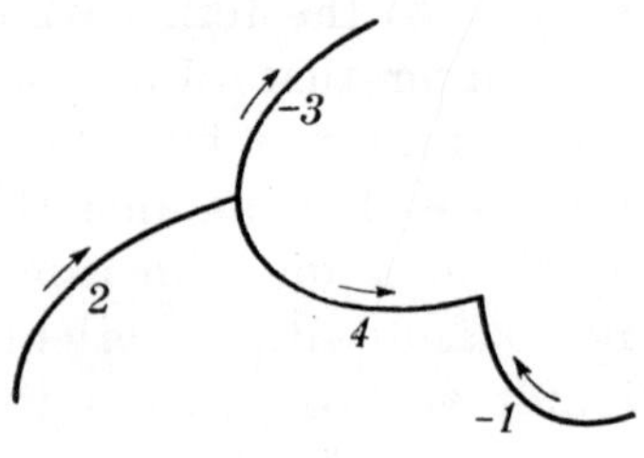

Fig. 45

A zero-dimensional positively oriented cell with support $\{q\}$ will be denoted by the symbol $\langle q\rangle$. Every zero-dimensional chain is of the form

$$\Gamma = a_1\langle q_1\rangle + \ldots + a_n\langle q_n\rangle \qquad (q_i \neq q_j \text{ for } i \neq j)\,.$$

It will be represented graphically so that the coefficients $a_1, \ldots, a_n$ will be appended to the points $q_1, \ldots, q_n$. For instance, if $q_1, \ldots, q_5$ are the points shown in figure 46, then figure 47 represents the zero-dimensional chain $\langle q_1\rangle - 2\langle q_2\rangle - 3\langle q_3\rangle + 10\langle q_4\rangle$.

Fig. 46

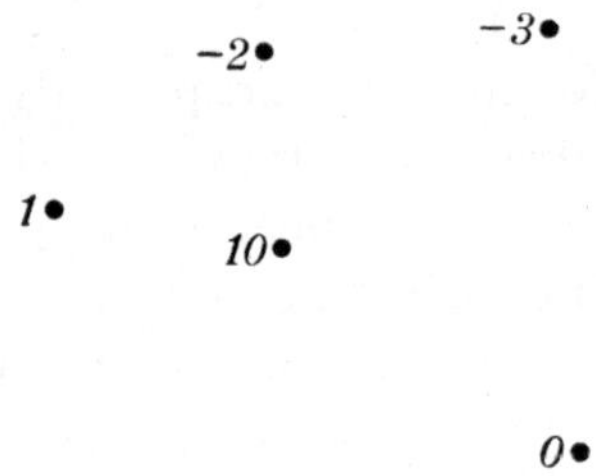

Fig. 47

The boundary of the one-dimensional oriented cell K shown in figure 48 is the zero-dimensional chain

(14) $$\mathrm{b}K = \langle q_2\rangle - \langle q_1\rangle = \langle q_2\rangle + (-1)\langle q_1\rangle$$

represented in figure 49.

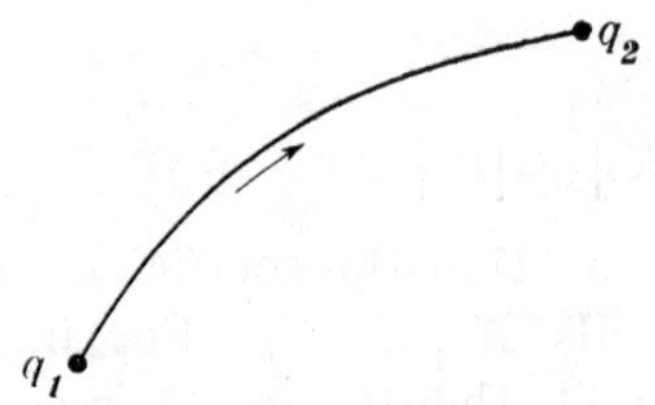

Fig. 48

Fig. 49

The zero-dimensional chain represented in figure 51 is the boundary of the one-dimensional chain in figure 50.

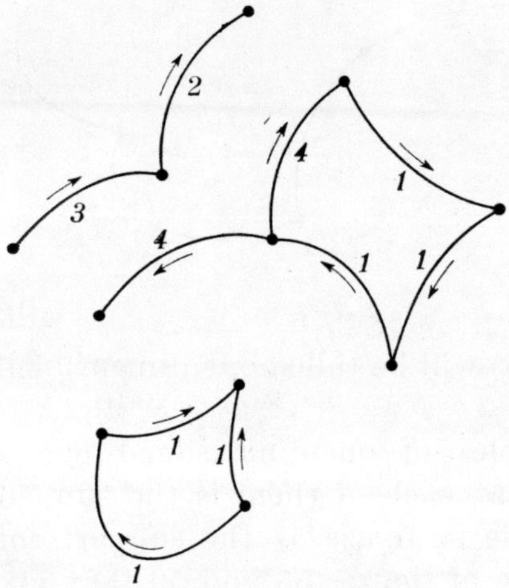

Fig. 50

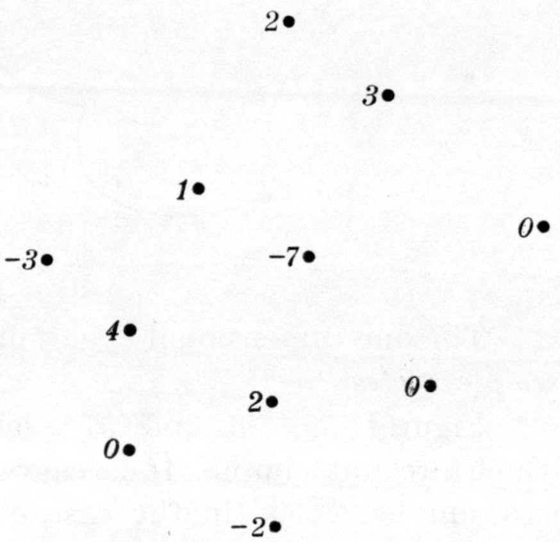

Fig. 51

The one-dimensional oriented cells in figure 52 are consistently oriented, but those in figure 53 are not.

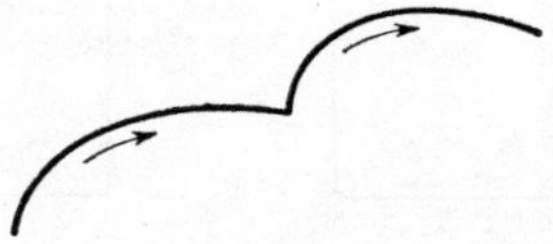

Fig. 52

Fig. 53

The simplest examples of one-dimensional cycles are chains

$$(15) \qquad \Gamma = K_1 + \dots + K_n \qquad (|K_i| \neq |K_j| \text{ for } i \neq j)$$

whose supports $|\Gamma| = |K_1| \cup \dots \cup |K_n|$ are closed curves—see figure 54 where a cycle Γ of this form is shown, and where the cycle $-\Gamma$, also of this form, is shown.

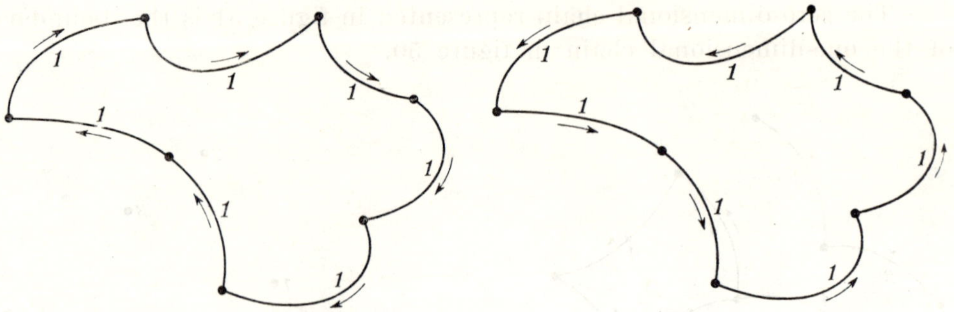

Fig. 54

The one-dimensional cycles of form (15) will be called one-dimensional *simple cycles.*

Figures 55, 56 and 57 yield examples of one-dimensional cycles which are not simple. It is easy to see that each of them is the sum of two simple cycles (in the case of the cycle in figure 57 the supports of the summands are disjoint, but in the case of figures 55 and 56 they are not disjoint). Analogously we can construct cycles which are sums of any finite number of simple cycles.

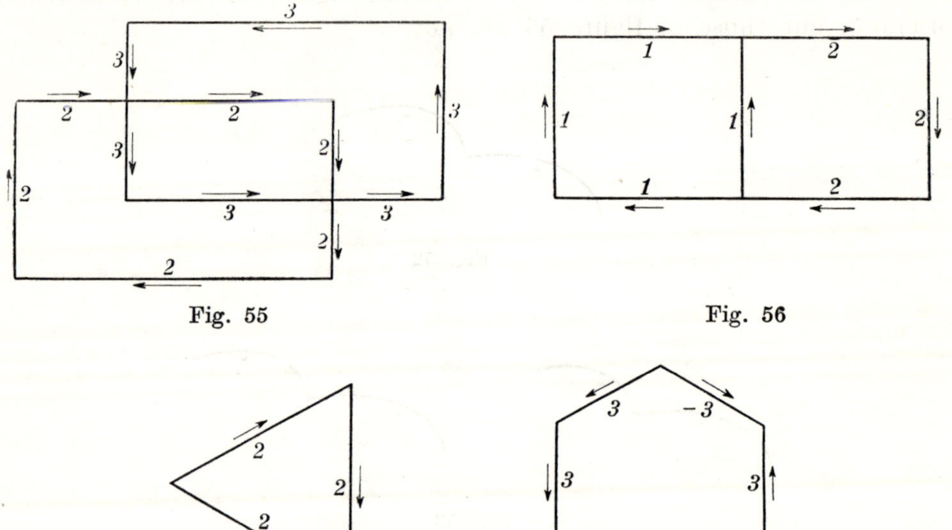

Fig. 55 Fig. 56

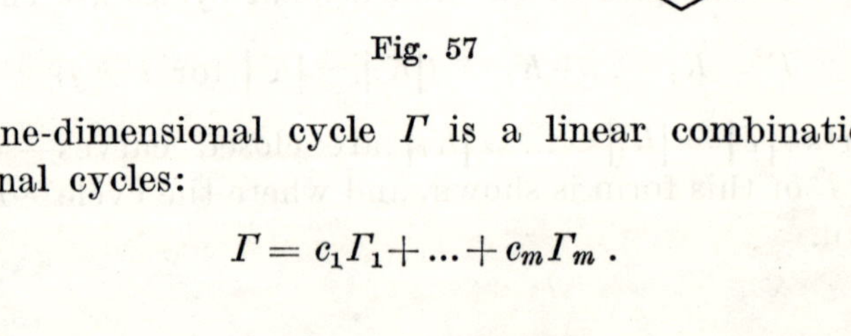

Fig. 57

Every one-dimensional cycle Γ is a linear combination of simple one-dimensional cycles:

(16) $$\Gamma = c_1\Gamma_1 + \ldots + c_m\Gamma_m\ .$$

The proof of this remark is by induction on n. We omit it because we shall not use this remark in what follows.

The two-dimensional oriented cell K with support C shown in figure 58, whose boundary is the one-dimensional cycle represented in figure 59, will be depicted graphically as in figure 60. The circle-like arrow in figure 60 shows the direction of the orientations of the one-dimensional faces, consistent with the orientation of K.

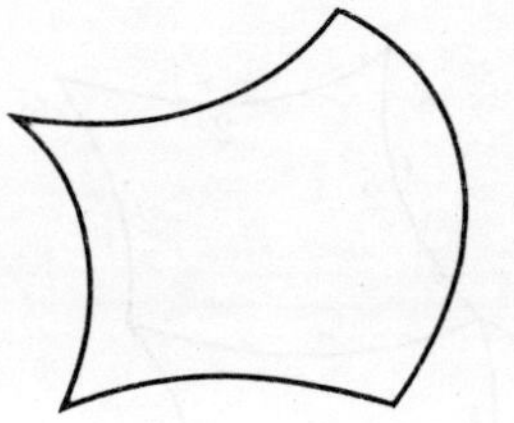

Fig. 58

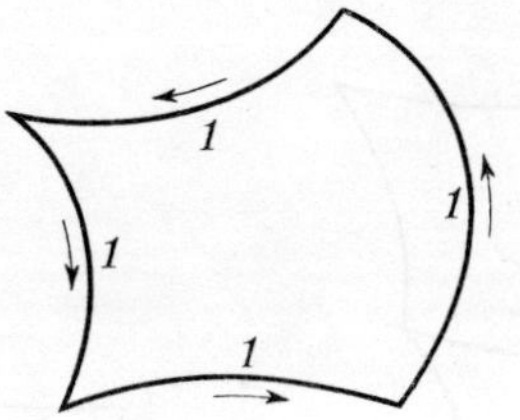

Fig. 59

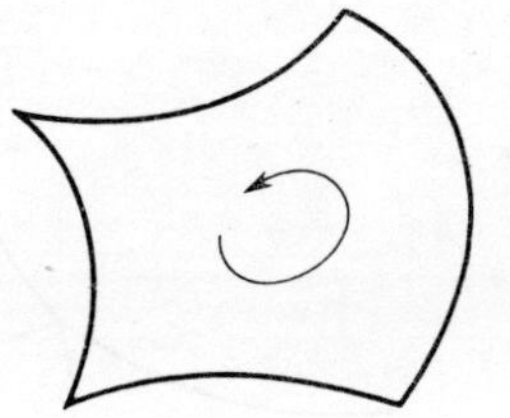

Fig. 60

Figure 61 represents the oriented cell $-K$. Its boundary is the one-dimensional cycle represented in figure 62.

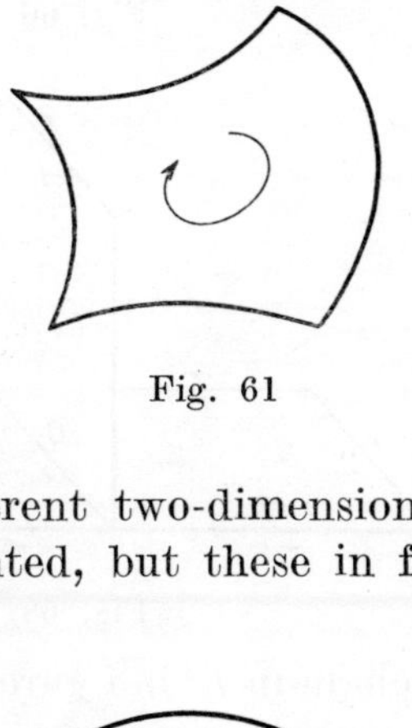

Fig. 61

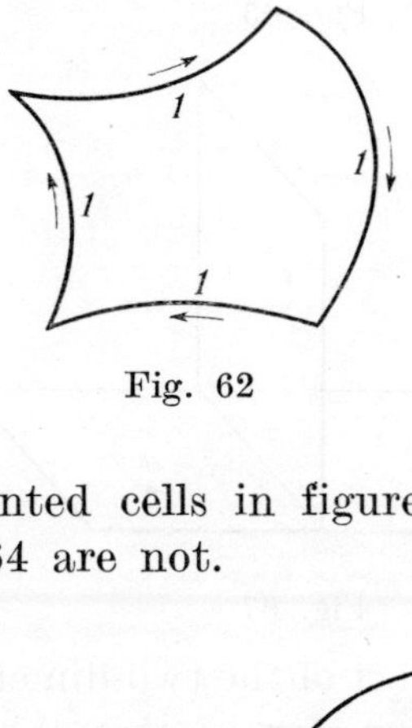

Fig. 62

The adherent two-dimensional oriented cells in figure 63 are consistently oriented, but these in figure 64 are not.

Fig. 63

Fig. 64

A two-dimensional chain $\Gamma = a_1 K_1 + \ldots + a_n K_n$ can be graphically represented by drawing the oriented cells $K_1, \ldots, K_n$ and writing the coefficients $a_1, \ldots, a_n$ close to the circle-like arrows showing the orientations of $K_1, \ldots, K_n$.

The reader can verify without difficulty that the boundary of the two-dimensional chain from figure 65 is the one-dimensional chain from figure 66, and that the boundary of the two-dimensional cycle (in R^3) from figure 67 is the one-dimensional chain from figure 68.

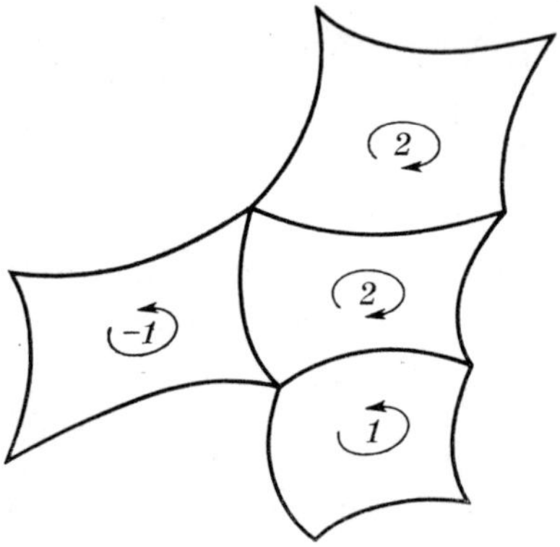

Fig. 65

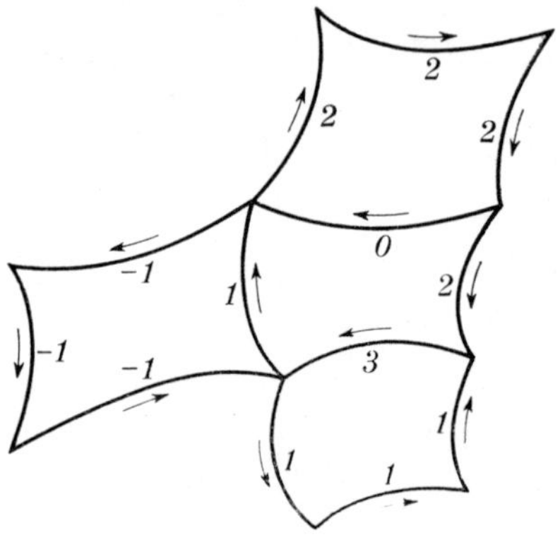

Fig. 66

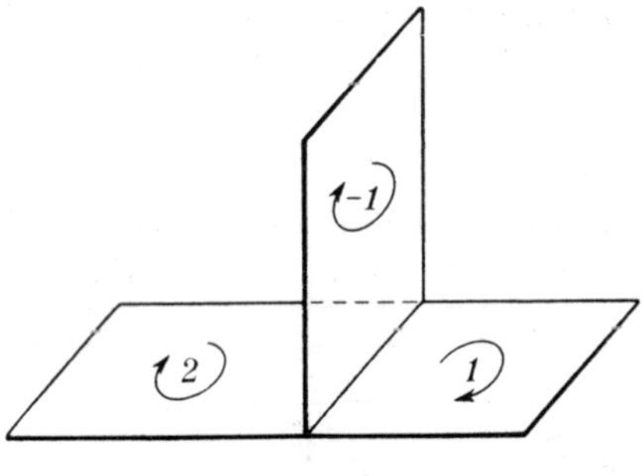

Fig. 67

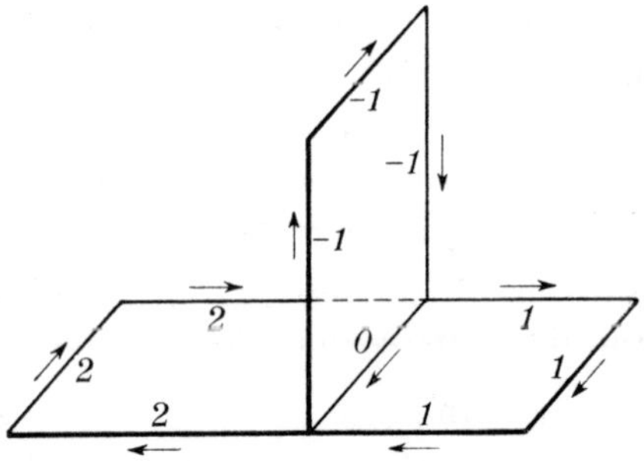

Fig. 68

The support of the two-dimensional chain in R^3 in figure 69 is a Möbius strip. The boundary of this chain is shown in figure 70.

From the intuitive point of view, a k-dimensional chain $\Gamma = a_1 K_1 + \ldots + a_n K_n$ in R^l can be visualized as the polyhedron $|\Gamma|$ whose cells are oriented in a certain way and are taken with multiplicities $a_1, \ldots, a_n$ respectively. This way of interpretation of the notion of chain is different from the formal definition of chain on p. 372. However it explains well the geometrical sense of the notion of chain. The geometrical sense of the notion of orientation of a cell (the choice of one of two possible fields of tangent unit k-vectors) can be understood rather easily. However it may seem strange, at first glance, why we take cells with multiplicities.

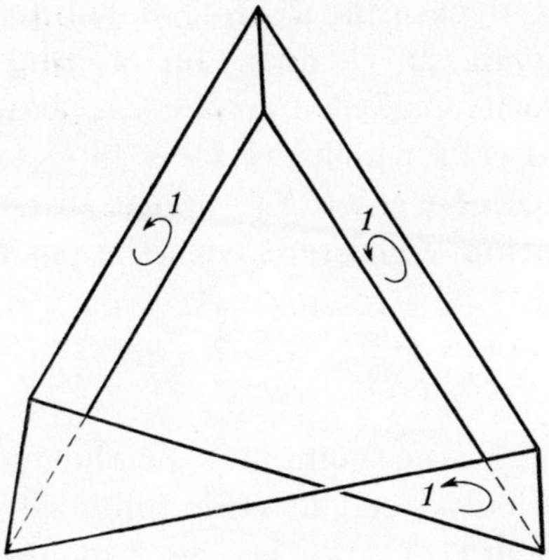

Fig. 69

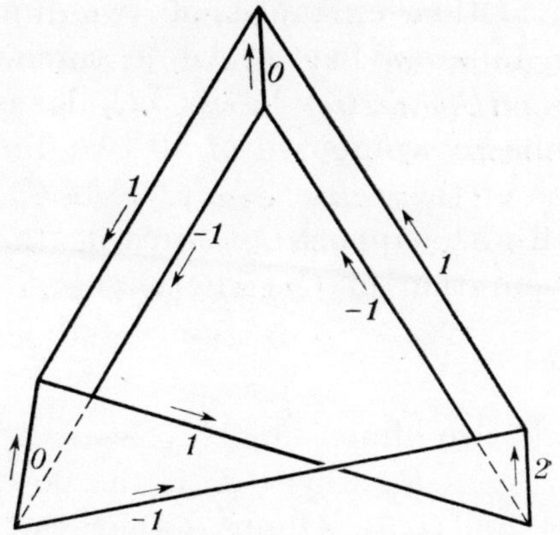

Fig. 70

One of the reasons why we do not restrict our investigation to chains of the form

(17) $\Gamma = K_1 + \ldots + K_n \qquad (|K_i| \neq |K_j| \text{ for } i \neq j)$

is that the boundary of chains of this form is not, in general, of this form (see the example in figures 69-70). In § 11 (4) an important formula from the theory of the integral will be proved, and the multiplicities of the cells in the chains under consideration play an important part in this formula. Also, in the case where the formula is applied for chains of the form (17), the multiplicities of the cells in the boundary bΓ play an essential part and cannot be ignored.

Now a few words about the notion of cycle.

The boundary of a three-dimensional oriented cell is the simplest example of a non-zero two-dimensional cycle. Figure 71 depicts the boundary of a positively oriented cube in R^3.

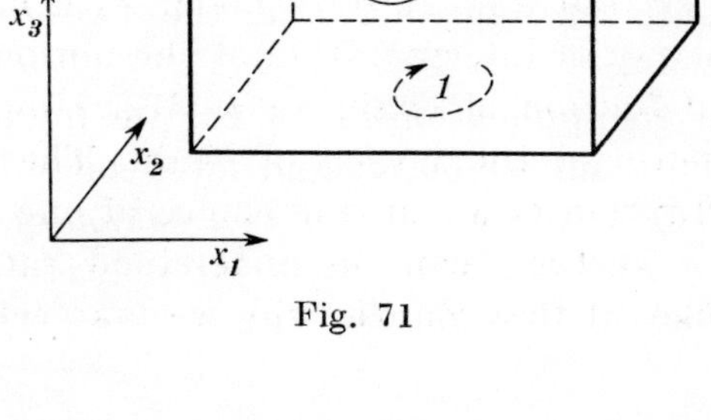

Fig. 71

Other examples of two-dimensional cycles in R^3 can be obtained as follows. Let P be a three-dimensional polyhedron in R^3 and let $D = \{C_1, \ldots, C_n\}$ be its cellular decomposition. Let $C_{0,1}, \ldots, C_{0,m}$ be the sequence composed of all two-dimensional cells which are faces of exactly one of the cells $C_1, \ldots, C_n$. If $C_{0,i}$ is a face of C_j, let $K_{0,i}$ be the oriented cell with support $C_{0,i}$ and with the orientation consistent with the positive orientation of C_j. The chain

$$\Gamma = K_{0,1} + \ldots + K_{0,m} \tag{18}$$

is a two-dimensional cycle in R^k, viz. it is the boundary of the chain $K_1 + \ldots + K_n$ where K_j is the positively oriented cell C_j (this follows from 6.4 and 6.5). Observe that the support $|\Gamma|$ of (18) is the topological boundary of the polyhedron $P \subset R^3$. Cycles of the form (18) are called *simple.*

In particular, the two-dimensional sphere (i.e. the topological boundary of a ball in R^3), ellipsoid and torus are supports of two-dimensional simple cycles. For a given cellular decomposition, each of these surface is the support of exactly two simple cycles which differ only by sign. Examples of non-simple two-dimensional cycles in R^3 can be obtained by an obvious generalization of the constructions from figures 55, 56 and 57. For instance, the union of a surfaces of two intersecting, or adherent or disjoint parallelepipeds is a support of certain two-dimensional cycles. Instead of the surfaces of parallelepipeds we can take spheres, ellipsoids, tori, surfaces of regular convex polyhedra, etc.

It follows from the above examples and from the definition of a cycle that the notion of a cycle is a precise formulation of what we would like to call a "closed surface" (the word "surface" has here an intuitive meaning, wider than that from V § 6, since we admit here such "surfaces" as the union of two intersecting spheres, tori, etc.). More precisely, the support of a two-dimensional cycle in R^3 is what we intuitively call a closed surface. The same is true for higher dimensions: the support of k-dimensional cycles is what should be called, from the intuitive point of view, a closed k-dimensional hypersurface (the word "hypersurface" has here an intuitive meaning wider than that in V § 6).

The intuitive interpretation of the notion of refinement from p. 374 is rather evident. The chains in figures 73 and 75 are refinements of the one-dimensional chain in figure 72 and the two-dimensional chain in figure 74, respectively.

Locking ahead to applications in § 13, we introduce the following definition. Let X be a subset of R^l, and let Γ be a k-dimensional cycle in R^l, $0 \leqslant k < l$, $|\Gamma| \subset X$. We say that Γ is a *boundary on* X and we write

$$\Gamma \sim 0 \text{ in } X \tag{19}$$

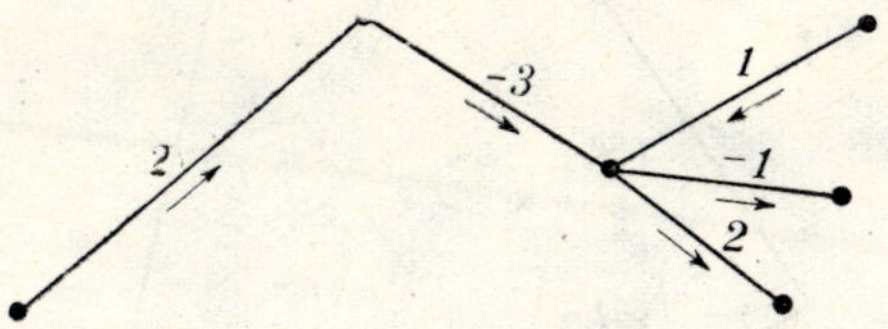

Fig. 72

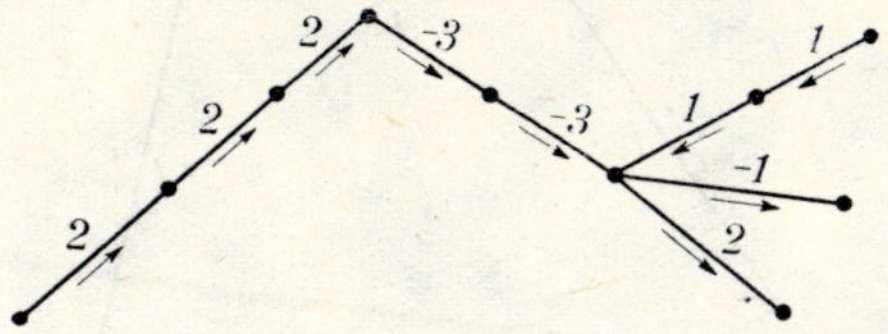

Fig. 73

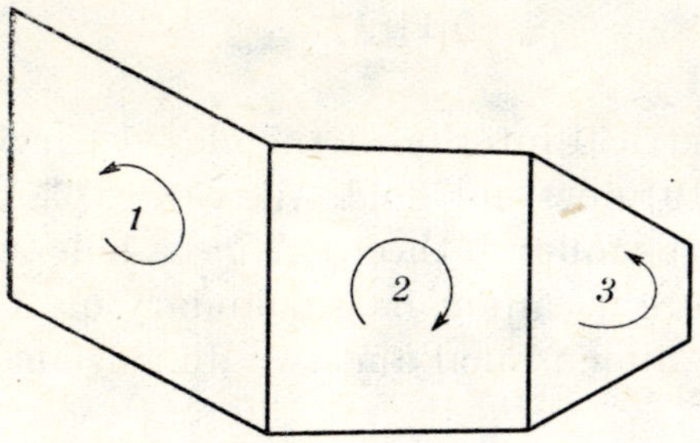

Fig. 74

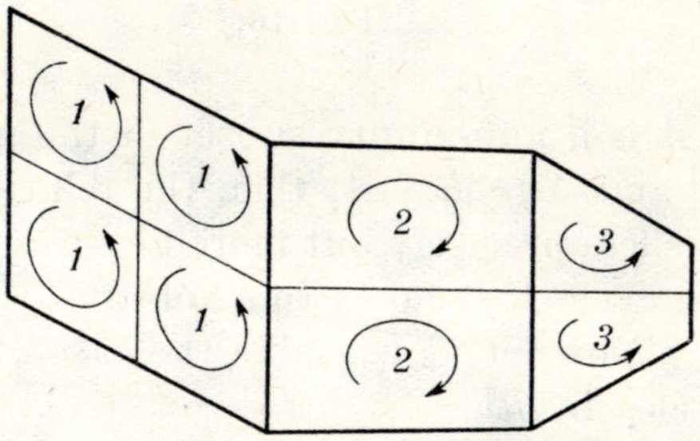

Fig. 75

if there exists a $(k+1)$-dimensional chain Γ_0 such that $|\Gamma_0| \subset X$ and the chains Γ and $\mathrm{b}\Gamma_0$ are equivalent (see the definition on p. 373), i.e.

$$\mathrm{b}\Gamma_0 \equiv \Gamma . \tag{20}$$

We shall try to explain the intuitive meaning of the relation (19). Let us consider, for simplicity, only such chains that the one-dimensional faces of their cells are segments, two-dimensional faces are plane quadrangles, etc.

Let us first examine the case where Γ is a simple one-dimensional cycle in R^2 (see figure 76). It is almost obvious that

$$\Gamma \sim 0 \text{ in } R^2 . \tag{21}$$

To ascertain (21), we must make a cellular decomposition D of the two-dimensional polyhedron P whose topological boundary is $|\Gamma|$, give the positive orientation to each of the two-dimensional cells of the decomposition D and then denote by Γ_0 the sum of the oriented cells so obtained. Then either $\Gamma \equiv \mathrm{b}\Gamma_0$ or $\Gamma \equiv \mathrm{b}-\Gamma_0$, which depends on the orientation (one of two possible) of the simple cycle Γ (see figure 77 where $\mathrm{b}-\Gamma_0 \equiv \Gamma$).

It is more difficult to show that (21) holds for every one-dimensional cycle on the plane R^2.

It is intuitively clear (we shall not give here a precise proof) that the above construction of the two-dimensional cycle Γ_0 satisfying (20) for a given simple one-dimensional cycle Γ is unique in this sense that the support $|\Gamma_0|$ has to fill the whole interior of the closed curve $|\Gamma|$. Thus,

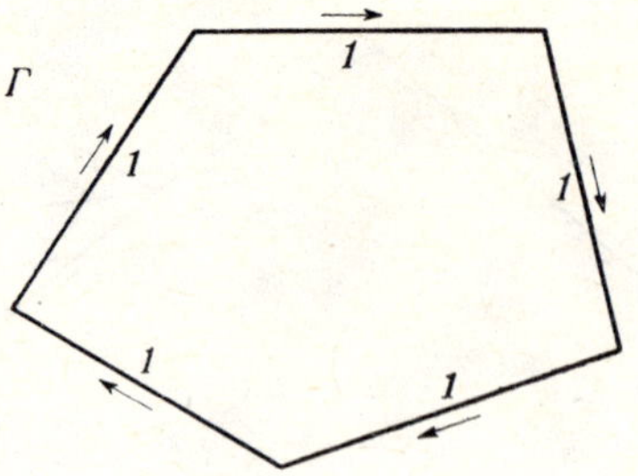

Fig. 76

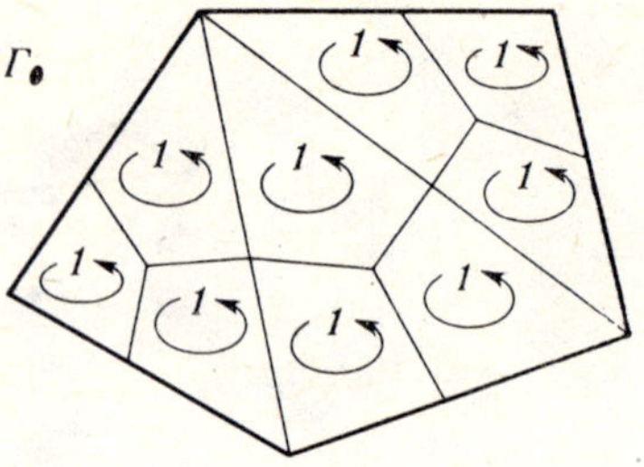

Fig. 77

if A is a non-empty subset of the geometrical interior of the closed curve $|\Gamma|$ and $X = R^2 - A$, then the relation (19) does not hold. We can explain this less precisely but more picturesquely as follows: the set X has a hole A, the cycle Γ encloses this hole and therefore it cannot be a boundary on X.

Now let us examine the case of one-dimensional and two-dimensional cycles in R^3.

It is easy to see that

(22) $$\Gamma \sim 0 \text{ in } R^3$$

for every two-dimensional simple cycle Γ in R^3. The construction of a three-dimensional chain Γ_0 in R^3 such that (20) holds is a repetition of the construction just described in the case of one-dimensional simple cycles in the plane. It is more difficult to be convinced that (22) holds for every simple one-dimensional cycle Γ in R^3. The support of Γ is a closed curve in R^3. We must stretch a two-dimensional polyhedron (a membrane) on this closed curve, take a cellular decomposition of the polyhedron, orient all the cells of the decomposition in a proper way and then define Γ_0 as the sum of the oriented cells. It is much more difficult to prove that (22) holds for every one-dimensional or two-dimensional cycle in R^3.

Now remove the ball

$$A = \{(x_1, x_2, x_3): x_1^2 + x_2^2 + x_3^2 \leqslant \varrho^2\}$$

from R^3, i.e. assume $X = R^3 - A$. It is easy to see that relation (19) does not hold for certain two-dimensional cycles Γ. For instance, if C is a cube such that $A \subset \operatorname{int} C$, and Γ is a simple two-dimensional cycle such that $|\Gamma|$ is the topological boundary of C, then (19) does not hold. The reason is that if (20) holds, then $|\Gamma_0|$ has to be the whole cube C, and therefore the inclusion $|\Gamma_0| \subset X$ is not true. On the other hand, it can be proved that (19) holds for every one-dimensional cycle, which can be easily imagined in the case of simple one-dimensional cycles.

Now remove from R^3 the infinite cylinder

$$B = \{(x_1, x_2, x_3): x_1^2 + x_2^2 \leqslant \varrho_0^2\},$$

i.e. assume $X = R^3 - B$. It is easy to see that (19) will hold for every two-dimensional simple cycle (in reality it holds for all two-dimensional cycles), but it does not hold for certain one-dimensional cycles. For instance, it does not hold for a simple cycle whose support is the set

$$|\Gamma| = \{(x_1, x_2, x_3)\colon (\max(|x_1|, |x_2|) = \varrho_1) \& (x_3 = 0)\}, \qquad \varrho_1 > \varrho_0 > 0$$

(see figure 78).

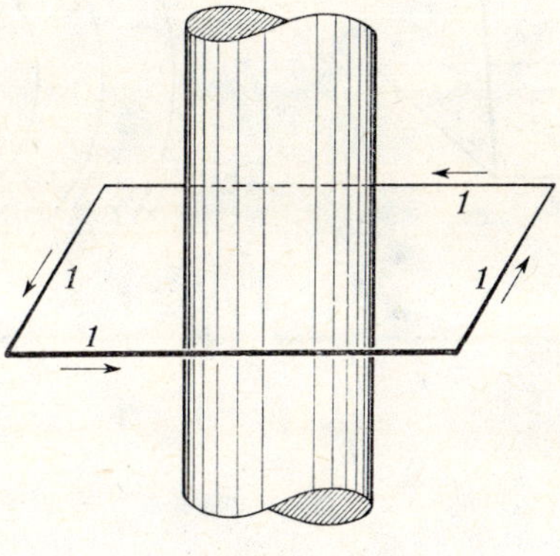

Fig. 78

In the first and in the second case, the reason that (19) does not always hold is, roughly speaking, the existence of "holes" in the set X which are "enclosed" by some cycles. The two last examples show that the "holes" can have completely different character.

The reader can easily guess the intuitive sense of the fact that, for a k-dimensional cycle Γ with $|\Gamma| \subset X \subset R^l$, the relation (19) does not hold: this a precise formulation of what it means intuitively that X has "holes" which are "enclosed" by Γ.

Additional information on this subject will also be given in X § 5.

EXERCISES

1. Calculate the boundaries of the one-dimensional chains from figures 79 and 80.

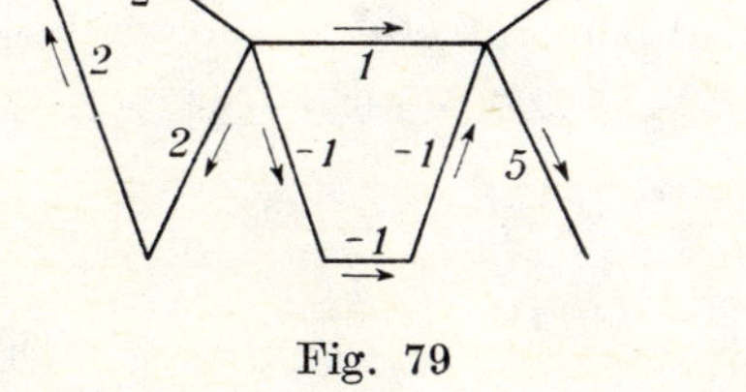

Fig. 79

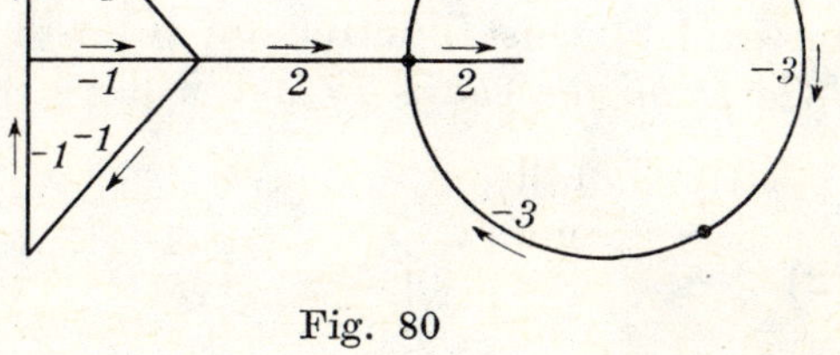

Fig. 80

2. Calculate the boundary of the two-dimensional chain in figure 81.

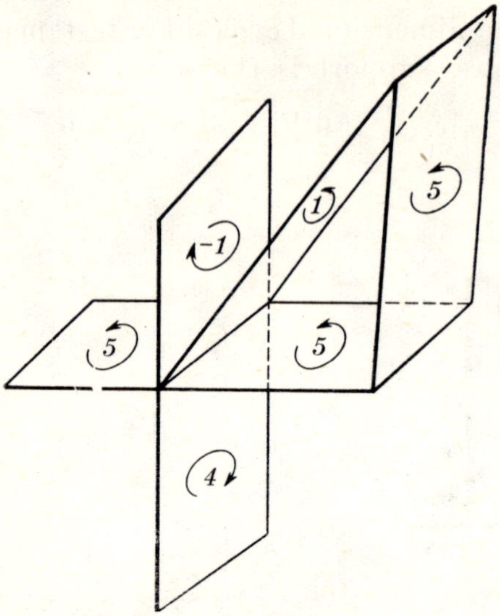

Fig. 81

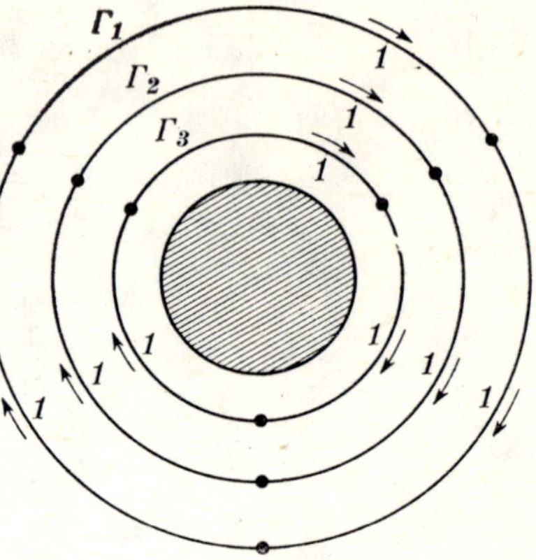

Fig. 82

3. Let $X = R^2 - \{p \in R^2 \colon |p - p_0| \leqslant 1\}$. None of the one-dimensional cycles Γ_1, Γ_2, Γ_3 in figure 82, whose supports are circles with radii >1 and center p_0, is a boundary on X. However the cycle $\Gamma_1 - 2\Gamma_2 + \Gamma_3$ is a boundary on X.

4. If a chain Γ_1 is a refinement of a chain Γ, then the chain $\mathrm{b}\Gamma_1$ is a refinement of the chain $\mathrm{b}\Gamma$. Hint: Use § 7, exercise 2.

§ 9. Integrals on chains. Differential forms. Let Γ be a k-dimensional chain in R^l $(0 \leqslant k \leqslant l)$ and let

$$\Omega = [\omega_{i_1, \ldots, i_k}]_{i_1, \ldots, i_k \leqslant l} \tag{1}$$

be a continuous skew-symmetric matrix field defined on the support $|\Gamma|$ of Γ. The scalar product $\Omega\Gamma$ is a bounded real function defined on the polyhedron $|\Gamma|$ and is measurable $\mu_{|\Gamma|}$. The last property follows from the fact that $\Omega\Gamma$ is continuous on the geometrical interiors of supports of oriented cells the linear combination of which is Γ. The union of all their $(k-1)$-dimensional faces has measure $\mu_{|\Gamma|}$ zero.

The real function $\dfrac{1}{k!}\Omega\Gamma$ is also bounded and measurable on $|\Gamma|$, therefore it is integrable on $|\Gamma|$ with respect to the measure $\mu_{|\Gamma|}$. The integral of this function on $|\Gamma|$ will be called the *integral of the matrix field Ω on the chain Γ* and will be denoted by the symbol $\int_\Gamma \Omega$.

By definition

$$\int_\Gamma \Omega = \int_{|\Gamma|} \frac{1}{k!}\Omega\Gamma d\mu_{|\Gamma|} = \frac{1}{k!}\int_{|\Gamma|} \Omega\Gamma d\mu_{|\Gamma|}\,. \tag{2}$$

The reason why the coefficient $1/k!$ is added was explained in II § 9, p. 85.

Often the polyhedron $|\Gamma|$ is not the domain of Ω but a subset of the domain of Ω. It is easy to guess then that $\int_\Gamma \Omega$ is the integral $\int_\Gamma \Omega_1$, where Ω_1 is the restriction of Ω to $|\Gamma|$, i.e. $\Omega_1 = \Omega| |\Gamma|$.

The chain Γ is a linear combination

$$\Gamma = a_1 K_1 + \ldots + a_n K_n \tag{3}$$

of oriented k-dimensional cells $K_1, \ldots, K_n$ whose supports $|K_1|, \ldots, |K_n|$ form a cellular decomposition of the polyhedron $|\Gamma|$. It follows from (2), (3) and § 5 (15) that

$$\int_\Gamma \Omega = a_1 \int_{K_1} \Omega + \ldots + a_n \int_{K_n} \Omega . \tag{4}$$

Thus the calculation of an integral on a chain can be reduced to the calculation of integrals on oriented cells.

If $K = \langle \Phi \rangle$, where Φ is a diffeomorphism on a k-dimensional closed interval $I \subset R^k$ ($k > 0$), then the orientation $T = K| |K|$ is given by the formula

$$T \circ \Phi = \frac{\Phi_{|1} \wedge \ldots \wedge \Phi_{|k}}{|\Phi'|} .$$

Consequently

$$\int_{\langle \Phi \rangle} \Omega = \frac{1}{k!} \int_I (\Omega \circ \Phi) \cdot (\Phi_{|1} \wedge \ldots \wedge \Phi_{|k}) , \tag{5}$$

where the integral on the right-hand side is with respect to the k-dimensional Lebesgue measure. Indeed, by 5.3,

$$k! \int_{\langle \Phi \rangle} \Omega = \int_{\Phi(I)} \Omega \cdot K = \int_{\Phi(I)} \Omega \cdot T = \int_I \big((\Omega \circ \Phi) \cdot (T \circ \Phi)\big) \cdot |\Phi'|$$

$$= \int_I (\Omega \circ \Phi) \cdot (\Phi_{|1} \wedge \ldots \wedge \Phi_{|k}) .$$

If $\Phi = (\varphi_1, \ldots, \varphi_l)$, then

$$\Phi_{|1} \wedge \ldots \wedge \Phi_{|k} = [J_{i_1, \ldots, i_k}]_{i_1, \ldots, i_k \leqslant l} ,$$

where $J_{i_1, \ldots, i_k}$ is the jacobian

$$J_{i_1, \ldots, i_k} = \begin{vmatrix} \varphi_{i_1|1} & \cdots & \varphi_{i_1|k} \\ \cdot & \cdots & \cdot \\ \varphi_{i_k|1} & \cdots & \varphi_{i_k|k} \end{vmatrix} .$$

If Ω is given by formula (1), then

$$(\Omega \circ \Phi)\cdot(\Phi_{|1} \wedge \ldots \wedge \Phi_{|k}) = \sum_{i_1,\ldots,i_k=1}^{l} (\omega_{i_1,\ldots,i_k} \circ \Phi)\cdot J_{i_1,\ldots,i_k}$$
$$= k! \sum_{i_1<i_2<\ldots<i_k} (\omega_{i_1,\ldots,i_k} \circ \Phi)\cdot J_{i_1,\ldots,i_k}\,.$$

Consequently,

$$\int_{\langle\Phi\rangle} \Omega = \frac{1}{k!}\int_I \sum_{i_1,\ldots,i_k=1}^{l} (\omega_{i_1,\ldots,i_k} \circ \Phi)\cdot J_{i_1,\ldots,i_k} \tag{6}$$
$$= \int_I \sum_{i_1<i_2<\ldots<i_k} (\omega_{i_1,\ldots,i_k} \circ \Phi)\cdot J_{i_1,\ldots,i_k}\,.$$

Since

$$\frac{1}{k!}(\Omega \circ \Phi)\cdot(\Phi_{|1} \wedge \ldots \wedge \Phi_{|k}) = (\Omega \circ \Phi)\cdot(\Phi_{|1} \times \ldots \times \Phi_{|k})$$

(see II § 10 (12)), we also have

$$\int_{\langle\Phi\rangle} \Omega = \int_I (\Omega \circ \Phi)\cdot(\Phi_{|1} \times \ldots \times \Phi_{|k})\,, \tag{7}$$

that is

$$\int_{\langle\Phi\rangle} \Omega = \int_I \sum_{i_1,\ldots,i_k=1}^{l} \omega_{i_1,\ldots,i_k}\varphi_{i_1|1}\cdots\varphi_{i_k|k}\,. \tag{8}$$

If $x_1, \ldots, x_k$ denote the coordinates of any point p in I, formulas (6), (8) can be written in the traditional notation as follows:

$$\int_{\langle\Phi\rangle} \Omega = \frac{1}{k!}\int\!\!\ldots\!\!\int_I \sum_{i_1,\ldots,i_k=1}^{l} \omega_{i_1,\ldots,i_k} \frac{\partial(\varphi_{i_1}, \ldots, \varphi_{i_k})}{\partial(x_1, \ldots, x_k)}\, dx_1 \ldots dx_k \tag{9}$$
$$= \int\!\!\ldots\!\!\int_I \sum_{i_1<i_2\ldots<i_k} \omega_{i_1,\ldots,i_k} \frac{\partial(\varphi_{i_1}, \ldots, \varphi_{i_k})}{\partial(x_1, \ldots, x_k)}\, dx_1 \ldots dx_k$$
$$= \int\!\!\ldots\!\!\int_I \sum_{i_1,\ldots,i_k=1}^{l} \omega_{i_1,\ldots,i_k}\cdot \frac{\partial\varphi_{i_1}}{\partial x_1} \ldots \frac{\partial\varphi_{i_k}}{\partial x_k}\, dx_1 \ldots dx_k\,.$$

In the case where $k = 1$, i.e. where the integrals under consideration are curvilinear, Ω is a vector field:

$$\Omega = [\omega_1, \ldots, \omega_l]$$

and

$$\Phi = (\varphi_1, \ldots, \varphi_l)$$

is a mapping on a one-dimensional interval $I = \langle a ; b \rangle$. Then formulas (5), (6), (7), (8) have the form

$$\int_{\langle\Phi\rangle} \Omega = \int_a^b (\Omega \circ \Phi)\cdot\Phi' = \int_a^b \Omega\big(\Phi(t)\big)\cdot\Phi'(t)\,dt \tag{10}$$

$$= \int_a^b \sum_{i=1}^{l} \omega_i\big(\Phi(t)\big)\varphi_i'(t)\,dt\,.$$

In the case where $k = 0$, a zero-dimensional chain Γ is a real function equal to 0 except for a finite set $|\Gamma| = \{q_1, \ldots, q_n\}$ ($q_i \neq q_j$ for $i \neq j$), and $\Omega = [\omega]$ is a real function ω on the set. Consequently (see § 1 (15))

$$\int_\Gamma \Omega = \omega(q_1)\Gamma(q_1) + \ldots + \omega(q_n)\Gamma(q_n)\,. \tag{11}$$

Let us return to the general case. It follows directly from the definition (2) that, for every real number a,

$$\int_{a\Gamma} \Omega = a \int_\Gamma \Omega = \int_\Gamma a\Omega\,. \tag{12}$$

If Ω_1 and Ω_2 are matrix fields of the form (1), then

$$\int_\Gamma (\Omega_1 + \Omega_2) = \int_\Gamma \Omega_1 + \int_\Gamma \Omega_2\,, \tag{13}$$

$$\int_\Gamma (\Omega_1 - \Omega_2) = \int_\Gamma \Omega_1 - \int_\Gamma \Omega_2\,. \tag{14}$$

If Γ_1, Γ_2 and $\Gamma_1 + \Gamma_2$ are k-dimensional chains, then

$$\int_{\Gamma_1+\Gamma_2} \Omega = \int_{\Gamma_1} \Omega + \int_{\Gamma_2} \Omega\,. \tag{15}$$

Similarly, if Γ_1, Γ_2 and $\Gamma_1 - \Gamma_2$ are k-dimensional chains, then

$$\int_{\Gamma_1-\Gamma_2} \Omega = \int_{\Gamma_1} \Omega - \int_{\Gamma_2} \Omega\,. \tag{16}$$

Let (1) be a matrix field on a set $G \subset R^l$. For every fixed point $p \in G$, the k-indexed l-dimensional matrix $\Omega(p)$ is skew-symmetric and, therefore, it determines a skew-symmetric multi-linear form F_p: $V_l^k \to R$, viz. the form

$$F_p(\mathfrak{a}_1, \ldots, \mathfrak{a}_k) = \Omega(p)\cdot(\mathfrak{a}_1 \times \ldots \times \mathfrak{a}_k) \tag{17}$$

$$= \frac{1}{k!}\Omega(p)\cdot(\mathfrak{a}_1 \wedge \ldots \wedge \mathfrak{a}_k) \quad \text{for} \quad \mathfrak{a}_1, \ldots, \mathfrak{a}_k \in V_l$$

(see II § 6, p. 70 and II 9.1). According to the general rule of notation, the symbol F will denote the function which assigns to every $p \in G$ the skew-symmetric form F_p. Every function F of this kind, i.e. every function which assigns to points $p \in G$ skew-symmetric k-linear forms F_p: $V_l^k \to R$ is called a *differential form* (more precisely, a *differential k-form*) on G. We see that every skew-symmetric k-indexed l-dimensional matrix field on G determines a differential k-form on G (see (17)). It is easy to verify that every differential k-form on G is determined by a unique skew-symmetric matrix field Ω on G. Namely, for every $p \in G$, $\Omega(p)$ is the matrix of coordinates of the k-linear form F_p. The matrix field Ω will be called the *matrix of the coordinates of the differential form* F. The natural correspondence between differential forms and skew-symmetric matrix fields is one-to-one. Roughly speaking, both notions express the same mathematical content.

We say that a differential form F on G is *continuous* (or *of class* C_m) if the matrix of the coordinates of F is a continuous matrix field (or a matrix field of class C_m).

By the *integral* $\int_\Gamma F$ *of a continuous differential k-form F on a k-dimensional chain Γ*, whose support is a subset of the domain G of F, we mean the number $\int_\Gamma \Omega$ where Ω is the matrix of the coordinates of F.

It follows from (7) and (17) that, in the case where $\Gamma = \langle\Phi\rangle$ and Φ is a diffeomorphism on a closed k-dimensional interval I,

$$(18)\qquad \int_{\langle\Phi\rangle} F = \int_I F_{\Phi(p)}(\Phi_{|1}(p), \ldots, \Phi_{|k}(p))\,dp\,,$$

the integral on the right-hand side being with respect to the k-dimensional Lebesgue measure.

By (4), if Γ is a chain (3), then

$$(19)\qquad \int_\Gamma F = a_1 \int_{K_1} F + \ldots + a_n \int_{K_n} F\,.$$

Identities (18), (19) enable us to reduce the calculation of $\int_\Gamma F$ on an arbitrary chain Γ to integration with respect to the Lebesgue measure.

By (12),

$$(20)\qquad \int_{a\Gamma} F = a \int_\Gamma F = \int_\Gamma aF\,.$$

If F_1 and F_2 are continuous differential forms, then by (13) and (14)

$$(21)\qquad \int_\Gamma (F_1 + F_2) = \int_\Gamma F_1 + \int_\Gamma F_2\,,$$

$$(22)\qquad \int_\Gamma (F_1 - F_2) = \int_\Gamma F_1 - \int_\Gamma F_2\,.$$

By (15), if Γ_1, Γ_2 and $\Gamma_1+\Gamma_2$ are k-dimensional chains, then

$$\int_{\Gamma_1+\Gamma_2} F = \int_{\Gamma_1} F + \int_{\Gamma_2} F. \tag{23}$$

By (16), if Γ_1, Γ_2 and $\Gamma_1-\Gamma_2$ are k-dimensional chains, then

$$\int_{\Gamma_1-\Gamma_2} F = \int_{\Gamma_1} F - \int_{\Gamma_2} F. \tag{24}$$

By a *differential zero-form* on G we shall mean any real function on G. By (11), if ω is a differential zero-form, and Γ is a zero-dimensional chain, $|\Gamma| = \{q_1, \ldots, q_n\}$ ($q_i \neq q_j$ for $i \neq j$), then

$$\int_{\Gamma} \omega = \omega(q_1)\Gamma(q_1) + \ldots + \omega(q_n)\Gamma(q_n). \tag{25}$$

Let F be a differential k-form on an open set $G \subset R^l$. For any vectors $\mathfrak{a}_1, \ldots, \mathfrak{a}_k$ in R^l the symbol $F(\mathfrak{a}_1, \ldots, \mathfrak{a}_k)$ will denote the real function on G defined by

$$F(\mathfrak{a}_1, \ldots, \mathfrak{a}_m)(p) = F_p(\mathfrak{a}_1, \ldots, \mathfrak{a}_m) \quad \text{for} \quad p \in G. \tag{26}$$

In other words, the value $F(\mathfrak{a}_1, \ldots, \mathfrak{a}_k)(p)$ of the function $F(\mathfrak{a}_1, \ldots, \mathfrak{a}_k)$ at a point $p \in G$ is equal to the value $F_p(\mathfrak{a}_1, \ldots, \mathfrak{a}_k)$ of the k-linear form F_p at $(\mathfrak{a}_1, \ldots, \mathfrak{a}_k) \in V_l^k$. If (1) is the matrix of coordinates of F and

$$\mathfrak{a}_1 = [a_{1,1}, \ldots, a_{1,l}],$$
$$\ldots\ldots\ldots\ldots$$
$$\mathfrak{a}_k = [a_{k,1}, \ldots, a_{k,l}],$$

then

$$F(\mathfrak{a}_1, \ldots, \mathfrak{a}_k) = \Omega\cdot(\mathfrak{a}_1 \times \ldots \times \mathfrak{a}_k) = \sum_{i_1,\ldots,i_k=1}^{l} \omega_{i_1,\ldots,i_k} a_{1,i_1} \ldots a_{k,i_k}. \tag{27}$$

Moreover

$$\omega_{i_1,\ldots,i_k} = F(\mathfrak{e}_{i_1}, \ldots, \mathfrak{e}_{i_k}) \tag{28}$$

where $\mathfrak{e}_1, \ldots, \mathfrak{e}_l$ are unit vectors of the coordinate axes in R^l. Hence it follows that F is continuous or of class C_m if and only if $F(\mathfrak{a}_1, \ldots, \mathfrak{a}_k)$ is continuous or of class C_m for any vectors $\mathfrak{a}_1, \ldots, \mathfrak{a}_k$.

EXERCISES

1. Let $I = \{(x_1, x_2, x_3): |x_i| \leqslant 1 \text{ for } i = 1, 2, 3\}$ and let $K_{i,\eta}$ be the two-dimensional oriented cell whose support is the face $I_{j,\eta}$ of I and whose orientation is consistent with the positive orientation of I (see figure 71). Calculate the integral $\int_{\Gamma}\Omega$ where $\Gamma = \sum_{j=1}^{3}\sum_{\eta=\pm 1} j\eta K_{j,\eta}$ and $\Omega = [\omega_{i,j}]_{i,j\leqslant 3}$ is the skew-symmetric matrix field such that

$$\omega_{2,3}(x_1, x_2, x_3) = x_1^2 + x_2^2 + x_3^2, \quad \omega_{3,1}(x_1, x_2, x_3) = x_1x_2x_3, \quad \omega_{1,2}(x_1, x_2, x_3) = x_1 + x_2^2 + x_3^3.$$

2. Let $K_1, \dots, K_{12}$ be the one-dimensional oriented cells whose supports are the one-dimensional closed faces of the cube I from exercise 1 and the orientation of each of the faces is equal to the unit vector of the parallel coordinate axis. Calculate the curvilinear integral $\int_\Gamma \Omega$ where

$$\Gamma = K_1 + \dots + K_{12} \quad \text{and} \quad \Omega(x_1, x_2, x_3) = [e^{x_1+x_2+x_3}, \sin \pi x_1, x_2^2 + x_3].$$

3. Let $p_1 = (3, 0, 1)$, $p_2 = (0, 4, 1)$, $p_3 = (-3, 1, 1)$, $p_4 = (3, 0, 0)$, $p_5 = (0, 4, 0)$, $p_6 = (-3, 0, 0)$, $p_7 = p_1$. Let K_i $(i = 1, \dots, 6)$ be the oriented one-dimensional cell such that its support is the segment joining p_i and p_{i+1}, and $\mathrm{b}K_i = \langle p_{i+1}\rangle - \langle p_i\rangle$. Calculate the integral $\int_\Gamma \Omega$ where

$$\Gamma = K_1 + 2K_2 + \dots + 6K_6 \quad \text{and} \quad \Omega(x_1, x_2, x_3) = [x_1, x_2 + x_3, x_1 + x_2 + x_3].$$

4. Let $\Gamma_1, \Gamma_2, \Gamma_3$ be the one-dimensional chains in R^2 shown in figure 82, whose supports are circles with radii $\varrho_1, \varrho_2, \varrho_3$ and with center $p_0 = (0, 0, 0)$. Calculate the integral $\int_\Gamma \Omega$ where $\Gamma = \Gamma_1 - \Gamma_2 + \Gamma_3$ and $\Omega(x_1, x_2) = [x_1 + x_2, x_1^2 + x_2^2]$.

5. Let $I = \{(x_1, \dots, x_k) \colon 0 \leqslant x_i \leqslant 1 \text{ for } i = 1, \dots, k\}$ and let $K_{i,\eta}$ be the oriented $(k-1)$-dimensional cell whose support is the face $I_{j,\eta}$ of I and whose orientation is consistent with the positive orientation of I. Calculate the integral $\int_\Gamma \Omega$ where

$$\Gamma = \sum_{j=1}^{k} \sum_{\eta=\pm 1} \eta K_{j,\eta}, \quad \Omega = [\omega_{i_1,\dots,i_{k-1}}]_{i_1,\dots,i_{k-1} \leqslant k},$$

$$\omega_{i_1,\dots,i_{k-1}}(x_1, \dots, x_k) = x_{i_1} \quad \text{when} \quad i_1 < i_2 < \dots < i_{k-1}.$$

§ 10. The skew derivative. Let F be a differential k-form of class C_1 on an open set $G \subset R^l$, $k < l$.

For any fixed point $p \in G$, let F'_p be defined by

$$(1) \qquad F'_p(\mathfrak{a}_1, \mathfrak{a}_2, \dots, \mathfrak{a}_{k+1}) = F(\mathfrak{a}_2, \dots, \mathfrak{a}_{k+1})'_{\mathfrak{a}_1}(p) \quad \text{for} \quad \mathfrak{a}_1, \dots, \mathfrak{a}_{k+1} \in V_l.$$

By definition, F'_p is a $(k+1)$-linear form on V_l^{k+1} whose value at $(\mathfrak{a}_1, \dots, \mathfrak{a}_{k+1}) \in V_l^{k+1}$ is equal to the directional derivative of the real function $F(\mathfrak{a}_2, \dots, \mathfrak{a}_{k+1})$ (see § 9 (26)) at the point p in the direction $\mathfrak{a}_1$. Moreover, F'_p is skew-symmetric in the last k variables (i.e. in $\mathfrak{a}_2, \dots, \mathfrak{a}_{k+1}$) by the skew symmetry of F. Thus the equation

$$(2) \qquad \mathrm{d}F_p = \mathrm{sk}_{1,k} F'_p$$

defines a skew-symmetric $(k+1)$-linear form $\mathrm{d}F_p$ on V_l^{k+1} (see II § 8 (12)). The function $\mathrm{d}F$ which assigns the skew-symmetric $(k+1)$-linear form $\mathrm{d}F_p$ to any $p \in G$ is a differential $(k+1)$-form on G called the *skew derivative* (or the *exterior derivative*) of F. By definition (see II § 8 (12)),

$$(3) \qquad \mathrm{d}F_p(\mathfrak{a}_1, \dots, \mathfrak{a}_{k+1}) = \sum_{j=1}^{k+1} (-1)^{j-1} F'_p(\mathfrak{a}_j, \mathfrak{a}_1, \dots, \mathfrak{a}_{j-1}, \mathfrak{a}_{j+1}, \dots, \mathfrak{a}_{k+1})$$

$$= \sum_{j=1}^{k+1} (-1)^{j-1} F(\mathfrak{a}_1, \dots, \mathfrak{a}_{j-1}, \mathfrak{a}_{j+1}, \dots, \mathfrak{a}_{k+1})'_{\mathfrak{a}_j}(p)$$

and consequently

$$(4)\quad \mathrm{d}F(\mathfrak{a}_1, \dots, \mathfrak{a}_{k+1}) = \sum_{j=1}^{k+1} (-1)^{j-1} F(\mathfrak{a}_1, \dots, \mathfrak{a}_{j-1}, \mathfrak{a}_{j+1}, \dots, \mathfrak{a}_{k+1})'_{\mathfrak{a}_j}$$
$$= F(\mathfrak{a}_2, \dots, \mathfrak{a}_{k+1})'_{\mathfrak{a}_1} - F(\mathfrak{a}_1, \mathfrak{a}_3, \dots, \mathfrak{a}_{k+1})'_{\mathfrak{a}_2} +$$
$$+ F(\mathfrak{a}_1, \mathfrak{a}_2, \mathfrak{a}_4, \dots, \mathfrak{a}_{k+1})'_{\mathfrak{a}_3} - \dots + (-1)^k F(\mathfrak{a}_1, \dots, \mathfrak{a}_k)'_{\mathfrak{a}_{k+1}}.$$

Thus the differential form $\mathrm{d}F$ is continuous. If F is of class C_m then $\mathrm{d}F$ is of class C_{m-1}, $m = 2, 3, \dots$

Let

$$(5)\quad \Omega = [\omega_{i_1,\dots,i_k}]_{i_1,\dots,i_k \leqslant l}$$

be the matrix of the coordinates of F. Since

$$F(\mathfrak{a}_1, \dots, \mathfrak{a}_{j-1}, \mathfrak{a}_{j+1}, \dots, \mathfrak{a}_{k+1}) = \Omega \cdot (\mathfrak{a}_1 \times \dots \times \mathfrak{a}_{j-1} \times \mathfrak{a}_{j+1} \times \dots \times \mathfrak{a}_{k+1})$$

by § 9 (27), we get by a simple verification

$$F(\mathfrak{a}_1, \dots, \mathfrak{a}_{j-1}, \mathfrak{a}_{j+1}, \dots, \mathfrak{a}_{k+1})'_{\mathfrak{a}_j} = \Omega'_{\mathfrak{a}_j} \cdot (\mathfrak{a}_1 \times \dots \times \mathfrak{a}_{j-1} \times \mathfrak{a}_{j+1} \times \dots \times \mathfrak{a}_{k+1})$$
$$= \Omega' \cdot (\mathfrak{a}_1 \times \dots \times \mathfrak{a}_{j-1} \times \mathfrak{a}_{j+1} \times \dots \times \mathfrak{a}_{k+1} \times \mathfrak{a}_j).$$

Consequently

$$(6)\quad \mathrm{d}F(\mathfrak{a}_1, \dots, \mathfrak{a}_{k+1}) = \sum_{j=1}^{k+1} (-1)^{j-1} \Omega' \cdot (\mathfrak{a}_1 \times \dots \times \mathfrak{a}_{j-1} \times \mathfrak{a}_{j+1} \times \dots \times \mathfrak{a}_{k+1} \times \mathfrak{a}_j).$$

Let

$$(7)\quad \mathrm{d}\Omega = [\omega^{\bullet}_{i_1,\dots,i_{k+1}}]_{i_1,\dots,i_{k+1} \leqslant l}$$

be the matrix of coordinates of $\mathrm{d}F$. By § 9 (28) (where F should be replaced by $\mathrm{d}F$, and k by $k+1$)

$$\omega^{\bullet}_{i_1,\dots,i_{k+1}} = \mathrm{d}F(\mathfrak{e}_{i_1}, \dots, \mathfrak{e}_{i_{k+1}})$$

where $\mathfrak{e}_1, \dots, \mathfrak{e}_l$ are the unit vectors of the coordinate axes in R^l. It follows from (6) and § 9 (28) by a simple calculation that

$$(8)\quad \omega^{\bullet}_{i_1,\dots,i_{k+1}} = \sum_{r=1}^{k+1} (-1)^{r-1} \omega_{i_1,\dots,i_{r-1},i_{r+1},\dots,i_{k+1}|i_r}$$
$$= \omega_{i_2,\dots,i_{k+1}|i_1} - \omega_{i_1,i_3,\dots,i_{k+1}|i_2} + \omega_{i_1,i_2,i_4,\dots,i_{k+1}|i_3} - \dots + (-1)^k \omega_{i_1,\dots,i_k|i_{k+1}}.$$

If $x_1, \dots, x_l$ denote the coordinates of any point q in G, the definition (8) can be written in the traditional notation as follows:

$$(8')\quad \omega^{\bullet}_{i_1,\dots i_{k+1}} = \sum_{r=1}^{k+1} (-1)^{r-1} \frac{\partial}{\partial x_{i_r}} \omega_{i_1,\dots,i_{r-1},i_{r+1},\dots,i_{k+1}}$$
$$= \frac{\partial}{\partial x_{i_1}} \omega_{i_2,\dots,i_{k+1}} - \frac{\partial}{\partial x_{i_2}} \omega_{i_1,i_3,\dots,i_{k+1}} +$$
$$+ \frac{\partial}{\partial x_{i_3}} \omega_{i_1,i_2,i_4,\dots,i_{k+1}} - \dots + (-1)^k \frac{\partial}{\partial x_{i_{k+1}}} \omega_{i_1,\dots,i_k}.$$

Since $\mathrm{d}\Omega$ is the matrix of coordinates of $\mathrm{d}F$, we have

$$\mathrm{d}F(\mathfrak{a}_1, \ldots, \mathfrak{a}_{k+1}) = \mathrm{d}\Omega\cdot(\mathfrak{a}_1\times\ldots\times\mathfrak{a}_{k+1})\,. \tag{9}$$

If follows from (9) and (6) that

$$\mathrm{d}\Omega\cdot(\mathfrak{a}_1\times\ldots\times\mathfrak{a}_{k+1}) = \sum_{j=1}^{k+1}(-1)^{j-1}\Omega'\cdot(\mathfrak{a}_1\times\ldots\times\mathfrak{a}_{j-1}\times\mathfrak{a}_{j+1}\times\ldots\times\mathfrak{a}_{k+1}\times\mathfrak{a}_j)\,. \tag{10}$$

If Ω is a skew-symmetric k-indexed l-dimensional matrix field (5) of class C_1, defined on an open set $G\subset R^l$, $0\leqslant k<l$, then the $(k+1)$-indexed l-dimensional matrix field $\mathrm{d}\Omega$ defined by (7)-(8) will be called the *skew derivative* (or the *exterior derivative*) of Ω. The matrix field $\mathrm{d}\Omega$ is skew-symmetric because it is the matrix of the coordinates of $\mathrm{d}F$ where F is the differential form whose matrix of the coordinates is Ω.

If $k=0$, i.e. $\Omega=[\omega]$ is a real function ω, then

$$\mathrm{d}\Omega = [\omega_{|i}]_{i\leqslant l} = [\omega_{|1}, \ldots, \omega_{|l}] = \omega'\,,$$

that is, the skew derivative coincides with the derivative of ω in the sense defined in IV § 3.

By the *dual skew derivative* of a skew-symmetric $\varkappa$-indexed matrix field of class C_1

$$\Lambda = [\lambda_{j_1,\ldots,j_\varkappa}]_{j_1,\ldots,j_\varkappa\leqslant l} \qquad (0<\varkappa<l) \tag{11}$$

defined on an open set $G\subset R^l$ we mean the skew-symmetric $(\varkappa-1)$-indexed matrix field

$$\delta\Lambda = [{}^{\bullet}\lambda_{j_1,\ldots,j_{\varkappa-1}}]_{j_1,\ldots,j_{\varkappa-1}\leqslant l}\,, \tag{12}$$

where

$${}^{\bullet}\lambda_{j_1,\ldots,j_{\varkappa-1}} = \sum_{j=1}^{l}\lambda_{j_1,\ldots,j_{\varkappa-1},j|j}\,. \tag{13}$$

In the traditional notation definition (13) can be written as

$${}^{\bullet}\lambda_{j_1,\ldots,j_{\varkappa-1}} = \sum_{j=1}^{l}\frac{\partial}{\partial x_j}\lambda_{j_1,\ldots,j_{\varkappa-1},j}\,. \tag{13'}$$

In the case where $\varkappa=1$, i.e. Λ is a vector field

$$\Lambda = [\lambda_1, \ldots, \lambda_l]\,,$$

the dual skew derivative $\delta\Lambda$ is a real function called the *divergence* of the vector field Λ and is denoted also by the symbol $\operatorname{div}\Lambda$. By definition,

$$\delta\Lambda = \operatorname{div}\Lambda = \lambda_{1|1}+\ldots+\lambda_{l|l}\,. \tag{14}$$

10.1. *The dual skew derivative* $\delta\delta\Lambda$ *of the dual derivative* $\delta\Lambda$ *of a* $\varkappa$*-indexed skew symmetrix matrix field* Λ ($\varkappa \geqslant 2$) *of class* C_2 *is the matrix field identically equal to zero:*

$$\delta\delta\Lambda = 0 .$$

By definition, $\Lambda = [^{\bullet\bullet}\lambda_{j_1,\ldots,j_{\varkappa-2}}]$, where

$$^{\bullet\bullet}\lambda_{j_1,\ldots,j_{\varkappa-2}} = \sum_{i=1}^{l}\sum_{j=1}^{l}\lambda_{j_1,\ldots,j_{\varkappa-2},i,j|j,i} .$$

It follows from the skew symmetry of Λ and from IV 4.1 that

$$^{\bullet\bullet}\lambda_{j_1,\ldots,j_{\varkappa-2}} = \sum_{i=1}^{l}\sum_{j=1}^{l} -\lambda_{j_1,\ldots,j_{\varkappa-2},j,i|j,i} = -^{\bullet\bullet}\lambda_{j_1,\ldots,j_{\varkappa-2}} .$$

Hence $^{\bullet\bullet}\lambda_{j_1,\ldots,j_{\varkappa-2}} = 0$.

According to the general convention assumed on p. 32 and pp. 76-78 and the notation introduced on p. 86, if Ω is a k-indexed skew-symmetric matrix field (1), then $\Omega^{\vee}$ denotes the *dual matrix field*, i.e. the matrix field whose value at a point $q \in G$ is the matrix

$$\Omega^{\vee}(q) = \Omega(q)^{\vee}$$

dual to $\Omega(q)$ (see II § 9, p. 86). By definition, $\Omega^{\vee}$ is an $(l-k)$-indexed skew-symmetric field of the same class as Ω, defined on the same open set G. Analogously we define the matrix field $\Lambda^{\wedge}$ for any given $\varkappa$-indexed skew-symmetric matrix field Λ. By definition, $\Lambda^{\wedge}$ is an $(l-\varkappa)$-indexed skew-symmetric matrix field of the same class as Λ and defined on the same open set G, such that for every $q \in G$

$$\Lambda^{\wedge}(q) = \Lambda(q)^{\wedge} .$$

10.2. *For every skew-symmetric matrix field* (5) *of class* C_1,

$$d\Omega = \big(\delta(\Omega^{\vee})\big)^{\wedge} .$$

For every skew-symmetric matrix field (11) *of class* C_1,

$$\delta\Lambda = \big(d(\Lambda^{\wedge})\big)^{\vee} .$$

It suffices to prove the first part of the theorem. The second part follows from the first where $\Omega = \Lambda^{\wedge}$. Indeed we get then

$$d(\Lambda^{\wedge}) = d\Omega = \big(\delta(\Omega^{\vee})\big)^{\wedge} = \big(\delta(\Lambda^{\wedge\vee})\big)^{\wedge} = (\delta\Lambda)^{\wedge}$$

since $\Lambda^{\wedge\vee} = \Lambda$ (see II § 9 (21)). Consequently

$$\delta\Lambda = (\delta\Lambda)^{\wedge\vee} = \big(d(\Lambda^{\wedge})\big)^{\vee} .$$

To prove the first part of 10.2 let as take in (11)

$$\varkappa = l-k \quad \text{and} \quad \Lambda = \Omega^{\vee} .$$

Since the matrix field Ω is k-indexed, the matrix field $\Omega^\vee$ is $(l-k)$-indexed, the matrix field $\delta(\Omega^\vee)$ is $(l-k-1)$-indexed, and the matrix field $(\delta(\Omega^\vee))^\wedge$ is $(k+1)$-indexed, i.e.

$$(\delta(\Omega^\vee))^\wedge = (\delta\Lambda)^\wedge = [\overline{\omega}_{i_1,\dots,i_{k+1}}]_{i_1,\dots,i_{k+1}\leqslant l}\,.$$

Let $i_1, \dots, i_{k+1}$ be a fixed sequence of different positive integers $\leqslant l$, and let $j_1, \dots, j_{l-k-1}$ be a fixed sequence formed of all the remaining positive integers $\leqslant l$. By (13) and the definition II § 9 (17)-(19)

$$\begin{aligned}\overline{\omega}_{i_1,\dots,i_{k+1}} &= \operatorname{sgn}(j_1, \dots, j_{l-k-1}, i_1, \dots, i_{k+1})\cdot{}^{\boldsymbol{\cdot}}\lambda_{j_1,\dots,j_{l-k-1}}\\ &= \sum_{j=1}^{l}\operatorname{sgn}(j_1, \dots, j_{l-k-1}, i_1, \dots, i_{k+1})\cdot\lambda_{j_1,\dots,j_{l-k-1},j|j}\,.\end{aligned}$$

If j is equal to one of the integers $j_1, \dots, j_{l-k-1}$, then $\lambda_{j_1,\dots,j_{l-k-1},j} = 0$ by the skew symmetry of Λ, and consequently $\lambda_{j_1,\dots,j_{l-k-1},j|j} = 0$. Hence it follows that in the last sum it suffices to consider only the summands where j is different from each of the numbers $j_1, \dots, j_{l-k-1}$, i.e. j is equal to one of the numbers $i_1, \dots, i_{k+1}$. Consequently

$$\overline{\omega}_{i_1,\dots,i_{k+1}} = \sum_{r=1}^{k+1}\operatorname{sgn}(j_1 \dots, j_{l-k-1}, i_1, \dots, i_{k+1})\cdot\lambda_{j_1,\dots,j_{l-k-1},i_r|i_r}\,.$$

Since $\Lambda = \Omega^\vee$, we have by II § 9 (18), (20)

$$\begin{aligned}&\lambda_{j_1,\dots,j_{l-k-1},i_r}\\ &\quad= \operatorname{sgn}(j_1, \dots, j_{l-k-1}, i_r, i_1, \dots, i_{r-1}, i_{r+1}, \dots, i_{k+1})\cdot\omega_{i_1,\dots,i_{r-1},i_{r+1},\dots,i_{k+1}}\end{aligned}$$

since the sequence $i_1, \dots, i_{r-1}, i_{r+1}, \dots, i_{k+1}$ contains all the positive integers $\leqslant l$ which do not appear in the sequence $j_1, \dots, j_{l-k-1}, i_r$. Substituting the obtained value of $\lambda_{j_1,\dots,j_{l-k-1},i_r}$ into the previous identity and using the identity

$$\begin{aligned}&\operatorname{sgn}(j_1, \dots, j_{l-k-1}, i_1, \dots, i_{k+1})\cdot\\ &\qquad\cdot\operatorname{sgn}(j_1, \dots, j_{l-k-1}, i_r, i_1, \dots, i_{r-1}, i_{r+1}, \dots, i_{k+1}) = (-1)^{r-1}\end{aligned}$$

we get

$$\overline{\omega}_{i_1,\dots,i_{k+1}} = \sum_{r=1}^{k+1}(-1)^{r-1}\omega_{i_1,\dots,i_{r-1},i_{r+1},\dots,i_{k+1}|i_r} = \omega^{\boldsymbol{\cdot}}_{i_1,\dots,i_{k+1}}\,.$$

If two integers among $i_1, \dots, i_{k+1}$ are equal, then $\overline{\omega}_{i_1,\dots,i_{k+1}} = 0$ by the skew symmetry of the matrix $(\delta\Lambda)^\wedge$. Then $\omega^{\boldsymbol{\cdot}}_{i_1,\dots,i_{k+1}} = 0$ also because if $i_{r_1} = i = i_{r_2}$ $(r_2 > r_1)$, then by the skew symmetry of Ω

$$\begin{aligned}\omega^{\boldsymbol{\cdot}}_{i_1,\dots,i_{k+1}} &= (-1)^{r_1-1}\omega_{i_1,\dots,i_{r_1-1},i_{r_1+1},\dots,i_{r_2-1},i,i_{r_2+1},\dots,i_{k+1}|i}+\\ &\qquad+(-1)^{r_2-1}\omega_{i_1,\dots,i_{r_1-1},i,i_{r_1+1},\dots,i_{r_2-1},i_{r_2+1},\dots,i_{k+1}|i} = 0\end{aligned}$$

since the sum of the two differentiated terms is equal to 0.

We have proved that always $\overline{\omega}_{i_1,\dots,i_{k+1}} = \omega^{\cdot}_{i_1,\dots,i_{k+1}}$, i.e. that the identity mentioned in the first part of 10.2 holds.

Observe that the identities mentioned in 10.2 can also be written

$$(15)\qquad (\mathrm{d}\Omega)^{\vee} = \delta(\Omega^{\vee}), \qquad (\delta\Lambda)^{\wedge} = \mathrm{d}(\Lambda^{\wedge})$$

(see II § 9 (21)).

10.3. *The skew derivative* $\mathrm{dd}\Omega$ *of the skew derivative* $\mathrm{d}\Omega$ *of a* k*-indexed matrix field* Ω $(k \leqslant l-2)$ *of class* C_2 *is the matrix field identically equal to zero*:

$$\mathrm{dd}\Omega = 0\,.$$

Assuming $\Lambda = \Omega^{\vee}$ we get from the first part of 10.2 and 10.1

$$\mathrm{dd}\Omega = \big(\delta((\delta\Lambda)^{\wedge\vee})\big)^{\wedge} = (\delta\delta\Lambda)^{\wedge} = 0^{\wedge} = 0$$

(see II § 9 (21)).

The following theorem is another formulation of 10.3.

10.4. *The skew derivative* $\mathrm{dd}F$ *of the skew derivative* $\mathrm{d}F$ *of a differential* k*-form* F $(k \leqslant l-2)$ *of class* C_2 *on an open set* $G \subset R^l$ *is identically equal to* 0:

$$\mathrm{dd}F = 0\,.$$

This means that for every $p \in G$ the multi-linear form $\mathrm{dd}F_p\colon V_l^{k+2} \to R$ is identically equal to 0.

To prove 10.4, let Ω be the matrix of coordinates of F. Then $\mathrm{d}\Omega$ is the matrix of coordinates of $\mathrm{d}F$, and $\mathrm{dd}\Omega$ is the matrix of coordinates of $\mathrm{dd}F$. Since $\mathrm{dd}\Omega = 0$, we have $\mathrm{dd}F = 0$.

Of course, Theorems 10.3 and 10.4 can be deduced directly from the definition of $\mathrm{d}\Omega$ and $\mathrm{d}F$.

EXERCISES

1. Calculate the skew derivative and the dual skew derivative of the skew-symmetric matrix field (5) where

$$\omega_{i_1,\dots,i_k}(x_1, \dots, x_l) = x_{i_k} \quad \text{if} \quad i_1 < i_2 < \dots < i_k\,.$$

2. Calculate the skew derivative and the dual skew derivative of the skew-symmetric matrix field (5) where

$$\omega_{i_1,\dots,i_k}(x_1, \dots, x_l) = \sum_{i=1}^{l} x_i - (x_{i_1} + \dots + x_{i_k}) \quad \text{for} \quad i_1 < i_2 < \dots < i_k\,.$$

3. Calculate the skew derivative and the dual skew derivative of Λ and $\Lambda^{\vee}$, where Λ is the vector field

$$\Lambda(x_1, \dots, x_l) = [x_1, x_2^2, \dots, x_l^l]\,.$$

4. Assuming the notation from IV § 3, exercise 6, verify that $\mathrm{d}\Omega = \nabla \wedge \Omega$ (see II § 9 (7)-(7′)). Find an analogous formula for $\delta\Lambda$.

5. Prove that for any function φ of class C_1 and a skew-symmetric matrix field Ω on an open set $G \subset R^l$, $\mathrm{d}(\varphi\Omega) = \varphi' \wedge \Omega + \varphi \mathrm{d}\Omega\,.$

Find the dual formula for $\delta(\varphi\Omega)$.

6. If $\Omega = \varphi_1' \wedge \ldots \wedge \varphi_k'$ is the skew product of the derivatives of real functions $\varphi_1, \ldots, \varphi_k$ defined on an open set $G \subset R^l$, then $\mathrm{d}\Omega = 0$.

If $\Lambda = \varphi_1' \vee \ldots \vee \varphi_\varkappa'$ is the dual skew product of the derivatives of real functions $\varphi_1, \ldots, \varphi_\varkappa$ of class C_1 defined on an open set $G \subset R^l$, then $\delta\Lambda = 0$.

Hint. Prove the first part by induction with respect to k. Use the formula given in exercise 5.

7. If $\varphi_1, \ldots, \varphi_{k-1}$ are real functions of class C_2 on an open set $G \subset R^l$ and

$$J_j = \begin{vmatrix} \varphi_{1|1} & \cdots & \varphi_{1|j-1} & \varphi_{1|j+1} & \cdots & \varphi_{1|k} \\ \cdot & \cdot & \cdot & \cdot & \cdot & \cdot \\ \varphi_{k-1|1} & \cdots & \varphi_{k-1|j-1} & \varphi_{k-1|j+1} & \cdots & \varphi_{k-1|k} \end{vmatrix},$$

then

$$\sum_{j=1}^{k} (-1)^{j-1} J_{j|j} = 0 .$$

§ 11. Stokes' theorem. This name is given to the following theorem:

11.1. *If F is a differential $(k-1)$-form of class C_1 on an open set $G \subset R^l$, and Γ is a k-dimensional chain, $|\Gamma| \subset G$, $1 \leqslant k \leqslant l$, then*

$$(1) \qquad \int_{\mathrm{b}\Gamma} F = \int_{\Gamma} \mathrm{d}F .$$

In other words,

If Ω is a skew-symmetric $(k-1)$-indexed l-dimensional matrix field of class C_1 on an open set $G \subset R^l$, and Γ is a k-dimensional chain, $|\Gamma| \subset G$, $1 \leqslant k \leqslant l$, then

$$(1') \qquad \int_{\mathrm{b}\Gamma} \Omega = \int_{\Gamma} \mathrm{d}\Omega .$$

To verify that the second formulation is equivalent to the first it suffices to take as Ω the matrix of coordinates of F. We shall prove Theorem 11.1 in the second formulation.

The chain Γ can be represented in the form

$$\Gamma = a_1 K_1 + \ldots + a_n K_n$$

where $K_1, \ldots, K_n$ are k-dimensional oriented cells and $\{|K_1|, \ldots, |K_n|\}$ is a cellular decomposition of the polyhedron $|\Gamma|$. Thus

$$\mathrm{b}\Gamma = a_1 \mathrm{b}K_1 + \ldots + a_n \mathrm{b}K_n ,$$

and

$$\int_{\mathrm{b}\Gamma} \Omega = a_1 \int_{\mathrm{b}K_1} \Omega + \ldots + a_n \int_{\mathrm{b}K_n} \Omega ,$$

$$\int_{\Gamma} \mathrm{d}\Omega = a_1 \int_{K_1} \mathrm{d}\Omega + \ldots + a_n \int_{K_n} \mathrm{d}\Omega .$$

Hence it follows that, in order to prove 11.1, it suffices to show that

(2) $$\int_{bK} \Omega = \int_K d\Omega$$

for every oriented k-dimensional cell K, $|K| \subset G$. Of course, (2) is a particular case of (1').

Consider first the case where $k = 1$. Then $\Omega = [\omega]$ is a real function ω of class C_1, and $d\Omega = \omega'$. The one-dimensional oriented cell K can be represented in the form $K = \langle \Phi \rangle$ where $\Phi: I \to R^l$ is a diffeomorphism on a closed interval $I = \langle a; b \rangle \subset R$. By § 8 (4), bK is a real function on R^l, defined by

$$bK(q) = \begin{cases} 1 & \text{if} \quad q = \Phi(b)\,, \\ -1 & \text{if} \quad q = \Phi(a)\,, \\ 0 & \text{elsewhere}\,, \end{cases}$$

and consequently (see § 9 (11) and § 9 (10))

$$\int_{bK} \Omega = \omega(\Phi(b)) - \omega(\Phi(a)) = \int_a^b (\omega \circ \Phi)' = \int_I (\omega' \circ \Phi)\cdot\Phi' = \int_K d\Omega\,,$$

that is, (2) is true.

Suppose now that $k > 1$. As above, to prove (2) we shall assume that $K = \langle \Phi \rangle$ where $\Phi = (\varphi_1, \dots, \varphi_l)$ is a diffeomorphism defined on a closed interval

$$I = \{(x_1, \dots, x_k)\colon \alpha_i \leqslant x_i \leqslant \beta_i \text{ for } i = 1, \dots, k\}\,.$$

By § 8 (4),

$$bK = \sum_{j=1}^{k} \sum_{\eta=\pm 1} (-1)^{j-1} \eta \langle \Phi_{j,\eta} \rangle$$

where $\Phi_{j,\eta}$ is the diffeomorphism defined by formula § 3 (22) on the interval

$$I_j = \{(x_1, \dots, x_{j-1}, x_{j+1}, \dots, x_k)\colon \alpha_i \leqslant x_i \leqslant \beta_i \text{ for } i \neq j\}\,.$$

By § 9 (7),

$$\int_{bK} \Omega = \sum_{j=1}^{k} \sum_{\eta=\pm 1} (-1)^{j-1} \eta \int_{\langle \Phi_{j,\eta} \rangle} \Omega = \sum_{j=1}^{k} (-1)^{j-1} \sum_{\eta=\pm 1} \eta \int_{I_j} f_{j,\eta}$$

where $f_{j,\eta}$ is the following scalar product

(3) $$f_{j,\eta}(p) = \Omega(\Phi_{j,\eta}(p)) \cdot (\Phi_{j,\eta|1}(p) \times \dots \times \Phi_{j,\eta|k-1}(p)) \quad \text{for} \quad p \in I_j\,.$$

On the other hand, by § 9 (7),

$$\int_K d\Omega = \int_I (d\Omega \circ \Phi) \cdot (\Phi'_{|1} \times \dots \times \Phi'_{|k})\,.$$

Thus Theorem 11.1 results from the following theorem where $f_{j,\eta}$ is defined by (3):

11.2. *For every mapping* Φ: $I \to G$ *of class* C_1,

$$\sum_{j=1}^{k}(-1)^{j-1}\sum_{\eta=\pm 1}\eta\int_{I_j} f_{j,\eta} = \int (\mathrm{d}\Omega\circ\Phi)\cdot(\Phi'_{|1}\times\ldots\times\Phi'_{|k})\,. \tag{4}$$

First we shall prove Theorem 11.2 under the hypothesis that Φ is of class C_2.

Let L denote the left-hand side of equation (4).

The main idea of the proof is as follows. For a fixed j, the expression $\sum_{\eta=\pm 1}\eta\int_{I_j} f_{j,\eta}$ is the increase of a real function of one real variable x_j on the interval $\langle\alpha_j;\beta_j\rangle$ and, consequently, it is equal to the integral, on this one-dimensional interval, of the derivative of this function. Using the Fubini theorem we state that L is the integral of a function f on I. By means of some algebraic and differential transformations of the formula for f we assert that f is equal to the function under the integral on the right-hand side of equation (4).

Fix an integer j. For every point $p=(x_1,\ldots,x_{j-1},x_{j+1},\ldots,x_k)\in I_j$ and for every number x_j in the interval $\langle\alpha_j;\beta_j\rangle$, let (p,x_j) denote the point $q=(x_1,\ldots,x_{j-1},x_j,x_{j+1},\ldots,x_k)$. We recall (see § 3 (22)) that for $p\in I_j$

$$\Phi_{j,\eta}(p)=\begin{cases}\Phi(p,\beta_j) & \text{if}\quad \eta=+1\,,\\ \Phi(p,\alpha_j) & \text{if}\quad \eta=-1\,.\end{cases}$$

Thus for $\eta=+1$ the vectors

$$\Phi_{j,\eta|1}(p)\,,\;\ldots,\;\Phi_{j,\eta|k-1}(p) \tag{5}$$

coincide respectively with the vectors

$$\Phi_{|1}(p,\beta_j),\;\ldots,\;\Phi_{|j-1}(p,\beta_j),\;\Phi_{|j+1}(p,\beta_j),\;\ldots,\;\Phi_{|k}(p,\beta_j)\,,$$

and for $\eta=-1$ the vectors (5) coincide respectively with the vectors

$$\Phi_{|1}(p,\alpha_j),\;\ldots,\;\Phi_{|j-1}(p,\alpha_j),\;\Phi_{|j+1}(p,\alpha_j),\;\ldots,\;\Phi_{|k}(p,\alpha_j)\,.$$

Assuming for $q=(p,x_j)\in I$

$$f_j(q)=f_j(p,x_j)=\Omega\big(\Phi(q)\big)\cdot\big(\Phi_{|1}(q)\times\ldots\times\Phi_{|j-1}(q)\times\Phi_{|j+1}(q)\times\ldots\times\Phi_{|k}(q)\big)\,,$$

we get

$$f_{j,\eta}(p)=\begin{cases}f_j(p,\beta_j) & \text{for}\quad \eta=+1\,,\\ f_j(p,\alpha_j) & \text{for}\quad \eta=-1\,.\end{cases}$$

By the general convention in IV § 2, the symbol $f_{j|j}$ denotes the partial derivative of f_j with respect to the jth coordinate. By the Fubini theorem,

$$\int_I f_{j|j} = \int_{I_j} dp \int_{\alpha_j}^{\beta_j} f_{j|j}(p, x_j)\,dx_j$$

$$= \int_{I_j} \big(f_j(p, \beta_j) - f_j(p, \alpha_j)\big)\,dp = \sum_{\eta=\pm 1} \eta \int_{I_j} f_{j,\eta}(p)\,dp\,.$$

Consequently,

$$L = \sum_{j=1}^{k} (-1)^{j-1} \int_I f_{j|j} = \int_I f$$

where

$$f = \sum_{j=1}^{k} (-1)^{j-1} f_{j|j}\,.$$

Thus

$$f = g + h$$

where

$$g = \sum_{j=1}^{k} (-1)^{j-1} (\Omega \circ \Phi)_{|j} \cdot (\Phi_{|1} \times \ldots \times \Phi_{|j-1} \times \Phi_{|j+1} \times \ldots \Phi_{|k})\,,$$

$$h = \sum_{j=1}^{k} (-1)^{j-1} (\Omega \circ \Phi) \cdot (\Phi_{|1} \times \ldots \times \Phi_{|j-1} \times \Phi_{|j+1} \times \ldots \times \Phi_{|k})_{|j}\,.$$

It is easy to see that

$$(\Omega \circ \Phi)_{|j} = (\Omega' \circ \Phi) \cdot \Phi_{|j}$$

and consequently (see II § 4 (23))

$$g = \sum_{j=1}^{k} (-1)^{j-1} (\Omega' \circ \Phi) \cdot (\Phi_{|1} \times \ldots \times \Phi_{|j-1} \times \Phi_{|j+1} \times \ldots \times \Phi_{|k} \times \Phi_{|j})$$

$$= (\mathrm{d}\Omega \circ \Phi) \cdot (\Phi_{|1} \times \ldots \times \Phi_{|k})$$

by § 10 (10) (where k should be replaced by $k-1$). On the other hand,

$$h = \sum_{j=1}^{k} \Big((-1)^{j-1} (\Omega \circ \Phi) \cdot \Big(\sum_{r=1}^{j-1} \Phi_{|1} \times \ldots \times \Phi_{|r-1} \times \Phi_{|r,j} \times$$

$$\times \Phi_{|r+1} \times \ldots \times \Phi_{|j-1} \times \Phi_{|j+1} \times \ldots \times \Phi_{|k} +$$

$$+ \sum_{r=j+1}^{k} \Phi_{|1} \times \ldots \times \Phi_{|j-1} \times \Phi_{|j+1} \times \ldots \times \Phi_{|r-1} \times \Phi_{|r,j} \times \Phi_{|r+1} \times \ldots \times \Phi_{|k} \Big) \Big) = 0$$

by II 8.1 (where $F(\mathfrak{a}_1, \ldots, \mathfrak{a}_{k-1}) = \Omega(\Phi(q))\cdot(\mathfrak{a}_1 \times \ldots \times \mathfrak{a}_{k-1})$; assume $\mathfrak{a}_j = \Phi_{|j}(q)$ and $\mathfrak{a} = \Phi_{|r,j}(q)$, $q \in I$).

Consequently

$$L = \int_I g = \int_I (\mathrm{d}\Omega \circ \Phi)\cdot(\Phi_{|1} \times \ldots \times \Phi_{|k}),$$

which we wished to prove.

Now we shall show that 11.2 is true under the hypothesis that Φ is of class C_1. The mapping Φ can be extended to a mapping of class C_1 defined on the whole space R^k with values in R^l. Applying Theorem VIII 7.2 to the coordinates of the extension, we infer that there exists a sequence of mappings $\Phi_n\colon I \to G$ of class C_2 (and even of class C_∞) such that

$$\Phi_n \to \Phi \quad \text{and} \quad \Phi_{n|i} \to \Phi_{|i} \quad (i = 1, \ldots, k)$$

uniformly on I (see also III 9.4 and III 6.4). For each of the mappings Φ_n the identity 11.2 holds. Passing to the limit and applying VII 6.5 we get identity 11.2 for the mapping Φ.

§ 12. Integrals on oriented polyhedra. The purpose of this section is to investigate a particular case of Stokes' Theorem 11.1 whose characteristic property is that it can be formulated without the notion of chain. This particular case appears often in applications, especially in the case of the plane R^2 and the space R^3. It can be formulated in two different versions—see Theorems 12.2 and 12.3 below.

Let P be a k-dimensional polyhedron in R^l ($1 \leqslant k \leqslant l$). A continuous matrix field T is called an *orientation* (or a *tangent orientation*) of P if there exists a cellular decomposition

(1) $$D = \{C_1, \ldots, C_n\} \quad (C_i \neq C_j \text{ for } i \neq j)$$

of P such that

1) T is defined on a subset of P containing the set $C_1^\circ \cup \ldots \cup C_n^\circ$;

2) for every point $p \in C_i^\circ$, $T(p)$ is a unit k-vector tangent to C_i° at p;

3) if a $(k-1)$-dimensional cell is a common face of exactly two cells in D, then the orientations of the cells determined by T are consistent.

If, moreover, the decomposition D has the property

4) every $(k-1)$-dimensional cell is the common face of at most two cells in D,

then we say that the orientation T is *proper*.

It is easy to guess that by the orientation of a cell C_i in D, *determined by a continuous matrix field* T satisfying conditions 1) and 2) we mean an orientation T_i of C_i such that $T_i(p) = T(p)$ for every $p \in C_i^\circ$. In other words, T_i is the only continuous extension of $T|C_i^\circ$ over C_i.

We say that a polyhedron P is *orientable* if it has a proper orientation.

According to a general definition in VII § 5, two orientations T_1 and T_2 of a polyhedron P are said to be *equivalent* if they differ only on a set of measure zero. Of course, by "measure" we understand here the measure μ_P on P.

If T is an orientation (a proper orientation) of a polyhedron P, then $-T$ is also an orientation (a proper orientation) of P which is not equivalent to T. It may happen that besides T and $-T$ the polyhedron P also has other essentially different orientations, i.e. orientations which are equivalent neither to T nor to $-T$. On the other hand, it is not difficult to formulate some connectedness conditions for a polyhedron P to have at most two non-equivalent orientations (see exercise 2).

It can be proved that every orientation T of a polyhedron P can be extended to an orientation defined on re$P$; the extension of T is equivalent to T. We shall not prove the existence of the extension because we shall not make any use of it. Orientations T will be used only to define certain integrals on P. For this purpose it suffices if T is defined almost everywhere on P. This is assured by condition 1).

Suppose P is orientable, and T and D satisfy conditions 1)-4). Let

$$D_0 = \{C_{0,1}, \dots, C_{0,m}\} \quad (C_{0,i} \neq C_{0,j} \text{ for } i \neq j) \tag{2}$$

be the set of all $(k-1)$-dimensional cells that are $(k-1)$-dimensional closed faces of exactly one of the cells in D. By definition, for every i there exists exactly one integer j such that

$$C_{0,i} \text{ is a } (k-1)\text{-dimensional face of } C_j\,. \tag{3}$$

The union of all the cells in D_0 is a $(k-1)$-dimensional polyhedron which will be called the *geometrical boundary* of P and be denoted by bP. By definition, D_0 is a cellular decomposition of bP. The set bP may be empty.

It can be proved that bP does not depend on D, i.e. that it is uniquely determined by P only. Namely, bP is the closure of the set of all points $q \in P$ that have a neighbourhood U in P which is homeomorphic with a k-dimensional half-ball

$$\{(x_1, \dots, x_k) \in R^k\colon (x_1^2 + \dots + x_k^2 < \varrho^2) \& (x_k \geqslant 0)\}\,.$$

We omit the proof of this statement because we shall not use it in what follows.

We say that an orientation τ of bP is *consistent* with the orientation T of P if the orientation of $C_{0,i}$ determined by τ $(i = 1, \dots, n)$ is consistent with the orientation of C_j determined by T. Of course, C_j is assumed to be the only cell in D satisfying (3), and τ is assumed to satisfy conditions 1)-3) where D is replaced by D_0.

12.1. *For every proper orientation T of an orientable k-dimensional polyhedron P $(k>0)$ there exists an orientation τ of its geometrical boundary* bP *such that T and τ are consistent. The consistent orientation τ of* bP *is determined by T uniquely up to equivalence.*

Let K_j be the oriented k-dimensional cell whose support is C_j and whose orientation is that determined by T $(j=1,\ldots,n)$ and let

(4) $$\Gamma = K_1+\ldots+K_n\,.$$

Let $K_{0,i}$ be the oriented $(k-1)$-dimensional face of K_j whose support is $C_{0,i}$, where i, j satisfy (3). Let

(5) $$\Gamma_0 = K_{0,1}+\ldots+K_{0,m}\,.$$

It follows from conditions 3) and 4) for T that

(6) $$\Gamma_0 = \mathrm{b}\Gamma\,.$$

Let P_0 denote the union of the geometrical interiors $C^\circ_{0,i}$ of all the cells $C_{0,i}$, $i=1,\ldots,m$.

The function $\tau = \Gamma_0|(C_{0,1}\cup\ldots\cup C_{0,m})$ is the required orientation of bP. In fact, it satisfies conditions analogous to 1) and 2). It follows from (6) and 8.1 that $\mathrm{b}\Gamma_0=0$. This proves that τ also satisfies 3). By definition, τ is continuous on each of the sets $C^\circ_{0,i}$. Since the sets are open in their union which is the domain of τ, the function τ is continuous. It follows directly from the definition of τ that τ is consistent with T.

To prove the last part of 12.1 let us suppose that τ is an orientation on bP satisfying conditions 1)-3) with D replaced by D_0. If τ is consistent with T, then

(7) $$\tau(p) = \Gamma_0(p) \quad \text{for} \quad p\in C^\circ_{0,i}, \quad i=1,\ldots,m\,.$$

This proves that τ is equivalent to the orientation defined in the proof of the first part of the theorem.

Observe that if orientations T, τ are consistent, then $-T$, $-\tau$ are also consistent, but T, $-\tau$ and $-T$, τ are not consistent (except the case where bP is empty).

12.2. *Let*

$$\Omega = [\omega_{i_1,\ldots,i_{k-1}}]_{i_1,\ldots,i_{k-1}\leqslant l}$$

be a skew-symmetric matrix field of class C_1 *defined on an open set $G\subset R^l$ $(1\leqslant k\leqslant l)$. Let T be a proper tangent orientation of an orientable polyhedron $P\subset G$, and let τ be a tangent orientation of the geometric boundary* bP *of P, such that T and τ are consistent. Then*

(8) $$\frac{1}{(k-1)!}\int\limits_{\mathrm{b}P}\Omega\cdot\tau = \frac{1}{k!}\int\limits_{P}\mathrm{d}\Omega\cdot T\,.$$

The integral on the left-hand side of (8) is with respect to the $(k-1)$-dimensional measure μ_{bP}, and the integral on the right-hand side is with respect to the k-dimensional measure μ_P.

Identity (8), called *Stokes' formula*, is a particular case of § 11 (1') where Γ is defined by (4). In fact,

$$\int_{\Gamma} \mathrm{d}\Omega = \frac{1}{k!} \int_{P} \mathrm{d}\Omega \cdot T \qquad \text{and} \qquad \int_{\mathrm{b}\Gamma} \Omega = \frac{1}{(k-1)!} \int_{\mathrm{b}P} \Omega \cdot \tau$$

by (6) and (7). Since the integrals on the left-hand sides of these identities are equal, (8) holds.

Of course, if bP is empty, we assume that the integral on bP is equal to 0. The same remark holds for other integrals on bP examined in this section.

By an *oriented k-dimensional polyhedron* $\mathcal{T}$ we mean a k-dimensional polyhedron P in R^l with a fixed orientation of P

$$T = [\xi_{i_1,\ldots,i_k}]_{i_1,\ldots,i_k \leqslant l} .$$

From the logical point of view $\mathcal{T}$ can be understood as the pair composed of P and T (see the analogous remarks on oriented cells at the beginning of § 6). Let f be a real function integrable on the *support* P of the oriented polyhedron $\mathcal{T}$. The integral

$$\int_P f\xi_{i_1,\ldots,i_k} \tag{9}$$

with respect to the measure μ_P is denoted in the traditional notation by the symbol

$$\int_{\mathcal{T}} \!\!\ldots\! \int f(x_1, \ldots, x_l)\, dx_{i_1} \ldots dx_{i_k} \tag{10}$$

(the sign $\int$ is repeated k times) or by the shorter symbol

$$\int_{\mathcal{T}} \!\!\ldots\! \int f dx_{i_1} \ldots dx_{i_k} \tag{10'}$$

where $x_1, \ldots, x_l$ are the coordinates of any point in R^l. Integrals (10)-(10') are called *oriented integrals*. Since

$$\xi_{i_1,\ldots,i_{r-1},i_{r+1},i_r,i_{r+2},\ldots,i_k} = -\xi_{i_1,\ldots,i_{r-1},i_r,i_{r+1},i_{r+2},\ldots,i_k}$$

we have

$$\int_{\mathcal{T}} \!\!\ldots\! \int f dx_{i_1} \ldots dx_{i_{r-1}} dx_{i_{r+1}} dx_{i_r} dx_{i_{r+2}} \ldots dx_{i_k}$$
$$= -\int_{\mathcal{T}} \!\!\ldots\! \int f dx_{i_1} \ldots dx_{i_{r-1}} dx_{i_r} dx_{i_{r+1}} dx_{i_{r+2}} \ldots dx_{i_k} .$$

We see that the ordering of the symbols $dx_{i_1}, \ldots, dx_{i_k}$ plays an essential part in integrals (10)-(10'). The expressions $dx_{i_1} \ldots dx_{i_k}$ in the integrals form, roughly speaking, a skew-symmetric matrix, i.e., for every permutation $r_1, \ldots, r_k$ of the integers $1, \ldots, k$,

$$dx_{i_{r_1}} \ldots dx_{i_{r_k}} = \operatorname{sgn}(r_1, \ldots, r_k)\, dx_{i_1} \ldots dx_{i_k} . \tag{11}$$

This purely formal identity is only an abbreviation of the identity

$$\int_{\mathcal{T}} \!\!\ldots\! \int f dx_{i_{r_1}} \ldots dx_{i_{r_k}} = \operatorname{sgn}(r_1, \ldots, r_k) \int_{\mathcal{T}} \!\!\ldots\! \int f dx_{i_1} \ldots dx_{i_k} \tag{11'}$$

which is valid for every integrable function f. Identity (11) suggests that it would be more reasonable to write $dx_{i_1} \wedge \ldots \wedge dx_{i_k}$ instead $dx_{i_1} \ldots dx_{i_k}$.

To explain the origin of the notation (10)-(10′) let us examine the case where $\mathfrak{T}$ is an oriented cell $\langle \Phi \rangle$ where $\Phi = (\varphi_1, \ldots, \varphi_l)$ is a diffeomorphism of an interval $I \subset R^k$ onto the support P of $\langle \Phi \rangle$. Let $u_1, \ldots, u_k$ denote the coordinates of any point in R^k. By § 6 (2′) the orientation T of $\mathfrak{T}$ is given by the formula

$$T(\Phi(u_1, \ldots, u_k)) = |\Phi'(u_1, \ldots, u_k)|^{-1} \left[\frac{\partial(\varphi_{i_1}, \ldots, \varphi_{i_k})}{\partial(u_1, \ldots, u_k)}\right],$$

that is,

$$\xi_{i_1, \ldots, i_k} = |\Phi'|^{-1} \cdot \frac{\partial(\varphi_{i_1}, \ldots, \varphi_{i_k})}{\partial(u_1, \ldots, u_k)} = |\Phi'|^{-1} \frac{\partial(x_{i_1}, \ldots, x_{i_k})}{\partial(u_1, \ldots, u_k)}.$$

By 5.3, for any integrable function f,

$$(12) \quad \int_{\mathfrak{T}} \cdots \int f(x_1, \ldots, x_l)\, dx_{i_1} \ldots dx_{i_k} = \int_I \cdots \int f(\Phi(u_1, \ldots, u_k)) \frac{\partial(x_{i_1}, \ldots, x_{i_k})}{\partial(u_1, \ldots, u_k)}\, du_1 \ldots du_k$$

where the integral on the right-hand side is with respect to the k-dimensional Lebesgue measure. In other words, the following formal identity holds:

$$(13) \quad dx_{i_1} \ldots dx_{i_k} = \frac{\partial(x_{i_1}, \ldots, x_{i_k})}{\partial(u_1, \ldots, u_k)}\, du_1 \ldots du_k.$$

The ordering of the symbols $du_1, \ldots, du_k$ on the right-hand side is of no consequences but the ordering of the symbols $dx_{i_1}, \ldots, dx_{i_k}$ on the left-hand side is significant. Thi, is an inconsistency in the notation but is easy to explain and remember: on the left-hand side of (12) we integrate on an oriented polyhedron, therefore the ordering of $dx_{i_1}, \ldots, dx_{i_k}$ plays an essential part; on the right-hand side of (12) we integrate on a non-oriented polyhedron, therefore the ordering of $du_1, \ldots, du_k$ is of no consequence.

The formal identity (13) is an analogue of the formal identity VIII § 5 (5′) for integration by substitution.

Let us come back to the general case where $\mathfrak{T}$ is an oriented k-dimensional polyhedron with a support $P \subset R^l$ and an orientation T. Let

$$\Omega = [\omega_{i_1, \ldots, i_k}]_{i_1, \ldots, i_k \leq l}$$

be a continuous matrix field defined at least on P. Using the notation (10) we get

$$(14) \quad \int_P \Omega T\, d\mu_P = \sum_{i_1, \ldots, i_k = 1}^{l} \int_{\mathfrak{T}} \cdots \int \omega_{i_1, \ldots, i_k} \xi_{i_1, \ldots, i_k} = \sum_{i_1, \ldots, i_k = 1}^{l} \int_{\mathfrak{T}} \cdots \int \omega_{i_1, \ldots, i_k}\, dx_{i_1} \ldots dx_{i_k}.$$

Therefore the integral (14) is denoted in the traditional notation by the symbol

$$(15) \quad \int_{\mathfrak{T}} \cdots \int \sum_{i_1, \ldots, i_k = 1}^{l} \omega_{i_1, \ldots, i_k}\, dx_{i_1} \ldots dx_{i_k}$$

or by

$$(15') \quad \int_{\mathfrak{T}} \cdots \int \sum_{i_1, \ldots, i_k = 1}^{l} \omega_{i_1, \ldots, i_k}(x_1, \ldots, x_l)\, dx_{i_1} \ldots dx_{i_k}.$$

The sign $\int$ is repeated k times in (15) and (15′). The ordering of $dx_{i_1}, \ldots, dx_{i_k}$ in *oriented integrals* (15)-(15′) is significant and the rule (11) is valid.

In the case where Ω is skew-symmetric,

$$\frac{1}{k!}\int_P \Omega T = \int_{\mathcal{T}} \cdots \int \sum_{i_1<\ldots<i_k} \omega_{i_1,\ldots,i_k} dx_{i_1} \ldots dx_{i_k} . \tag{16}$$

We get identity (16) from (15) by grouping terms which differ only by a permutation of lower indices and by selecting from each such group the term with an increasing sequence of lower indices (see II § 9 (12)). Sometimes it is convenient to select another term than that with an increasing sequence of lower indices. See e.g. formula (22) below for surface integrals in R^3.

Let us consider now the case where $k = 1$. Then

$$\Omega = [\omega_1, \ldots, \omega_l] \tag{17}$$

is a vector field defined at least on a one-dimensional polyhedron $P \subset R^l$, and T is an orientation of P, i.e. a continuous unit vector field defined on P except at a finite number of points (see condition 1) on p. 404). Let, as previously, $\mathcal{T}$ denote the oriented polyhedron with support P and orientation T. In the traditional notation the curvilinear integral $\int_P \Omega T$ (with respect to the one-dimensional measure μ_P) is denoted by the symbol

$$\int_{\mathcal{T}} \omega_1 dx_1 + \ldots + \omega_l dx_l \tag{18}$$

or by the symbol

$$\int_{\mathcal{T}} \omega_1(x_1, \ldots, x_l)\, dx_1 + \ldots + \omega_l(x_1, \ldots, x_l)\, dx_l . \tag{18'}$$

In other words (see the traditional notation on p. 362),

$$\int_P \Omega T\, ds = \int_{\mathcal{T}} \sum_{i=1}^{l} \omega_i dx_i . \tag{19}$$

Assume now that Ω is a skew-symmetric matrix field

$$\Omega = [\omega_{i_1,\ldots,i_{k-1}}]_{i_1,\ldots,i_{k-1} \leqslant l}$$

of class C_1, defined on an open set $G \subset R^l$. Let $\mathcal{T}$ be an oriented k-dimensional polyhedron with support $P \subset G$ and proper orientation T. Let $\mathrm{b}\mathcal{T}$ denote the oriented $(k-1)$-dimensional polyhedron with support $\mathrm{b}P$ and orientation τ consistent with T. Using the traditional notation, Stokes' formula (8) can be written as follows

$$\int_{\mathrm{b}\mathcal{T}} \cdots \int \sum_{i_1<\ldots<i_{k-1}} \omega_{i_1,\ldots,i_{k-1}} dx_{i_1} \ldots dx_{i_{k-1}} = \int_{\mathcal{T}} \cdots \int \sum_{i_1<\ldots<i_k} \omega^{\bullet}_{i_1,\ldots,i_k} dx_{i_1} \ldots dx_{i_k} , \tag{20}$$

where as always $\omega^{\bullet}_{i_1,\ldots,i_k}$ are coordinates of the skew derivative $\mathrm{d}\Omega$, i.e.

$$\omega^{\bullet}_{i_1,\ldots,i_k} = \sum_{r=1}^{k} (-1)^{r-1} \frac{\partial}{\partial x_{i_r}} \omega_{i_1,\ldots,i_{r-1},i_{r+1},\ldots,i_k} .$$

Let us discuss now the case where $k = 2$. Then Ω is a vector field (17) of class C_1 defined on G and

$$\omega^{\bullet}_{i,j} = \frac{\partial}{\partial x_i} \omega_j - \frac{\partial}{\partial x_j} \omega_i .$$

Stokes' formula then has the form

$$\int_{b\mathfrak{T}} \omega_1 dx_1 + \ldots + \omega_l dx_l = \frac{1}{2} \iint_{\mathfrak{T}} \sum_{i,j=1}^{l} \omega_{i,j}^{\bullet} dx_i dx_j , \tag{21}$$

that is,

$$\int_{b\mathfrak{T}} \omega_1 dx_1 + \ldots + \omega_l dx_l = \iint_{\mathfrak{T}} \sum_{i<j} \omega_{i,j}^{\bullet} dx_i dx_j . \tag{21'}$$

Assume additionally that $l = 3$. Since $\omega_{1,3}^{\bullet} = -\omega_{3,1}^{\bullet}$ and $dx_1 dx_3 = -dx_3 dx_1$, the formula (21′) can now be written in the form

$$\int_{b\mathfrak{T}} \omega_1 dx_1 + \omega_2 dx_2 + \omega_3 dx_3 \tag{22}$$

$$= \iint_{\mathfrak{T}} \left(\frac{\partial}{\partial x_2}\omega_3 - \frac{\partial}{\partial x_3}\omega_2\right) dx_2 dx_3 + \left(\frac{\partial}{\partial x_3}\omega_1 - \frac{\partial}{\partial x_1}\omega_3\right) dx_3 dx_1 + \left(\frac{\partial}{\partial x_1}\omega_2 - \frac{\partial}{\partial x_2}\omega_1\right) dx_1 dx_2 .$$

This is the *classical Stokes formula* for R^3.

By a *dual orientation* (or *orthogonal orientation* or *normal orientation*) of a k-dimensional polyhedron $P \subset R^l$ we mean any matrix field N of the form

$$N = T^{\vee} \tag{23}$$

where T is a tangent orientation of P. If T is proper, then N is also said to be *proper*.

In other words, N is an orthogonal (proper) orientation of P if and only if $T = N^{\wedge}$ is a tangent (proper) orientation of P in the sense defined on p. 404.

It follows from (23) and II 10.2 that if N is an orthogonal orientation of P, then $N(p)$ is a unit $(l-k)$-vector orthogonal to C_i° at any point $p \in C_i^{\circ}$ (for notation, see p. 404). Thus an orthogonal orientation of P is a unit $(l-k)$-vector field defined almost everywhere on P.

If N is an orthogonal orientation of P, then $-N = (-T)^{\vee}$ is the *opposite* orthogonal orientation of P.

Suppose that P is orientable and N is proper. Let ν be an orthogonal orientation of bP. We say that the orthogonal orientations N and ν are *consistent* if the tangent orientations $N^{\wedge}$ and $\nu^{\wedge}$ are consistent. It follows from the definition that if N and ν are consistent, then $-N$ and $-\nu$ are also consistent but N and $-\nu$ are not consistent, and $-N$ and ν are not consistent (except the case where bP is empty).

12.3. *Let*

$$\Lambda = [\lambda_{i_1,\ldots,i_{l-k+1}}]_{i_1,\ldots,i_{l-k+1} \leqslant l}$$

be a skew-symmetric matrix field of class C_1 *defined on an open set* $G \subset R^l$ $(1 \leqslant k \leqslant l)$. *Let* N *be a proper orthogonal orientation of an orientable* *polyhedron $P \subset G$ and let ν be an orthogonal orientation of the geometrical boundary* bP *of P, such that N and ν are consistent. Then*

$$\frac{1}{(l-k+1)!}\int_{\mathrm{b}P} \Lambda\cdot\nu = \frac{1}{(l-k)!}\int_{P} \delta\Lambda\cdot N\,. \tag{24}$$

Of course, the integral on the left-hand side is with respect to the $(k-1)$-dimensional measure $\mu_{\mathrm{b}P}$, and the integral on the right-hand side is with respect to the k-dimensional measure μ_P.

Identity (24) will be called the *dual Stokes formula.*

(24) follows directly from (8), where

$$T = N^{\wedge}, \qquad \tau = \nu^{\wedge}, \qquad \Omega = \Lambda^{\wedge}, \tag{25}$$

that is,

$$N = T^{\vee}, \qquad \nu = \tau^{\vee}, \qquad \Lambda = \Omega^{\vee}. \tag{25'}$$

Indeed,

$$(\delta\Lambda)^{\wedge} = \mathrm{d}(\Lambda^{\wedge}) = \mathrm{d}\Omega$$

by § 10 (15) and consequently, by II § 9 (24),

$$\frac{1}{(l-k+1)!}\,\Lambda\nu = \frac{1}{(k-1)!}\,\Lambda^{\wedge}\cdot\nu^{\wedge} = \frac{1}{(k-1)!}\cdot\Omega\cdot\tau\,,$$

$$\frac{1}{(l-k)!}\,\delta\Lambda\cdot N = \frac{1}{k!}(\delta\Lambda)^{\wedge}\cdot N^{\wedge} = \frac{1}{k!}\mathrm{d}\Omega\cdot T.$$

Hence

$$\frac{1}{(l-k+1)!}\int_{\mathrm{b}P}\Lambda\nu = \frac{1}{(k-1)!}\int_{\mathrm{b}P}\Omega\tau\,, \qquad \frac{1}{(l-k)!}\int_{P}\delta\Lambda\cdot N = \frac{1}{k!}\int_{P}\mathrm{d}\Omega\cdot T, \tag{26}$$

which proves (24) by (8).

Now we shall discuss the case where $l = k$. We recall that there are only two k-dimensional unit k-vectors, namely

$$\mathrm{sgn} = [\mathrm{sgn}(i_1, \ldots, i_k)]_{i_1,\ldots,i_k\leqslant k}$$

and $-$sgn.

For any k-dimensional polyhedron $P \subset R^k$ the matrix field

$$T(p) = \mathrm{sgn} \quad \text{for all } p \in P$$

is a proper orientation of P. Indeed, as (1) we can take an arbitrary cellular decomposition D of P. Conditions 1) and 2) are obvious. Condition 3) follows from 6.4. Condition 4) follows from 6.5.

The proper orientation T just defined is said to be *positive.* The proper orientation $-T$ is said to be *negative.*

We see that every k-dimensional polyhedron $P \subset R^k$ is orientable.

A tangent orientation of the geometrical boundary $\mathrm{b}P$ of P is said to be *positive* (*negative*) if it is consistent with the positive (negative) orientation of P. Thus if τ is a positive orientation of $\mathrm{b}P$, then $-\tau$ is a negative orientation of $\mathrm{b}P$, and conversely.

An orthogonal orientation ν of $\mathrm{b}P$ is said to be *exterior* (*interior*) if $\nu^\wedge$ is a positive (negative) tangent orientation of $\mathrm{b}P$, i.e. if ν is of the form $\nu = \tau^\vee$ where τ is a positive (negative) tangent orientation of $\mathrm{b}P$. An orthogonal orientation ν is exterior if and only if $-\nu$ is interior.

Let ν be an exterior orthogonal orientation of $\mathrm{b}P$. Let q be a point in the geometrical interior $C^\circ_{0,i}$ of a $(k-1)$-dimensional cell $C_{0,i}$ of the decomposition D_0 of $\mathrm{b}P$ (see p. 405). The cell $C_{0,i}$ is the closed face of exactly one cell C_j of the decomposition D of P (see (3)). By 6.3, the vector $\nu(q)$ is orthogonal to $C^\circ_{0,i}$ and out-tangent to C_j. Consequently the unit vector $\nu(q)$ is also orthogonal to $\mathrm{reb}P$ and out-tangent to P.

Similarly, if ν is an interior orthogonal orientation, then, for the same point q, the unit vector $\nu(q)$ is orthogonal to $\mathrm{reb}P$ and in-tangent to P.

By II § 9 (29), if T is the positive orientation of P, then the dual orientation $N = T^\vee$ is the constant function equal to 1. If ν is an exterior orthogonal orientation, then $\nu^\wedge$ is consistent with T and, consequently, ν is consistent with $N = 1$.

The following *Gauss-Ostrogradski theorem* is a particular case of 12.3.

12.4. *If A is a k-dimensional vector field of class C_1 defined on an open set $G \subset R^k$ and ν is an exterior orthogonal orientation of the geometrical boundary $\mathrm{b}P$ of a k-dimensional polyhedron $P \subset G$, then*

$$\int_{\mathrm{b}P} A\nu = \int_P \delta A\,, \tag{27}$$

that is,

$$\int_{\mathrm{b}P} A\nu = \int_P \operatorname{div} A\,. \tag{27'}$$

Of course, the integrals on the left-hand side are with respect to the $(k-1)$-dimensional measure $\mu_{\mathrm{b}P}$, and the integrals on the right-hand side are with respect to the k-dimensional measure μ_P, i.e. with respect to the k-dimensional Lebesgue measure in R^k.

Now we shall formulate some particular cases of Theorems 12.2 and 12.4.

Assume in 12.2 that $k = l = 2$, T is the positive orientation of a two-dimensional polyhedron $P \subset R^2$, τ is a positive orientation of the boundary $\mathrm{b}P$ of P, and

$$\Omega = [\omega_1, \omega_2]$$

is a vector field of class C_1. Since $\dot{\omega_{1,2}} = \omega_{2|1} - \omega_{1|2}$, identity (8) has then the following form

(29) $$\int_{bP} \Omega \cdot \tau = \int_{P} (\omega_{2|1} - \omega_{1|2}) \,.$$

The integral on the left-hand side is with respect to the two-dimensional Lebesgue measure, and the integral on the right-hand side is with respect to the one-dimensional measure on bP.

Identity (29) is called *Green's formula*. Remember that τ is here the positive tangent orientation.

Figures 83 and 84 depict the positive and negative tangent orientation of the boundary of a two-dimensional polyhedron in the plane. With the numeration of the coordinate axes in these figures, the orientation of the boundary in fig. 83 is positive, and that in fig. 84 is negative.

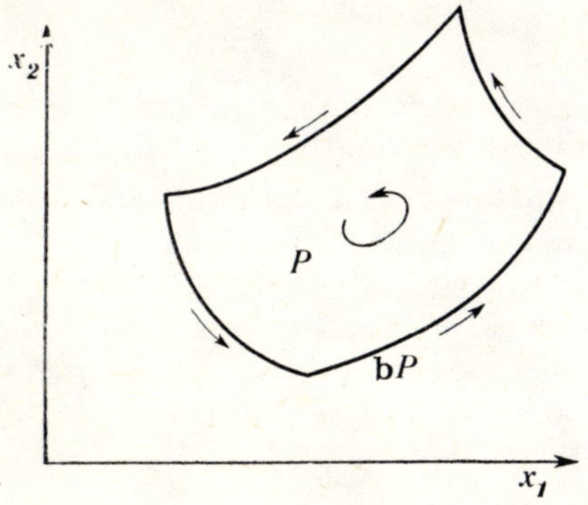

Fig. 83

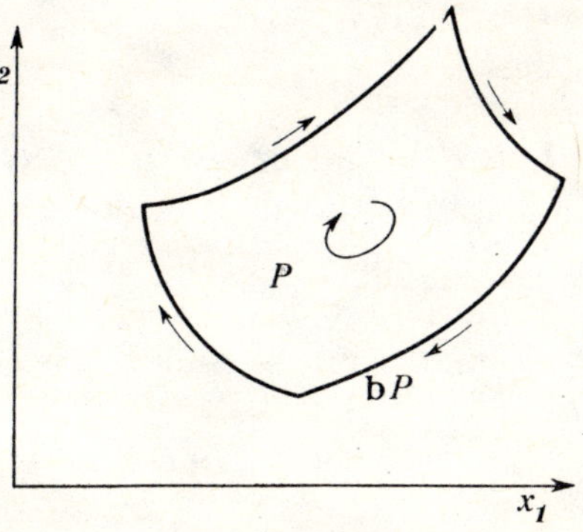

Fig. 84

Let $b\mathcal{T}$ denote the one-dimensional oriented polyhedron whose support is bP and whose orientation is the positive orientation of bP. The oriented polyhedron $b\mathcal{T}$ is called the *positively oriented boundary* of the two-dimensional polyhedron $P \subset R^2$. Green's formula is written in the traditional notation as follows

(30) $$\int_{b\mathcal{T}} \omega_1(x_1, x_2)\, dx_1 + \omega_2(x_1, x_2)\, dx_2 = \iint_{P} \left(\frac{\partial}{\partial x_1} \omega_2(x_1, x_2) - \frac{\partial}{\partial x_2} \omega_1(x_1, x_2) \right) dx_1 dx_2$$

or

(30′) $$\int_{b\mathcal{T}} \Omega \tau \, ds = \iint_{P} \left(\frac{\partial \omega_2}{\partial x_1} - \frac{\partial \omega_1}{\partial x_2} \right) dx_1 dx_2 \,.$$

Now take $k = 2$ in Theorem 12.4. In this case $\Lambda = [\lambda_1, \lambda_2]$ and

$$\operatorname{div} \Lambda = \lambda_{1|1} + \lambda_{2|2} \,.$$

Formula (27) now has the form

(31) $$\int_{bP} \Lambda \nu = \int_{P} (\lambda_{1|1} + \lambda_{2|2}) \,.$$

This particular case of the Gauss-Ostrogradski formula will be called the *dual Green formula*. Remember that ν is here an exterior orthogonal orientation of the boundary bP of a two-dimensional polyhedron $P \subset R^2$.

In the traditional notation the dual Green formula is written as follows

$$\int_{\mathrm{b}P} A\nu\, ds = \iint_P \left(\frac{\partial \lambda_1}{\partial x_1} + \frac{\partial \lambda_2}{\partial x_2}\right) dx_1 dx_2 . \tag{32}$$

Assume now that

$$\Omega = [\omega_1, \omega_2, \omega_3]$$

is a vector field of class C_1 defined on an open set $G \subset R^3$. By the *rotation* of the vector field Ω we shall mean the following vector field on G:

$$\mathrm{rot}\,\Omega = [\omega_{3|2} - \omega_{2|3}, \omega_{1|3} - \omega_{3|1}, \omega_{2|1} - \omega_{1|2}] . \tag{33}$$

Observe that

$$\mathrm{rot}\,\Omega = (\mathrm{d}\Omega)^{\vee} . \tag{33'}$$

Let P be an orientable two-dimensional polyhedron contained in G, let N be an orthogonal orientation of P and let τ be a tangent orientation of bP. We shall say that the orientations N and τ are *consistent* if the tangent orientations $T = N^{\wedge}$ and τ are consistent.

Assume that N and τ are consistent. If $T = N^{\wedge}$, i.e. $N = T^{\vee}$, then

$$\frac{1}{2!}\mathrm{d}\Omega \cdot T = \frac{1}{1!}(\mathrm{d}\Omega)^{\vee} \cdot T^{\vee} = \mathrm{rot}\,\Omega \cdot N .$$

Consequently, by (8),

$$\int_{\mathrm{b}P} \Omega \cdot \tau = \int_P \mathrm{rot}\,\Omega \cdot N . \tag{34}$$

In the traditional notation the identity is written as follows:

$$\iint_{\mathrm{b}P} \Omega \cdot \tau\, ds = \iint_P \mathrm{rot}\,\Omega \cdot N\, d\sigma \tag{34'}$$

(see § 5 (17)). Identities (34), (34′) are another formulation of the classical Stokes, formula (22).

Let us recall how the consistency of N and τ can be verified.

Let q be a point in the geometric interior $C^{\circ}_{0,i}$ of a cell $C_{0,i}$ in the cellular decomposition D_0 of the geometrical boundary bP of P. The cell $C_{0,i}$ is a one-dimensional face of exactly one two-dimensional cell C_j in a cellular decomposition of P (see (3)). We may suppose that the tangent orientation $T = N^{\wedge}$ and the orthogonal orientation N of P are defined on $C^{\circ}_{0,i}$ (if not, we can extend them with preservation of continuity). Let $\mathfrak{b}$ be a vector out-tangent to C_j at q. The orientation T and τ are consistent if $T(q) = c\mathfrak{b}\wedge\tau(q)$ where $c > 0$ (see § 6 (4)). Hence it follows that N and ν are consistent if $N(q) = c\mathfrak{b}\vee\tau(q)$ where $c > 0$ (see figure 85). This should hold for every point q in the geometrical interiors of the cells of the cellular decomposition D_0 of bP. We recall that $\mathfrak{b}\vee\tau(q)$ is the vector product of the vectors $\mathfrak{b}$ and $\tau(q)$.

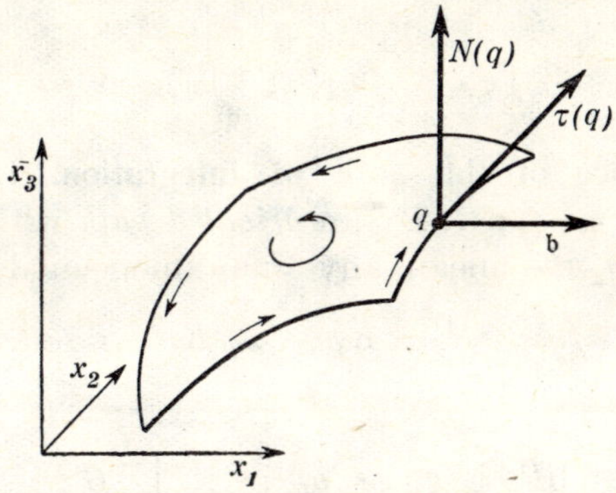

Fig. 85

EXERCISES

1. Let P be the union of three triangles such that any two of them have only one common vertex and the intersection of all three triangles is empty. How many orientations does the two-dimensional polyhedron P have? Is P orientable?

2. We say that a k-dimensional polyhedron $P \subset R^l$ is *regularly connected* if every cellular decomposition (1) of P has the following property. For any two cells C_r and C_s in the decomposition (1) there exists a sequence $C_{i_1}, \ldots, C_{i_m}$ of cells in (1) such that $C_{i_1} = C_r$, $C_{i_m} = C_s$ and the intersection $C_{i_j} \cap C_{i_{j+1}}$ $(j = 1, \ldots, m-1)$ is the common $(k-1)$-dimensional closed face of C_{i_j} and $C_{i_{j+1}}$ but is not a face of any other cell in (1).

Prove that every regularly connected polyhedron P has at most two non-equivalent orientations which differ only by sign.

3. Calculate the integral

$$\iint_{\mathfrak{T}} (x_2^2 + x_3^2)\, dx_2\, dx_3 + (x_3^2 + x_1^2)\, dx_3\, dx_1 + (x_1^2 + x_2^2)\, dx_1\, dx_2$$

where $\mathfrak{T}$ is the positively oriented torus from V § 6, pp. 213-214.

4. Calculate the integral

$$\iint_{\mathfrak{S}} (x_2 + x_3)\, dx_2\, dx_3 + (x_3 + x_1)\, dx_3\, dx_1 + (x_1 + x_2)\, dx_1\, dx_2$$

where $\mathfrak{S}$ is the positively oriented surface of the pyramid with vertices (0, 0, 0), (1, 0, 0), (0, 1, 0), (0, 0, 1).

5. Prove that there exists no orientation on the Möbius strip from § 5, exercise 3.

6. Let φ and ψ be functions of class C_2 on an open set $G \subset R^2$, let P be a two-dimensional polyhedron contained in G, let bP be its boundary, and let ν be an exterior orthogonal orientation of bP. Prove the following formulas

$$\iint_{\mathrm{b}P} \varphi'\psi' = \int_{\mathrm{b}P} \varphi\psi'_\nu - \iint_P \varphi\Delta\psi\,, \qquad \int_{\mathrm{b}P} (\varphi\psi'_\nu - \psi\varphi'_\nu) = \iint_P (\varphi\Delta\psi - \psi\Delta\varphi)\,,$$

where Δ is the differential operator defined in IV § 4, exercise 2. The symbol $\int\int$ denotes the integral with respect to the Lebesgue measure, and the symbol $\int$ denotes the integral with respect to the one-dimensional measure on bP. The symbol φ'_ν denotes the function whose value at a point p is the directional derivative $\varphi'_{\nu(p)}(p)$.

7. A function φ of class C_2, defined on an open set $G \subset R^2$ is said to be *harmonic* if $\Delta\varphi = 0$.

Verify that if φ and ψ are harmonic on G, and if P and ν have the same meaning as in exercise 6, then

$$\int_{\mathrm{b}P} \varphi\psi'_\nu = \int_{\mathrm{b}P} \psi\varphi'_\nu , \qquad \int_{\mathrm{b}P} \varphi'_\nu = 0 .$$

§ 13. Independence of the path of integration. Let G be an open subset of R^l and let $q_1, q_2 \in G$. By an *oriented path* with the *initial point* q_1 and the *end point* q_2 we mean any one-dimensional chain

$$\Gamma = K_1 + \ldots + K_n \tag{1}$$

such that

$$\mathrm{b}\Gamma = \langle q_2\rangle - \langle q_1\rangle ; \qquad |\Gamma| \subset G ;$$

$$|K_i| \cap |K_{i+1}| = \{p_i\} \text{ is a one-point set}, \qquad i = 1, \ldots, n-1;$$

$$|K_i| \cap |K_j| = 0 \quad \text{if} \quad j > i+1 .$$

It follows from the definition that

$$\mathrm{b}K_1 = \langle p_1\rangle - \langle q_1\rangle , \qquad \mathrm{b}K_i = \langle p_i\rangle - \langle p_{i-1}\rangle \quad \text{for} \quad i = 2, \ldots, n-1 ,$$

$$\mathrm{b}K_n = \langle q_2\rangle - \langle p_{n-1}\rangle .$$

Figure 86 gives an example of a path with an initial point q_1 and an end point q_2.

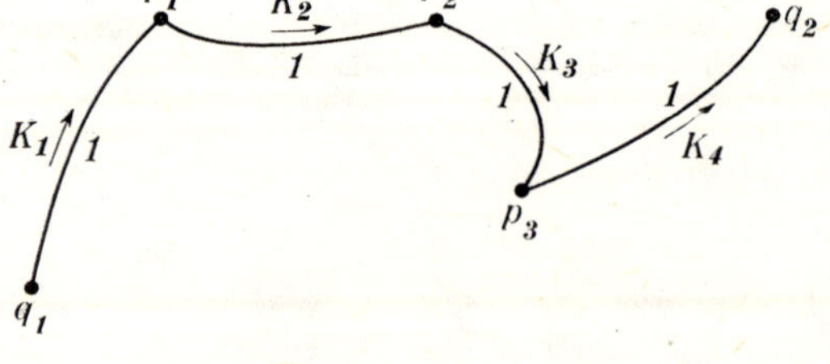

Fig. 86

The support $|\Gamma|$ of a path Γ is an arc with end points q_1, q_2. The arc $|\Gamma|$ is the union of a finite number of one-dimensional cells (regular arcs) $|K_1|, \ldots, |K_n|$.

If all the cells $|K_1|, \ldots, |K_n|$ are segments, then Γ is called an *oriented polygonal path.*

In what follows we shall always assume that the open set G is connected. By III 11.7 this implies that, for any points q_1, q_2 in G, there exists a polygonal path with the initial point q_1 and the end point q_2.

All chains under consideration are assumed to have supports contained in G.

Let $\Omega = [\omega_1, \ldots, \omega_l]$ be a fixed continuous vector field on G. We are interested in determining under what conditions

(2)
$$\int_{\Gamma_1} \Omega = \int_{\Gamma_2} \Omega$$

for any paths Γ_1, Γ_2 in G with the same initial point and the same end point. In other words, we shall discuss the question: when does the integral $\int_\Gamma \Omega$ depend only on the initial point q_1 and the end point q_2 of the integration path Γ. If (2) holds for any paths Γ_1, Γ_2 in G with the same initial point and the same end point, we say that the *curvilinear integral* $\int_\Gamma \Omega$ *does not depend on the path of integration*. In the case where (2) holds for any polygonal paths Γ_1, Γ_2 in G with the same initial point and the same end point, we say that the *curvilinear integral* $\int_\Gamma \Omega$ *does not depend on the polygonal path of integration.*

13.1. *The following conditions are equivalent:*

(a) *the curvilinear integral* $\int_\Gamma \Omega$ *does not depend on the path of integration;*

(a′) *the curvilinear integral* $\int_\Gamma \Omega$ *does not depend on the polygonal path of integration;*

(b) *for any one-dimensional chains* Γ_1 *and* Γ_2,

$$\textit{if } \mathrm{b}\Gamma_1 = \mathrm{b}\Gamma_2, \textit{ then } \int_{\Gamma_1} \Omega = \int_{\Gamma_2} \Omega\,;$$

(c) $\int_\Gamma \Omega = 0$ *for every one-dimensional cycle* Γ;

(d) *there exists a function* ω *of class* C_1 *on* G *such that* $\omega' = \Omega$ (*i.e.* $\mathrm{d}\omega = \Omega$).

It suffices to show the implications (d)$\Rightarrow$(b), (b)$\Rightarrow$(a), (a)$\Rightarrow$(a′), (a′)$\Rightarrow$(d), (b)$\Rightarrow$(c), (c)$\Rightarrow$(a′).

(d)$\Rightarrow$(b). By Stokes' Theorem 11.1 (with Ω replaced by ω, and $\mathrm{d}\Omega$ by Ω), if $\omega' = \Omega$, then

(3)
$$\int_\Gamma \Omega = \int_{\mathrm{b}\Gamma} \omega\,.$$

Thus the integral $\int_\Gamma \Omega$ depends only on the boundary $\mathrm{b}\Gamma$ of Γ. More precisely, if $\mathrm{b}\Gamma_1 = \mathrm{b}\Gamma_2$, then $\int_{\Gamma_1} \Omega = \int_{\mathrm{b}\Gamma_1} \omega = \int_{\mathrm{b}\Gamma_2} \omega = \int_{\Gamma_2} \Omega$.

(a) is a particular case of (b) and (a′) is a particular case of (a). Thus the implications (b) ⇒ (a) and (a) ⇒ (a′) are true.

(a′) ⇒ (d). For any points $q_1, q_2 \in G$ let

$$\int_{q_1}^{q_2} \Omega = \int_{\Gamma} \Omega \,, \tag{4}$$

where Γ is a polygonal path with the initial point q_1 and the end point q_2. By (a′), the integral on the right-hand side of (4) does not depend on Γ. Thus the expression on the left-hand side is well defined. We assume that (4) is equal to 0 if $q_1 = q_2$. It is easy to prove that

$$\int_{q_1}^{q_2} \Omega + \int_{q_2}^{q_3} \Omega = \int_{q_1}^{q_3} \Omega \,. \tag{5}$$

Let q_0 be a fixed point in G. For any point $q \in G$, let

$$\omega(q) = \int_{q_0}^{q} \Omega \,.$$

It follows from (4) and (5) that

$$\omega(q + \Delta t \cdot \mathfrak{e}_i) - \omega(q) = \int_{q}^{q+\Delta t \cdot \mathfrak{e}_i} \Omega = \int_0^{\Delta t} \Omega(q + t\mathfrak{e}_i) \cdot \mathfrak{e}_i \, dt = \int_0^{\Delta t} \omega_i(q + t\mathfrak{e}_i)\, dt \,,$$

where $\mathfrak{e}_i$ is the unit vector of the ith coordinate axis in R^l. Consequently

$$\omega_{|i}(q) = \lim_{\Delta t \to 0} \frac{\omega(q + \Delta t \cdot \mathfrak{e}_i) - \omega(q)}{\Delta t} = \omega_i(q) \,, \quad \text{that is,} \quad \omega' = \Omega \,.$$

(b) ⇒ (c). It suffices to take $\Gamma_1 = \Gamma$ and $\Gamma_2 = 0$ in (b).

(c) ⇒ (a′). Let Γ_1 and Γ_2 be polygonal paths with the same initial point q_1 and the same end point q_2. Besides the points q_1 and q_2, the supports $|\Gamma_1|$ and $|\Gamma_2|$ can have other common points, and even common segments. Without any loss of generality we may assume that the common segments are supports of oriented cells in both the chains Γ_1 and Γ_2, and that all other common points are 0-dimensional faces of cells in both the chains Γ_1 and Γ_2 (for if not, we can replace Γ_1 and Γ_2 by their refinements satisfying the above conditions; the passage to refinements changes the values of Γ_1 and Γ_2 only at a finite number of points and, therefore, does not change the integrals $\int_{\Gamma_1} \Omega$ and $\int_{\Gamma_2} \Omega$—see § 7, p. 374). Under this hypothesis, $\Gamma_1 - \Gamma_2$ is also a chain, and even a cycle, because $\mathrm{b}(\Gamma_1 - \Gamma_2) = \mathrm{b}\Gamma_1 - \mathrm{b}\Gamma_2 = 0$. Consequently $\int_{\Gamma_1} \Omega - \int_{\Gamma_2} \Omega = \int_{\Gamma_1 - \Gamma_2} \Omega = 0$.

13.2. *If Ω is a vector field of class C_1 and if the curvilinear integral $\int_{\Gamma} \Omega$ does not depend on the path of integration, then $\mathrm{d}\Omega = 0$.*

Indeed, it follows from 13.1 (d), that $\Omega = \mathrm{d}\omega$ where ω is a real function (i.e. a 0-indexed matrix field) of class C_2. By 10.3, $\mathrm{d}\Omega = \mathrm{dd}\omega = 0$. Note that it is not necessary to refer to the general Theorem 10.3 because in our case $\omega_i = \omega_{|i}$ and consequently $\mathrm{d}\Omega = [\omega^{\cdot}_{i,j}]$, where

$$\omega^{\cdot}_{i,j} = \omega_{j|i} - \omega_{i|j} = \omega_{|j,i} - \omega_{|i,j} = 0$$

(see IV 4.1).

The condition $\mathrm{d}\Omega = 0$ is necessary for the curvilinear integral $\int_\Gamma \Omega$ to be independent of the path of integration. However it is not sufficient. For instance, let $l = 2$, let $G = R^2 - \{(0, 0)\}$ and let

$$\Omega(x_1, x_2) = \left[-\frac{x_2}{x_1^2 + x_2^2}, \frac{x_1}{x_1^2 + x_2^2}\right] \quad \text{for} \quad (x_1, x_2) \in G .$$

It is easy to verify that $\mathrm{d}\Omega = 0$. In every open interval $J \subset G$ there is a continuous single-valued branch ω of the multi-valued function $\operatorname{arctg}(x_2/x_1)$ and $\Omega|J = \omega'$. However, there exists no function ω defined in the whole set G such that $\omega' = \Omega$. For such a function would differ by a constant from a continuous single-valued branch of $\operatorname{arctg}(x_2/x_1)$. This is impossible since $\operatorname{arctg}(x_2/x_1)$ does not have any continuous single-valued branch in the whole set G.

Only the following partial inverse of Theorem 13.2 is true.

13.3. *If* $\mathrm{d}\Omega = 0$ *and if*

(6) $\quad \Gamma \sim 0$ *in* G *for every one-dimensional cycle* Γ *with* $|\Gamma| \subset G$,

then the curvilinear integral $\int_\Gamma \Omega$ *does not depend on the path of integration.*

Indeed, if Γ is a cycle in G, then it follows from (6) that $\Gamma \equiv \mathrm{b}\Gamma_0$ for a two-dimensional cycle Γ_0 with $|\Gamma_0| \subset G$. By Stokes' Theorem 11.1 (where Γ should be now replaced by Γ_0)

$$\int_\Gamma \Omega = \int_{\mathrm{b}\Gamma_0} \Omega = \int_{\Gamma_0} \mathrm{d}\Omega = 0 ,$$

i.e. condition (c) in 13.1 is satisfied.

It can be proved that if Γ is a one-dimensional cycle in G and if the condition

$$\Gamma \sim 0 \text{ in } G$$

is not satisfied, then there exists a vector field Ω of class C_1 on G such that $\int_\Gamma \Omega \neq 0$. Thus, in this case, the curvilinear integral of Ω depends on the path of integration.

Theorems 13.1-13.3 have multi-dimensional generalizations. To formulate them, let us now suppose that G is an open subset of R^l and that

(7) $$\Omega = [\omega_{i_1,\ldots,i_k}]_{i_1,\ldots,i_k \leq l}$$

is a skew-symmetric continuous matrix field on G $(1 \leqslant k \leqslant l)$. Then the following three conditions analogous to (b), (c), (d) in 13.1 are equivalent:

1° for any k-dimensional cycles Γ_1, Γ_2 with supports contained in G,

$$\text{if } \mathrm{b}\Gamma_1 = \mathrm{b}\Gamma_2, \text{ then } \int_{\Gamma_1} \Omega = \int_{\Gamma_2} \Omega;$$

2° $\int_{\Gamma} \Omega = 0$ for every k-dimensional cycle Γ with $|\Gamma| \subset G$;

3° there is a $(k-1)$-indexed l-dimensional skew-symmetric matrix field Ω_0 of class C_1 on G such that $\Omega = \mathrm{d}\Omega_0$.

If Ω has properties 1°-3°, we say that the *integral* $\int_{\Gamma} \Omega$ *does not depend on the path of integration*

If $\int_{\Gamma} \Omega$ does not depend on the path of integration and Ω is of class C_1, then $\mathrm{d}\Omega = 0$. Conversely, if $\mathrm{d}\Omega = 0$ and if

(8) $\quad \Gamma \sim 0$ in G for every k-dimensional cycle Γ with $|\Gamma| \subset G$,

then the integral $\int_{\Gamma} \Omega$ does not depend on the path of integration. Moreover, if Γ is a k-dimensional cycle in G such that the condition

$$\Gamma \sim 0 \quad \text{in } G$$

is not satisfied, then there exists a matrix field (7) of class C_1 such that $\int_{\Gamma} \Omega \neq 0$.

Additional information on this subject will be given in X § 5.

EXERCISES

1. Let $p_1, \ldots, p_n$ be different points in E^2, $p_i = (x_i, y_i)$. Let Γ_i be a positively oriented simple chain whose support is the circle $\{p \in E^2 \colon |p - p_i| = \varrho\}$. The number ϱ is so small that none of the points p_j $(j \neq i)$ lies inside this circle. Let Ω be a vector field of class C_1 on the open set $G = E^2 - \{p_1, \ldots, p_n\}$ such that $\mathrm{d}\Omega = 0$. Prove that the number

$$c_i = \int_{\Gamma_i} \Omega$$

does not depend on the radius ϱ. The number c_i remains unchanged if Γ_i is replaced by the positively oriented simple cycle whose support is the boundary of an interval (or triangle, or a polygon) which contains p_i inside, but none of the points p_j $(j \neq i)$.

2. Under the same notation and hypotheses as in exercise 1, prove that if Γ_1 and Γ_2 are any paths in G with the same initial point q_1 and the same end point q_2, then

$$\int_{\Gamma_1} \Omega = \int_{\Gamma_2} \Omega + \sum_{i=1}^{n} m_i c_i \quad \text{where the } m_i \text{ are integers}.$$

3. Calculate the numbers c_i in exercise 1 in the case where

$$\Omega(x, y) = \left[-\sum_{i=1}^{n} a_i \frac{y - y_i}{(x - x_i)^2 + (y - y_i)^2}, \sum_{i=1}^{n} a_i \frac{x - x_i}{(x - x_i)^2 + (y - y_i)^2}\right].$$

CHAPTER X

SUPPLEMENTARY INFORMATION

§ 1. Linear spaces. By a *linear space* we mean a non-void set E in which there are defined two operations:

addition that assigns to every pair $u, v \in E$ their sum $u+v \in E$, and

multiplication by scalars that assigns to any $a \in R$ and $u \in E$ their product $au \in E$,

in such a way that the following conditions are satisfied:

(l_1) $u+v = v+u$,

(l_2) $u+(v+w) = (u+v)+w$,

(l_3) if $u+v = u+w$, then $v = w$,

(l_4) $a(u+v) = au+av$,

(l_5) $(a+b)u = au+bu$,

(l_6) $a(bu) = (ab)u$,

(l_7) $1 \cdot u = u$.

It is easy to verify that the element $0u$ does not depend on $u \in E$. It is denoted by 0 and called the *zero element* of E. It satisfies the following identity

$$u+0 = u \quad \text{for every } u \in E.$$

The element $-u = (-1)u$ satisfies the equation

$$u+(-u) = 0 .$$

Instead of $v+(-u)$ we write $v-u$. The element $v-u$ is called the *difference* of v and u.

Roughly speaking, the operations addition $u+v$, *subtraction* $u-v$ and multiplication au have the same properties as the analogous operations on vectors in R^k. The set V_k of all vectors in R^k is the simplest example of a linear space.

The reader familiar with group theory can easily verify that a linear space with the addition $u+v$ is an abelian group.

Generalizing a definition from II § 1, we say that elements $u_1, \ldots, u_n$ in E are *linearly independent* if

$$a_1 u_1 + \ldots + a_n u_n = 0 \quad \text{implies} \quad a_1 = \ldots = a_n = 0.$$

By the *dimension* of a linear space E we mean the maximal number of linearly independent elements in E. The dimension can be finite or infinite.

We admit the case where the linear space E under consideration has only one element: the zero element. By the dimension of this space we understand the number zero.

Let E be a linear space whose dimension is a finite integer k. By a *basis* of E we mean any sequence $u_1, \ldots, u_k$ of linearly independent elements in E. Every element $u \in E$ is a linear combination of elements of the basis

$$u = a_1 u_1 + \ldots + a_k u_k \tag{1}$$

and the representation (1) is unique.

We say that a mapping f from a linear space E into another linear space E' is *linear* if

$$f(u+v) = f(u) + f(v), \quad f(au) = af(u).$$

Then also

$$f(u-v) = f(u) - f(v).$$

If E' is the linear space of all real numbers, then f is often called a *linear form*, or a *linear functional.*

If a linear mapping $f\colon E \to E'$ is one-to-one and maps E onto E', then f is said to be a *linear isomorphism.* We say that E and E' are *isomorphic* if there is a linear isomorphism from E onto E'.

For instance, if $u_1, \ldots, u_k$ is a basis of a finite-dimensional linear space E, then the mapping

$$f(a_1 u_1 + \ldots + a_k u_k) = [a_1, \ldots, a_k]$$

is a linear isomorphism from E onto V_k.

A non-empty subset E' of a linear space E is said to be a *linear subspace* of E if it is a linear space under the same operations restricted to E', i.e. if

$$u+v \in E' \quad \text{and} \quad au \in E' \quad \text{for all } u, v \in E' \text{ and } a \in R.$$

Let E' be a linear subspace of a linear space E. For any $u, v \in E$ we shall write

$$u \sim v \quad \text{if} \quad u - v \in E'. \tag{2}$$

It follows easily from the linearity of E' that the relation $\sim$ is reflexive, symmetric, and transitive, i.e.

$$u \sim u;$$

$$\text{if } u \sim v, \text{ then } v \sim u;$$

$$\text{if } u \sim v \text{ and } v \sim w, \text{ then } u \sim w.$$

Hence it follows that the whole set E decomposes into disjoint sets, called the *equivalence classes* of $\sim$, such that two elements u, v are in the same equivalence class if and only if $u \sim v$. The equivalence class containing a given element u will be denoted by $u^{\sim}$. The set of all equivalence classes $u^{\sim}$ is denoted by E/E' and is called the *quotient space* of E and E'. The quotient space E/E' is a linear space with the following definition of the operations:

$$u^{\sim} + v^{\sim} = (u+v)^{\sim}, \qquad a \cdot u^{\sim} = (au)^{\sim}.$$

It can be proved that the result of the operations on elements $u^{\sim}$, $v^{\sim}$ in E/E' does not depend on the choice of the representatives $u, v \in E$.

Observe that elements $u_1^{\sim}, \ldots, u_n^{\sim}$ are linearly independent in E/E' if and only if the condition

(3) $$a_1 u_1 + \ldots + a_n u_n \sim 0 \quad \text{implies} \quad a_1 = \ldots = a_n = 0.$$

If the linear space E is finite-dimensional, then each of its subspaces E' is finite-dimensional and the quotient space E/E' is finite-dimensional.

In this book we have examined many finite-dimensional and infinite-dimensional linear spaces. The set of all matrices of a fixed type is an example of a finite-dimensional linear space. The set of all continuous matrix field (or matrix fields of class C_m) of a fixed type defined on an open set $G \subset R^l$ is an example of an infinite-dimensional linear space.

§ 2. Calculus in Banach spaces. By a *normed space* we mean a linear space E on which there is defined a non-negative real function $\|u\|$ in such a way that

(n_1) $\|u\| = 0$ if and only if $u = 0$;

(n_2) $\|u+v\| \leqslant \|u\| + \|v\|$;

(n_3) $\|au\| = |a| \cdot \|u\|$.

The number $\|u\|$ is called the *norm* of u.

Every normed space is a metric space, the distance between two elements u, v in E being defined as the norm of their difference: $|u-v| = \|u-v\|$. Conditions III § 1 (1)-(3) follow easily from (n_1)-(n_3).

If the normed space is complete, we say that it is a *Banach space.*

For instance, the linear space of all matrices of a fixed type, with the norm defined in II § 4 (11), is a Banach space (see II § 4 (12)-(13) and III § 2, p. 102). This Banach space is finite-dimensional. In particular, the set V_k of all vectors in R^k is a Banach space (see II § 1 (34)-(37) and III § 2, p. 102). Identifying points in R^k with vectors in V_k which have the same coordinates, we can consider the k-dimensional Euclidean space R^k as a Banach space.

Many interesting examples of Banach spaces are provided by the theory of the Lebesgue integral from Chapter VII.

Let μ be a measure on a σ-algebra of subsets of a space X, and let r be a fixed real number, $1 \leqslant r < \infty$. By $L^r(X, \mu)$ we shall denote the set of all measurable functions f defined on X, such that the number

$$\|f\| = \Big(\int_X |f|^r d\mu\Big)^{1/r} \tag{1}$$

is finite. Equivalent functions, i.e. functions which differ only on a set of measure zero, are identified one with another.

One can prove that $L^r(X, \mu)$ is a linear space, that (1) is a norm in this space and that $L^r(X, \mu)$ is complete, i.e. it is a Banach space.

The notion of Banach space is one of the fundamental notions of modern mathematics. The hypothesis of completeness is very essential; without this hypothesis, basic theorems on Banach spaces fail. In particular, the Banach spaces $L^r(X, \mu)$ play a very important part in many domains of modern mathematical analysis, e.g. in the theory of orthogonal series, of Fourier series and Fourier transforms, in the theory of integral equations, etc. First of all, the completeness of $L^r(X, \mu)$ implies important consequences.

The completeness of $L^r(X, \mu)$ is caused by the wide generality of the Lebesgue integral, i.e. by the fact that the Lebesgue integral is defined for a very large class of functions. In the historical development of mathematics, the theory of the Lebesgue integral was preceded by other theories of integration, much less general than the Lebesgue theory. As an example we mention here the theory of the Riemann integral which permits us (in the case where X is a closed interval in R^k) to define the integral of any function continuous at every point except at a set of Lebesgue measure zero. If the notion of the Lebesgue integral is replaced in the definition of $L^r(X, \mu)$ by the narrower notion of the Riemann integral, the normed space $L^r(X, \mu)$ fails to be complete, and many interesting and deep theorems from modern analysis have to be abandoned. Adjoining an enormous number of non-continuous functions to the theory of the integral by Lebesgue proved to be a very fruitful idea.

The space $L^1(X, \mu)$ of all integrable functions and the space $L^2(X, \mu)$ of all functions whose squares are integrable play a special role among the spaces $L^r(X, \mu)$. The space $L^2(X, \mu)$ has many interesting properties, especially from a geometrical point of view. Besides the norm, one can define in the space the scalar product of two functions f, g; it is the integral

$$\int_X fg\,d\mu\,.$$

The scalar product has, roughly speaking, the same properties as the scalar product of vectors in a Euclidean space. For this reason the geometry of the space $L^2(X, \mu)$ is, roughly speaking, the same as the geometry of R^k.

On one hand, the theory of the Lebesgue integral yields interesting examples of Banach spaces. On the other hand, the theory of Banach spaces enables us to generalize the notion of the Lebesgue integral. Namely, one can introduce the notion of the integral of functions whose values are in a Banach space E and which are defined on a set X with a measure μ. The definition of the integral from Chapter VII cannot be applied directly because it makes an essential use of the ordering of real numbers. It must be replaced by another definition which in the case of $E = R$ yields the same class of integrable functions. We shall quote below one such definition. For simplicity, we shall consider only the case where the measure μ under consideration is finite.

A function $f\colon X \to E$ will be called *simple* if it assumes only a finite number of values, say

$$u_1, \dots, u_m \in E \qquad (u_i \neq u_j \text{ for } i \neq j)$$

and the sets

$$A_i = \{p \in X\colon f(p) = u_i\} \qquad (i = 1, \dots, m)$$

are measurable. The *integral of the simple function* f is the element

$$\int_X f\,d\mu = \sum_{i=1}^{m} \mu(A_i)\cdot u_i$$

in E.

We shall say that a function $f\colon X \to E$ is *integrable* on X if it is the limit of a sequence (f_n) of simple functions such that

$$\int_X \|f_n(p) - f_m(p)\|\,d\mu(p) \to 0 \qquad \text{for} \qquad n, m \to \infty\,.$$

Then the sequence $(\int_X f_n\,d\mu)$ of elements in E converges in the metric space E, and its limit does not depend on the choice of (f_n). We define

$$\int_X f\,d\mu = \lim_{n\to\infty} \int_X f_n\,d\mu\,.$$

All the theorems from the theory of the integral in Chapter VII remain true except those connected with the ordering of values of integrated functions.

The reader interested in this subject is referred to the books [2], [8], [13].

Also the theory of differentiation can be generalized to the case of Banach spaces.

Let Φ be a mapping from an open subset G of a Banach space E into another Banach space E'. A linear mapping $F\colon E \to E'$ is said to be the *differential* of Φ at a point $u \in G$ provided that

$$\lim_{v\to 0} \frac{\Phi(u+v)-\Phi(u)-F(v)}{\|v\|} = 0 .$$

For every $v \in E$ the element $F(v)$ is then the *derivative of* Φ *at* u *in the direction* v, i.e. $F(v) = \Phi'_v(u)$, where

$$\Phi'_v(u) = \lim_{\Delta t\to 0} \frac{\Phi(u+\Delta t\cdot v)-\Phi(u)}{\Delta t} \quad \text{for} \quad v \in E .$$

We say that the mapping $\Phi\colon G \to E'$ is *differentiable* at a point $u \in G$ if it has a differential at u. We say that the mapping Φ is *differentiable* if it is differentiable at every point $u \in G$.

No coordinate system is given in the Banach space E under consideration. Therefore one cannot introduce the notion of partial derivative. It is also impossible to introduce the notion of scalar derivative or of modulus of derivative.

In the case of Euclidean spaces there is a natural one-to-one correspondence between rectangular matrices and linear mappings (see II 7.1). Therefore it does not matter of whether we speak about linear mappings or about the corresponding matrices. In order to accentuate the use of the natural coordinate systems in Euclidean spaces we presented the theory of differentiation in Euclidean spaces using the language of the theory of matrices. In the case of arbitrary Banach spaces there is no substitute for the notion of a matrix. Therefore the whole exposition of the theory of differentiability should be formulated in the language of linear mappings. In particular, the notion of derivative (at a fixed point $u \in G$) defined in Chapters IV and V as a matrix should be replaced by the notion of differential. The notion of the derivative should now be replaced by the notion of a function which, to every $u \in G$, assigns the differential of Φ at u. We see that the notion of the derivative in the case of arbitrary Banach spaces is more complicated than in the case of Euclidean spaces. However, almost all theorems in Chapter V remain true for differentiable mappings between Banach spaces after a necessary reformulation, i.e. a translation from the language of the theory of matrices into the language of linear mappings. The reader interested in this subject is referred to the books [1] and [8].

The reader interested in the theory of differentiation of real functions under weaker hypotheses than those in this book is referred to the book [6].

§ 3. Measures. The general notion of measure introduced in VI § 3 plays a fundamental part in the theory of probability. From the modern point of view, the probability is a measure defined on a σ-algebra of subsets of a set X such that the whole space X has measure 1.

In IX § 1 and § 5 we introduced the notion of k-dimensional measure on a regular k-dimensional hypersurface $H \subset R^l$ and on a k-dimensional polyhedron $P \subset R^l$, $k \leqslant l$. The notion of the k-dimensional measure was introduced, one can say, for every hypersurface H or polyhedron P separately. However, it is possible to introduce a general notion of k-dimensional measure in the whole space R^l for a very large class of subsets. Moreover, k can here be an arbitrary positive number (not necessarily an integer).

For this purpose, if $A \subset R^l$ and $\varepsilon > 0$, let

$$|A|_{k,\varepsilon} = \inf\Big\{\sum_{n=1}^{\infty}\varrho_n^k\colon \big(A \subset \bigcup_{n=1}^{\infty} \mathrm{B}(p_n, \varrho_n)\big)\&(\varrho_n < \varepsilon \text{ for } n = 1, 2, \ldots)\Big\}.$$

In other words, we consider all coverings of A by open balls with radii $< \varepsilon$. To every covering we assign the sum of all kth powers of the radii. The number $|A|_{k,\varepsilon}$ is the greatest lower bound of the sums.

By definition,

$$\text{if } \varepsilon_1 > \varepsilon_2, \text{ then } |A|_{k,\varepsilon_1} \leqslant |A|_{k,\varepsilon_2}.$$

Thus the limit

$$|A|_k = \lim_{\varepsilon \to 0} |A|_{k,\varepsilon}$$

exists (it is finite or infinite).

The set function $|A|_k$ has the properties of the Lebesgue outer measure formulated in VI 5.1-5.2 and, consequently, it also has the properties in VI §§ 5-6. Following a definition in VI § 6 we define the class $\mathcal{M}_k$ of k-measurable sets $A \subset R^l$ as follows

$$\mathcal{M}_k = \{A \subset R^l\colon \bigwedge_{Z \subset R^l} (|Z \cap A|_k + |Z - A|_k = |Z|_k)\}.$$

Just as in the case of the Lebesgue outer measure one proves that $\mathcal{M}_k$ is a σ-algebra of subsets of R^l, and that the set function μ_k defined by

$$\mu_k(A) = |A|_k \quad \text{for} \quad A \in \mathcal{M}_k$$

is a complete measure on $\mathcal{M}_k$. The measure μ_k is called the *k-dimensional Hausdorff measure.* One can prove that in the case where k is a positive integer $\leqslant l$ and A is a subset of a regular k-dimensional hypersurface $H \subset R^l$, then

$$A \in \mathcal{M}_H \text{ if and only if } A \in \mathcal{M}_k$$

and

$$\mu_H(A) = c_k \mu_k(A)$$

where the constant c_k is equal to the k-dimensional measure $V_k(1)$ of a ball with radius 1 in R^k (see VIII § 6, p. 317). The reader interested in this subject is referred to the paper [7].

Observe that the theorems given in VI § 5-6 can be generalized as follows. Let X be an arbitrary set. By a *Carathéodory outer measure* in X we mean any non-negative extended real function ν defined on the class of all subsets of X, such that

(a) $\nu(0) = 0$;

(b) if $A \subset \bigcup_{n=1}^{\infty} A_n$, then $\nu(A) \leqslant \sum_{n=1}^{\infty} \nu(A_n)$.

Conditions (a) and (b) are the subject of Theorems VI 5.1-5.2. All the remaining theorems in VI § 5-6 are true in the case of an arbitrary Carathéodory outer measure. In particular, the class of sets

$$\mathcal{M} = \Big\{A: \bigwedge_{Z \subset X} \big(\nu(Z \cap A) + \nu(Z - A) = \nu(Z)\big)\Big\}$$

is a σ-algebra of subsets of X and the set function μ defined by

$$\mu(A) = \nu(A) \quad \text{for} \quad A \in \mathcal{M}$$

(i.e. $\mu = \nu|\mathcal{M}$) is a complete measure. The theorem quoted, called the *Carathéodory theorem,* is often used to construct various interesting measures. The proof of the Carathéodory theorem is the same as that of its particular case in VI § 6.

The following theorem can be obtained as an application of the Carathéodory theorem: Let φ be a real non-decreasing function defined on a bounded or unbounded interval X of real numbers (open, closed, or open from one side). There exists exactly one measure μ_φ on the σ-algebra of all Borel subsets of X such that

$$\mu_\varphi(\langle a; b\rangle) = \varphi(b+) - \varphi(a-)$$

for every closed interval $\langle a; b\rangle \subset X$. The symbols $\varphi(b+)$ and $\varphi(a-)$ denote here the right-hand and the left-hand limit of φ at the points b or a, respectively.

The integral $\int_A f d\mu_\varphi$ is usually denoted by the symbol

$$\int_A f d\varphi \quad \text{or} \quad \int_A f(x)\, d\varphi(x) \tag{1}$$

and is called the *Lebesgue-Stieltjes integral.* This integral is a very convenient tool, e.g., in theory of probability. More generally, one defines the integral (1) in the case where φ is the difference of two monotonic functions (i.e. φ is a *function of bounded variation*). The reader interested in this subject is referred to the book [12].

The Fubini theorem can be generalized to the case of arbitrary measures as follows.

Let μ and ν be σ-finite measures defined on σ-algebras $\mathcal{M}$ and $\mathcal{N}$ of subsets of spaces X and Y respectively. There exists exactly one measure λ defined on the smallest σ-algebra (of subsets of $X \times Y$) containing all the sets

$$A \times B \quad \text{where} \quad A \in \mathcal{M} \text{ and } B \in \mathcal{N},$$

such that

$$\lambda(A \times B) = \mu(A)\nu(B) \quad \text{for} \quad A \in \mathcal{M} \text{ and } B \in \mathcal{N}.$$

For every λ-measurable function f defined on $X \times Y$ and non-negative or integrable,

$$\begin{aligned}\int_{X\times Y} f(p, q)\, d\lambda(p, q) &= \int_X \Big(\int_Y f(p, q)\, d\nu(q)\Big) d\mu(p) \\ &= \int_Y \Big(\int_X f(p, q)\, d\mu(p)\Big) d\nu(q).\end{aligned}$$

The interpretation of this formula is the same as in the case of the Fubini theorem proved in VIII § 2. Of course, p denotes an arbitrary point in X, and q denotes an arbitrary point in Y.

The Fubini theorem can be generalized to the case of an arbitrary (finite or infinite) number of measures. The reader interested in this subject is referred to the books [5] and [12].

§ 4. Homology groups of a polyhedron. Let P be an r-dimensional polyhedron in R^l and let D be a fixed cellular decomposition of P. By a *k-dimensional D-chain* we shall mean any chain

$$\Gamma = a_1 K_1 + \ldots + a_n K_n \tag{1}$$

such that the cells $|K_1|, \ldots, |K_n|$ are k-dimensional closed faces of cells in the decomposition D, $0 \leqslant k \leqslant r$. By a *$k$-dimensional D-cycle* we shall mean a k-dimensional D-chain which is a cycle, i.e. whose boundary is equal to 0.

Observe that the set of all k-dimensional D-chains is a linear space. The set $\mathcal{C}_k$ of all k-dimensional D-cycles is a linear subset of this space. The set $\mathcal{B}_k$ of all k-dimensional D-cycles which are boundaries of $(k+1)$-dimensional D-chains is a linear subspace of $\mathcal{C}_k$, $k < r$. For $k = r$ we assume that $\mathcal{B}_k$ is the subspace of $\mathcal{C}_k$ composed only of the zero element.

For any cycles $\Gamma_1, \Gamma_2 \in \mathcal{C}_k$ we shall write

$$\Gamma_1 \sim \Gamma_2 \tag{2}$$

if $\Gamma_1 - \Gamma_2 \in \mathcal{B}_k$. Then we shall say that the cycles Γ_1, Γ_2 are *homologous* (more precisely, *D-homologous*).

By § 1, p. 423, we can form the quotient linear space

$$\mathcal{H}_k(P) = \mathcal{C}_k / \mathcal{B}_k$$

which will be called the *k-th homology group of P* $(0 \leqslant k \leqslant r)$.

We say "group" instead of "linear space" since in many problems only the group structure of $\mathcal{H}_k(P)$ plays an essential part. Moreover, the notion of the kth homology group is often generalized in such a way that the coefficients $a_1, \ldots, a_n$ in chains (1) are taken from a fixed abelian group which yields, as a final result, a group, not a linear space.

According to the convention on p. 423, the element h of $\mathcal{H}_k(P)$ which is determined by a cycle $\Gamma \in \mathcal{C}_k$ will be denoted by $\Gamma^{\sim}$.

Since the linear space of all k-dimensional D-chains has a finite dimension (equal to the number of all k-dimensional closed faces of cells in D), the linear space $\mathcal{C}_k$ has finite dimension, too. Consequently the quotient linear space $\mathcal{H}_k(P)$ has finite dimension. The dimension of the linear space $\mathcal{C}_k$ and the dimension of the linear space $\mathcal{B}_k$ depend on the cellular decomposition D under consideration. However, the dimension of $\mathcal{H}_k(P)$ does not depend on the decomposition D (the proof is omitted). The dimension of the linear space $\mathcal{H}_k(P)$ is called the *k-th Betti number* of the polyhedron P.

One can prove that the Betti numbers are topological invariants of P, i.e. that homeomorphic polyhedra have the same Betti numbers.

Now we shall try to explain the intuitive meaning of the Betti numbers.

We shall say that cycles $\Gamma_1, \ldots, \Gamma_m \in \mathcal{C}_k$ are *homologously independent* if the elements $\Gamma_1^{\sim}, \ldots, \Gamma_m^{\sim}$ of $\mathcal{H}_k(P)$ are linearly independent (see § 1, pp. 422-423), i.e. if the relation

$$c_1\Gamma_1 + \ldots + c_m\Gamma_m \sim 0 \quad \text{implies} \quad c_1 = \ldots = c_m = 0 .$$

Thus the kth Betti number is the maximal number of homologously independent k-dimensional D-cycles.

We shall illustrate this notion by a few examples.

If P is the two-dimensional polyhedron on the plane in figure 87 (a rectangle with two squares removed), then the cycles Γ_1, Γ_2 in figure 88 are homologously independent (each of them encloses a different hole). One can prove that for every cycle $\Gamma \in \mathfrak{C}_1$ there exist numbers c_1 and c_2 such that

$$\Gamma \sim c_1\Gamma_1 + c_2\Gamma_2, \quad \text{i.e.} \quad \Gamma^\sim = c_1\Gamma_1^\sim + c_2\Gamma_2^\sim .$$

Thus the maximal number of homologously independent one-dimensional cycles of P is equal to 2, i.e. the first Betti number is equal to 2.

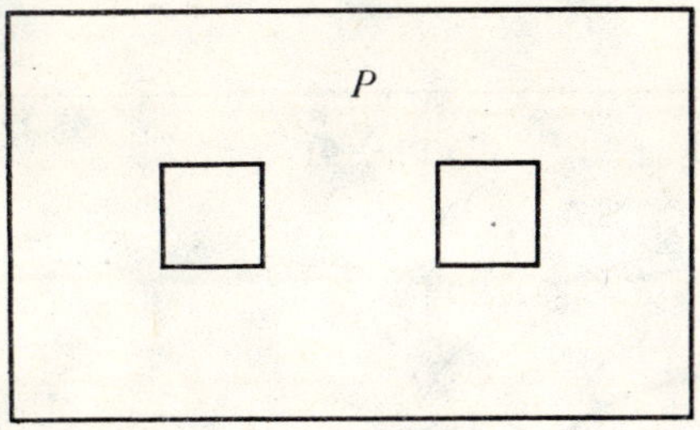

Fig. 87

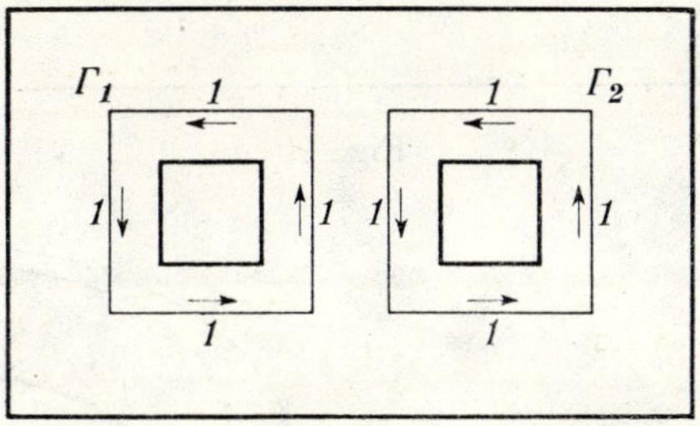

Fig. 88

Instead of the cycles in figure 88 we can take other cycles, suitably chosen, e.g. the cycles in figure 89.

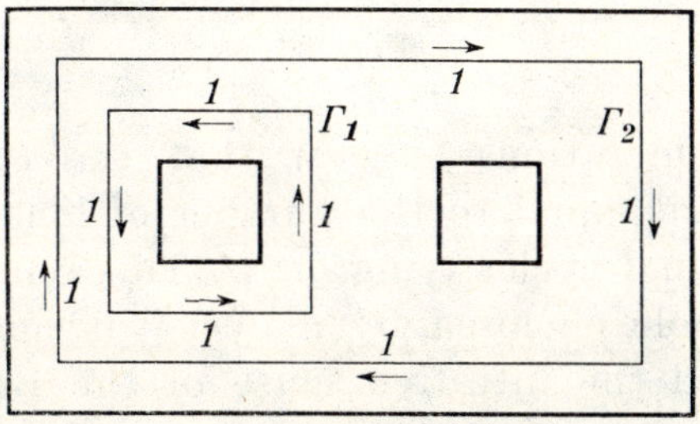

Fig. 89

If P is the two-dimensional polyhedron in the plane from figure 90 (a rectangle with three squares removed), then the cycles $\Gamma_1, \Gamma_2, \Gamma_3$ shown in figure 91 are homologously independent, and for any one-dimensional cycle $\Gamma \in \mathfrak{C}_1$ there exist numbers c_1, c_2, c_3 such that $\Gamma \sim c_1\Gamma_1 + c_2\Gamma_2 + c_3\Gamma_3$. Thus the first Betti number of P is equal to 3.

In both the examples the first Betti number of P is equal to the number of holes enclosed by one-dimensional cycles. A similar conclusion can be obtained when simple examples from the space R^3 are taken into consideration. For instance, if P is a two-dimensional sphere, then the first Betti number of P is 0, and the second is 1. If P is a torus (see V

§ 6, p. 214), then the second Betti number is 1 (the only two-dimensional cycles are of the form $c\Gamma$, where Γ is the cycle determined by an orientation of the torus). The first Betti number of the torus is 2. A maximal set of homologously independent one-dimensional cycles on the torus is shown in figure 92.

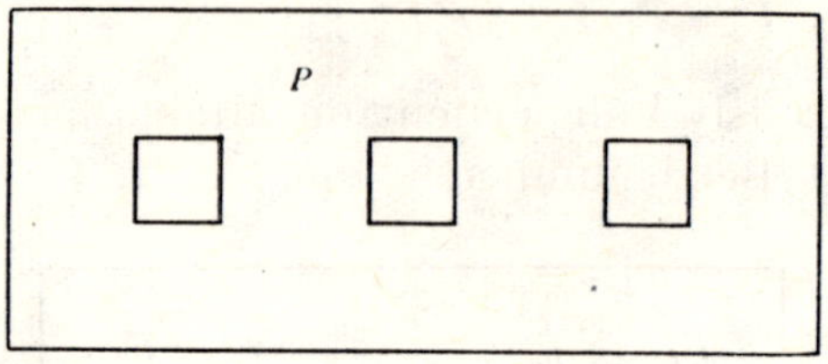

Fig. 90

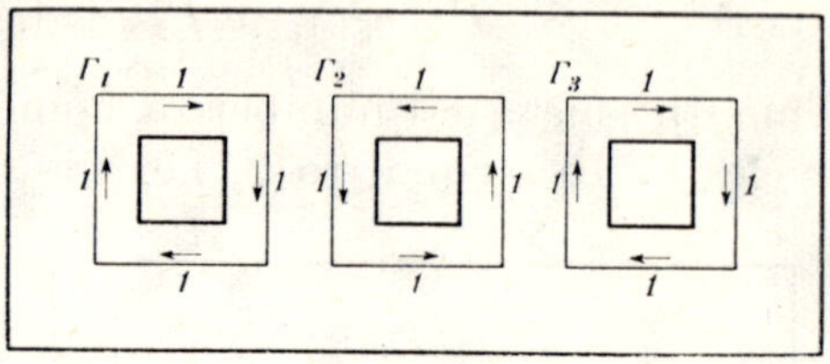

Fig. 91

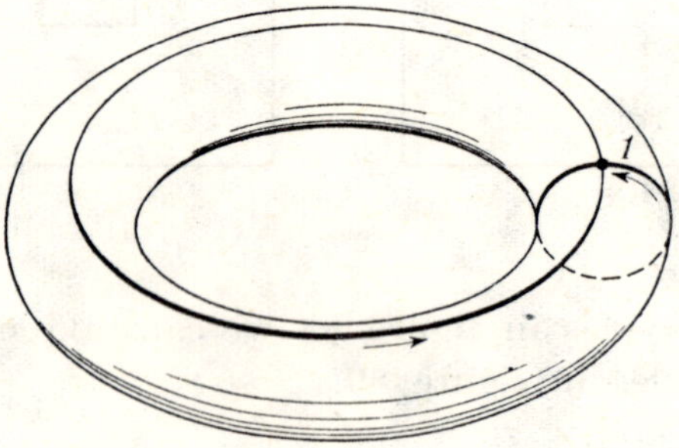

Fig. 92

The above considerations suggest that the kth Betti number of a polyhedron $P \subset R^l$ is equal to the number of holes in P which are enclosed by k-dimensional cycles lying on P. This remark is not a theorem (the expression "a hole enclosed by a cycle" is not defined). However, it explains quite well the intuitive sense of the Betti number and the homology group.

Observe that the zero Betti number of a polyhedron P is equal to the number of components of the set P (in the sense defined in III § 11) diminished by 1.

§ 5. Homology groups of an open set. The de Rham theorem. The notion of homology groups of an open subset of a Euclidean space can be introduced in a way similar to that for polyhedra. For this purpose it is convenient to treat an open set as an infinite polyhedron, i.e. as the union of infinitely many cells.

More precisely, by a *cellular decomposition* of a non-empty open subset G of R^l we shall mean a countable set

(1) $$D = \{C_1, C_2, \ldots\}$$

of l-dimensional cells, such that

(2) $$G = C_1 \cup C_2 \cup \dots ;$$

(3) if $i \neq j$, then the set $C_i \cap C_j$ is either empty or is a common r-dimensional closed face of the cells C_i and C_j $(r = 0, \dots, l-1)$;

(4) every point $q \in G$ has a neighbourhood U such that U intersects only a finite number of cells in D.

One can prove that every open set G has a cellular decomposition. In what follows D denotes a fixed cellular decomposition of G.

The following definitions are analogous to those in § 4.

By a *k-dimensional D-chain* we shall mean any chain

$$\Gamma = a_1 K_1 + \dots + a_n K_n$$

such that the cells $|K_1|, \dots, |K_n|$ are k-dimensional closed faces of cells in D. The set of all k-dimensional D-chains is an infinite-dimensional linear space. The set $\mathfrak{C}_k$ of all k-dimensional D-cycles (i.e. D-chains which are cycles) is a linear subspace of the linear space of all D-chains, i.e. it is a linear space. The set $\mathfrak{B}_k$ of all cycles in $\mathfrak{C}_k$ which are boundaries of D-cycles is a linear subspace of $\mathfrak{C}_k$ (in the case where $k = l$, $\mathfrak{C}_l$ and $\mathfrak{B}_l$ are composed only of the zero chain). The quotient linear space (finite- or infinite-dimensional)

$$\mathcal{H}_k(G) = \mathfrak{C}_k / \mathfrak{B}_k \qquad (k = 0, 1, \dots, l)$$

is called the *k-th homology group of the open set G*, and the dimension of $\mathcal{H}_k(G)$ is called the *k-th Betti number of G*. If $\Gamma \in \mathfrak{C}_k$, then $\Gamma^{\sim}$ denotes the element $h \in \mathcal{H}_k(G)$ determined by Γ.

We also assume the same definition of homologously independent cycles and the notation § 4 (2).

Of course, $\mathfrak{C}_k$ and $\mathfrak{B}_k$ depend on the choice of the cellular decomposition D. One can prove that $\mathcal{H}_k(G)$ is independent of D. The proof is too complicated to quote here. However, it is easy to deduce from theorems quoted in this book that $\mathcal{H}_k(G)$ is invariant under diffeomorphisms. More precisely, every diffeomorphism $\Phi\colon G \to R^l$ induces in a natural way an isomorphism from $\mathcal{H}_k(G)$ onto $\mathcal{H}_k(\Phi(G))$ since the cellular decomposition D of G determines a corresponding cellular decomposition $D' = \{\Phi(C_1), \Phi(C_2), \dots\}$ of $\Phi(G)$. It is much more difficult to prove that the homology groups (and, consequently, also the Betti numbers) are invariant under homeomorphisms.

We introduced here the notion of homology groups of G in order to be able to formulate the de Rham theorem which is an important supplement to the result quoted at the end of IX § 13.

For this purpose let us denote by $\mathcal{M}_k$ the linear space of all k-indexed l-dimensional skew-symmetric matrix fields Ω of class C_1 defined in G, such that $d\Omega = 0$ $(0 < k < l)$. Let $\mathcal{T}_k$ denote the linear subspace of $\mathcal{M}_k$ composed of all those Ω which are skew derivatives of $(k-1)$-indexed l-dimensional skew-symmetric matrix fields of class C_1, defined in G. For any $\Omega_1, \Omega_2 \in \mathcal{M}_k$ we shall write

$$\Omega_1 \sim \Omega_2 \text{ if and only if } \Omega_1 - \Omega_2 \in \mathcal{T}_k .$$

Let $\mathcal{S}_k(G) = \mathcal{M}_k/\mathcal{T}_k$. According to § 1, p. 423, the element $\chi \in \mathcal{S}_k(G)$, which is determined by $\Omega \in \mathcal{M}_k$, will be denoted by $\Omega^\sim$.

For any $\Gamma, \Gamma_1 \in \mathcal{C}_k$ and $\Omega \in \mathcal{M}_k$

$$\text{if } \Gamma \sim \Gamma_1, \text{ then } \int_\Gamma \Omega = \int_{\Gamma_1} \Omega . \tag{5}$$

Indeed, $\Gamma - \Gamma_1 \equiv b\Gamma_0$ where Γ_0 is a $(k+1)$-dimensional D-chain. By Stokes' theorem,

$$\int_\Gamma \Omega - \int_{\Gamma_1} \Omega = \int_{b\Gamma_0} \Omega = \int_{\Gamma_0} d\Omega = \int_{\Gamma_0} 0 = 0 .$$

For any $\Omega, \Omega_1 \in \mathcal{M}_k$ and $\Gamma \in \mathcal{C}_k$,

$$\text{if } \Omega \sim \Omega_1, \text{ then } \int_\Gamma \Omega = \int_\Gamma \Omega_1 . \tag{6}$$

Indeed, $\Omega - \Omega_1 \equiv d\Omega_0$ for a $(k-1)$-indexed skew-symmetric matrix field Ω_0 of class C_1 defined on G. By Stokes' theorem,

$$\int_\Gamma \Omega - \int_\Gamma \Omega_1 = \int_\Gamma d\Omega_0 = \int_{b\Gamma} \Omega_0 = \int_0 \Omega_0 = 0 .$$

Now let $h \in \mathcal{H}_k(G)$ and $\chi \in \mathcal{S}_k(G)$. By the integral

$$\int_h \chi \tag{7}$$

we shall mean the real number $\int_\Gamma \Omega$ where $h = \Gamma^\sim$ and $\chi = \Omega^\sim$. It follows from (5) and (6) that the number (7) does not depend on the choice of the representative Γ of h or the representative Ω of χ. It follows easily from known properties of the integral of a skew-symmetric matrix field on a chain (see IX § 9 (10)-(14)) that

$$\int_{ah} \chi = a \int_h \chi = \int_h a\chi , \tag{8}$$

$$\int_h \chi + \int_{h_1} \chi = \int_{h+h_1} \chi , \tag{9}$$

$$\int_h \chi + \int_h \chi_1 = \int_h (\chi + \chi_1) . \tag{10}$$

It follows from (8)-(9) that, for every fixed $\chi \in \mathcal{S}_k(G)$, the integral (7), considered as a function of $h \in \mathcal{H}_k(G)$, is a linear real function on $\mathcal{H}_k(G)$. The *de Rham theorem* states, roughly speaking, that every linear real function F on $\mathcal{H}_k(G)$, i.e. every function $F\colon \mathcal{H}_k(G) \to R$ such that

$$F(ah) = aF(h)\,, \qquad F(h+h_1) = F(h)+F(h_1) \qquad (h, h_1 \in \mathcal{H}_k(G))$$

is of the form (7), i.e. there exists a $\chi \in \mathcal{S}_k(G)$ such that

$$F(h) = \int_h \chi \qquad \text{for every } h \in \mathcal{H}_k(G)\,.$$

The reader who is interested in the de Rham theorem is referred to the books [9], [11], [14].

In this book, in the part concerning integrals on polyhedra and chains, we assumed the notion of interval as the fundamental notion with the help of which we defined further notions: cell, oriented cell, polyhedron, chain. This procedure is not typical. As a rule, the notion of simplex is taken as the fundamental notion rather than that of interval.

§ 6. Differentiable manifolds. In V § 6 an introduction to the theory of k-dimensional hypersurfaces in the l-dimensional Euclidean space has been presented. The exposition presented referred always to the "exterior" space R^l containing the hypersurfaces under examination. However, it is possible to develop the theory of hypersurfaces in another "intrinsic" manner without referring to the exterior Euclidean space. Moreover, the notion of hypersurface can be replaced by another more general notion which is also a natural multi-dimensional generalization of the intuitive notion of surface, and which can be introduced without referring to the exterior Euclidean space. This notion, called a *differentiable manifold*, is defined as a metric space satisfying certain additional conditions which ensure that the manifold is locally diffeomorphic with an open subset of a Euclidean space.

By a *k-dimensional differentiable manifold of class* C_m ($m = 1, 2, \ldots, \infty$) we mean a non-void separable metric space H and a set $\mathfrak{D}$ of mappings such that

(a) every $\Phi \in \mathfrak{D}$ is a homeomorphism from an open set $G_\Phi \subset R^k$ onto an open subset $U_\Phi = \Phi(G_\Phi)$ of H;

(b) H is the union of all the open sets U_Φ ($\Phi \in \mathfrak{D}$);

(c) for any $\Phi, \Psi \in \mathfrak{D}$, if $U_\Phi \cap U_\Psi \neq 0$, then the homeomorphism $\Phi^{-1} \circ \Psi$ from the open set $\Psi^{-1}(U_\Phi) \subset R^k$ onto the open set $\Phi^{-1}(U_\Psi) \subset R^k$ is of class C_m.

Interchanging Φ and Ψ in (c) we infer that the mapping $(\Phi^{-1} \circ \Psi)^{-1} = \Psi^{-1} \circ \Phi$ is of class C_m. Thus for any $\Phi, \Psi \in \mathfrak{D}$ the homeomorphism $\Phi^{-1} \circ \Psi$ is a diffeomorphism of class C_m.

The space R^k is the simplest example of a k-dimensional manifold (of class C_∞). As $\mathfrak{D}$ we take here the set composed only of the identity mapping in R^k.

Every k-dimensional hypersurface $H \subset R^l$ of class C_m is another example of a k-dimensional manifold of class C_m. As $\mathfrak{D}$ we should take now the class of all diffeomorphisms Φ: $G \to H$ ($G \subset R^k$) of class C_m. Condition (c) follows from Theorem V 5.4 on three mappings.

Thus the notion of a differentiable manifold of class C_m is a generalization of the notion of a hypersurface of class C_m.

A large part of the theory of differentiation can be generalized, in a natural way, to differentiable manifolds. However, the generalization often requires us to change some definitions and to replace some notions by others which are different from the purely formal point of view, but are equivalent from the point of view of their content and their intention.

For instance, a real function φ defined on an open subset of a k-dimensional manifold of class C_m is said to be *of class* C_r ($1 \leqslant r \leqslant m$) if the composite mapping $\varphi \circ \Phi$ is of class C_r (in the sense defined in IV § 5) for every $\Phi \in \mathfrak{D}$.

The restriction $r \leqslant m$ is often inconvenient. To omit it, let us assume that $m = \infty$, i.e. that the k-dimensional manifold H under consideration is of class C_∞.

In V § 6 we have defined the notion of a vector tangent to a k-dimensional regular hypersurface $H \subset R^l$ at a point q. The set $V(q)$ of all vectors tangent to H at q is a k-dimensional linear space; more precisely, it is a k-dimensional linear subspace of the linear space V_l. In general, the spaces $V(q)$ of tangent vectors corresponding to different points q are different. However, all are subspaces of one linear space V_l which enables us to compare vectors from different spaces $V(q)$. For instance, we can say that vectors $\mathfrak{a} \in V(q)$ and $\mathfrak{b} \in V(p)$, $q \neq p$, are equal or parallel, etc.

In the case of an arbitrary k-dimensional differentiable manifold we can also define the k-dimensional linear space $V(q)$ of all vectors tangent to H at q. However, the linear spaces $V(q)$ are not now subspaces of a common linear space. Consequently it is not possible to compare vectors from different spaces $V(q)$, $V(p)$, $q \neq p$. The definition of tangent vectors can be introduced in various ways. We shall quote here only one of the known definitions. It is based on the one-to-one natural correspondence between tangent vectors and directional derivatives on hypersurfaces (see V § 3 exercise 13 and V § 6 exercise 13).

For this purpose, let $C(q)$ be the set of all functions φ of class C_∞ defined on a neighbourhood (depending on φ) of a fixed point q of a k-dimensional differentiable manifold H. By a *vector tangent to* H *at* q we shall mean any function $\mathfrak{w}$: $C(q) \to R$ such that

$$\mathfrak{w}(\varphi+\psi) = \mathfrak{w}(\varphi)+\mathfrak{w}(\psi)\,, \qquad \mathfrak{w}(a\varphi) = a\mathfrak{w}(\varphi)\,,$$
$$\mathfrak{w}(\varphi\psi) = \mathfrak{w}(\varphi)\psi(q)+\varphi(q)\mathfrak{w}(\psi)$$

for all $\varphi, \psi \in C(q)$. It follows from the definition that the set $V(q)$ of all vectors tangent to H at q is a linear space with respect to ordinary addition and multiplication by a scalar of functions $\mathfrak{w}$ in $V(q)$. One can prove that $V(q)$ is k-dimensional.

It follows from exercises 12-13 in V § 6, in the case where H is a hypersurface of class C_∞ in a Euclidean space, that each function $\mathfrak{w}$ satisfying the above conditions is of the form

$$\mathfrak{w}(\varphi) = \varphi'_{\mathfrak{a}}(q)$$

where $\mathfrak{a}$ is a vector tangent to H at q in the sense defined in V § 6, and $\varphi'_{\mathfrak{a}}(q)$ is the derivative of φ at q in the direction $\mathfrak{a}$. Moreover, the mutual correspondence between $\mathfrak{w}$ and $\mathfrak{a}$ is one-to-one. We see that the abstract vector $\mathfrak{w}$ just defined should be identified with the corresponding ordinary vector $\mathfrak{a}$ tangent to H at q in the sense of V § 6.

As we stated before, a large part of the theory of differentiation can be generalized to differentiable manifolds after a suitable modification of notions and definitions. In particular, we can generalize the notion of skew derivative, Stokes' theorem, the de Rham theorem, etc. The reader interested in this subject is referred to the books [11] and [14].

On the other hand, every k-dimensional manifold can be imbedded—with preservation of differential properties—into the Euclidean space R^{2k+1}. This means that, in practice, every differentiable manifold can be considered as a regular hypersurface, of the same class, in a Euclidean space. A proof of the imbedding theorem can be found, e.g., in the book [14].

REFERENCES

[1] J. Dieudonné, *Foundations of Modern Analysis*, New York–London 1960.

[2] N. Dunford and J. T. Schwartz, *Linear Operators*, Part I: *General Theory*, New York 1958.

[3] W. H. Fleming, *Functions of Several Variables*, Reading 1965.

[4] C. Goffman, *Calculus of Several Variables*, New York 1965.

[5] P. R. Halmos, *Measure Theory*, Toronto–New York–London 1950.

[6] O. Haupt, G. Aumann and Ch. Y. Pauc, *Differential- und Integralrechnung*, Vol. I, Berlin 1948; vol. II, Berlin 1950; vol. III, Berlin 1955.

[7] F. Hausdorff, *Dimension und äusseres Mass*, Mathematische Annalen 79 (1919), pp. 157-179.

[8] E. Hille and R. S. Phillips, *Functional Analysis and Semi-groups*, Providence 1957.

[9] W. V. D. Hodge, *The Theory and Applications of Harmonic Integrals*, Cambridge 1941.

[10] C. Kuratowski, *Introduction to Calculus*, Oxford–London–New York–Paris 1961.

[11] G. W. de Rham, *Variétés différentiables*, Paris 1955.

[12] R. Sikorski, *Funkcje rzeczywiste* (*Real functions*; in Polish), vol. I, Warsaw 1958; vol. II, Warsaw 1959.

[13] A. E. Taylor, *Introduction to Functional Analysis*, New York 1958.

[14] H. Whitney, *Geometric Integration Theory*, Princeton 1957.

LIST OF SYMBOLS

CHAPTER I

CHAPTER II

CHAPTER III

CHAPTER IV

Symbol	Description	Page
$\overset{(m)}{\varphi}_{x_{j_1},\ldots,x_{j_m}}$, $\varphi_{x_{j_1},\ldots,x_{j_m}}$, $\dfrac{\partial^m\varphi}{\partial x_{j_m}\ldots\partial x_{j_1}}$, $\dfrac{\partial}{\partial x_{j_m}}\cdots\dfrac{\partial}{\partial x_{j_1}}\varphi$, $\dfrac{\partial^m z}{\partial x_{j_m}\ldots\partial x_{j_1}}$, $\dfrac{\partial}{\partial x_{j_m}}\cdots\dfrac{\partial}{\partial x_{j_1}}z$, $\overset{(m)}{\varphi}_{x_{j_1},\ldots,x_{j_m}}(x_1,\ldots,x_k)$, $\varphi_{x_{j_1},\ldots,x_{j_m}}(x_1,\ldots,x_k)$, $\dfrac{\partial^m}{\partial x_{j_m}\ldots\partial x_{j_1}}\varphi(x_1,\ldots,x_k)$	the traditional notation for partial derivatives of order m of a function $z = \varphi(x_1, \ldots, x_k)$ and for their values at a point	148, 149
$\dfrac{\partial^3\varphi}{\partial x_1^3}$, $\dfrac{\partial^4}{\partial x_1^4}\dfrac{\partial^2}{\partial x_2^2}\varphi$ etc.,	the traditional notation for partial derivatives with multiple differentiation with respect to some variables	149

CHAPTER V

Symbol	Description	Page
$\Phi'_{\mathfrak{a}}$, $\Phi'_{\mathfrak{a}}(p)$	the directional derivative of a mapping, a vector field or a matrix field in a direction $\mathfrak{a}$, and its value at a point p	166
$\Phi_{\mid j}$	the partial derivative of a mapping, a vector field or a matrix field with respect to the jth coordinate (variable)	167
Φ'_{x_j}, Φ_{x_j}, $\dfrac{\partial\Phi}{\partial x_j}$, $\dfrac{\partial}{\partial x_j}\Phi$, $\dfrac{\partial}{\partial x_j}\Phi(x_1,\ldots,x_k)$, $\Phi'_{x_j}(x_1,\ldots,x_k)$, $\Phi_{x_j}(x_1,\ldots,x_k)$	the traditional notation for partial derivatives of a mapping, vector field or matrix field and for their values at a point	167
Φ', $\Phi'(p)$	the derivative of a mapping, vector fields or matrix field and its value at a point	168
Φ', $\Phi'(t)$	the derivative of a mapping of a set of real numbers and its value at a point t	169, 170
$\Phi_{\mid\mathbf{j}}$, $(\varphi_1,\ldots,\varphi_l)_{\mid\mathbf{j}}$, $(\varphi_1,\ldots,\varphi_l)_{\mid(j,\ldots,j_\varkappa)}$	the derivative of a mapping $\Phi = (\varphi_1, \ldots, \varphi_l)$ (or of a vector field) with respect to $\mathbf{j}$th coordinates, $\mathbf{j} = (j_1, \ldots, j_\varkappa)$	172
$/(\varphi_1,\ldots,\varphi_l)_{\mid\mathbf{j}}/$, $/\Phi_{\mid\mathbf{j}}/$, $/(\varphi_1,\ldots,\varphi_l)_{\mid(j_1,\ldots,j_l)}/$, $/\Phi_{\mid(j_1,\ldots,j_l)}/$	the determinant (the jacobian) of all partial derivatives of the coordinates $\varphi_1, \ldots, \varphi_l$ of a mapping (or a vector field) Φ with respect to $\mathbf{j}$th coordinates, $\mathbf{j} = (j_1, \ldots, j_l)$	173
$/\Phi'/$	the scalar derivative (the jacobian) of a mapping or a vector field Φ	173
$\lvert\Phi'\rvert$	the modulus of the derivative of a mapping or a vector field Φ	173
$\dfrac{\partial(\varphi_1,\ldots,\varphi_r)}{\partial(x_{j_1},\ldots,x_{j_r})}$, $\dfrac{\partial(z_1,\ldots,z_r)}{\partial(x_{j_1},\ldots,x_{j_r})}$	the traditional notation for jacobians	174, 175

CHAPTER X

SUBJECT INDEX